ENVIRONMENTAL REMEDIATION TECHNOLOGIES, REGULATIONS AND SAFETY

HANDBOOK OF REMEDIATION FOR COMPLEX ENVIRONMENTAL PROBLEMS

ENVIRONMENTAL REMEDIATION TECHNOLOGIES, REGULATIONS AND SAFETY

Additional books and e-books in this series can be found
on Nova's website under the Series tab.

ENVIRONMENTAL REMEDIATION TECHNOLOGIES, REGULATIONS AND SAFETY

HANDBOOK OF REMEDIATION FOR COMPLEX ENVIRONMENTAL PROBLEMS

SADIA AMEEN
M. SHAHEER AKHTAR
AND
HYUNG SHIK SHIN
EDITORS

nova science publishers
New York

NOTICE TO THE READER

Library of Congress Cataloging-in-Publication Data

ISBN: 978-1-53615-118-3

Published by Nova Science Publishers, Inc. † New York

CONTENTS

PREFACE

Environmental issues are of great concerns and would keep increasing in the next few decades. The field of environmental science has broadened and expanded greatly. The demand for clean environment in our current society has increased with large-scale economic developments and population growth. It is crucial to build clean energy systems in order to solve the environmental issues. Taking care of the environment today is a sustainable act for the generations of tomorrow. The use of automobile is responsible for much of the air pollution that affects cities and runoff from lawns, city streets, and farms is an important source of water pollution. Many new environmental problems have also arisen. Climate change, the AIDS pandemic, genetically modified crops, and concerns about the loss of biodiversity have become core issues. The process of developing new techniques as well as improving older types of technologies is still ongoing. The most important thing is to understand that remediation actions need to be justified and optimized the adopted actions and must do more good than harm. The end results should always be a balance between risks, costs, benefits and remediation viability.

The central theme of the book is interrelatedness. In this book, each part opens with a new author's essay highlighting their work for contribution towards environmental remediation. There has been significant progress toward solving some environmental problems, progress in others has been slow, and new threats or concerns have emerged. To help the reader evaluate progress and recognize how difficult some problems are, chapter presents charts, graphs, and figures to help illustrate the scope of environmental issues, the text's main focus is on identifying major issues and giving appropriate examples to illustrate the complex interactions that are characteristic of all environmental problems. Sustainability is also discussed in the chapters on energy use. We need to note that the number of disciplines involved in the Research for environmental remediation has increased dramatically. Thus, in addition to traditional approaches, varieties of new themes like photo-active nanomaterials, sustainable energy for environmental remediation have been introduced to achieve the goals of analyzing methodology on the macro and nano-scales.

It is necessary to comment briefly on the coverage of the literature. *Handbook of Remediation for Complex Environmental Problems* highlights the fundamentals of solar energy conversion, natural and artificial photosynthetic systems, nanotechnology and nanoscience, and the application of nanoscience and nanotechnology in environmental remediation. The unique properties of nanoscale materials, such as nanoparticles, nanotubes, nanowires, nanofibers, nanocomposites, nanopores, and nanofilms, allow them to design the next generation of photosynthetic devices in environmental applications. Of course, not all materials are covered equally, in many cases, the level of detailed elaboration was determined by their significance and interest shown in that class of materials. The editors are very pleased to present this book for researchers working in active nanomaterials, environment friendly molecules, catalytic materials and their environment applications, and also will be an excellent text book for advanced study in the universities from fundamental points of views. During our work on this book we tried to cover the field more or less completely. We are thankful for the author's hard work and contributions, copyright owners, and reviewers's comments and suggestions. Without all the above support, it would not have been possible for us to publish this book. We wish to express our gratitude to the staff of IK publishers for their patience during the development of this project and for encouraging us during the various stages of preparation.

Editors
Sadia Ameen, PhD
M. Shaheer Akhtar, PhD
Hyung-Shik Shin, PhD

In: Handbook of Remediation ...
Editors: Sadia Ameen, M. Shaheer Akhtar et al.
ISBN: 978-1-53615-118-3

Chapter 1

DEVELOPMENT OF THIAZOLO [5,4-D]THIAZOLE-BASED NOVEL ORGANIC CHROMOPHORES AND THEIR PHOTOVOLTAIC APPLICATIONS

***M. Nazim*[1], *Sadia Ameen*[2,*], *M. Shaheer Akhtar*[3] *and Hyung-Shik Shin*[1,†]**

[1]Energy Materials and Surface Science Laboratory,
Solar Energy Research Center, School of Chemical Engineering,
Chonbuk National University, Jeonju, Republic of Korea
[2]Advanced Materials and Devices Laboratory,
Jeongeup Industry-Academic Cooperation Support Centre,
Chonbuk National University, Republic of Korea
[3]New and Renewable Energy Material Development Center (NewREC),
Chonbuk National University, Jeonbuk, Republic of Korea

ABSTRACT

Organic solar cells have recently received a great deal of interest owing the advantages of inexpensive materials, easy to process and high power conversion efficiency. OPV materials in bulk heterojunction (BHJ) solar cells basically consist of a phase-separated mixture of a p-type conjugated polymer or small organic molecule as a donor and the fullerene derivative; phenyl-C_{61}-butyric acid methyl ester ($PC_{61}BM$) as an acceptor material. Organic chromophores display numerous advantages such as low bandgap, ease and reproducibility of synthesis, ease of purification process and high solubility in common organic solvents. Various π-conjugated small molecules of D-A-D type architecture have

* Corresponding Author's Email: sadiaameen@jbnu.ac.kr.
† Corresponding Author's Email hsshin@jbnu.ac.kr.

been designed to improve the light harvesting ability, charge-separation property, energy levels and photovoltaic properties. Thiazolo [5, 4-*d*]thiazole is a rigid and fused heterocyclic ring moiety which ensures a coplanar structure of a π-conjugated system and a strong π-π stacking ability. Thiazolo [5,4-*d*]thiazole as electron-acceptor unit has aroused much interest in efficient small organic molecules of D–A–D or A–D–A type and accomplishes good photovoltaic properties. The both side terminal n-hexyl side chains in these molecules lead to good solubility in common organic solvents, enhances the energy levels and induces self-assembly behavior. From the morphological analysis, the presence of different liquid-crystal phases affects the film-morphology of the devices during the solution-processed fabrication and leads to self-organization behavior. Due to the incorporation of triphenyl amine donor, small molecule presents high absorption, good film morphology and the solubility which eventually lead to better performances of small molecule organic solar cells (SMOSCs). The overall improved performance of the devices might be attributed to their improved absorption, electrochemical-properties and the presence of strong electron-withdrawing furan moieties. Moreover, the presence of furan spacer in small molecule in blend thin film substantially improves the interface area for an exciton dissociation, which results in the high photocurrent density and PCE of the devices. The enhanced J_{SC} and PCEs are attributed to the presence of terminal alkyl units and the homogenous-miscibility of the blend thin film which assists the ultrafast intermolecular-charge transfer from donor to acceptor. This chapter includes some of the promising electron-donor organic small molecules and their applications in organic solar cells. The synthesis methods, morphology, physical and chemical properties of the materials have shown that the further architecting of these organic materials would surely their applications in solar cells and other opto-electronic devices due to enhanced optical, electrochemical and photovoltaic properties.

Keywords: small organic molecule, solar cells, electrochemical properties, photocurrent density, energy conversion efficiency

1. Introduction

1.1. Renewable Energy

Recently, the increasing demand of clean energy resources has led to the dramatic growth in the research, development, and manufacturing of cost-effective energy sources which play an important role for the development and progress of human beings. However, with the rapid depletion of the conventional energy sources and increasing energy demand, worldwide primary energy consumption has grown by ~1.8%, as reported in 2012 [1]. Due to various environmental protection issues, the clean fuel technologies and new and renewable energy resources are being intensively pursued and investigated. Therefore, shifting towards renewable energy resources could help to reduce the greenhouse gas emissions, improves the climate impacts, and ensures a reliable, timely, and cost-effective form of energy.

The renewable energy sources are continually replenished by nature and derived directly such as thermal, photo-chemical, and photo-electric or used indirectly like wind, hydropower, and photosynthetic energy stored in biomass or taken from other natural mechanisms of the environment such as geothermal and tidal energy. Renewable energy does not include energy resources derived from fossil fuels, waste products from fossil sources, or waste products from inorganic sources [2]. Renewable energy is the energy collected from resources which are naturally replenished, as sunlight, wind, rain, tides, waves, and geothermal heat [3]. There are conventional energy resources (i.e., oil, coal, uranium etc.) with some damaging effects on the atmosphere. The plants are not adequate to absorb the huge amount of carbon dioxide which releases into the atmosphere due to the burning of fossil fuels [4].

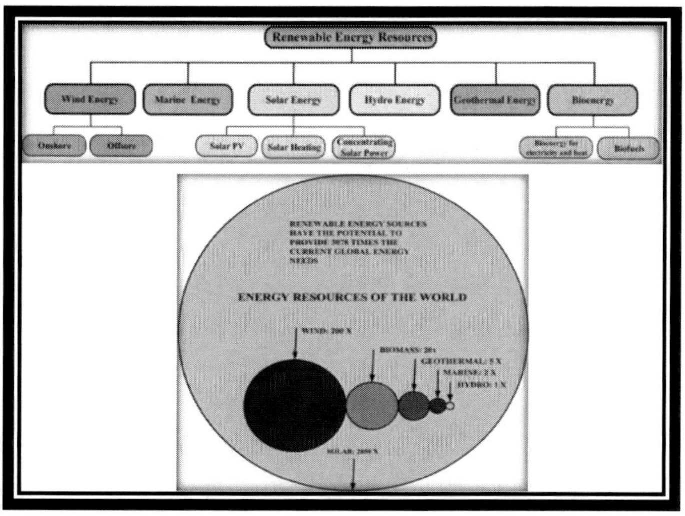

Figure 1. An overview of different types of renewable energy resources of world (http://www.sciencedirect.com/science/article/pii/S1364032114005656).

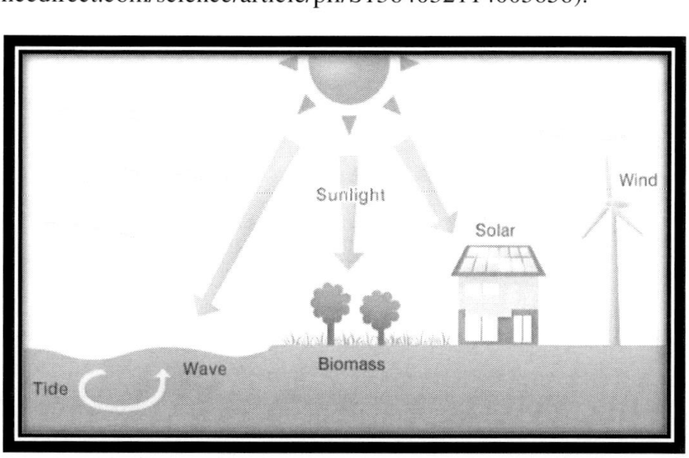

Figure 2. Various types of renewable energy resources (https://www.mhi-global.com/discover/earth/issue/history/future/renewable/about.html).

Hence, the increased concentration of carbon dioxide in the atmosphere considerably adds to the greenhouse effect which results in the increase of global surface temperature [5]. The sources of renewable energy (Figure 1) are widely available in the environment and will last forever. The renewable energy sources have a great advantage that they are eco-friendly to the environment because of negligible emissions of carbon dioxide (CO_2) gas.

1.2. Types of Renewable Energy

- Solar energy
- Wind energy
- Biomass energy
- Hydroelectric power
- Geothermal energy

The worldwide energy consumption would be expected to increase with increasing world population and hence, higher energy demand in the future. Among all the renewable energy forms (Figure 2), solar energy has shown its advantages and potential for the generation of huge amount of electricity. Solar radiation amounts to 3.8 million EJ/year, which is approximately 10,000 times more than current energy needs [6]. From the energy conservation and environmental protection perspective, it is needed to convert solar radiation directly into electrical energy by the photovoltaic devices. The annual global energy consumption in just one hour is less than the amount of solar energy reaching to the Earth's surface. The consumption of solar energy is increasing worldwide in a great extent and could be used in two ways; one is solar cell that uses sunlight to irradiate the semiconductors and generates electricity. Other one is solar-thermal systems deployed for supplying the hot water, heating and cooling. The efficiency of solar-thermal systems is ≥40%, which is superior to solar cells. In the solar-thermal systems, the heat energy stored during the day could be used at night.

Photovoltaics (PV) are a simple and elegant method of harnessing the solar energy (Figure 3). Inorganic solar cells, such as conventional silicon solar cells and heterojunction solar cells, are relatively mature technologies, and the performance of these devices is approaching to the record efficiency 24.7% for crystalline silicon solar cells [7]. Despite being unfavorable effects on environment, the fossil fuels are also the key source of energy with around 86% in the world [8-10]. The PV devices could be easily made in the cost-effective way as compared to other energy technologies of fossil fuels. Solar cells are a robust, reliable and long lasting source of energy because they directly convert the solar radiation into electricity without creating noise pollution or harm to the environment. Every day, sunlight is provided to the earth in huge amount as compared to other energy resources such as oil, coal or uranium. Generally, sun supplies energy to the earth about 3 × 10^{24} Joules, which is ~10000 times than the total global energy consumption in a year [11, 12].

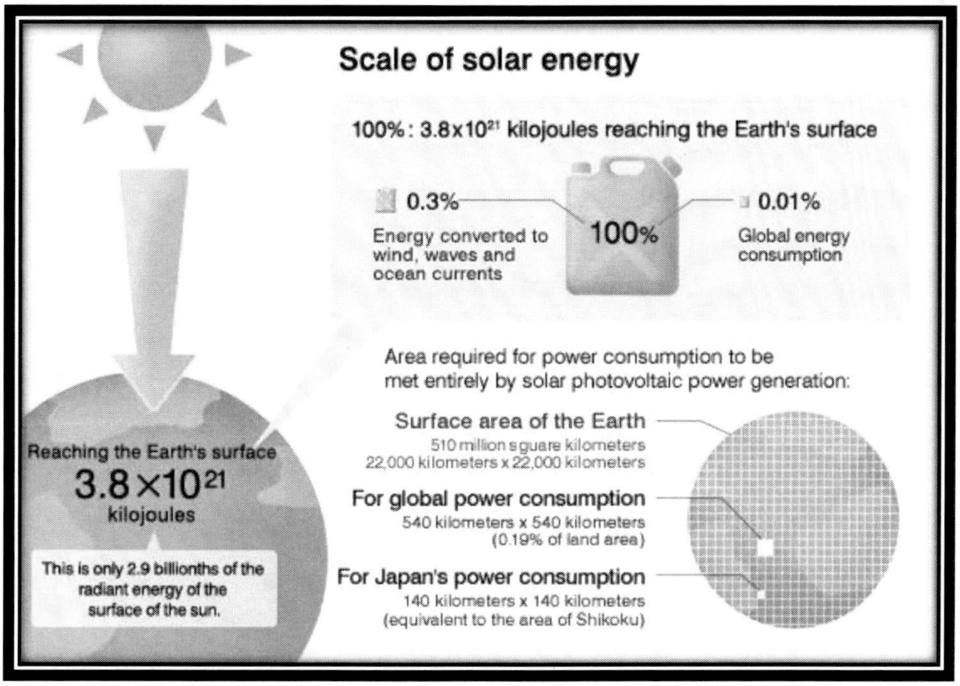

Figure 3. Sunlight and its application in solar energy resources (https://www.mhi-global.com/discover/earth/issue/history/future/renewable/solar.html).

1.3. Principle of Solar Cells: Photovoltaic Effect

In progress of the photovoltaic cell (Figure 4), silicon is the most widely used semiconductor material because silicon atoms have four valence electrons. In the crystal structure, each silicon atom shares four valence electrons with other Si atom and hence, creates covalent bonds resulting in a tetrahedral (Td) lattice structure. When semiconductor materials exposed to light, some of the photons of light are absorbed by the semiconductor material which causes a number of free electrons in thc semiconductor crystal lattice and hence, produces the electricity. This phenomenon is known photovoltaic effect (Figure 5a). The band gap is the minimum change in energy which is required to excite the electron so that it could participate in the conduction of electricity [13]. The electrons in excited state (conduction band) are free to move through the semiconductor materials. Consequently, the excitation of an electron into the conduction band results in not only an "electron" in the conduction band but also a "hole" in the valence band. Thus, both the electron and hole could participate in the conduction and are called "charge carriers" [14].

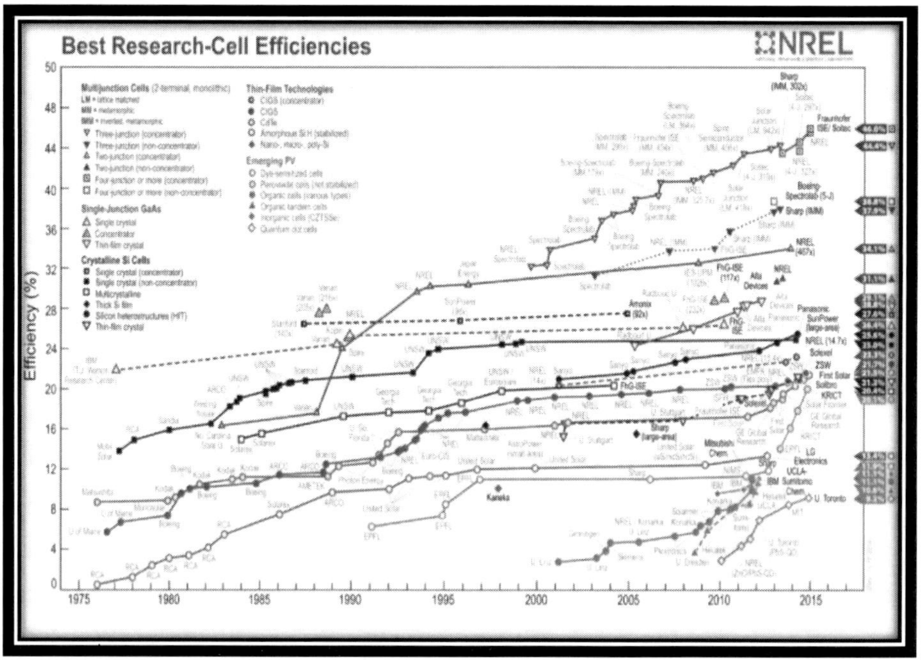

Figure 4. Progress of various types of renewable energy sources (http://solarlove.org/first-solar-breaks-solar-cell-efficiency-record-cdte-solar-cells-yet/).

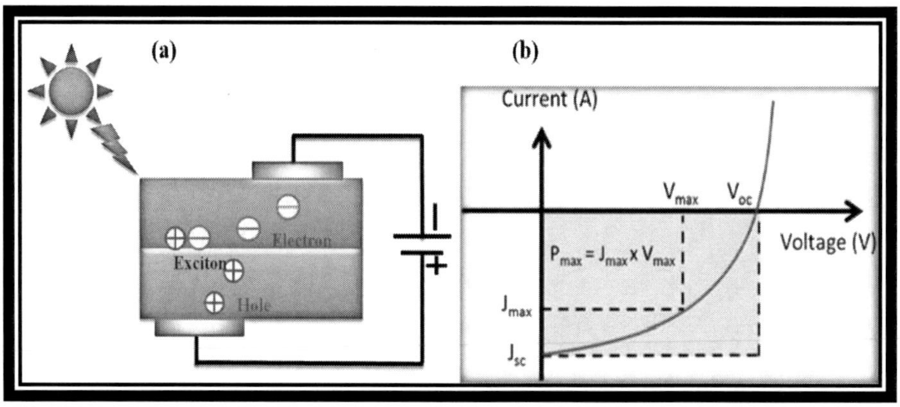

Figure 5. (a) Photovoltaic effect of solar cells (b) Characterstics of a typical solar cell (http://pubs.rsc.org/en/Content/ArticleHtml/2011/PY/c1py00275a).

1.4. Characteristics of Solar Cells

Generally, the solar cell devices are characterized by various parameters (Figure 5 b) to calculate the performances. The most important parameters that are widely used to determine the performance of solar cell devices are open-circuit voltage (V_{OC}), short-circuit current (J_{SC}), fill factor (FF), and power conversion efficiency (PCE).

1.4.1. Open Circuit Voltage (VoC)

Open circuit voltage (Voc) is simply defined as the maximum voltage available from a solar cell at zero current. In other words, when there is no current flow, the potential difference between the two terminals of solar cells under illumination is called Voc. Hence, open-circuit voltage is a measure of the amount of recombination in the devices. Hence, Voc depends on the saturation current of the solar cell devices and the light-generated current.

1.4.2. Short-Circuit Current Density (J_{SC})

Short-circuit current is defined as the current produced by solar cells under illumination without application of any external potential. The short-circuit current density (J_{SC}) is the current through the solar cell when the voltage across the solar cell is zero. The short-circuit current is originated due to the generation and collection of light-generated charge carriers. For an ideal solar cell, the short-circuit current and the light-generated current are identical. Therefore, the short-circuit current is the largest current which may be drawn from the solar cells. For the solar cell devices, it is better to state the maximum current-density in place of maximum current only because the current generation depends upon the surface area of the solar cells. The maximum current-density or short-circuit current density is the ratio of maximum or short-circuit current to exposed surface area of the cells.

1.4.3. Fill Factor (FF)

The fill factor (FF) of a solar cell determines the maximum power of the devices. Generally, FF is defined as the ratio of the maximum power from the solar cells to the product of V_{oc} and J_{sc}. Graphically; the FF is a measure of the "squareness" of the solar cell and is also the area of the largest rectangle which fits in the IV curve. Owing to the presence of various recombinations and loss mechanisms, the real FF value remains lower than the expected. The FF can be calculated by the following formula:-

$$P_{max} = J_{sc} V_{oc} FF \quad (1)$$

$$FF = J_{max} V_{max}/J_{sc} V_{oc} \quad (2)$$

where J_{max} is maximum short circuit current density and V_{max} is the maximum open circuit voltage of the devices.

1.4.4. Power Conversion Efficiency (η)

The power conversion efficiency (PCE) is more commonly known as the efficiency of solar cells which is the ratio of the maximum power (P_{max}) generated by the solar cells to the incident radiant power (P_{in}). Efficiency is defined as the ratio of energy output from the solar cell to input energy from the sun. In addition, the efficiency depends on the spectrum and intensity of the incident sunlight and the temperature of the solar cells. Hence, the conditions of solar cells could be controlled in order to compare the performance of devices.

$$\eta = V_{oc}.J_{sc}.FF/P_{in} \quad (3)$$

where V_{oc} is the open circuit voltage, J_{sc} is the short circuit current density, FF is the fill factor and P_{in} is the incident radiant power.

1.5. Types of Solar Cells

Semiconducting materials are much useful because their properties could be manipulated and controlled rather easily by the addition of impurities through the process known as "doping." The material's conductivity could be controlled by applying an electric or magnetic field, through light exposure or heat, or by the mechanical deformation of the devices (Figure 6). The semiconductor is a solid chemical-element that could conduct electricity under the specified conditions through different techniques. Semiconductor devices could display the useful properties of flowing current in one direction, shows variable resistances, and the sensitivity towards light or heat. Organic molecules are needed for the construction of solar cells which exhibits a high optical absorption-coefficient in the solution [15]. In solar cell devices, organic materials could be used as a sensitizer or electron-donor or electron-acceptor. The organic materials might be used as polymers (macromolecules) or small molecules as photoactive materials in solar cell applications [16, 17].

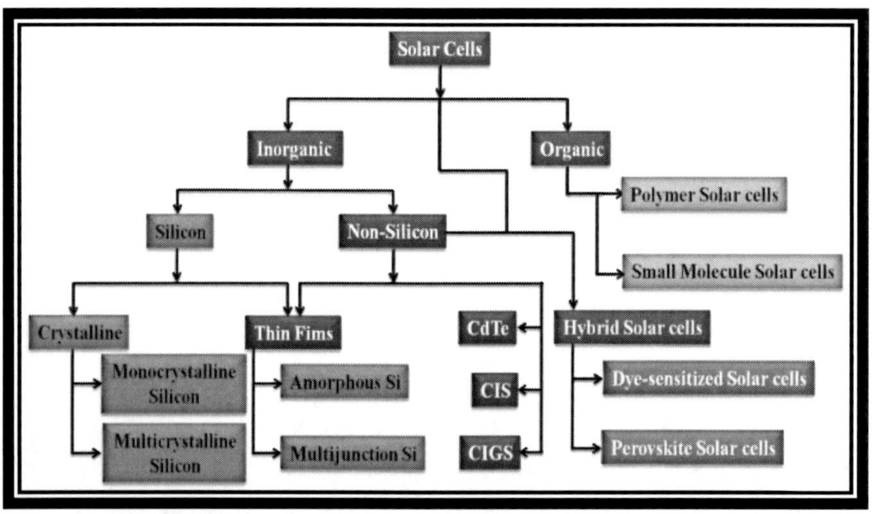

Figure 6. Model presentation of various types of solar cell technologies.

1.5.1. Inorganic Solar Cells

The semiconductor materials with doping from III group elements like boron, aluminum etc are deposited on silicon wafers and thus, creates positively charged free holes which are called as "p-type" semiconductor [18]. On the other hand, the semiconductors doped with the V group elements as nitrogen, phosphorus etc. contains free electrons and are known as "n-type" semiconductor. The p-n junction is generally formed by combining the p-type and n-type semiconducting materials which is responsible for the conversion of sunlight into the electricity [19]. When p- and n-type semiconductors are brought together, the electrons near to the junction could easily diffuse into the p-type region and creates an electric-field that opposes the flow of electrons [20]. At equilibrium state, there is no net flow of charges across the junction and that region between the two materials is called as the "depletion region." Silicon semiconductor absorbs photons in the visible-region to excite the electrons from valance-band to the conduction-band and leaves hole in the valance-band [21].

1.5.2. Organic Solar Cells

On the light absorption, the excitation of electron generates an electron-hole pair which is tightly bound by coulomb-attraction forces and called as exciton. After diffusion towards the donor-acceptor (D-A) interface, the exciton dissociates in the presence of an external electric-field with sufficient energy of molecular orbitals [22]. Since, the exciton-diffusion length ranges from 1 nm to 10 nm which is much shorter than the device thickness, exciton-diffusion limits the charge-carrier generation through the charge-recombination. The incorporation of the buffer-layers such as; PEDOT:PSS, ZnO etc ensures that the charge-carriers move towards the electrode in OPVs and thus, utilizes the use of organic compounds in various optoelectronic-devices [23, 24].

The solar cell devices which are fabricated by the vacuum-deposition process contain copper-phthalocyanine as the hole-transporting material and perylene as an electron-transporting material [25]. However, the effective charge-separation limits the planar heterojunction (PHJ) devices owing to short exciton-diffusion length (<10 nm) resulting in lower performance of the devices. However, use of fullerene ester ($PC_{60}BM$) as electron-acceptor has significantly improved the power conversion efficiency (PCE) up to ~4.2% [26, 27]. A novel hybrid heterojunction developed by sandwiching a mixed layer of donor and acceptor materials between pure donor and acceptor layers achieved an efficiency of ~5.0% [28]. Due to different electron-affinity and the ionization-energies, the electrostatic-attractions are generated at the interface between the two organic layers [29, 30]. However, the bilayer structure is limited by the short exciton-diffusion lengths. So, the thin films of organic photovoltaics must ascertain that the created excitons should diffuse to reach the donor-acceptor (D-A) interface which results in a limited optical-density and hence, restricts the efficiency of soalr cell devices [31-33]. As compared to a planar-heterojunction (bilayer architecture), the bulk-heterojunction (BHJ) architecture provides a large surface area for the charge-separation and hence, reduces the charge-recombination which improved performances [34, 35]. Heeger et al. developed the BHJ concept of the blend thin films to refer a bicontinuous, interpenetrating network of donor (D) and acceptor (A) phases [36]. The nanostructural morphology of BHJ devices plays a crucial role for the solar cell performance. The good dispersion of the organic chromophore results in poor intermolecular charge-transfer (ICT) transitions [37].

1.5.3. Polymer Organic Solar Cells

Off late, organic semiconducting polymers as donor materials display a great attention in the solution-processed fabrication of organic solar cells owing to their high performance [38-40]. In polymer solar cells, poly-3-hexylthiophene (P3HT) was utilized as the standard material due to its absorption and self-organization properties which resulted in the improvement of charge-carrier mobilites [41]. The thermal or solvent vapor annealing of the devices significantly enhanced the self-assembly property of the polymers and thus, resulted in better performance of up to 5% [42-44]. One of the restricting factors of P3HT is the maximum absorption at ~530 nm in the solid state, while a high-lying HOMO of ~5.2 eV value is considered as the retarding factor for the photovoltaic applications. For low band gap polymers, light harvesting over a broader absorption range is possible to achieve high V_{OC} values. Due to the introduction of $PC_{71}BM$ acceptor, a significant enhancement in the absorption spectra of the solar cell devices with efficiency >8% was reported [45].

1.5.4. Small Molecule Organic Solar Cells

The organic chromophores have been reported with various alkyl chains on pyrrole moiety to control the geometry of molecules in the solid state. The enhanced charge-transport and the photocurrent density of solar cell devices significantly improved the efficiency upto 2.59% [46, 47]. The squaraine-based small molecules have achieved a high crystallinity and hole mobility by the replacement of ketone on the squarine moiety with dicyanovinyl group [48]. The indigo-derivatives building blocks have been employed in solution-processed BHJ organic solar cells among all organic dyes. The (donor-acceptor-donor, D-A-D) type isoindigo-based oligothiophenes shows similar HOMO energy level of ~5.5 eV. After annealing, the blended thin films with $PC_{61}BMs$ (D-A-D type) have achieved more crystalline domains than acceptor-donor-acceptor (A-D-A type) [49]. Theoretically, the maximum efficiency for OPVs could be ~20%. The difference in the experimental and the theoretical performances of solar cells could be explained by the following losses: a) the incident light is absorbed by the materials that is not involved in the charge carrier production is called as optical loss; b) when excitons fail to reach at D-A interface i.e., exciton loss; c) the charge-carriers recombines before their split i.e., recombination loss; d) the charge-carrier mobility is insufficient to get charges towards electrodes i.e., collection loss of the devices. Out of which, the greatest loss mechanism is exciton recombination at D-A interface resulting from the mismatched energy levels between two organic materials.

1.5.5. Hybrid Solar Cells

In recent years, hybrid solar cells HSCs) are considered as possible cost-effective way for energy harvesting and its conversion. The hybrid solar cells are generally lightweight, relatively thin, strong light-absorbing, and solution-processable which makes it very attractive for the large-scale flexible solar cell fabrication. Hybrid solar cells (HSCs) essentially consist of organic molecules/polymers and inorganic semiconducting nanocrystals as electron donors or acceptors in the photoactive layer. The best HSCs device so far is based on CdSe fabricated by using a mixture of pyridine-treated nanorods and quantum dots together with Poly[2,6-(4,4-bis-(2-ethylhexyl)-4H-cyclopenta [2,1-b;3,4-b′]dithiophene)-alt-4,7(2,1,3-benzothiadiazole)] (PCPDTBT) as donor with ~3.6% efficiency [50]. Inorganic nanocrystals offer several attractive properties such as a tunable band gap, higher intrinsic charge mobility, the possibility of multiple exciton generation if the band gap is sufficiently small.

1.5.6. Dye-Sensitized Solar Cells

The dye-sensitized solar cells (DSSCs) are one of the most promising alternatives to the silicon solar cells in terms of efficiency and the fabrication process. Gratzel reported the first dye sensitized nanocrystalline solar cell (Figure 7) with energy conversion efficiency of ~7.1% and incident photon-to-electrical current conversion efficiency (IPCE) was ~80% [8]. The simple structure and low cost technology have further improved the efficiency up to ~10% level needed for the commercial use [51]. The dye sensitizer gets excited on exposure to sunlight and an electron is injected into the conduction band of the mesoporous oxide (TiO_2) film. The excited electrons diffuse to anode and utilizes at the external load before being collected by the electrolyte at cathode. The conducting glass is used as substrate to enhance the electrical conductivity and the light transmittance. The semiconductor electrode is a layer of nanocrystalline titanium dioxide (TiO_2) thin film (thickness ~5–30 nm) which plays an important role in the electron transfer process. The film morphology and porosity of TiO2 layer are crucial factors that determine the amount of dye absorbed on the surface which provides a large reaction sites for the dye molecules to harvest the incident light [52]. Recently, a number of dyes have been synthesized and some of them have already been commercialized as N3, N719 and Z907 dye.

Hence, the ideal sensitizer has some requirements that guide effective molecular engineering: a) the sensitizer should absorb all the incident light below the near-IR wavelength; b) it must carry a carboxylate or phosphonate group to anchor on the surface of semiconductor oxide; c) the lowest unoccupied molecular orbital (LUMO) of the sensitizer must match to the edge of the conduction band of the oxide to avoid the energetic potential losses during the electron transfer; d) the highest occupied orbital (HOMO) of the sensitizer must be low to accept electron from an electrolyte or a hole conducting material [53, 54].

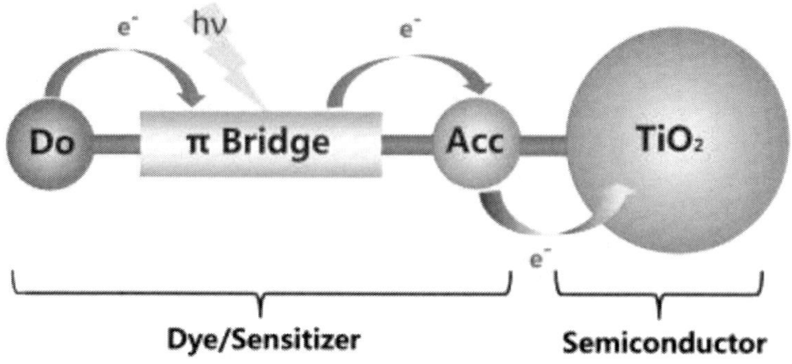

Figure 7. Model design structure of organic dyes for DSSCs (http://onlinelibrary.wiley.com/doi/10.1002/anie.200804709/epdf).

In addition, the desirable dyes have long-term operational stability; firm graft on the semiconductor surface, redox potential should be high enough to facilitate the regeneration reaction with a redox mediator. The iodide and triiodide (I/I_3) redox couple is the most commonly used in the liquid electrolyte, with other solid-state or quasi-solid electrolytes as organic hole-transport materials (HTMs) and polymer gels along with platinum as the cathode [55-57].

1.5.7. Perovskite Solar Cells

Perovskites is the naming for any material which acquires the similar crystal structure as calcium titanate or ABX_3. Different materials adopt this structure, with different properties, such as insulation, antiferromagnetic, piezoelectric, thermoelectric, semiconducting, conducting, and superconducting [58]. The semiconducting perovskites could be fabricated from dried solutions of the precursor salts which interestingly applicable for printable electronics [59, 60]. The perovskite solar cells (PSCs) (Figure 8a) as light absorbers have arisen from the dye-sensitized solar cells (DSSCs) [8, 61].

DSSCs consist of three main components, like mesoporous n-type TiO_2 which is sensitized by light-absorbing organic dye and filled with a redox-active liquid electrolyte. The porous TiO_2 layer is required to provide a sufficient surface area to adsorb the organic dye for the absorption of the incident sunlight. However, for DSSCs, the 10 μm thick films are required to complete light absorption of the organic dyes. However, for solid-state DSSCs (ssDSSCs), the film thickness limits to < 2 μm [62]. Hence, inorganic absorbers as quantum dots or extremely thin semiconductor absorber layers should be applied as an alternative to enable complete light absorption in the thin films and the additional possibilities for photoactivity into the near-infrared (NIR) region [63-67]. To find out more advanced light absorbers than organic dyes, T. Miyasaka and co-workers reported the first perovskite-sensitized solar cells and employed $CH_3NH_3PbI_3$ and $CH_3NH_3PbBr_3$ light absorbers with either an iodide/triiodide redox couple or a polypyrrole carbon black composite hole conductor and measured PCEs between 0.4 and 2% for solid-state and liquid electrolyte cells, respectively [68, 69]. However, the first perovskite-sensitized solar cell reported in 2009 where the $CH_3NH_3PbI_3$ absorber resulted in PCE of ~3.5% employed the iodide/triiodide redox couple [70]. During the past two decades, inorganic–organic hybrid perovskite is investigated enthusiastically due to its excellent optical and electronic properties [71, 72]. The hybrid perovskite has the formula ABX_3, where X is a halogen ion (I−, Br−, Cl−), A is an organic ammonium ion (e.g., $CH_3NH_3^+$) and B is Pb^{2+} or Sn^{2+} ion. The corners of the cube are occupied by the organic cations (Figure 8b). Therefore, a three-dimensional (3D) framework could form with the arrays of the corner-sharing metal halide octahedra, resulting in a narrow band gap [73]. Generally, perovskite ($CH_3NH_3PbI_3$) is obtained from a solution of PbI_2 and methylammonium iodine (MAI), while $CH_3NH_3{-}xClx$, is obtained from $PbCl_2$ and MAI with a 1:3 molar ratio. The most popular $CH_3NH_3PbI_3$ and $CH_3NH_3PbI_3{-}xClx$ has a direct band gap of ~1.55 eV which corresponds to an absorption onset of ~800 nm and intense absorption of the whole visible region [74-76]. Moreover, the exciton binding energy of the perovskites is very small as compared to that of organic semiconductors which corresponds to the long exciton-diffusion length of 100–1000 nm and lifetime of ~100 ns [77-80]. The high performances of PSCs enable the intrinsically favorable optoelectronic properties of lead halide perovskites as high absorption coefficients and sharp absorption edges [81]. Each fabrication process produces a different perovskite film morphology (grain size/shape, film roughness, surface coverage, etc.) which significantly influences the optoelectronic properties of the thin film, e.g., carrier diffusion lengths, charge carrier mobilities, etc. One of the most crucial issues for PSCs is to control the film morphology of the perovskite [82]. The spin-coating method involves a precursor solution of lead halide with certain amount of an organic ammonium halide where perovskite crystallites form and grow during the solvent evaporation. Then, the thermal annealing at high temperature (~100°C) is required for the complete crystallization of the perovskite and the removal of the residual solvent impurities [83, 84]. The perovskite crystallites have some defects and the crystal grain boundaries which act as the traps of the charge carriers and aggravate the charge-recombination. The hysteresis effect causes inaccuracy in the cell efficiency evaluation is related to the crystallinity and interfaces of the perovskite [85].

The PSCs have two types of architectures, the meso-structured device and the planar devices. The perovskite crystallites growth process differs significantly in both types of the devices. The meso-structured device employs a layer of ~100 nm thick mesoporous TiO_2 or Al_2O_3 which greatly benefits the crystallization of the perovskite at the mesostructure surface. During the film formation, pore filling and uniformity of the perovskite layer are main concerns for such devices. Moreover, a capping layer of perovskite on the mesoscopic structure is found to be beneficial for the device performance [86-89].

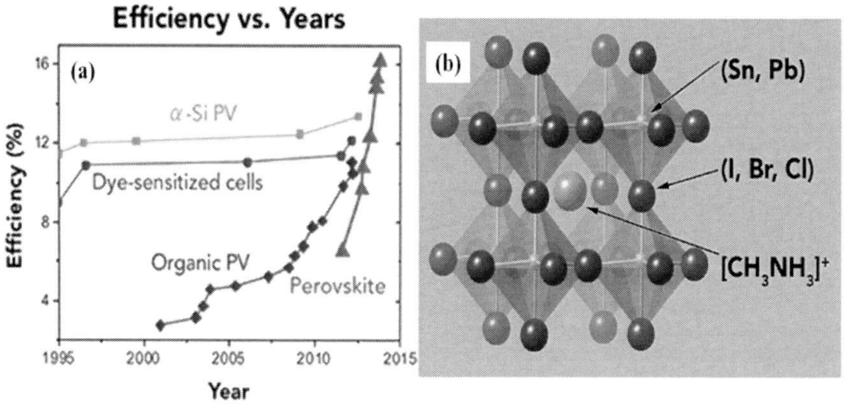

Figure 8. (a) Progress growth of various types solar cell technologies (b) Crystal structure of the perovskite (http://www.depts.ttu.edu/coe/publications/envision/snapshots/fan.php).

The planar device structure contains the perovskite which is directly deposited on a compact n-type inorganic oxide or p-type PEDOT: PSS film covered by a solid organic HTM or fullerene electron acceptor, respectively. However, it is quite difficult to control the morphology of the perovskite films due to lack of scaffolds [90-92]. The perovskite layer with big crystal grain size is crucial for the planar devices to acquire efficient light-absorption and reduce the energy loss in the charge-transport and the charge-collection processes [93, 94].

1.6. Principle of Organic Solar Cells

The electron is excited from the ground state to the excited state upon the irradiation of sunlight. An electron-hole pairs (exciton) are created which diffused on the D-A interface are, as shown in Figure 9. The exciton dissociates at the interface to release the charge carriers of the opposite electrodes [95]. The exciton could only be dissociated when there is an electronic offset of 0.3–0.4 eV between the LUMO levels of donor and acceptor materials. The free charge-carriers migrate within the domain of the acceptor material towards the metal electrode by hopping-transport, while the holes diffuse in the donor material towards the front electrode [96]. The operational mechanism of OPVs is illustrated and could be summarized in the following steps:

Step 1: Light Absorption and excitation of electrons.
Step 2: Exciton (electron-hole pair) formation and diffusion.
Step 3: Charge separation.
Step 4: Charge collection.

Symmetric systems of D-A-D type architecture have attracted a huge attraction where donor and acceptor units are connected by various spacer groups as thiophene, furan, phenyl or vinylene, etc. [97]. The push-and-pull approach is based on the incorporation of electron-rich and electron-deficient unit in the alternating fashion along with the molecular backbone. Owing to push-pull driving forces, the electron-delocalization form the quinoid resonating structures which effectively leads to small band gap materials [98, 99]. The D-A approach is the best tool to control the energy levels and the band gaps of organic molecules and avoids the necessity of interchain effects or generating insoluble for rigidly planar polymer backbones. By extending the oligothiophene length, a red-shift of the absorption and an increase in the HOMO energy upto -5.26 eV was observed despite a rough film-morphology of devices with the overall efficiency upto 0.8% [100]. Small molecular photovoltaic materials have attracted considerable attention, due to their advantages over their polymer counterparts, such as well-defined molecular structure, definite molecular weight, and high purity without batch to batch variations [101, 102].

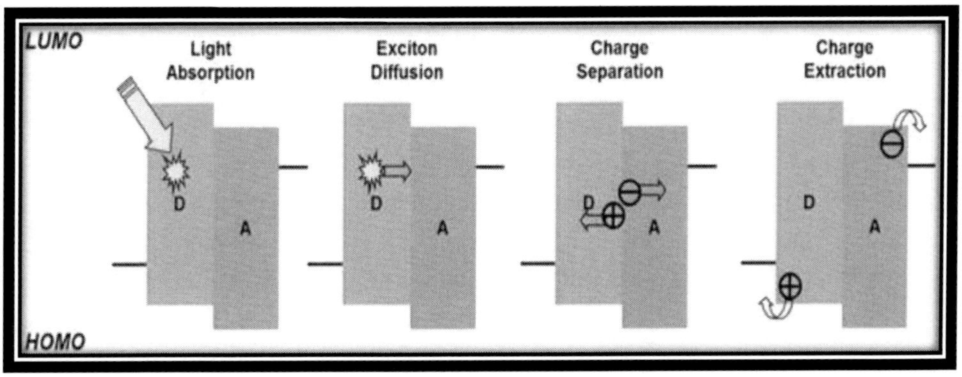

Figure 9. Principle of typical organic solar cell (http://www.sigmaaldrich.com/materials-science/organic-electronics/opv-tutorial.html).

2. THIAZOLO [5,4-*D*]THIAZOLE-BASED ORGANIC CHROMOPHORES

Thiazole derivatives display electron-accepting properties bccause of the presence of electron-withdrawing imine (-C=N-) group in the five membered ring structure. Organic chromophore based on thiazole derivatives exhibited high electron or hole mobilities and applied as n- or p-type semiconductors in OFETs [103-106]. Thiazole-based orgaqnic materials were utilized as electron-donors and achieved high performance in solution-processed fabrication of OPVs [107-110]. The well-known heterocycles, thiazolo[5,4-*d*]thiazole has fused ring rigid planar structure with high oxidative stability and high charge-carrier mobility. The thiazolo[5,4-*d*]thiazole derivatives were easily synthesized and recognized as a good semiconducting material in organic pgotovoltaics (OPVs) [111-114]. Therefore, the introduction of the thiazolo[5,4-*d*]thiazole moiety as an acceptor in the

D-A alternating copolymers might generate an efficient photovoltaic properties along with low band gap and suitable energy levels [115-117]. Thiophene-based organic chromophores with thiazolothiazole unit have shown several advantages for the applications in FETs with reduced band gap [118-120]. The rigid planar geometry leads to the efficient intermolecular π-π interactions. Earlier explored as biological-activity, these molecules have focused their applications for semi-conducting materials [121-124].

2.1. Thiazolo[5,4-*d*]Thiazole-Based Organic Chromophore (RTzR) Containing Terminal Alkyl Thiophenes as Electron-Donor Moiety

2.1.1. Synthesis and Characterizations of Thiazolo [5,4-d]Thiazole-Based Organic Chromophore (RTzR)

Organic conjugated chromophores are considered as potential alternatives for low-band gap organic conjugated polymers in solution-processed fabrication of organic solar cells (OSC) [125-128]. In solution-processed solar cells, organic chromophores are highly explored active layer materials due to their advantages of light weight, ease of synthesis and the purification process etc. [129, 130]. The fullerene derivative, [6,6]-phenyl-C_{61}-butyric acid methyl ester ($PC_{61}BM$) has contributed a crucial role in achieving the high performance OSCs [131, 132]. Small molecule organic solar cells (SMOSCs) of donor-acceptor-donor (D-A-D) type architecture have received huge interests due to their easy synthesis, ease of fabrication, good solubility and hence, display high performance [133, 134]. The synthesized organic chromophore, RTzR shows good thermal stability as analyzed by thermogravimetric analysis. It reveals the decomposition temperature (T_d) point of the RTzR is ~327°C with ~5% of weight loss [Figure 10 (a)]. The low glass transition temperature (T_g) at ~1.2°C of the organic chromophore suggests the flexibility of RTzR as evidenced by differential thermal calorimetric analysis, as shown in the inset of Figure 10(a). The liquid-crystal (LC) property of RTzR might be induced by the presence of the terminal alkyl chains [135]. RTzR has melting transition at ~58°C which forms a LC smectic C phase. RTzR converts into new LC smectic A phase at ~90°C through LC-LC transitions [136, 137]. The phase changes into LC nematic phase at ~138°C and finally reaches to ~170°C i.e., transition into isotropic liquid phase of RTzR.

The chloroform solution (1×10^{-4} M) of RTzR displays an absorption band at λmax ≈ 447 nm in an ultraviolet-visible (UV-Vis) spectroscopy [Figure 10(b)] and the molar absorption coefficient (ε) is calculated as ~$1.78 \times 10^4\ M^{-1}\ cm^{-1}$. The absorption properties indicate an intramolecular charge transfer (ICT) interaction in donor and acceptor moieties [138]. The slight red shifting in thin film spectra of RTzR than chloroform solution might be due to aggregate formation of the molecules. It is noted that RTzR thin film has uniform film thickness which enhances the intermolecular-electron delocalization and a self-assembly behavior [139]. The optical band gap (E_g) of ~2.21 eV was calculated by the absorption edge of thin film of RTzR (Table 1).

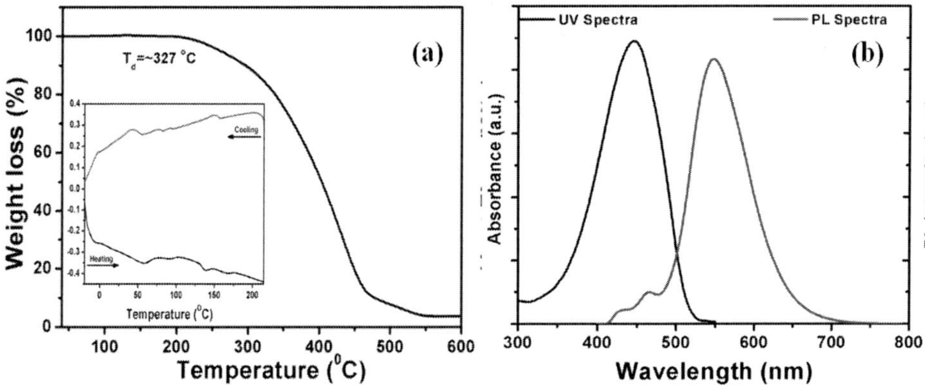

Figure 10. (a) TGA plot with DSC plot (inset) of RTzR at a scan rate of 10 °C/min (b) UV-vis spectra (Black line) and PL spectra (Red line) of RTzR in chloroform solution. Reprinted with permission from M. Nazim, 2013, Chem. Phys. Lett., (2013) © 2013 Elsevier Ltd.

Table 1. Optical and electrochemical properties of RTzR

Small molecule	$\lambda_{max}{}^{a}$ (nm)	$\lambda_{max}{}^{b}$ (nm)	HOMO[c] (eV)	LUMO[d] (eV)	$E_g{}^{e}$ (eV)	$E_g{}^{f}$ (eV)
RTzR	447	452	-5.49	-3.63	1.86	2.21

[a]Absorption in chloroform solution [b]Absorption of thin film on ITO [c]Estimated from the onset of oxidation wave of cyclic voltammogram, [d]Estimated from the onset of reduction wave of cyclic voltammogram, [e]Electrochemical band gap calculated from cyclic voltammogram, [f]Optical band gap calculated from the onset of the UV-vis spectra of the film.

Reprinted with permission from [M. Nazim, 2013], *Chem. Phys. Lett.*, (2013) © 2013 Elsevier Ltd

The photoluminescence (PL) spectroscopy [Figure 10(b)] has been performed in solution and the thin film state for the charge transfer process. The emission band at ~547 nm is recorded in chloroform solution of RTzR. Additionally, RTzR thin film displays strong quenching properties upon blending with $PC_{60}BM$ and hence, facilitates charge-transfer from donor to acceptor.

The electrochemical properties of RTzR thin film [Figure 11(a)] was studied by cyclic voltammetry (CV) in 0.1 M acetonitrile solution of tetra butyl ammonium hexaflouro-phosphate $[nBu_4N]^+[PF_6]^-$ at the potential scan rate of 50 mV/s. The reversible oxidation and reduction peaks are situated at the onset values, $E_{ox} = +1.09 \pm 0.02$ eV and $E_{red} = -0.77 \pm 0.02$ eV, respectively. Hence, RTzR thin film exhibits a good energy level calculated as HOMO value –5.49 eV and LUMO value –3.63 eV from the onset. Noticeably, the difference between HOMO and LUMO is quite lower than that of the optical band gap, indicating the fast delocalization of charges in the film [140, 141]. Generally, mass spectra represent a relationship between the mass of ion and the number of elementary charges that it carries. The mass spectra [Figure 11(b)] of RTzR has shown a strong molecular ion peak at m/z = 802.12 along with some fragment peaks at m/z values at 482.43, 554.60 etc.

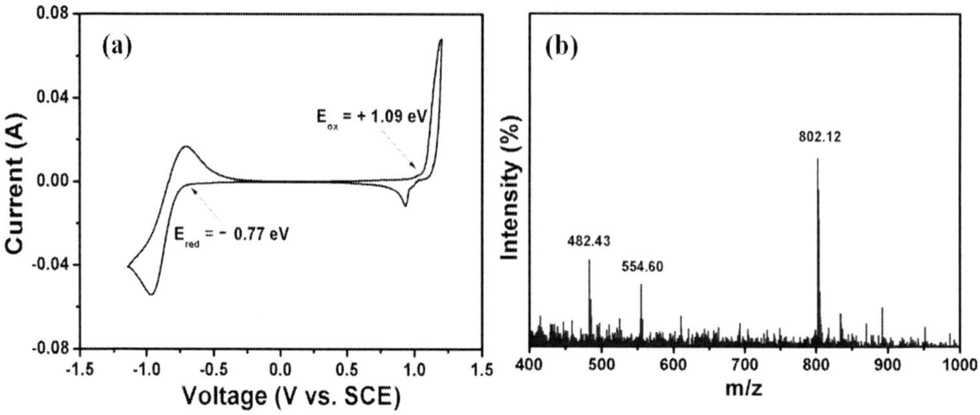

Figure 11. (a) Cyclic voltammetry spectra of RTzR thin film in 0.1 mol/L acetonitrile solution with a scan rate of 50 mV s^{-1} of $[nBu_4N]^+[PF_6]^-$ (b) Mass spectra of RTzR. Reprinted with permission from M. Nazim, 2013, Chem. Phys. Lett., (2013) © 2013 Elsevier Ltd.

2.1.2. Photovoltaic Applications of Thiazolo[5,4-D]Thiazole Based Organic Small Molecule (RTzR)

The synthesized organic chromophore has been applied for the fabrication of solution-processed SMOSCs. The RTzR based SMOSCs has a typical configuration of ITO/PEDOT:PSS/RTzR:$PC_{60}BM$/Ag with the thicknesses of PEDOT:PSS layer as ~50 nm, the RTzR layer of ~80 nm, and Ag layer of ~80-100 nm. The current density-voltage (J-V) curves [Figure 12(a)] of the fabricated SMOSCs was recorded under 100 mW/cm^2 (1.5 AM). The SMOSC fabricated with RTzR:$PC_{60}BM$ active layer (1:1, w/w) ratio achieves a good PCE of ~1.57% with high J_{sc} of ~7.85 mA/cm^2 and V_{oc} of ~0.650 V. However, a low PCE of ~1.20% with low V_{oc} of ~0.424 V are obtained in SMOSCs fabricated with RTzR:$PC_{60}BM$ (1:2, w/w) active layer (Table 2). The J_{sc} is related to the absorption of RTzR:$PC_{60}BM$ active layer (1:1, w/w) ratio which might be due to the large excitons generation and easy exciton-dissociation at interface [142]. The film morphology of the photoactive layers is a crucial factor for achieving a high performance. Hence, the AFM images [Figure 12(b)] are studied for analyzing the morphology of the blend thin film of RTzR:$PC_{60}BM$ active layer ratios before and after annealing.

The RTzR:$PC_{60}BM$ active layer (1:1, w/w) ratio exhibits a rough film morphology as compared to RTzR:$PC_{60}BM$ (1:2, w/w) ratio. The surface roughness is decreased after annealing the devices at 90 °C. Generally, the nanoscale phase separation is responsible for the large interface area for exciton dissociation in the blend thin film [143]. Contrary, the low FF could be obtained from a high rough surface of the blend film which might increase the series resistance of the blend layer/ITO and results in the recombination [144, 145].

Table 2. Summary of J-V curves of the RTzR based fabricated devices in SMOSCs

RTzR:$PC_{60}BM$	Photovoltaic parameters			
	J_{SC} (mA/cm^2)	V_{OC} (V)	FF	PCE (%)
1:1, w/w	**7.85**	**0.650**	**0.31**	**1.57**
1:2, w/w	**7.59**	**0.424**	**0.37**	**1.20**

Reprinted with permission from [M. Nazim, 2013], *Chem. Phys. Lett.*, (2013) © 2013 Elsevier Ltd

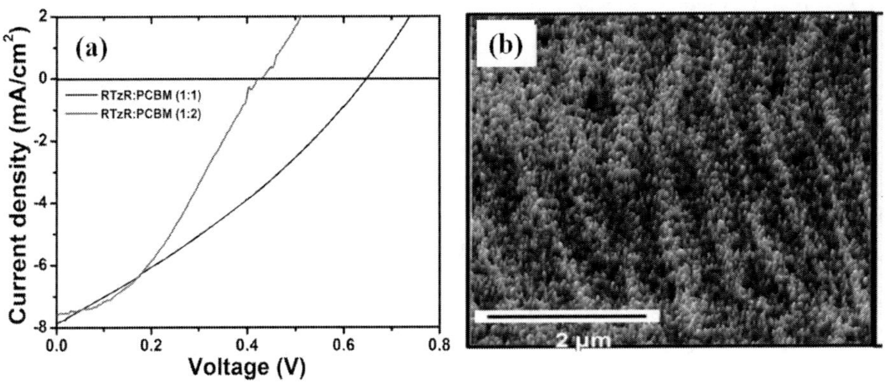

Figure 12. (a) J-V curves of fabricated organic solar cell devices of RTzR:$PC_{60}BM$ active layers (b) Topographic AFM image of RTzR:$PC_{60}BM$ active layer (1:1, w/w) ratio thin film. Reprinted with permission from M. Nazim, 2013, Chem. Phys. Lett., (2013) © 2013 Elsevier Ltd.

2.2. Thiazolo [5,4-*d*]Thiazole-Based Organic Chromophore (TPTzR) Containing Triphenyl Amine (TPA) and Terminal Alkyl Thiophenes as Electron-Donor Moiety

2.2.1. Synthesis and Characterizations of Thiazolo[5,4-d]Thiazole-Based Organic Chromophore (TPTzR)

In OPV devices, chromophore-based organic solar cells (OSCs), received a lot of interest as an alternative and cost-effective source to the renewable energy [146-148]. The D–A–D type organic chromophores display numerous advantages such as low band gap, ease and reproducibility of synthesis, easy purification process and a good solubility in common organic solvents and uniform film morphology which results in high performance of the devices [149-152]. The triphenylamine (TPA)-based SMOSCs have shown promising results because of its high mobility, good solubility and 3D type propeller structure [153-156]. In addition, TPA-based chromophores were synthesized with different acceptor units in linear or star-shaped manner to achieve good performances [157, 158]. Generally, the fused thiazolo[5,4-*d*]thiazole heterocyclic rings show resonating structures which stabilizes the structure and increases the co-planarity of the moleculer backbone [159,160]. Thiazolo [5,4-*d*]thiazole as an acceptor moiety has much importance for the designing of various types of chromophores owing to their good photoactive properties for the solution-processed fabrication [161-163]. This reveals the connection of the solubility with alkyl chains and the absorption enhancement via extending the conjugation length due to the presence of TPA unit that reduces the optical band gap of organic chromophores.

The thermogravimetry analysis [Figure 13(a)] is used to reveal the thermal stability of the organic chromophores. TPTzR shows ~5% weight loss at the ~415 °C which is considered as a good property for the OPVs. TPTzR has observed a low glass transition temperature (T_g) at -42°C, suggesting the elastomeric nature of the small molecule by DSC studies [Figure 13(a) inset] [164]. The alkyl chain in the TPTzR might induce the liquid-crystal (LC) properties and hence, self-assembly behavior. Additionally, TPTzR has melting transitions at ~165°C, at ~183°C and ~194°C, which are associated to the smectic'C' and smectic 'A' phases and nematic phases, respectively. Moreover, smectic 'C' phase changes into nematic phase and then, turns into a completely isotropic liquid phase at ~247°C [165]. In UV-Vis spectra [Figure 13(b)], TPTzR records good optical properties in chloroform solution (1×10^{-4} M) and a strong absorption band at ~474 nm is observed due to the intramolecular charge transfer (ICT) between D and A units [166].

The replacement of alkyl-bithiophene unit by TPA donor extends the conjugation length for the electron-delocalization due to stronge electron-donating ability of TPA. The high molar absorption coefficient (ε) of ~2.08×10^4 M^{-1} cm^{-1} is also calculated for TPTzR in solution which facilitates the morphology of devices [167, 168].

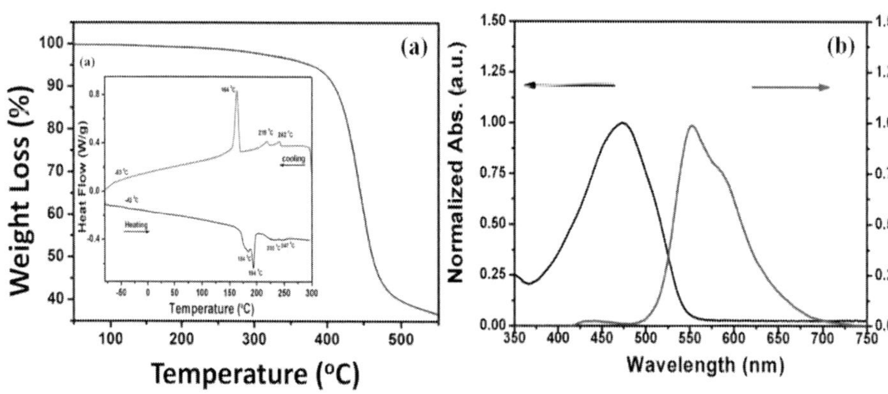

Figure 13. (a) TGA and DSC plots (inset) of TPTzR with scan rates of 10°C/min (b) UV-vis spectra (Black Line) and PL spectra (Red Line) of TPTzR in chloroform solution at room temperature. Reprinted with permission from M. Nazim, 2014, Synth. Met., 187 (2014) 178 © 2014 Elsevier Ltd.

Photoluminescence spectra [Figure 13(b)] of TPTzR in chloroform solution obtains a strong emission band at ~553 nm which shows a strong quenching behavior in the TPTzR:$PC_{60}BM$ active layer. It has been reported that a larger Stokes shift attributes to a smaller overlap area between the absorption and emission spectra. Herein, TPTzR thin film presents a significant stokes shift with difference of ~76 nm as compared to TPTzR solution, representing the excited-state intramolecular interaction between donor and acceptor molecules. Thus, the significant energy difference and stroke shift might result from the good film morphology of TPTzR with $PC_{60}BM$ [169].

Table 3. Optical and electrochemical properties of TPTzR

Small molecule	λ_{max}^{a} (nm)	λ_{max}^{b} (nm)	HOMO^{c} (eV)	LUMO^{d} (eV)	Eg^{e} (eV)	Eg^{f} (eV)
TPTzR	474	365, 515	–6.28	–3.41	2.87	1.98

a Absorption in chloroform solution bAbsorption of thin film on ITO cEstimated from the onset of oxidation wave of cyclic voltammogram, dEstimated from the onset of reduction wave of cyclic voltammogram, eElectrochemical band gap calculated from cyclic voltammogram, fOptical band gap calculated from the onset of the UV-vis spectra of the film.

Reprinted with permission from [M. Nazim, 2014], *Synth. Met.*, 187 (2014) 178 © 2014 Elsevier Ltd

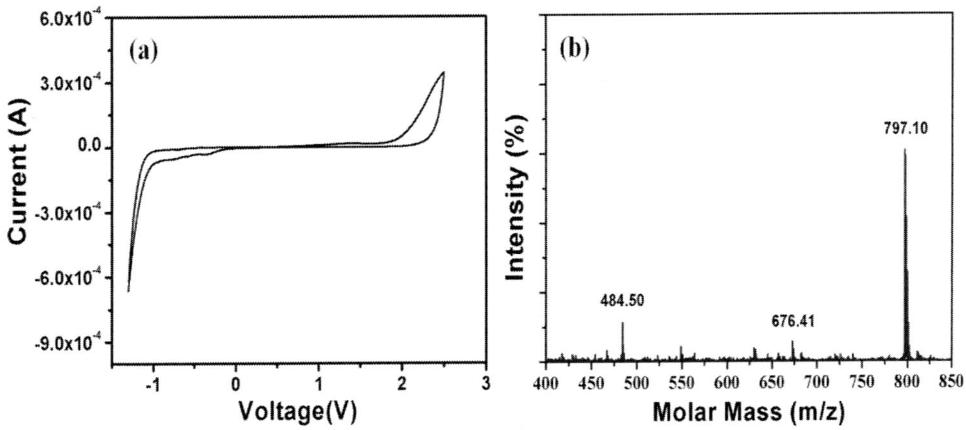

Figure 14. (a) Cyclic voltammogram of TPTzR thin film in 0.1M acetonitrile solution of $[^{n}Bu_4N]^{+}[PF_6]^{-}$ (b) Mass spectra of the organic chromophore, TPTzR. Reprinted with permission from M. Nazim, 2014, Synth. Met., 187 (2014) 178 © 2014 Elsevier Ltd.

The electrochemical analysis [Figure 14(a)] of TPTzR thin film investigated by the cyclic voltammetry (CV) by using 0.1 M acetonitrile solution of tetra butyl-ammonium hexaflouro-phosphate $[nBu_4N]^{+}[PF_6]^{-}$ at a scan rate of 50 mV. The onset values of oxidation and reduction peaks are observed at E_{ox} = +1.88 ± 0.02 eV and E_{red} = –0.99 ± 0.02 eV, respectively which leads to be match up with organic solar cell devices. The incorporation of TPA unit, TPTzR exhibits absorption spectra with a clear red-shift and down-shifted HOMO and LUMO energy levels resulting in the reduced optical band gap (Table 3). The electrochemical reduction requires a higher energy than LUMO which makes the electrochemical band gap higher than the optical band gap of the solar cell devices [170]. In the mass spectrometer, mostly molecules have a single charge, so the m/z value is equivalent to mass itself. Hence, the mass spectra [Figure 14(b)] of RTzR has shown a strong molecular ion peak at m/z = 797.10 along with some fragment peaks at m/z values at 484.50, 676.41 etc. The molecular ion peak is always a radical cation, but the fragment ions might either be even-electron cations or odd-electron radical cations, depending on the neutral fragment lost from the molecule.

2.2.2. *Photovoltaic Applications of Thiazolo[5,4-d]Thiazole Based Organic Chromophore (TPTzR)*

As a photoactive material, TPTzR was applied for the solution-processed fabrication of SMOSCs with the fabricated SMOSC device structure as- ITO/PEDOT:PSS/ TPTzR:PC_{60}BM/Ag. TPTzR based solar cell devices were fabricated on ITO-coated glass substrates. Firstly, ITO-coated glass substrates were cleaned with detergent, ultrasonicated in deionized water, acetone and isopropyl alcohol, subsequently dried overnight. Poly (3,4-ethylene-dioxythiophene):poly(styrenesulfonate) [(PEDOT:PSS)] films were spin-coated on ITO substrates with the scan rate of ~4000 rpm for 60 sec. The layer was annealed at 130°C for 10 min. Additionally, the photoactive layers of different ratios of TPTzR:PC_{60}BM (1:1, 1:2 and 1:3, w/w) were deposited by spin-coating, and finally a top Ag layer was deposited by thermal evaporation. The performances of the fabricated cells of TPTzR:PC_{60}BM ratios are evaluated by current-density versus voltage (J-V) curves [Figure 15(a)] under 100 mW/cm^2 light intensity (1.5 AM). The PCE of ~2.43% with a J_{sc} of ~12.05 mA/cm^2 and V_{oc} of ~0.691 V are attained by TPTzR:PC_{60}BM (1:2, w/w) ratio. However, SMOSCs fabricated with TPTzR:PC_{60}BM of 1:1 and 1:3, (w/w) ratios exhibit the low PCE (Table 4) of ~1.09% and ~1.57%, respectively.

Table 4. Summary of photovoltaic properties of TPTzR based fabricated SMOSCs

Blend film	Photovoltaic parameters			
TPTzR:PCBM	J_{SC} (mA/cm^2)	V_{OC} (V)	FF	PCE (%)
TPTzR (1:1, w/w)	5.22	0.646	0.32	1.09
TPTzR (1:2, w/w)	12.05	0.691	0.30	2.43
TPTzR (1:3, w/w)	7.52	0.651	0.31	1.57

The high performance is generally attributed to a good miscibility of TPTzR and PC_{60}BM resulting in uniform morphology of the devices. Introduction of TPA donor unit has considerably extended the conjugation length and hence, enhanced the absorption edge towards higher wavelength [143]. The presence of alkyl chain along with TPA donor in TPTzR might favor a quick intramolecular charge transfer (ICT) which suppresses the charge-recombination. The surface film morphology is important to determine the performance of solar cell devices. The atomic force microscopy (AFM) analysis has been performed to investigate the morphological properties of the blend TPTzR:PC_{60}BM films formed by spin-coating the chlorobenzene solution. The homogenous film morphology [Figure 15(b)] with low roughness is found with the nanoscale phase separation which suggests a good blend or miscibility of TPTzR with $PC_{60}BM$. The TPTzR:$PC_{60}BM$ blend film provides a large interface area to diffuse and dissociate the excitons which significantly, enhances the performance of TPTzR based solar cell devices [144].

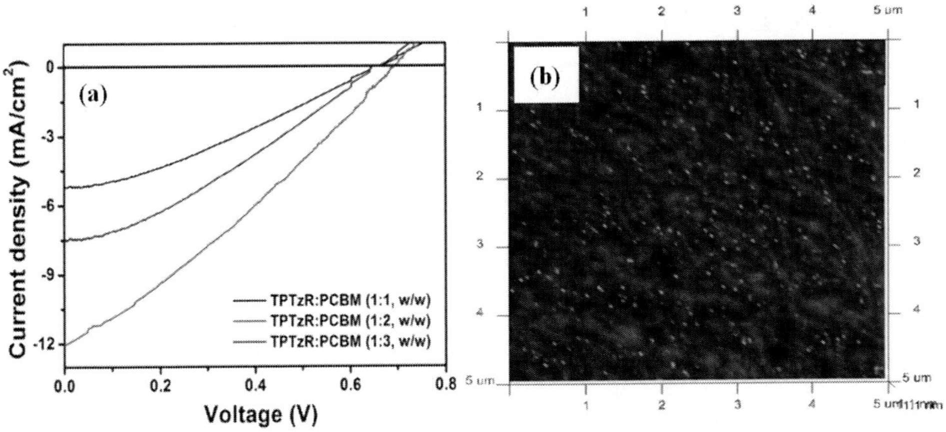

Figure 15. (a) J-V curves of the SMOSC devices with different ratio of TPTzR:$PC_{60}BM$ (1:1, 1:2 and 1:3 w/w) ratios (b) Topographic AFM image of TPTzR:$PC_{60}BM$ active layer (1:2, w/w) ratio film. Reprinted with permission from M. Nazim, 2014, Synth. Met., 187 (2014) 178 © 2014 Elsevier Ltd.

2.3. Thiazolo [5,4-*d*]Thiazole-Based Organic Chromophore (RFTzR) Contains Terminal Alkyl Thiophenes as Electron-Donor Moiety and Furan as Spacer

2.3.1. Synthesis and Characterizations of Thiazolo [5,4-d]Thiazole-Based Organic Chromophore (RFTzR)

Solution-processed small molecule organic solar cells (SMOSCs) are the potential alternatives to the conventional photovoltaic devices due to the definite molecular structures, small molecular weight, easy synthetic and purification techniques [145, 146]. Organic π-conjugated chromophores are considered as a good alternative to low band-gap organic conjugated polymers in solution-processed bulk-heterojunction (BHJ) organic solar cells (OSCs) [147, 148]. Various fabrication parameters as D-A composition ratio, film-thickness, and annealing temperatures etc. significantly influenced the performance [149, 150]. As the electro-negativity of oxygen atom is higher than sulfur atom which reduces the HOMO energy level of the organic chromophore by the introduction of heterocyclic moieties with oxygen. The heterocyclic unit, furan, has been considered as a good replacement for thiophene owing to its good electron-withdrawing nature, and tunale HOMO and LUMO energy levels for organic conjugated chromophore and dyes [151, 152]. The small size of oxygen as compared to sulfur leads significant effect on the molecular rigidity and the geometry with less steric hindrance [153, 154]. The furan-based materials are found to be biodegradable and are obtained from the natural sources directly. These materials are electron-rich, show higher fluorescence with better molecular-packing and better processibility as compared to their thiophene-counterparts. RFTzR shows ~5% weight loss at ~394°C which suggests a good thermal stability of the devices [Figure 16(a)]. The synthesized organic chromophore exhibits the melting transitions at ~205°C and ~225°C which suggest the existence of liquid-crystalline property as shown by DSC analysis [Figure 16(a) inset]. RFTzR displays crystalline transitions at ~222°C and ~187°C during the cooling cycle [155]. The self-assembly behavior is the electrostatic interactions due to the result of π-π staking ability and the hydrogen bonding nature of the organic materials [156].

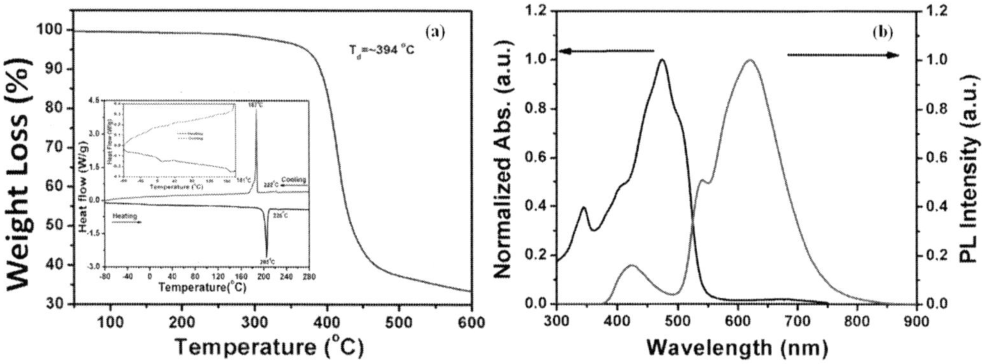

Figure 16. (a) TGA and DSC thermograms (inset) of RFTzR with scan rates of 10°C/min (b) UV-vis spectra (Black Line) and PL spectra (Red Line) of RFTzR in chloroform solution at room temperature. Reprinted with permission from M. Nazim, 2015, RSC Adv., 5 (2015) 6286 © 2015 Royal Society of Chemistry Ltd.

Table 5. Optical and Electrochemical properties of the furan-bridged small molecule, RFTzR

Small molecule	$\lambda_{max}{}^{a}$ (nm)	$\lambda_{max}{}^{b}$ (nm)	HOMO^{c} (eV)	LUMO^{d} (eV)	$E_g{}^{e}$(eV)
RFTzR	473	484	-5.36	-3.14	2.18

a Absorption in chloroform solution bAbsorption of thin film on ITO cEstimated from the onset of oxidation potential from the formula: HOMO = –(4.4 + Eoxonset) (eV), dEstimated from the formula: LUMO = Egopt + HOMO, eOptical band gap calculated from the onset of the UV-vis spectra of the film from the formula: Egopt = 1240/(λonset)film.

Reprinted with permission from [M. Nazim, 2015], *RSC Adv.*, 5 (2015) 6286 © 2015 Royal Society of Chemistry Ltd.

RFTzR in chloroform (1 × 10^{-4} M) [Figure 16(b)] shows an absorption band at $\lambda_{max} \approx$ 473 nm due to π-π* transition which is red shifted in thin film state as studied by UV-Vis absorption spectroscopy. Hence, RFTzR obtains a higher absorption edge which leads to a low optical band gap than that of previously reported RTzR [157]. RFTzR displays a molar absorption coefficient (ε) ~2.0 × 10^{4} $M^{-1}cm^{-1}$ in the solution at maxima absorption which suggests an efficient intramolecular charge transfer (ICT) transition. The red shifting occurs due to the strong molecular ordered packing through a planar molecular structure in the solid state film [158,159]. The film thickness of RFTzR is estimated as ~60 nm which could adsorb enough photons and delivers the homogenous thin film morphology. The molecular geometry of RFTzR permits the electronic interaction between the acceptor and donor moieties. RFTzR exhibits relatively a small optical band gap of ~2.18 eV as compared to its thiophene counterparts [160, 161].

The D-A charge transfer process of RFTzR has been analyzed in the thin film state and the blend films with $PC_{60}BM$ layer by the photoluminescence spectroscopy [Figure 16(b)]. RFTzR film displays a strong emission band at ~625 nm (Table 5) in chloroform solution which might induce exciton-diffusion towards the D-A interface and facilitates the exciton-dissociation. RFTzR:$PC_{60}BM$ blend active layer film (2:1, w/w) ratio displays the complete quenching which suggests an efficient charge transport during the solution-processed fabrication [162, 163]. RFTzR thin film is utilized in cyclic voltammetry studies [Figure 17(a)] in which 0.1 M acetonitrile solution of tetra butyl ammonium hexaflouro phosphate $[nBu_4N]^+[PF_6]^-$ at a potential scan rate of 100 mV/s. Furan-bridged organic chromophore exhibits a low electrochemical-stability owing to irreversible oxidation potential peak which might influence the photovoltaic properties. The oxidation potential peak is situated at the onset values of $E_{ox}^{onset} = +\,0.96 \pm 0.02$ eV. The HOMO value of –5.36 eV is obtained by using the equation:

$$HOMO = -(4.4 + E_{ox}^{onset})\ (eV) \qquad (4)$$

The LUMO energy level is estimated by using the following formula:

$$LUMO = E_g^{opt} + HOMO \qquad (5)$$

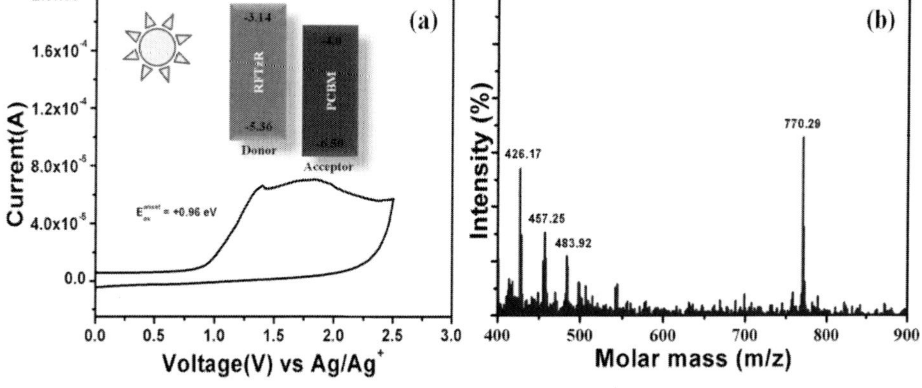

Figure 17. (a) Cyclic voltammogram of RFTzR thin film in 0.1M acetonitrile solution of $[^nBu_4N]^+[PF_6]^-$ (b) Mass spectra of the organic chromophore, RFTzR. Reprinted with permission from M. Nazim, 2015, RSC Adv., 5 (2015) 6286 © 2015 Royal Society of Chemistry Ltd.

Generally, the electron transfer requires a difference in LUMO level of donor by ~0.3 eV to LUMO energy level of the acceptor molecule. For efficient donor to acceptor electron transfer a LUMO-LUMO offset of 0.3–0.4 eV is necessary for an exciton-dissociation [164, 165]. The substitution of thiophene from furan unit and alkyl chains at both the terminal ends might improve the ICT transition and hence, induces the self-assembly behavior RFTzR and improves the efficiency [166].

In mass spectrometry, the molecular ion peak is always a radical cation, while various fragment ion peaks might be either even-electron cations or odd-electron radical cations which depend upon the loss of the neutral fragments from the molecule. The mass spectra generally represents a single charge on the molecules, hence, m/z value is equivalent to the mass itself. The mass spectra [Figure 17(b)] of RFTzR has shown a strong molecular ion peak at m/z = 770.29 along with some fragment peaks at m/z values at 426.17, 457.25 and 483.92 etc.

2.3.2. Photovoltaic Applications of Thiazolo[5,4-d]Thiazole Based Organic Small Molecule (RFTzR)

For solution-processed fabrication, RFTzR is mixed in different ratio with $PC_{60}BM$ in chlorobenzene solution. The architecture of solar cell devices is ITO/PEDOT:PSS/ RFTzR:$PC_{60}BM$/Ag.

Table 6. Photovoltaic parameters of RFTzR based fabricated devices

RFTzR: $PC_{60}BM$	Photovoltaic parameters			
	J_{SC} (mA/cm^2)	V_{OC} (V)	FF	PCE (%)
1:1, w/w	6.38	0.667	0.40	1.72
2:1, w/w	10.13	0.756	0.34	2.72
3:1, w/w	8.29	0.723	0.32	1.94

The ITO substrates were cleaned with detergent, deionized water, acetone and isopropyl alcohol using ultrasonication, and dried in vacuum oven. PEDOT:PSS as a buffer layer (thickness ~80 nm) was deposited by spin-coating with the scan rate of ~4000 rpm for 60 s which was annealed at 130°C for 10 min. The photoactive layers (thickness ~60 nm) of RFTzR:$PC_{60}BM$ (1:1, 2:1 and 3:1, w/w) ratios were deposited above the PEDOT:PSS/ITO films by spin-coating, and finally, top Ag layer of ~100 nm thickness was deposited by thermal evaporation.

The photovoltaic parameters are evaluated by the current density-voltage (J-V) measurements [Figure 18(a)] under the 1 sun light (100 mW/cm^2, 1.5 AM). The PCE of ~2.72% is achieved with RFTzR:$PC_{60}BM$ (2:1, w/w) ratio whereas, the other fabricated device ratios (Table 6) exhibit a lower efficiency of ~1.72% for RFTzR:$PC_{60}BM$ (1:1, w/w) and ~1.94% for RFTzR:$PC_{60}BM$ (3:1, w/w) ratios. The fabricated RFTzR:$PC_{60}BM$ (2:1, w/w) ratio presents a J_{sc} of ~10.13 mA/cm^2 due to higher absorption along with V_{oc} of ~0.756 V. The decrease in V_{oc} with increased concentration of RFTzR in blend layer ratio might relate to the surface morphologies of the devices [167]. RFTzR:$PC_{60}BM$ (1:1, w/w) blend film exhibits a high roughness indicating the low homogeneous film between RFTzR and $PC_{60}BM$ which might result in low V_{oc}.

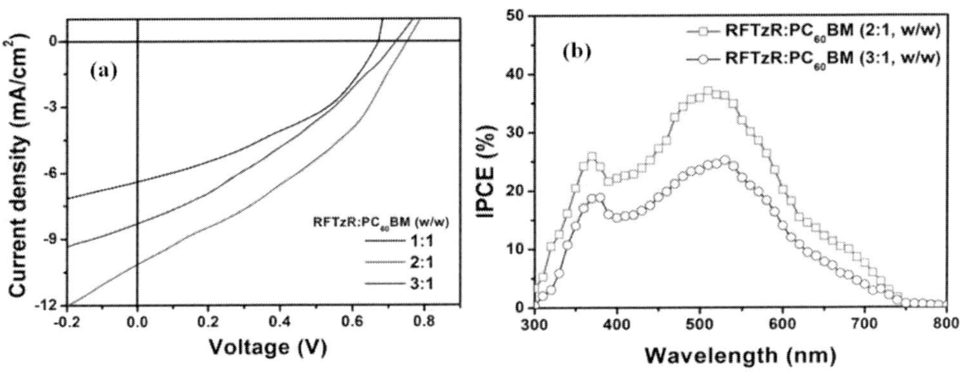

Figure 18. (a) J-V curves of fabricated SMOSCs with RFTzR:$PC_{60}BM$ (1:1, 2:1 and 3:1, w/w) ratios (b) IPCE spectra of the RFTzR:$PC_{60}BM$ at 2:1 and 3:1 (w/w) ratios. Reprinted with permission from M. Nazim, 2015, RSC Adv., 5 (2015) 6286 © 2015 Royal Society of Chemistry Ltd.

The improved J_{sc} and V_{oc} might be explained by the complete ICT due to the incorporation of furan unit in the molecular backbone. Additonally, the introduction of furan spacer might be an effective hole generator owing to its less resonance energy (16 kcal mol^{-1}) than the thiophene as spacer (29 kcal mol^{-1}). The small size of oxygen in furan might induce the planarity and improves the film morphology which enhances the photocurrent density and the performances of solar cell devices [168]. Additionally, the presence of two terminal alkyl chains might facilitate the charge transfer rate and increases the photocurrent density of the devices. In devices, a low V_{OC} and FF might be related to the fast electron transfer and a high recombination rate at the D-A interface of the active layer of the device [169].

The IPCE spectra [Figure 18(b)] are examined to study the contribution of incident-photons to the photocurrent for the fabricated OSC devices. Solar cell devices of RFTzR:$PC_{60}BM$ (2:1, w/w) ratio records the maximum IPCE of ~37% in the broad absorption wavelengths between ~450 to ~650 nm while devices with RFTzR:$PC_{60}BM$ (3:1, w/w) shows low IPCE of ~25% in absorption. The AFM surface morphology (Figure 19) displays a homogeneous behavior of RFTzR:$PC_{60}BM$ blend active layer (1:1, 2:1, 3:1, w/w) ratios, as studied by using the atomic force microscopy (AFM). The samples were prepared by spin-coating of films on the ITO/PEDOT:PSS for better film morphology [170, 171]. The smooth film morphology with the nanoscale phase separation is responsible for a large D-A interface area for an exciton dissociation. The blended film RFTzR:$PC_{60}BM$

(2:1, w/w) exhibits homogeneous and smooth film morphology (Figure 19 (d) with low roughness (R_{rms} = 1.94 nm), suggesting the better mixing of RFTzR with $PC_{60}BM$. The other blended films of RFTzR:$PC_{60}BM$ (1:1 and 3:1, w/w) ratios record a high roughness of R_{rms} = 2.73 nm and R_{rms} = 2.02 nm, respectively. The results show RFTzR:$PC_{60}BM$ active layer (2:1, w/w) ratio is optimized for the best performance of organic solar cell devices [172]. The surface morphology reveals that RFTzR:$PC_{60}BM$ blend film (2:1, w/w) ratio depicts a low surface roughness value (~1.94 nm) as compared to other RFTzR:$PC_{60}BM$ ratios (1:1 and 3:1, w/w) which provides enough surface area. The resistance of different semiconductor layers of the devices in addition to the metal semiconductor layer is defined as series-resistance.

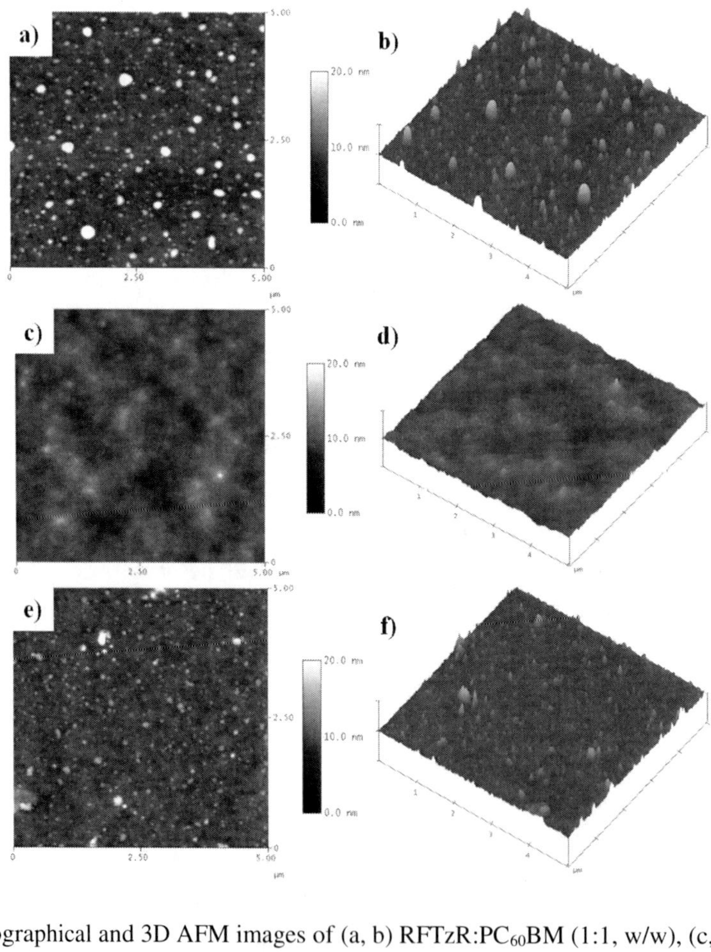

Figure 19. Topographical and 3D AFM images of (a, b) RFTzR:$PC_{60}BM$ (1:1, w/w), (c, d) RFTzR:$PC_{60}BM$ (2:1, w/w), and (e, f) RFTzR:$PC_{60}BM$ (3:1, w/w) ratios. Reprinted with permission from M. Nazim, 2015, RSC Adv., 5 (2015) 6286 © 2015 Royal Society of Chemistry Ltd.

The low FF is related to the increase in the series-resistance of RFTzR:$PC_{60}BM$/ITO, resulting in the high recombination rate [173,174]. The RFTzR:$PC_{60}BM$ (1:1, w/w) ratio has a high FF which is related to a low series resistance (6.16 $\Omega.cm^2$) as compared to other OSC devices. RFTzR:$PC_{60}BM$ (2:1, w/w) ratio with high series-resistances (~9.18 $\Omega.cm^2$)

and RFTzR:$PC_{60}BM$ (3:1, w/w) with ~11.42 Ω.cm^2. RFTZR:$PC_{60}BM$ active layer (2:1, w/w) ratio suggests that RFTzR has a good miscibility with $PC_{60}BM$ in the blended films in phase-segregation process to form a bicontinuous network for the efficient carrier-collection. The presence of furan spacer with the homogeneous blended film of RFTzR:$PC_{60}BM$ (2:1, w/w) ratio improves the D-A interface area, resulting in the high performance of the devices [175, 176].

CONCLUSION

By the generation of photogenerated carriers, the organic solar cells are the promising photovoltaic devices by converting the solar radiation. Organic materials have shown the potential applications due to various advantages of low cost, high efficiency and ease of fabrication process etc. OSCs are efficient photovoltaic technology with the possibility of improvements in the device structures. Untill now, organic chromophores have presented the highest solar efficiency (η) of ~20% for the hybrid solar cells. Thus, novel D-A-D type thiazolo [5,4-*d*]thiazole-based organic chromophores, named RTzR, RFTzR and TPTzR are synthesized by Suzuki cross-coupling reaction. The synthesized chromophores are efficiently utilized as electron donor materials for the solution processed BHJ-SMOSCs. RTzR contains thiazolothiazole-core acceptor and terminal alkyl thiophenes as donor. RTzR shows a good absorption spectrum which reveals the intermolecular interactions. The fabricated devices with RTzR:$PC_{60}BM$ active layer (1:1, w/w) shows a PCE of ~1.57% with J_{SC} of ~7.85 mA/cm^2 under 100 mW/cm^2 (1.5 AM). Another novel D_1-A-D_2 type organic chromophore, TPTzR with thiazolo [5,4-*d*]thiazole-core as acceptor and triphenylamine (TPA) and bithiophenes as donor units is synthesized via Suzuki cross-coupling reaction and applied for organic solar cells. The strong red shift in the absorption of TPTzR is due to the presence of strong electron-donating TPA unit. The solution-processed devices with TPTzR:$PC_{60}BM$ (1:2, w/w) ratio shows a high performance of ~2.43% with a high J_{sc} of ~12.05 mA/cm^2 owing to the uniform film morphology of the fabricated devices. Additionally, a novel furan-bridged thiazolo[5,4-*d*]thiazole based π-conjugated organic chromophore, RFTzR is synthesized and utilized for the solution-processed fabrication of OSCs. The synthesized RFTzR exhibits an excellent solubility in common organic solvent. RFTzR shows the HOMO and LUMO energy levels of −5.36 eV and−3.14 eV, respectively. The fabricated devices with RFTzR:$PC_{60}BM$ (2:1, w/w) ratio exhibits relatively a high performance of ~2.72% with a high J_{SC} of ~10.13 mA/cm^2 and V_{OC} of 0.756 V. The better efficiency of furan-bridged organic chromophore might attribute to a good miscibility with $PC_{60}BM$ and hence, the formation of the uniform film morphology of the devices.

REFERENCES

[1] *BP Statistical Review of World Energy*, bp.com/statistical review; June 2013 [accessed on 13/8/2013].

[2] http://www.treia.org [accessed on 6/6/2013]

[3] A. Brown, S. Müller, Z. Dobrotková "Renewable energy: markets and prospects by technology" *Int Energy Agency Inf Paper* (2011).

[4] O. Ellabbana, H. Abu-Rubb and F. Blaabjerg, Renewable energy resources, current status, future prospects and their enabling technology, *Renew. Sustain. Energy Rev.* 39 (2014) 748.

[5] United Nations Environment Programme (UNEP), *Global environment outlook (GEO year book 2004/5)*, Web site: www.unep.org/geo/yearbook.

[6] Intergovernmental Panel on Climate Changes (IPCC), *Third assessment report-climate changes 2001*, Web site: www.meto.gov.uk.

[7] M. A. Hasan, K. Sumathy, Photovoltaic thermal module concepts and their performance analysis: a review, *Renew. Sustain. Energy Rev.* 14 (2010) 1845.

[8] G. Dennler, M. C. Scharber,C. J. Brabec, Polymer-fullerene bulk heterojunction solar cells, *Adv. Mater.* 21 (2009) 1323.

[9] B. O' Regan, M. Grätzel, A low cost, high efficiency solar cell based on dye-sensitized colloidal TiO_2 films. *Nature*, 353 (1991) 737.

[10] H. Y. Chen, J. Hou, S. Zhang, Y. Liang, G. Yang, Y. Yang, L. Yu, Y. Wu, G. Li, Polymer solar cells with enhanced open-circuit voltage and efficiency, *Nat. Photon.* 3 (2009) 649.

[11] J. C. Hummelen, B. W. Knight, F. Lepeq, F. Wudl, J. Yao, C. L. Wilkins, Preparation and characterization of fulleroid and methanofullerene derivatives, *J. Org. Chem.* 60 (1995) 532.

[12] Y. J. Cheng, S. H. Yang, C. S. Hsu, Synthesis of conjugated polymers for organic solar cells, *Chem. Rev.* 109 (2009) 5868.

[13] A. W. Hains, Z. Liang, M. A. Woodhouse, B. A. Gregg, Molecular semiconductors in organic photovoltaic cells, *Chem. Rev.* 110 (2010) 6689.

[14] H. Hoppe, N. S. Sariciftci, Organic solar cells: an overview, *J. Mater. Res.* 19 (2004) 1924.

[15] S. Gunes, H. Neugebauer, N. S. Sariciftci, Conjugated polymer-based organic solar cells, *Chemical Reviews* 107 (2007) 1324.

[16] B. Walker, C. Kim, T. Q. Nguyen, Small molecule solution-processed bulk heterojunction solar cells, *Chem. Mater.* 23 (2011) 470.

[17] N. S. Sariciftci, L. Smilowitz, A. J. Heeger, F. Wudl, Photoinduced electron transfer from a conducting polymer to Buckminsterfullerene, *Science* 258 (1992) 1474.

[18] A. Facchetti, π-conjugated polymers for organic electronics and photovoltaic cell applications, *Chem. Mater.* 23 (2011) 733.

[19] R. C. Haddon, A. S. Perel, R. C. Morris, T. T. M. Palstra, A. F. Hebard, R. M. Fleming, C_{60} thin film transistors, *Appl. Phys. Lett.* 67 (1995) 121.

[20] M. A. Green, K. Emery, D. L. King, S. Igari, W. Warta, Solar cell efficiency tables (version 29), *Prog. Photovolt.* 13 (2005) 49.

[21] X. Guo, N. Zhou,S. J. Lou,J. Smith, D. B. Tice, J. W. Hennek, R. P. Ortiz, J. T. López Navarrete, S. Li, J. Strzalka, L. X. Chen, R. P. H. Chang, A. Facchetti, T. J. Marks, Polymer solar cells with enhanced fill factors, *Nat. Photon.* 7 (2013) 825.

[22] L. Lu, T. Xu, W. Chen, E. S. Landry, L. Yu, Ternary blend polymer solar cells with enhanced power conversion efficiency, *Nat. Photon.* 8 (2014) 716.

[23] F. C. Krebs, Fabrication and processing of polymer solar cells: a review of printing and coating techniques, *Sol. Ener. Mater. Sol. Cells* 93 (2009) 394.

[24] Y. Kim, S. Cook, S. M. Tuladhar, S. A. Choulis, J. Nelson, J. R. Durrant, D. D. C. Bradley, M. Giles, I. McCulloch, C. S. Ha, M. Ree, A strong regioregularity effect in self-organizing conjugated polymer films and high-efficiency polythiophene:fullerene solar cells, *Nat. Mater.* 5 (2006) 197.

[25] M. C. J. M. Vissenberg, P. W. M. Blom, Transient hole transport in poly(-p-phenylene vinylene) LEDs, *Synth. Met.* 102 (1999) 1053.

[26] C. W. Tang, Two-layer organic photovoltaic cell, *Appl. Phys. Lett.* 48 (1986) 183.

[27] C. W. Tang, S. A. VanSlyke, Organic electroluminescent diode, *Appl. Phys. Lett.* 51 (1987) 913.

[28] G. Yu, J. Gao, J. C. Hummelen, F. Wudl, A. J. Heeger, Polymer photovoltaic cells: enhanced efficiencies via a network of internal donor acceptor heterojunctions, *Science* 270 (1995) 1789.

[29] J. G. Xue, B. P. Rand, S. Uchida, S. R. Forrest, A hybrid planar-mixed molecular heterojunction photovoltaic cell, *Adv. Mater.* 17 (2005) 66.

[30] A. L. Ayzner, C. J. Tassone, S. H. Tolbert, B. J. Schwartz, Reappraising the need for bulk heterojunctions in polymer-fullerene photovoltaics: the role of carrier transport in all solution processed P3HT/PCBM bilayer solar cells, *J. Phys. Chem. C* 113 (2009) 20050.

[31] K. Walzer, B. Maennig, M. Pfeiffer, K. Leo, Highly efficient organic devices based on electrically doped transport layers, *Chem. Rev.* 107 (2007) 1233.

[32] M. A. Green, K. Emery, Y. Hishikawa, W. Warta, Solar cell efficiency tables (version 37) *Prog. Photov. Res. & Appl.* 19 (2011) 84.

[33] A. J. Parnell, A. D. F. Dunbar, A. J. Pearson, P. A. Staniec, A. J. C. Dennison, H. Hamamatsu, M. W. A. Skoda, D. G. Lidzey, R. A. L. Jones, Depletion of PCBM at the cathode interface in P3HT/PCBM thin films as quantified via neutron reflectivity measurements, *Adv. Mater.* 22 (2010) 2444.

[34] J. J. M. Halls, R. H. Friend, M. D. Archer, R. D. Hill, Ed. *Clean electricity from photovoltaics.* (2001) London: Imperial College Press. pp. 377. ISBN 1860941613.

[35] G. Li, V. Shrotriya, J. Huang, Y. Yao, T. Moriarty, K. Emery, Y. Yang, High-efficiency solution processable polymer photovoltaic cells by self-organization of polymer blends, *Nat. Mater.* 4 (2005) 864.

[36] C. Weiran, X. Jiangeng, Recent progress in organic photovoltaic: device architecture and optical design, *Energy Environ. Sci.*, 7 (2014) 2123.

[37] J. Peet, J. Y. Kim, N. E. Coates, W. L. Ma, D. Moses, A. J. Heeger, G. C. Bazan, Efficiency enhancement in low-bandgap polymer solar cells by processing with alkane dithiole, *Nat. Mater.* 6 (2007) 497.

[38] A. J. Heeger, 25th anniversary article: Bulk heterojunction solar cells: understanding the mechanism of operation. *Adv. Mater.* 26 (2014) 10.

[39] F. Yang, M. Shtein, S. R. Forrest, Controlled growth of a molecular bulk heterojunction photovoltaic cells, *Nat. Mater.*, 4 (2005) 37.

[40] G. Dennler, M. C. Scharber, C. J. Brabec, Polymer-fullerene bulk heterojunction solar cells, *Adv. Mater.* 21 (2009) 1323.

[41] G. Li, R. Zhu, Y. Yang, Polymer solar cells, *Nat. Photon.* 6 (2012) 153.

[42] N. M. Kronenberg, M. Deppisch, F. Wurthner, H. W. A. Lademann, K. Deing, K. Meerholz, Bulk heterojunction organic solar cells based on merocyanine colorants, *Chem. Commun.* (2008) 6489.

[43] H. Burckstummer, N. M. Kronenberg, M. Gsanger, M. Stolte, K. Meerholz, F. Wurthner, Tailored merocyanine dyes for solution-processed BHJ solar cells, *J. Mater. Chem.* 20 (2010) 240.

[44] F. Silvestri, M. D. Irwin, L. Beverina, A. Facchetti, G. A. Pagani, T. J. Marks, Efficient squaraine-based solution processable bulk-heterojunction solar cells, *J. Am. Chem. Soc.* 130 (2008) 17640.

[45] D. Bagnis, L. Beverina, H. Huang, F. Silvestri, Y. Yao, H. Yan, G. A. Pagani, T. J. Marks, A. Facchetti, Marked alkyl-vs alkenyl substitutent effects on squaraine dye solid-state structure, carrier mobility, and bulk heterojunction solar cell efficiency, *J. Am. Chem. Soc.* 132 (2010) 4074.

[46] A. Sharenko, C. M. Proctor, T. S. van der Poll, Z. B. Henson, T. Q. Nguyen, G. C. Bazan, A high performing solution processed small molecule perylene diimide bulk heterojunction solar cell, *Adv. Mater.* 25 (2013) 4403.

[47] M. C. Scharber, D. Mühlbacher, M. Koppe, P. Denk, C. Waldauf, A. J. Heeger, C. J. Brabec, Design rules for donors in bulk heterojunction solar cells-towards 10% energy conversion efficiency, *Adv. Mater.* 18 (2006) 789.

[48] L. J. A. Koster, V. D. Mihailetchi, P. W. M. Blom, Ultimate efficiency of polymer/fullerene bulk heterojunction solar cells, *Appl. Phys. Lett.* 88 (2006) 093511.

[49] Y. Sun, G. C. Welch, W. L. Leong, C. J. Takacs, G. C. Bazan, A. J. Heeger, Solution processed small molecule solar cells with 6.7% efficiency, *Nat. Mater.* 11 (2012) 44.

[50] Q. Shi, P. Cheng, Y. Li, X. Zhan, A solution processable D_A-D molecule based on thiazolothiazole for high performance organic solar cells, *Adv. Energy Mater.* 2 (2012) 63.

[51] K. F. Jeltsch, M. Schadel, J. B. Bonekamp, P. Niyamakom, F. Rauscher, H. W. A. Lademann, I. Dumsch, S. Allard, U. Scherf, K. Meerholz, Efficiency enhanced hybrid solar cells using a blend of quantum dots and nanorods, *Adv. Funct. Mater.* 22 (2012), 397.

[52] F. Gao, Y. Wang, D. Shi, J. Zhang, M. Wang, X. Jing, et al. Enhance the optical absorptivity of nanocrystalline TiO_2 film with high molar extinction coefficient ruthenium sensitizers for high performance dye-sensitized solar cells, *J. Am. Chem. Soc.*,130 (2008) 10720.

[53] C. Sima, C. Grigoriu, S. Antohe. Comparison of the dye-sensitized solar cells performances based on transparent conductive ITO and FTO, *Thin Solid Films* (2010), 519 (2):595.

[54] A. Hagfeldt, M. Gratzel, Molecular photovoltaics, *Accounts of Chemical Research* 33(2000), 269.

[55] B. O' Regan, F. Lenzmann, R. Muis, J. Wienk, A solid state sye sensitized solar cell fabricated with pressure treated P25-TiO2 and CuSCN: analysis of pore filling and IV characteristics, *Chem. Mat.* 14 (2002) 5023.

[56] T. Taguchi, X. T. Zhang, I. Sutanto, K. Tokuhiro, T. N. Rao, H. Watanabe, T. Nakamori, M. Uragami, A. Fujishima, Improving the performance of solid-state dye-sensitized solar cell using MgO-coated TiO_2 nanoporous film, *Chem. Comm.* 19 (2003) 2480.

[57] J. Hagen, W. Schaffrath, P. Otschik, R. Fink, A. Bacher, H. W. Schmidt, D. Haarer, Novel hybrid solar cells consisting of inorganic nanoparticles and an organic hole transport material, *Syn. Metals* 8 9(1997) 215.

[58] A. C. Arango, L. R. Johnson, V. N. Bliznyuk, Z. Schlesinger, S.A. Carter, H. H. Horhold, Efficient titanium oxide/conjugated polymer photovoltaic for solar energy conversion, *Adv. Mat.* 12 (2000) 1689.

[59] J. G. Bednorz, K. A. Muller, Possible high T_c superconductivity in the Ba-La-Cu-O system, *B. Phys: Condens. Matter*, 64 (1986) 189.

[60] Z. Cheng, J. Lin, Layered organic-iorganic hybrid perovskite: structure, optical properties, film preparation, patterning and templating engineering, *Cryst. Eng. Comm.*, 12 (2010) 2646.

[61] D. B. Mitzi, S. Wang, C. A. Feild, C. A. Chess, A. M. Guloy, Conducting layered organic-iorganic halides containing <110>-oriented perovskite sheet. *Science* 267(1995) 1473.

[62] B. E. Hardin, H. J. Snaith, M. D. McGehee, The renaissance of dye-sensitized solar cells, *Nat. Photonics* 6 (2012) 162.

[63] H. J. Snaith, L. Schmidt-Mende, Advances in liquid-electrolyte and solid-state dye-sensitized solar cells, *Adv. Mater.*, 19 (2007) 3187.

[64] H. J. Snaith, A. Stavrinadis, P. Docampo, A. A. R. Watt, Lead-sulphide quantum dot sensitization of tin oxide based hybrid solar cells, Solar Cells. *Sol. Ener.*, 85 (2011) 1283.

[65] P. V. J. Kamat, Quantum dot solar cells, semiconductor nanocrystals as light harvesters, *Phys. Chem. C* 112 (2008) 18737.

[66] H. Lee, H. C. Leventis, S. J. Moon, P. Chen, S. Ito, S. A. Haque, T. Torres, F. Nuesch, T. Geiger, S. M. Zakeeruddin, M. Gratzel, M. K. Nazeeruddin, PbS and CdS quantum dot-sensitized solid stste solar cells: old concepts, new results, *Adv. Funct. Mater.* 19 (2009) 2735.

[67] Y. Itzhaik, O. Niitsoo, M .Page, G. J. Hodes, Sb_2S_3-sensitized nanoporous TiO_2 solar cells, *Phys. Chem. C* 113 (2009) 4254.

[68] P. V. Kamat, J. Phys. Quantum dot solar cells, the next big thing in photovoltaics, *Chem. Lett.*, 4(2013) 908.

[69] A. Kojima, K. Teshima, Y. Shirai, T. Miyasaka, Hetrojunction modification for highly efficient organic iorganic perovskite solar cells. Modification for highly efficient organic-inorganic. *210th ECS Meeting*, Cancun, Mexico, Oct. 29 Nov. 3, 2006.

[70] A. Kojima, K. Teshima, Y. Shirai, T. Miyasaka, Novel photoelectrochemical cell with mesoscopic electrodes sensitized by lead-halide compounds (11). In *Proc. 214th ECS Meeting* (ECS, 2014).

[71] A. Kojima, K. Teshima, Y. Shirai, T. J. Miyasaka, Organometal halide perovskite as visible light sensitizers for photovoltaic cells, *Am. Chem. Soc.*, 131 (2009) 6050.

[72] D. B. Mitzi, Solution-processed inorganic semiconductors, *J. Mater. Chem.*, 14 (2004) 2355.

[73] (a) C. C. Stoumpos, C. D. Malliakas, M. G. Kanatzidis, semiconducting Tin and lead iodide perovskite with organic cations: phase transitions, high mobilities, and near infrared photoluminescent properties, *Inorg. Chem.*, 52 (2013) 9019. (b) W.J. Yin, J.H. Yang, J. Kang, Y. Yan, S.H. Wei, Halide perovskite materials for solar cells: a theoretical review, *J. Mater. Chem. A* (2015), DOI: 10.1039/ C4TA05033A.

[74] P. Gao, M. Grätzel, M. K. Nazeeruddin, organohalide lead perovskites for photovoltaic applications, *Energy Environ. Sci.* 7(2014) 2448.

[75] T. Baikie, Y. Fang, J. M. Kadro, M. Schreyer, F. Wei, S. G. Mhaisalkar, M. Grätzel, T. J. White, Synthesis and crystal chemistry of the hybrid perovskite $(CH_3NH_3)PbI_3$ for solid-state sensitized solar cell applications, *J. Mater. Chem. A*, 1(2013) 5628.

[76] S. A. Bretscheider, J. Weickert, J. A. Dorman, L. Schmidt-Mende, Research update: physical and electrical charcteristics of lead halide perovskites for solar cell applications, *APL Mater.* 2 (2014) 040701.

[77] S. Colella, E. Mosconi, P. Fedeli, A. Listorti, F. Gazza, F. Orlandi, P. Ferro, T. Besagni, A. Rizzo, G. Calestani, G. Gigli, F. D. Angelis, R. Mosca, $MAPbI_{3-x}Cl_x$ mixed halide perovskite for hybrid solar cells: the role of chloride as dopant on the transport and structural properties, *Chem. Mater.*, 25 (2013) 4613.

[78] S. D. Stranks, G. E. Eperon, G. Grancini, C. Menelaou, M. J. P. Alcocer, T. Leijtens, L. M. Herz, A. Petrozza, H. J. Snaith, Electron-hole diffusion lengths exceeding 1 micrometer in an organometal trihalide perovskite absorber, *Science*, 342 (2013) 341.

[79] G. Xing, N. Mathews, S. Sun, S. S. Lim, Y. M. Lam, M. Grätzel, S. Mhaisalkar, T. C. Sum, Long-range balanced electron and hole transport length in organic-inorganic $CH_3NH_3PbI_3$, *Science*, 342 (2013) 344.

[80] C. S. Ponseca Jr., T. J. Savenije, M. Abdellah, K. Zheng, A. Yartsev, T. Pascher, T. Harlang, P. Chabera, T. Pullerits, A. Stepanov, J. P. Wolf, V. Sundström, organometal halide perovskite solar cell materials rationalized: ultrafast charge generation, high and microsecond long balanced mobilities, and slow recombination, *J. Am. Chem. Soc.*, 136 (2014) 5189.

[81] Q. Lin, A. Armin, R. C. R. Nagiri, P. L. Burn, P. Meredith, Electro-optics of perovskite solar cells, *Nat. Photonics*, 9 (2015) 106.

[82] D. Wolf, S. Holovsky, J. Moon, S. J. Löper, P. Niesen, B. Ledinsky, M. Haug, F. J. Yum, J. H. Ballif, Organometallic halide perovskites: sharp optical absorption edge and its relation to photovoltaic performance, *J. Phys. Chem. Lett.*, 5(2014) 1035.

[83] (a) H.S. Kim, N. G. Park, Parameters Affecting I–V Hysteresis of CH3NH3PbI3 Perovskite Solar Cells: Effects of Perovskite Crystal Size and Mesoporous TiO_2 Layer, J. Phys. Chem. Lett., 5 (2014) 2927. (b) E. L. Unger, E. T. Hoke, C. D. Bailie, W. H. Nguyen, A. R. Bowring, T. Heumüller, M. G. Christoforo, M. D. McGehee, Hysteresis and transient behavior in current-voltage measurements of hybrid-perovskite absorber solar cells, *Energy Environ. Sci.*, 7 (2014) 3690.

[84] H. J. Snaith, A. Abate, J. M. Ball, G. E. Eperon, T. Leijtens, N. K. Noel, S. D. Stranks, J. T.W. Wang, K. Wojciechowski, W. Zhang, Anomalous hysteresis in perovskite solar cells, *J. Phys. Chem. Lett.*, 5(2014) 1511.

[85] R. S. Sanchez, V. Gonzalez-Pedro, J.-W. Lee, N. G. Park, Y. S. Kang, I. Mora-Sero, J. Bisquert, Slow dynamic processes in lead halide perovskite solar cells, characteristic times and hysteresis, *J. Phys. Chem. Lett.* 5 (2014) 2357.

[86] J. Wei, Y. Zhao, H. Li, G. Li, J. Pan, D. Xu, Q. Zhao, D. Yu, Hysteresis analysis based on the ferroelectric effect in hybrid perovskite solar cells, *J. Phys. Chem. Lett.*, 5 (2014) 3937.

[87] J. M. Ball, M. M. Lee, A. Hey, H. J. Snaith, Low-temperature processed meso-superstructured to thin-film perovskite solar cells, *Energy Environ. Sci.*, 6 (2013) 1739.

[88] N. Li, H. Dong, H. Dong, J. Li, W. Li, G. Niu, X. Guo, Z. Wu, L. Wang, Multifunctional perovskite capping layers in hybrid solar cells, *J. Mater. Chem. A*, 2 (2014) 14973.

[89] T. Leijtens, B. Lauber, G. E. Eperson, S. D. Stranks and H. J. Snaith, The importance of perovskite pore filling in organometal mixed halide sensitized TiO_2-based solar cells, *J. Phys. Chem. Lett.* 5 (2014) 1096.

[90] J. H. Im, I. H. Jang, N. Pellet, M. Grätzel, N. G. Park, Growth of $CH_3NH_3PbI_3$ cuboids with controlled size for high-efficiency perovskite solar cells, *Nat. Nanotechnol.* 9 (2014) 927.

[91] G. E. Eperon, V. M. Burlakov, P. Docampo, A. Goriely, H. J. Snaith, Morphological control for high performance, solution-processed planar heterojunction perovskite solar cells, *Adv. Funct. Mater.* 24 (2014) 151.

[92] Z. Xiao, Q. Dong, C. Bi, Y. Shao, Y. Yuan, J. Huang, Solvent annealing of perovskite-induced crystal growth for photovoltaic device efficiency enhancement, *Adv. Mater.* 26(2014) 6503.

[93] P.W. Liang, C.-Y. Liao, C. C. Chueh, F. Zuo, S. T. Williams, X. K. Xin, J. Lin, A. K. Y. Jen, Additive enhanced crystallization of solution-processed perovskite for highly efficient planr-heterojunction solar cells, *Adv. Mater.* 26(2014) 3748.

[94] Y. Zhao, K. Zhu, CH_3NH_3Cl-assisted one-step solution growth of $CH_3NH_3PbI_3$: structure, charge-carrier dynamic and photovoltaic of perovskite solar cells, *J. Phys. Chem. C.* 118 (2014) 9412.

[95] M. Saliba, K. W. Tan, H. Sai, D. T. Moore, T. Scott, W. Zhang, L. A. Estroff, U. Wiesner, H. J. Snaith, Influence of thermal processing protocol upon the crystallization and photovoltaic performance of organic-inorganic lead trihalide perovskites, *J. Phys. Chem. C.*, 118 (2014) 17171.

[96] A. Hadipour, B. de Boer, J. Wildeman, F. B. Kooistra, J. C. Hummelen, M. G. R. Turbiez, M. M. Wienk, R. A. J. Janssen, P. W. M. Blom, Solution-processed organic tandem solar cells, *Adv. Funct. Mater.* 16 (2006) 1897.

[97] A. Facchetti, Y. Deng, A. Wang, Y. Koide, H. Sirringhaus, T. J. Marks, R. H. Friend, Tuning the semiconducting properties of sexithiophene by α,ω-substitution-α,ω-diperfluorohexylsexithiophene: the first n-type sexithiophene for thin-film transistors, *Angew. Chem. Int. Ed.*, 39 (2000) 4547.

[98] R. C. Knighton, A. J. Hallett, B. M. Kariuki, S. J. A. Pope, A one-step synthesis towards new ligands based on aryl-functionalised thiazolo[5,4-d]thiazole chromophores, *Tetrahedron Lett.*, 51 (2010) 5419.

[99] I. H. Jung, Y. K. Jung, J. Lee, J. H. Park, Y. H. Woo, J. I. Lee, H. Y. Chu, H. K. Shim, Synthesis and electroluminescent properties of fluorine-based copolymers containing electron-withdrawing thiazole derivatives, *J. Polym. Sci. Part A: Polym. Chem.* 46 (2008) 7148.

[100] Y. Lin, P. Cheng, Y. Liu, Q. Shi, W. Hu, Y. Li, X. Zhan, Small molecules based on bithiazole for solution-processed organic solar cells, *Org. Electron.* 13 (2012) 673.

[101] U. Mitschke, T. Debaerdemaeker, P. Bauerle, Structure-proprty relationships in mixed oligoheterocycles based on end capped oligothiophenes, *Eur. J. Org. Chem.*, 2000 (2000) 425.

[102] X. Zhan, D. Zhu, Conjugated polymers for high-efficiency organic photovoltaics, *Polym. Chem.* 1 (2010) 409.

[103] A. B. Koren, M. D. Curtis, A. H. Francis, J. W. Kampf, Intermolecular interactions in π-stacked conjugated molecules, synthesis, structure, and spectral characterization of alkyl bithiazole oligomers, *J. Am. Chem. Soc.*, 125 (2003) 5040.

[104] Q. Peng, J. B. Peng, E. T. Kang, K. G. Neoh, Y. Cao, Synthesis and electroluminescent properties of copolymers based on fluorine and 2,5-di(2-hexyloxyphenyl) thiazolothiazole, *Macromolecules* 38 (2005) 7292.

[105] *J. Ephraim Ber.* 94 (1891) 1027.

[106] I. H. Jung, J. Yu, E. Jeong, J. Kim, S. Kwon, H. Kong, K. Lee, H. Y. Woo, H. K. Shim, synthesis and photovoltaic properties of cyclopentadithiophene based low bandgap copolymers that contain electron-withdrawing thiazole derivatives, *Chem. Eur. J.* 16 (2010) 3743.

[107] J. R. Johnson, R. Ketcham, Thiazolothiazoles. I. the reaction of aromatic aldehydes with dithiooxamide, *J. Am. Chem. Soc.* 82 (1960) 2719.

[108] M. Mamada, J. I. Nishida, D. Kumaki, S. Tokito, Y. Yamashita, n-type organic field-effect transistors with high electron mobilities based on thiazole-thiazolothiazole conjugated molecules, *Chem. Mater.* 19 (2007) 5404.

[109] Q. Fang, B. Jiang, B. Xu, W. Wang, F. Yu, X. Wu, Preparation and properties of new soluble π-conjugated polymers containing a fumaronitrile unit in the main chain, *Macromol. Rapid Commun.* 25 (2004) 1429.

[110] E. H. L. Falcao, Naraso, R. K. Feller, G. Wu, F. Wudl, A. K. Cheetham, Hybrid organic-inorganic framework structures: influence of cation size on metal-oxygen-metal connuctivity in the alkaline earth thiazolothiazoledicarboxylates, *Inorg. Chem.* 47 (2008) 8336.

[111] H. C. Yeh, S. J. Yeh, C. T. Chen, Readily synthesized arylamino fumaronitrile for non-doped red organic light-emitting diodes, *Chem. Commun.*, (2003) 2632.

[112] W. C. Wu, H. C. Yeh, L. H. Chan, C. T. Chen, Red organic light emitting diodes with a non-doping amorphous red emitter, *Adv. Mater.*, 14 (2002) 1072.

[113] B. J. Chen, X. W. Sun, X. Q. Lin, C. S. Lee, S. T. Lee, Improved luminescent efficiency of a red organic dye with modified molecular structure, *Mater. Sci. Eng. B*, 100 (2003) 59.

[114] H. C. Yeh, W. C. Wu, Y. S. Wen, D. C. Dai, J. K. Wang, C. T. Chen, Derivative of α,β-dicyanostilbene: convenient precursor for the synthesis of diphenylmaeimide

compounds, E-Z isomerization, crystal structure and solid state fluorescence, *J. Org. Chem.*, 69 (2004) 6455.

[115] M. Zhao, C. Zhang, C. Stern, A. G. M. Barrett, B. M. Hoffman, Synthesis and properties of dimetallic M^1[Pz[-M^2[Schiff Base] *Inorg. Chem.* 43 (2004) 3377.

[116] S. I. Vagin, M. Hanack, Synthesis and spectroscopic properties of non-symmetrical benzo-annulated porphyrazines and their metal complexes, *Eur. J. Org. Chem.* (2002) 2859.

[117] H. Miwa, K. Ishii, N. Kobayashi, Electronic structures of zinc and palladium tetraazaporphyrin derivatives controlled by fused benzo rings, *Chem. Eur. J.* 10 (2004) 4422.

[118] W. Zhang, Z. He, L. Mu, Y. Zou, Y. Wang,S. Zhao, Red non-doped electroluminescent dyes based on arylamino fumaronitrile derivatives, *Dyes Pigm.*, 85 (2010) 86.

[119] C. Q. Ma, M. Fonrodona, M. C. Schikora, M. M. Wienk, R. A. J. Janssen, P. Baurle, Solution-processed bulk-heterojunction solar cells based on monodisperse dendritic oligothiophenes, *Adv. Funct. Mater.* 18 (2008) 3323.

[120] S. Roquet, R. Bettignies, P. Leriche, A. Cravino, J. Roncali, Three-dimensional tetra(oligothienyl)silanes as donor material for organic solar cells, *J. Mater. Chem.* 16 (2006) 3040.

[121] N. Kopidakis, W. J. Mitchell, J. Lagemaat, D. S. Ginley, G. Rumbles, S. Shaheen, W. L. Rance, Bulk heterojunction organic photovoltaic devices based on phenyl-cored thiophene dendrimers, *Appl. Phys. Lett.* 89 (2006) 103524.

[122] A. Petrella, J. Cremer, L. De Cola, P. Bauerle, R. M. Williams, Charge transfer processes in conjugated triarylamine-oligothiophene-perylenemonoimide dendrimers, *J. Phys. Chem. A* 109 (2005) 11687.

[123] H. Fan, H. Shang, Y. Li, X. Zhan, Enhancement of optical absorption in thin film organic solar cells through the excitation of plasmonic modes in metallic gratings, *Appl. Phys. Lett.* 97 (2010) 133302.

[124] W. Li, C. Du, F. Li, Y. Zhou, M. Fahlman, Z. Bo, F. Zhang, Benzothiadiazole-based linear and star molecules: design, synthesis, and their application in bulk heterojunction organic solar cells, *Chem. Mater.*, 21 (2009) 5327.

[125] J. Cremer, P. Bauerle, Star-shaped perylene-oligothiophene-triphenylamine hybrid systems for photovoltaic applications, *J. Mater. Chem.*, 16 (2006) 874.

[126] S. Paek, N. Cho, K. Song, M. J. Jun, J. K. Lee, J. Ko, Efficient organic semiconductors containing fluorine-substituted benzothiadiazole for solution-processed small molecule organic solar cells, *J. Phys. Chem. C* 116 (2012) 23205.

[127] A. A. Mishra, P. Bauerle, Small molecule organic semiconductors on the move: promises for future solar energy technology, *Angew. Chem. Int. Ed.* 51 (2012) 2020.

[128] Y. Li, Q. Guo, Z. Li, J. Pei, W. Tian, Solution processable D-A small molecules for bulk-heterojunction solar cells, *Energy Environ. Sci.* 3 (2010) 1427.

[129] S. S. Palayangoda, X. Cai, R. M. Adhikari, D. C. Neckers, Carbazole-based donor-acceptor compounds: highly fluorescent organic nanoparticles, *Org. Lett.* 10 (2008) 281.

[130] J. Zhou, X. Wan, Y. Liu, Y. Zuo, Z. Li, G. He, G. Long, W. Ni, C. Li, X. Su, Y. Chen, Small molecules based on benzo[1,2-b:4,5-b]dithiophene unit for high performance solution-processed organic solar cells, *J. Am. Chem. Soc.* 134 (2012) 16345.

[131] X. Liu, Y. Sun, L. A. Perez, W. Wen, M. F. Toney, A. J. Heeger, G. C. Bazan, Narrow band gap conjugated chromophores with extended molecular lengths, *J. Am. Chem. Soc.* 134 (2012) 20609.

[132] Y. Huang, E. J. Kramer, A. J. Heeger, G. C. Bazan, Bulk heterojunction solar cells: morphology and performance relationships, *Chem. Rev.*, 114 (2014) 7006.

[133] Q. Sun, K. Park, L. Dai, Liquid crystalline polymers for efficient bilayer-heterojunction solar cells, *J. Phys. Chem. C* 113 (2009) 7892.

[134] X. Wang, S. Chen, Y. Sun, M. Zhang, Y. Li, X. Li, H. Wang, A furan-bridged D-π-A copolymer HOMO level: synthesis and application in polymer solar cells, *Polym. Chem.* 2 (2011) 2872.

[135] M. Zhang, X. Guo, Y. Li, Synthesis and characterization of a copolymer based on thiazolothiazole and dithienosilole for polymer solar cells, *Adv. Energy Mater.* 1 (2011) 557.

[136] J. S. Park, Y. H. Kim, M. Song, C. H. Kim, M. A. Karim, J. W. Lee, Y. -S. Gal, P. Kumar, S. W. Kang, S. H. Jin, Synthesis and photovoltaic properties of side-chain liquid crystal click polymers for dye sensitized solar cells application, *Macromol. Chem. Phys.* 211 (2010) 2464.

[137] M. Melucci, L. Favaretto, A. Zanelli, M. Cavallini, A. Bongini, P. Maccagnani, P. Ostoja, G. Derue, R. Lazzaroni, G. Barbarella, Thiophene-benzothiadiazole co-oligomers: synthesis, optoelectronic properties, electrical characterization, and thin film patterning, *Adv. Funct. Mater.* 20 (2010) 445.

[138] K. Yao, Y. Chen, F. Li, X. Li, X. Ren, H. Wang, T. Liu, Mesogens mediated self-assembly in applications of bulk heterojunction solar cells based on a conjugated polymer with narrow band gap, *Macromolecules* 44 (2011) 2698.

[139] N. Crivillers, L. Favaretto, A. Zanelli, I. Manet, M. Treier, V. Morandi, M. Gazzano, P. Samorì, M. Melucci, Self-assembly and electrical properties of a novel heptameric thiophene-benothiadiazole based architecture, *Chem. Commun.* 48 (2012) 12162.

[140] W. Li, K. H. Hendriks, A. Furlan, W. S. Roelofs, M. M. Wienk, R. A. Janssen, Universal correlation between fibril width and quantum efficiency in diketopyrrolopyrrole-based polymer solar cells, *J. Am. Chem. Soc.*, 135 (2013) 18942.

[141] J. M. Halls, C. A. Walsh, N. C. Greenham, E.A. Marseglia, R. H. Friend, S. C. Moratti, A. B. Holmes, Efficient photodiodes from interpenetrating polymer networks, *Nature* 376 (1995) 498.
[142] C. M. Cardona, W. Li, A. E. Kaifer, D. Stockdale, G. C. Bazan, Electrochemical considerations for determining absolute frontier orbital energy levels of conjugated polymers for solar cells application, *Adv. Mater.* 23 (2011) 2367.
[143] L. Xue, L. Liu, Q. Gao, S. Wen, J. He, W. Tian, Planar-diffused photovoltaic device based on the MEH-PPV/PCBM system prepared by solution process, *Sol. Energy Mater. Sol. Cells* 93 (2009) 501.
[144] M. Ali, M. Abbas, S. K. Shah, R. Tuerhong, A. Generosi, B. Paci, L. Hirsch, R. Gunnella, Realization of solution processed multi-layer bulk heterojunction organic solar cells by electro-spray deposition, *Org. Electron.* 13 (2012) 2130.
[145] Z. Li, Q. Dong, B. Xu, H. Li, S. Wen, J. Pei, S. Yao, H. Lu, P. Li, W. Tian, New amorphous small molecules synthesis, characterization and their application in bulk heterojunction solar cells, *Sol. Energy Mater. Sol. Cells* 95 (2011) 2272.
[146] B. Walker, A. B. Tamayo, D. T. Duong, X. D. Dang, C. Kim, J. Granstrom, T. Q. Nguyen, A systematic approach to solvent selection based on cohesive energy densities in a molecular bulk heterojunction system, *Adv. Energy. Mater.* 1 (2011) 221.
[147] Z. Du, W. Chen, Y. Chen, S. Qiao, X. Bao, S. Wen, M. Sun, L. Han, R. Yang, High efficiency solution-processed two-dimensional small molecule organic solar cell obtained via low-temperature thermal annealing, *J. Mater. Chem. A* 2 (2014)15904.
[148] S. H. Park, A. Roy, S. Beaupre, S. Cho, N. Coates, J. S. Moon, D. Moses, M. Leclerc, K. Lee, A. J. Heeger, Bulk heterojunction solar cells with internal quantum efficiency approaching 100%, *Nat. Photon.* 3 (2009) 297.
[149] M. A. Green, K. Emery, Y. Hishikawa, W. Warta, E. D. Dunlop, Solar cell efficiency tables, *Prog. Photo. Res. Appl.* 23 (2015) 1.
[150] T. S. Van der Poll, J. A. Love, T. Q. Nguyen, G. C. Bazan, Non-basic high performance molecular for solution-processed organic solar cells, *Adv. Mater.*, 24 (2012) 3646.
[151] J. Mei, K. R. Graham, R. Stadler, J. R. Reynolds, Synthesis of isoindigo-based oligothiophenes for molecular bulk heterojunction solar cells, *Org. Lett.*, 12 (2010) 660.
[152] J. Zhou, X. Wan, Y. Liu, Y. Zuo, Z. Li, G. He, G. Long, W. Ni, C. Li, X. Su, Y. Chen, Small molecules based on benzo[1,2-b:4,5-b']dithiophene unit for high-performance solution-processed organic solar cells, *J. Am. Chem. Soc.*, 134(2012) 16345.
[153] L. G. Kaake, G. C. Welch, D. Moses, G. C. Bazan, A. J. Heeger, Influence of processing additives on charge-transfer time scales and sound velocity in organic bulk heterojunction films, *J. Phys. Chem. Lett.*, 3 (2012) 1253.

[154] S. Shen, L. Gao, C. He, Z. Zhang, Q. Sun, Y. Li, A star-shaped oligothiophene with triphenylamine as core and octyl cyanoacetate as end groups for solution-processed organic solar cells, *Org. Electron.*, 14 (2013) 875.

[155] Y. Liang, B. Peng, J. Liang, Z. Tao, J. Chen, Triphenylamine-based dyes bearing functionalized 3.4-propylenedioxythiophene linkers with enhanced performance for dye-sensitized solar cells, *Org. Lett.*, 12 (2010) 1204.

[156] Y. Yang, J. Zhang, Y. Zhou, G. J. Zhao, C. He, Y. F. Li, M. Andersson, O. Inganas, F. L. Zhang, Solution-processable organic molecule with triphenylamine core and two benzothiadiazole-thiophene arms for photovoltaic application, *J. Phys. Chem.C*, 114 (2010) 3701.

[157] J. A. Mikroyannidis, M. M. Stylianakis, P. Suresh, P. Balraju, G. D. Sharma, Low band gap vinylene compounds with triphenylamine and benzothiadiazole segments for use in photovoltaic cells, *Org. Electron.*, 10 (2009) 1320.

[158] Z. Li, J. Pei, Y. Li, B. Xu, M. Deng, Z. Liu, H. Li, H. Lu, Q. Li, W. Tian, Synthesis and photovoltaic properties of solution processable small molecules containing 2-pyran-4-ylidenemalononitrile and oligothiophene moieties, *J. Phys. Chem. C*, 114 (2010) 18270.

[159] Y. Zhang, J. Liu, T. Q. Nguyen, Photoresponse of donor/acceptor blends in organic transistors: A tool for understanding field-assisted charge separation in small molecule bulk heterojunction solar cells, *ACS Appl. Mater. Inter.*, 5 (2013) 2347.

[160] M. Zhang, X. Guo, Y. Li, Synthesis and characterization of a copolymer based on thiazolothiazole and dithienosilole for polymer solar cells, *Adv. Energy Mater.*, 1 (2011) 557.

[161] Y. Lin, H. Fan, Y. Li, X. Zhan, Thiazole-based organic semiconductors for organic electronics, *Adv. Mater.*, 24 (2012) 3087.

[162] M. Nazim, S. Ameen, M. S. Akhtar, Y. S. Lee, H. S. Shin, Novel thiazolothiazole based linear chromophore for small molecule organic solar cells, *Chem. Phys. Lett.*, 574 (2013) 89.

[163] I. H. Jung, J. Yu, J. Jeong, J. Kim, S. Kwon, H. Kong, K. Lee, H. Y. Woo,H. K. Shim, Synthesis and photovoltaic properties of cyclopentasithiophene-based low-bandgap copolymers that contain electron-withdrawing thiazole derivatives, *Chem. Eur. J.*, 16 (2010) 3743.

[164] P. Cheng, Q. Shi, Y. Lin, Y. Li, X. Zhan, Evolved structure of thiazolothiazole based small molecules towards enhanced efficiency in organic solar cells, *Org. Electron.*, 14 (2013) 599.

[165] S. K. Lee, J. M. Cho, Y. Goo, W. S. Shin, J. C. Lee, W. H. Lee, I. N. Kang, H. K. Shim, S. J. Moon, Synthesis and characterization of a thiazolo[5,4-d]thiazole-based copolymer for high performance polymer solar cells, *Chem. Commun.*, 47 (2011) 1791.

[166] K. Yao, Y. Chen, F. Li, X. Li, X. Ren, H. Wang, T. Liu, Mesogens mediated self-assembly in applications of bulk heterjunction solar cells based on a conjugated polymer with narrow band gap, *Macromolecules* 44 (2011) 2698.

[167] H. J. Song, S. M. Lee, J. Y. Lee, B. H. Choi, D. K. Moon, The synthesis and electroluminescent properties of dithienyquinacridone-based copolymers for white light-emitting diodes, *Synth. Met.*, 161 (2011) 2451.

[168] P. Ding, C. C. Chu, B. Liu, B. Peng, Y. P. Zou, Y. H. He, K. C. Zhou, C. S. Hsu, A high-mobility low bandgap copolymer for efficient solar cells, *Macro. Chem. Phys.*, 211 (2010) 2555.

[169] Y. Chen, C. Li, P. Zhang, Y. Li, X. Yang, L. Chen, Y. Tu, Solution-processable tetrazine and oligothiophene based linear A–D–A small molecules: Synthesis, hierarchical structure and photovoltaic properties, *Org. Electron.*, 14 (2013) 1424.

[170] Z. Zeng, C. S. Garoufalis, S. Baskoutas, G. Bester, Electronic and optical properties of ZnO quantum dots under hydrostatic pressure, *Phys. Rev. B* 87 (2013) 125302.

[171] B. Walker, X. Han, C. Kim, A. Sellinger, T. Q. Nguyen, Solution-processed organic solar cells from dye molecules: an investigation of diketopyrrole: vinazene heterojunctions, *ACS Appl. Mater. Inter.*, 4 (2012) 244.

[172] J. S. Moon, C. J. Takacs, Y. Sun, A. J. Heeger, Spontaneous formation of bulk heterojunction nanostructures: multiple routes to equivalent morphologies, *Nano Lett.* 11 (2011) 1036.

[173] K. Wang, B. Guo, Z. Xu, X. Guo, M. Zhang, Y. Li, Solution-processable organic molecule for high performance organic solar cells with low acceptor content, *ACS Appl. Mater. Interfaces* 7 (2015) 24686.

[174] S. Wang, L. Hall, V. V. Diev, R. Haiges, G. Wei, X. Xiao, P. I. Djurovich, S. R. Forrest, M. E. Thompson, N,N-diarylanilinosquaraines and their application to organic photovoltaics, *Chem. Mater.*, 23 (2011) 4789.

[175] A. C. Grimsdale, K. Leok Chan, R. E. Martin, P. G. Jokisz, A. B. Holmes, Synthesis of light-emitting conjugated polymers for applications in electroluminescent devices, *Chem. Rev.* 109 (2009) 897.

[176] A. Guerrero, S. Loser, G. Garcia-Belmonte, C. J. Bruns, J. Smith, H. Miyauchi, S. I. Stupp, J. Bisquert, T. J. Marks, Solution-processed small molecule: fullerene bulk-heterojunction solar cells: impedance spectroscopy deduced bulk and interfacial limits to fill factors, *Phys. Chem. Chem. Phys.* 15 (2013) 16456.

[177] G. Wei, R. R. Lunt, K. Sun, S. Wang, M. E. Thompson, S. R. Forrest, Efficient, ordered bulk heterojunction nanocrystalline solar cells by annealing of ultrathin squaraine thin films, *Nano Lett.*, 10 (2010) 3555.

In: Handbook of Remediation ... ISBN: 978-1-53615-118-3
Editors: Sadia Ameen, M. Shaheer Akhtar et al.

Chapter 2

NANOENGINEERING ASPECTS FOR ENVIRONMENTAL REMEDIATION

Ved Vati Singh*

Department of Chemistry, Indian Institute of Technology Delhi, Delhi, India

ABSTRACT

The clean up or removal of contaminants from the environment is called "environmental remediation." Environmental pollution has become a global menace due to gradual increase in the industrialization, constant urbanization, and the changing life style of the people. Cleaning up the environment from various sources of pollution is imperative to protect the ecological and general public health. The environmental remediation encompasses all the remediation methods employed to reduce the environmental and/or human health risks due to environmental pollution. Nanotechnology could play an important role in reducing the existing environmental problems. Nano-remediation of environment addresses the possible solutions of the existing environmental problems using nanotechnology and the preventive measures for the future.

This chapter illustrates the comprehensive knowledge about the existing development and research activities on the environmental remediation using nanotechnology. Various types of the treatments and remediation are discussed by exploring different technologies/nanostructured materials from the contaminated waste water, surface water, ground water, mine tailings, pollute atmosphere, air and soil. The chapter also features a section with nano-sensing of heavy metals and the poisonous gases. We believe that this chapter will serve as a reference guide to provide the background information for the innovations of sustainable solutions for the environmental remediation.

Keywords: environmental remediation, waste water, hazardous gases, nanostructured materials, heavy metals

* Corresponding Author's Email: singhved30@gmail.com.

1. INTRODUCTION

The heavy industrialization and the resulting industrial by products (hazardous wastes, poisonous gas fumes and smokes) have severe effects on the environment. Currently, the conventional technologies, such as adsorption, biological and chemical oxidation, and incineration, etc. are used for the treatment of all types of organic and toxic wastes. In addition, the extensive research is being carried out to design the efficient and cost-effective techniques for treating the environmental pollution. The water remediation techniques could be broadly classified as: *ex-situ* and *in-situ*. The treatment of the pollution after the removal of the contaminated soil, sediments, or water from the polluted site is known as *ex-situ* technique, while the clean-up of the contamination when it is still inside the ground, without the need of off-site treatment falls under *in-situ* technique. Most of the conventional remediation methods being used are based on classical *ex-situ* (pump-and-treat) approaches that involve the removal or extraction of the contaminated soil, ground water, and sediments, followed by their treatments. However, the *ex-situ* techniques are limited due to their high cost, permissible cleanup standards, and the problem of disposal of the concentrated hazardous residues. The *in-situ* techniques overcome these limitations upto some extents.

Most common the traditional environmental remediation techniques are being used to treat the large volumes of water intended for the municipal water supplies including carbon adsorption, air stripping, oxidation through ozonation or chlorination, ultrafiltration and sedimentation [1]. These approaches are generally used for the large scale pollutant removal. On small scale, activated carbon adsorbents are being used to remove the organic contaminants. Scientific and the technical methods to mitigate the environmental pollution rely on many approaches that vary for the soil, water and air. The choices of remediation approach depend on the complexity and nature of the contaminated media and economic costs of the treatment [2]. The rapid growth of the nanotechnology offers immense opportunities for the environmental applications. The environmental nanotechnology is considered to play a key role in the shaping of current environmental engineering and science. It has stimulated the development and use of novel and cost effective technologies for pollution detection, monitoring and the remediations [3]. Nanomaterials of extremely small size (1-10 nm) with large surface-to-volume ratios have an excellent chemical reactivity as compared to the bulk materials and unique functionalities to treat the pollutants. These materials are excellent adsorbents, catalysts and sensors due to their large specific surface areas and a high reactivity. Furthermore, the nanomaterials enhance the redox and the photocatalytic properties [4]. The unique properties of the nanoparticles as efficient catalysts, tunable photoactivity and increased strength make them desirable for several practical applications. Various natural and engineered nanomaterials including chitosan, silver nanoparticles (nAg), photocatalytic TiO_2, and carbon nanotubes (CNT) have shown strong antimicrobial properties, while others have shown proficiency in the detection of pesticides and heavy metals (e.g., cadmium, copper, lead, mercury, arsenic, etc. [5].

Nanoremediation is the term used to describe various techniques and the methods to clean up the contaminated sites using engineered nanomaterials. Even in the natural environment, nanomaterials of various shapes/morphologies/forms have a significant impact on maintaining water and an air quality. The high mobility of the nanomaterials in solution makes it possible for a quick scanning of entire volume with the small amounts of the nanomaterials. Thus, the pollutants adsorbed onto the nanomaterials could be easily removed by applying mild gravitational (centrifugal) or magnetic (in the case of magnetic nanoparticles) forces. According to the recent data, currently there are approximately 70 sites around the world that are using nanoremediation techniques either at pilot or full study scales [6].

Fresh drinking water sources are being increasingly scarce due to overconsumption and the contamination. Scientists have begun to consider sea water as an alternating source for the drinking water. However, sea water contains too much salt for human consumption. The nano-filters based on carbon nanotube membranes have shown a great potential for desalination, to convert the salty water into drinking water and to make it free from the hazardous substances. These methods have also proved to be cost effective. Air pollution is another area where nanotechnology has shown great potential. Nano-filtration techniques for air, similar to the water purification methods, could be used in the buildings to purify indoor air, and applied to automobile tailpipes and factory smokestacks to separate out contaminants as well as prevent them from entering the atmosphere. Finally, the nano-sensors are developed to detect toxic gas leaks at extremely low concentrations and to detect water contaminants. In this chapter, nano-remediation techniques for water and air will be discussed separately.

2. Environment Remediation Techniques

The three main ways to discuss the remediation are biological, physical and chemical processes.

2.1. Biological Treatments

Biodegradation is defined as the breakdown of organic compounds by the living organisms, resulting in the formation of carbon dioxide and water or methane [7]. Biodegradation is catalyzed by the microorganisms like bacteria, fungi, and microfauna (e.g., protozoans, some worms, and some insects), etc. leading to a specific and non-specific degradation of the substances. Detail discussion about bioremediation is beyond the scope of this chapter.

2.2. Physical Treatments

The most common form of the physical remediation technologies utilized today are air sparging/stripping, dual phase vacuum extraction, pump and treat and soil washing. Soil vapor extraction involves the installation of wells and pipes in the soil, through which soil contaminants are extracted in vapor form. Soil washing involves the physical removal of the contaminated soil, followed by their treatment at a plant on or off-site. After the treatment of the contamination, the soil is returned to the site. Soil washing is a water-based process for scrubbing soil *ex-situ*. The process removes the contaminants either by dissolving or suspending them in the wash solution (attained by chemical manipulation of pH for a period of time). Soil washing is a proven technology for the remediation of radionuclides, heavy metals and hydrocarbons.

2.3. Chemical Treatments

The process of using chemicals to clean-up pollutants from the contaminated media is known as chemical remediation. The key chemical remediation methods available are solvent extraction, chemical oxidation, chemical immobilization, critical fluid, *in-situ* catalyzed peroxide remediation and photo-degradation. Solvent extraction is normally used for the clean-up of the chemicals like PCBs that do not dissolve in water and therefore, tends to stick (sorb) to soil and sediments. Chemical oxidation involves pumping of oxidants into the ground where they are mixed with the harmful chemicals and helps to break the chemicals down into harmless substances such as water and carbon dioxide. Chemical remediation methods help to keep the costs of remediation low.

3. Environment Remediation by Chemical Degradation

Chemical degradation is one of the widely used environmental remediation techniques. The various chemical degradation methods are ozone/UV radiation/H_2O_2 oxidation, supercritical water oxidation, photocatalytic and sonochemical degradation, electron beam process, Fenton method, electrochemical method, solvated electron reduction, permeable reactive barriers of iron and other zero-valent metals, and enzymatic treatment methods.

Detail discussion about these techniques is out of scope of this chapter. Therefore, author is presenting a brief introduction about some of these techniques.

The use of O_3/UV/H_2O_2 generally involves an oxidation/photolysis process to remove the foreign matters which are present in the water. This oxidation process is destructive type, and also termed as "photocatalytic degradations" that means an involvement of photons and a catalyst. The oxidation process generates hydroxyl free radicals in-situ, used for the degradation of pollutants. Phenols and many pesticides react rapidly with these hydroxyl radicals whereas, the organochlorine compounds are less reactive. In principle, the photocatalytic reaction occurs in the adsorbed phase and the overall process could be formally divided into five steps, 1) The transfer of the reactants from the fluid phase to the surface, 2) The adsorption of at least one of the reactants, 3) The reaction in the adsorbed phase, 4) The desorption of the product(s), and 5) The removal of the product(s) from the interfacial region. The adsorption and desorption rates depends on the temperature and the affecting rate of overall photocatalytic reaction.

In the supercritical water oxidation (SCWO) process, an aqueous waste stream and oxygen are brought together in a heated pressurized reactor, operating above the critical point of water (374°C, 22.1 MPa or 218 atm). Under these conditions, the solubility of water is reversed with increase in inorganic solubility and decreased in inorganic solubility, while the viscosity diminishes to attain a gas-like feature that overall enhances the mass transfer properties of the water. These unique properties of hot pressurized water allows oxygen and organics to be contacted in a single phase in which the oxidation of the organics proceeds rapidly. At 400 - 650°C and 3750 psi, SCWO could be used to achieve the complete oxidation of many organic compounds with destruction rate efficiencies of ~ 99.99% or higher. It is said that at the supercritical temperature and pressure, metal oxide nanoparticles are produced by the hydrothermal synthesis process. The SCWO has been successfully employed to reduce the TOC of waste water from 37480 ppm to 200 ppm after the reaction [8].

Fenton's reagent is a solution of hydrogen peroxide with ferrous iron as a catalyst that is used to oxidize the contaminants (organic compounds such as trichloroethylene and tetrachloroethylene) in waste water [9]. During the Fenton reaction, hydrogen peroxide oxidizes Iron (II) to Iron (III), forming a hydroxyl radical and hydroxide ion. With another molecule of hydrogen peroxide, Iron (III) is reduced back to Iron (II) with the formation of hydroperoxyl radical and a proton. The free radical generated in this process then engaged in the secondary reaction. Oxidation of free radical by this method is rapid and exothermic and results in the oxidation of contaminants to primarily carbon dioxide and water [10].

$$Fe^{2+} + H_2O_2 + H^+ \rightarrow Fe^{3+} + HO^- + H_2O \tag{3.1}$$

$$Fe^{3+} + H_2O_2 \rightarrow Fe^{2+} + HOO^- + H^+ \tag{3.2}$$

$$Fe^0 + NO_3^- + 2H^+ \rightarrow Fe^{2+} + H_2O + NO_2^- \quad (4.5)$$

$$NO_3^- + 6H_2O + 8e^- \rightarrow NH_3 + 9OH^- \quad (4.6)$$

Oxides of poly-valent metals like Fe (III), Al (III), Ti (IV) and Zr (IV) show ligand sorption properties through the formation of inner-sphere complexes. Furthermore, the hydrated Fe (III) oxide (HFO) is inexpensive, readily available and is chemically stable over a wide pH range, making it lucrative for the practical applications. Iron (III) oxides have high sorption affinity towards both As (V) or arsenates and As (III) or arsenites [18]. Hybrid-ion-exchange (HIX) could be used for the simultaneous removal of arsenate and perchlorate by the impregnation of nano-crystaline iron hydroxide nanoparticle onto a strong base ion-exchange (IX) resin [19]. Carboxy methylcellulose (CMC) stabilizes ZVI nanoparticles reduced Cr (VI) in both an aqueous and soil media. The stabilized ZVI nanoparticle is more effective than the non-stabilized one for the removal of Cr (VI). CMC-stabilized ZVI nanoparticles catalyst enhances the perchlorate reduction rate in saline. The different environmental remediation using iron-based nanoparticles is summarized in Table 1.

4.1.3. Bimetallic Nanoparticles

ZVI based environmental remediation techniques as discussed above, even though it is one of the latest innovations, encounters various challenges with the implementation of zero-valent iron based techniques as given below:

- Production and accumulation of chlorinated by-products in the form of toxic chlorinated hydrocarbons due to a low reactivity of iron powder.
- Iron reactivity decreases over time due to the formation of surface passive layers or the precipitation of metal hydroxides and metal carbonates on the iron surface.
- Although, nZVI is efficient for enhancing the rate of reaction and reducing the formation of by-products [20], overtime it become less stable, and prone to air oxidation and water hydrolysis. This leads to a significant reduction in the efficiency of nZVI particles in case of large-scale environmental remediation.

The deposition of small amounts of second metals such as Pd, Pt, Ag, Ni and Cu on iron (bimetallic) has shown to be beneficial for accelerating the reaction rate. Relative potential difference of iron deposited with any of the above mentioned metals drive the electron from iron to that metal [21]. Bimetallic nanoparticles of ZVI and Al are very efficient for the removal of chlorinated organic solvents, nitrate and heavy metals, such as Cr^{6+} and Cu^{2+}, and to degrade CCl_4 by a factor of 6 as well as perchlorate ions from the waste water in comparison to micro sized Fe ZVI [22]. Bimetallic nanoparticles of Pd-Fe/Au provide a higher degradation of CCl_4 than one-dimensional ZVI nanoparticles due to the catalytic effects of the metals (Pd or Ni) through a direct hydrogen reduction and the prevention of oxide formation at the iron surface. These catalysts are considerably more active than Pd NPs.

Table 1. Iron based nanoremediation for environment

Type of Nanoparticles	Target material	Type of remediation
Iron Nanoparticles	Ni(II)	Metal removal
Zero valant iron nanoparticles	Nitrate	Nutrient removal
	Humic acid	Organic removal
	Arsenic(V)	Metal removal
Iron sulfide nanoparticles	Lindane	
BCF	Arsenate and Arsenite	Metal removal
CMC stabilized ZVI nanoparticle	Perchlorate	Salt removal
Colloidal zero-valent powder iron	Herbicide Molinate	Biological removal

Table 2. Removal of pollutants using bimetallic nanoparticles

Type of Bimetallic Nanoparticles	Target material	Type of remediation
Ni/Fe nanoparticles	Trichloroethene	Organic removal
	Carbon tetrachloride	-
Pd/Fe Nanoparticles	Trichloroethene	Organic removal
	Tetrachloroethene	Organic removal
Cu/Fe Nanoparticles	Nitrate	Nutrient removal

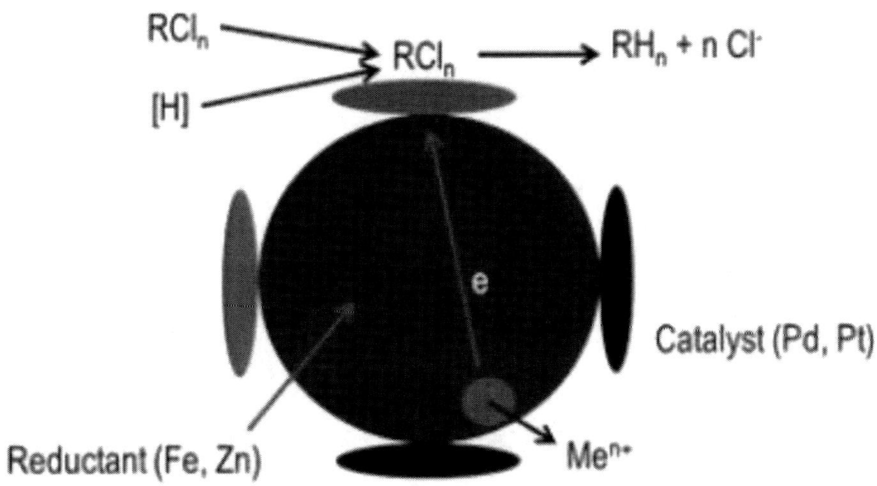

Reprinted with the permission from W. X. Zhang, C. B. Wang, H. L. Lien, *Catal. Today* 40 (1998) 378.

Figure 1. Removal of chlorinated solvent using bimetallic nanoparticle.

Polymer membrane immobilized nanoparticles have two advantages i.e., 1) it controls growth of zero-valent iron metal nanoparticles, and 2) the localization and higher concentration of organic substances in the membrane domain leads to a significant enhancement of the reaction rates [23]. Figure 1 and Figure 2 shows the scheme for the removal of chlorinated solvent and the nitrate reduction reaction at Cu/Fe system using bimetallic nanoparticles respectively.

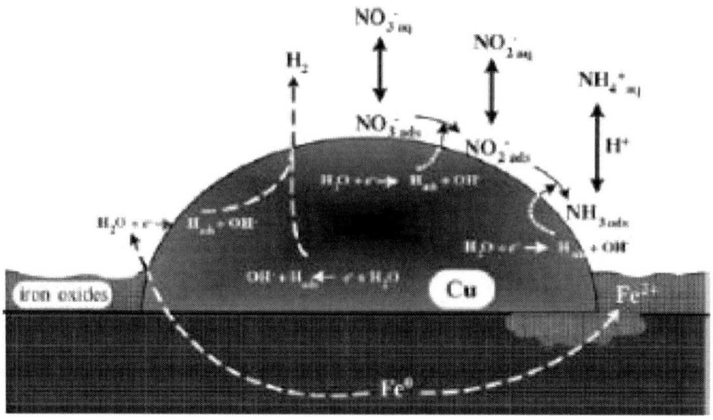

Reprinted with the permission from Y. H. Liouet al. *J. Hazardous Materials B* 127 (2005) 102.

Figure 2. Scheme of the nitrate reduction reaction at Cu/Fe system.

4.1.4. Titanium Dioxide (TiO_2) Based Nanoparticles

Titanium dioxide (TiO_2) is one of the most popular semiconducting materials used in various applications due to its photocatalytic, energy converting, electronic and gas sensing properties. It is available in three different polymorphs; Rutile, Anatase and Brookite (Figure 3) [24]. Semiconducting materials are of two types; one is adopted with donor atoms that provides electron to the conduction band as charge carrier, while another one is adopted with accepter atoms that takes electron from valance band and leaves behind the holes.

A B C

Reprinted with the permission from http://ruby.colorado.edu/~smyth/min/tio2.html.

Figure 3. Crystal structures A) rutile, B) anatase, C) brookite.

Large bandgap of TiO_2 (~ 3.2 eV) requires UV excitation to induce the charge separation within the particle. The semiconducting properties of TiO_2 are mainly responsible for the removal of various organic pollutants with certain limitations due to a rapid recombination of the photo-generated electron-hole pairs and the non-selectivity of the system [25]. Nevertheless, TiO_2 is widely applied in heterogeneous photocatalysis to decompose a host of the organic pollutants, such as phenolic compounds, metal ethylene diamine tetra acetate complexes, airborne microbes and odorous chemicals. Anatase and Rutile forms of TiO_2 nanoparticles are being used for the wet oxidation of phenols, considered to be one of the toxic materials found in municipal and wastewaters, by the hydrothermal treatment of microemulsions and their photocatalytic activities [26]. The advantage of such treatment is the effects of the surfactant to water ratio on the size of particles. The main reactions involved for phenol degradation are [26].

$$TiO_2 + h\upsilon \rightarrow e^- + TiO_2(h)^+ \quad (4.7)$$

$$TiO_2(h) + +H_2O(ads) \rightarrow TiO_2 + H^+ + OH^\bullet \quad (4.8)$$

$$\text{Phenol} + OH^\bullet \rightarrow \text{intermediate products (e.g., benzoquinone)} \quad (4.9)$$

$$TiO_2\ (h)^+ + \text{Intermediate products} \rightarrow CO_2 + H_2O + TiO_2 \quad (4.10)$$

In the presence of ultraviolet (UV radiation at 320 – 400 nm wavelength) radiation, nanostructured TiO_2 photocatalyst are used for the degradation of halogenated compounds, the removal of dyes *via* photocatalytic oxidation, the removal of metals and some disinfection processes for the drinking water and waste water [27], and the degradation of organic pollutants in water and air [28]. TiO_2 nanoparticles are more efficient than the immobilized catalysts, although their separation and recycling is quite expensive. The chelating agents at the surface of titanium dioxide (TiO_2) carry out the surface modification, which changes the energetic situation of the surface states, leading to the considerable alteration in their chemistry. The surface modification of TiO_2 with noble metal ions leads to decrease in the band gap that benefits the electron transfer from the valence band to the conduction band and facilitates the formation of the oxidative species, like OH [29]. The noble metal doped modified TiO_2 photocatalysts have capability to attract the cationic dyes, such as rhodamine B (RB) due to a larger number of electron traps. Band gap of TiO_2 could be further reduced with its surface modification with non-metals like N, F, S and C. This shows better absorption in the visible light than in UV radiation [30]. Substitution of oxygen for nitrogen in the TiO_2 lattice plays an important role in its visible light photo activity. Mixing of the N(2p) states with the O(2p) states also reduces the band gap in N doped TiO_2. Ag/TiO_2 or Au/TiO_2 shows a great potential as a photo catalyst due to its photo reactivity and response to the visible light. TiO_2 nanoparticles are also being widely used to reduce NO_x back to N_2 or oxidized into NO_2 and HNO_3. The oxidation of nitric acid converts it into a raw material like fertilizers, etc.

Table 3 summarizes the application of TiO_2 nanomaterials for the environmental remediation.

Table 3. TiO_2 nanomaterials for environmental remediation

Type of nanomaterial	Target material	Type of remediation
One dimensional TiO_2	Lead	Metal removal
TiO_3 NPs doped with Ag	2,4,6 Trichlorophenol	Organic removal
	Waterborne pathogenic virus in drinking water	Disinfection
Nanofibres of TiO_3 doped with Ag	Methylene blue,	Dye removal
	NO_x and toluene in air,	Air treatment
	S. aureus and *E. Coli* in water	Disinfection
TiO_3 NPs doped with Bi^{3+}	Nitrate	Nutrient removal
TiO_3 NPs doped with Cu/Au/Ag	Nitrate	Nutrient removal
TiO_3 thin film/composite material with Al_2O_3	Methylene blue, creatinine, biological toxins (microcystin LR), *E. Coli*	Dye removal/destruction of biological toxins and inactivation of bacteria
Fe doped TiO_2	Arsenic	Metal removal
Arginine- modified TiO_2	Nitrobenzene	Organic removal
Anatase TiO_2	Phenol	Organic removal

Photocatalytic inactivation of bacteria [31] and viruses enhance in TiO_2 doped with Ag. It is observed that silver doped TiO_2 particles are much more antibacterial effective than pure silver nanoparticles of a similar size [32]. Depending on the particle size, wavelength and the intensity of the radiation, the concentration of TiO_2 required to kill the bacteria varies between 100 and 1000 ppm [32]. The antibacterial activity of TiO_2 depends on ROS production, especially hydroxyl free radicals and the peroxide, formed under UVA irradiation *via* oxidative and the reductive pathways, respectively [33]. In this mechanism, silver facilitates electron–hole separation and also enhances the surface area for the adsorption [34]. TiO_2 nanoparticles are also capable of killing the viruses like polio [35] the hepatitis B [36], herpes simplex [37] and MS_2 bacteriophage [38].

Ultrasonic degradation technology is being used extensively to treat some organic pollutants. In this technique, heat-treated TiO_2 powder is being synthesized with low power ultrasound irradiation source. It is used for the decomposition of parathion with the nanometer. Rutile titanium dioxide (TiO_2) powder acts as the sonocatalyst after the treatment of high-temperature activation. Addition of an activated carbon (AC) to TiO_2 is another appropriate way to enhance the photocatalytic efficiency of TiO_2. Here, activated carbon acts as an adsorption trap for the organic pollutant and then immediately degrades through a mass transfer of the organic substances to the photoactivated TiO_2 on the surface. Advantages of using activated carbon are:

- Rate of photo-oxidation is enhanced due to the high concentration environment of the target organic substances around the loaded TiO_2 particles.
- The organic substances are oxidized on the photocatalyst surfaces via adsorption. The resulting toxic intermediates are adsorbed and oxidize and not released in the air to cause secondary pollution.
- High adsorbed ability of the hybrid photocatalysts for the organic substances is maintained for a long time because adsorbed substances are finally oxidized to give CO_2. Photo-efficiency of the hybrid catalysts depends upon the amount of TiO_2 in catalysts.

4.2. Nanofilteration by Membranes and Nanofiber Media

Membrane technology like microfiltration (MF), ultrafiltration (UF), nanofiltration (NF) and the reverse osmosis (RO) (all driven by pressure) has been applied for the water treatment, reuse and desalination systems all over the world. Membrane technology via reverse osmosis is very well known for the wastewater treatment and desalination. The principle of this method is the pressure difference across the membrane to overcome the osmotic pressure gradient. The smaller water molecules are literally pushed through the membrane while the large solute species are retained [39]. Among the different classes of membranes, reverse osmosis (RO) filtration is a well-known for the desalination of seawater, while the ultrafiltration (UF) is a well-established for the fractionation of natural organic matter (NOM). Membrane permeability in nanofiltration (NF) process is between RO and UF. Natural organic matters (NOM) such as humic acid and fulvic acid are widely distributed in soil, natural water and sediments. These materials include the mixture of the degradation products of plant and animal residues [40]. Membrane TiO_2-modified photocatalytic oxidation (PCO) has been focused on the NOM removal.

Nanofiltration (NF) removes the low molecular weight solutes like salts, glucose, lactose and micro pollutants from the contaminated water [41], the hardness (i.e., Ca (II)) and natural organic matter. However, the limitation of both RO and NF processes is the requirement of high pressures (100 – 1000 psi). Conversely, UF and MF membranes require lower pressures (5 – 60 psi) but fail to retain dissolved ions and the organic solutes. Both cellulose acetate and polyamide could be used to form NF membranes. In addition, other polymers (e.g., polyvinyl alcohol (PVA) and sulfonated polysulfone) and inorganic materials (e.g., some metal oxides) could also be used for NF membrane synthesis. The adsorption of the pollutants onto the membrane could be (1) physical in nature that is completely a reversible process, or (2) chemical in nature that is irreversible when strong chemical bonds are present, such as in polymerization, or reversible in case of weak secondary chemical bonds, such as hydrogen bonds and complexation or (3) both physical and chemical in nature. NF is capable of removing hardness, natural organic matter, particles and a number of other organic and inorganic substances *via* one single treatment. Water hardness is caused by the calcium and magnesium ions, while strontium and barium are rarely occur in the substantial concentrations. NF membranes could reject bivalent ions in significantly high amounts [42]. Mass transport through a nanofiltration membrane could be achieved by the two mechanisms; diffusion and convection. When the transmembrane pressure is high, the diffusion mechanism becomes less influential than the convection mass transport mechanism and therefore, a better retention of Ca^{2+} ions is found. Electrostatic interactions between the negatively charged membrane and the charged species appear as an important feature over the size effect and salts containing divalent sulfate anions are better rejected by the membrane than salts with a monovalent chloride anion. However, the differences in the retention of the examined chlorides are mostly controlled by the size of the cations. The total hardness of the surface water could be reduced by ~ 85.2%. The rejection of both divalent dibasic arsenate, $HAsO_4^{2-}$ and sulfate (SO_4^{2-}) ions was found to be notably higher than the rejection of chloride ions by the charged NF membranes due to the stronger electrostatic repulsion of the former ions than the repulsion of the lesser charged, monovalent chloride ion. This shows that the charge exclusion influences the retention of ions by the negatively charged NF membranes [43]. The high rejection of anti-scalants by the membranes takes place during the NF process due to their high molecular weight and negative charge [43]. The NF process is also used for the removal of natural organic matter (NOM) from water. NOM is a polydisperse mixture of individual particles in natural water originating from the degraded and partly resynthesized plant residues. Humic and fulvic acid seem to contribute to the natural color of the water, which becomes visible if the concentration of the dissolved organic carbon (DOC) is higher than 5 ppm [43]. The fouling mechanisms are adsorption, precipitation, gel formation and an interaction with multivalent ions. Thus, a pre-treatment with a coagulant are suggested to prevent these adverse effects. The combination of coagulation process and membrane filtration significantly reduces the necessary amount of coagulant required, while reducing the turbidity and DOC. Moreover, the coagulation is observed to provide an efficient hygienic barrier in the combination with membranes [44]. However, the removal of NOM has been significantly affected by the type of coagulant, coagulation conditions, type of membrane, filtration conditions and the characteristics of the treated water. In addition, cartridge filtration and activated carbon adsorption could also be used to prevent the membrane fouling by NOM [43]. Many pesticides and organohalide compounds have been found to be carcinogenic, even in very low concentrations. The removal of micropollutants is traditionally done by an activated carbon adsorption. This method is quite expensive when the large fractions of NOM are present, as the adsorption sites are occupied by the NOM due to its high concentration [45]. The NOM are assumed to enhance the adsorption of pesticides onto the membranes surface along with increased size exclusion and electrostatic repulsion observed during the NF process. The rejection of pesticides is reasonably high and correlates positively with the order of their molecular sizes. Important factors that must be considered for the selection of an appropriate membrane are (1) the molecular weight cutoff (MWCO) that is expressed in Dalton, (2) the membrane porosity, (3) the degree of membrane desalination, and (4) the charge of the membrane. The removal of viruses and bacteria is extremely important in drinking water purification. However, the removal of protozoa, giardia and cryptosporidium from the surface water sources is main priority for several government authorities and private companies dealing with drinking water. Traditional chlorination is still often applied as a disinfection method, but the disadvantages of chlorination are found to be (1) the formation of disinfection byproducts (DBPs) and (2) the resistance of cryptosporidium to chlorine and chloramines. Membrane filtration might facilitate an improvement in the disinfection process because of its extra barrier for viruses and bacteria [45]. Filtration processes by a membrane with carbon nanotube walls is another viable option. The filtration membrane consisted of a hollow cylinder with radially aligned carbon nanotube (CNT) walls [48]. Efficiently conducted the removal of *E. coli* from the drinking water using the CNT aligned membrane. Membranes that have CNTs as pores could be used in desalination and demineralization. Those tubes act as the pores in the membrane. A membrane filter possessing both super hydrophobicity and superoleophilicity was synthesized from the vertically aligned multiwalled carbon nanotubes on a stainless steel mesh for the possible separation of oil and water. Both super hydrophobicity and superoleophilicity could be obtained due to the dual scale structure and needle like nanotube geometry of the mesh with microscale pores combined with the low surface energy [46]. The nanotube filter could separate diesel and water layers and even surfactant-stabilized emulsions. The successful phase separation of the high viscosity lubricating oil and water emulsions has also been carried out. The separation mechanism could be readily expanded to a variety of different hydrophobic and oleophilic liquids, such as brewery waste water that might have a high content of organic matter in terms of COD. Polymeric membranes have many advantages, such as their easy formation, the selective transfer of the chemical species and their relatively low cost [47]. However, the inorganic membranes provide better resistance to the chemical attack, good mechanical strength and a high tolerance of extreme pH conditions and oxidation, thus are currently in tough competition with the organic membranes for the commercial applications. Inorganic membranes are composed of metal oxides that have higher durability in many water operating conditions. The decreased separation loads are other attractive features that promote inorganic membranes as chemical reactors in many established and novel reaction systems. NF are widely used for the industrial waste water treatments, such as (1) color removal and the recovery of salts from the waste water in the textile industry, (2) recovery of heavy metals and eluents from the waste water in the metal processing plants, (3) the removal of organics and recovery of proteins, enzymes and aromatic compounds from the waste water in the food processing industry, (4) the reclamation and reuse of water from the waste water in the pulp and paper industry, and (5) the removal of organics and multivalent ions from the waste water at the landfill leachate plants [43].

Nanofibers are defined on the bases of their effective diameter (roughly 1 - 100 nm). They behave like a nanomembrane due to its capability of removing the contaminants of size less than 0.1 µm. The nanofibers of carbon, ceramic and others synthesized by using fine spinning techniques, shows very good results as nanofilters (especially for the air filtration [48]. Microorganisms enter the filter with the airborne particulate matter and build up on the filter surface. They are usually resistant to the addition of antimicrobial functionality agents. At low levels, silver and silver oxides are safe and effective antimicrobial agents. Ag^+ bind with the sulfhydryl group on the cell membrane or bacterial DNA that prevent microorganism's proliferation [49]. The uniform porous channels between the nanofibers facilitate Hg^{2+} transportation in the adsorption process [50]. Polymer nanofiber ion exchangers have better performance due to extremely rapid kinetics, high ion-exchange capacities, and a higher swelling behavior as compared to the other media, because of their high surface area, porosity and capillary filters [51].

Combination filters have better capability of performing two separation tasks at once like solid-gas separation and adsorption of a gaseous impunity. These filters uses chemically active embedded in the filter medium material. The most common chemically active agent is activated carbon. Nanofiber filters based on silver zeolite technology are useful for the range of adsorptive and chemical treatments and antimicrobial action. Porous polymeric membranes have their intrinsic limitations of low flux rates and high fouling in case of liquid filters due to their geometric structure of the pores, the corresponding pore size distribution and the formation of undesirable macro voids across the whole membrane layer.

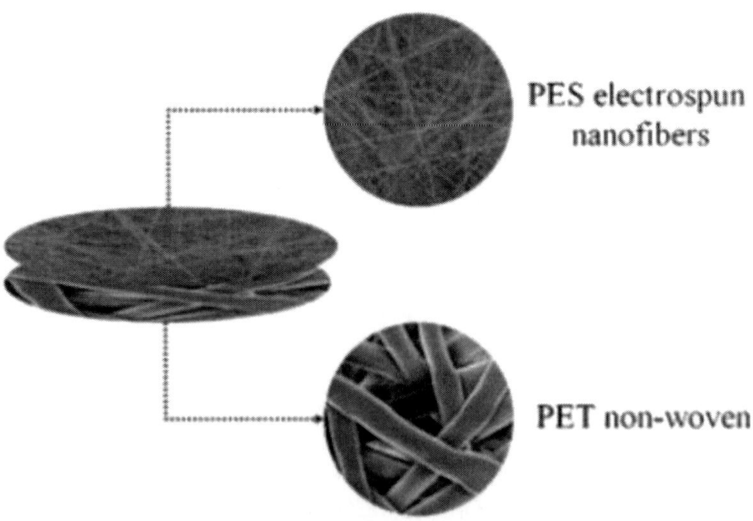

Reprinted with the permission from S. S. Homaeigohar, K. Buhr, K. Ebert, *J. Membr. Sci.* 365 (2010) 68.

Figure 4. PES-PET composite membrane structure.

Electrospun nanofibrous membranes are capable to overcome some of these limitations. Polyethersulfone (PES) electrospun nanofiber membranes are useful for liquid filtration. To increase the mechanical strength of the PES fiber, a polyethylene terephthalate) (PET) non-woven sub-layer is also being used. The composite membrane of polyethersulfone and polyethylene terephthalate (PES–PET, shown in Figure 4) possesses high initial flux.

4.3. Environment Remediation by Carbon Nanotubes (CNTs)

Nanoscale activated carbon has advantages over conventional materials due to larger surface area, unique structure and the electronic properties. Thus, activated carbon, extremely popular among the adsorbents has various sources like coconut coir, jute stick, rice husk, etc. Adsorption methods for wastewater treatment include specific ion exchangers/extractants and a combination of adsorption with catalytic treatment methods, redox and the magnetic methods. Carbon nanotube clusters are unique due to their ability to remove bacteria from water by an adsorption method. It is one of the preferred methods to remove the toxic contaminants from contaminated water and natural gas contaminated air. Carbon nanotubes/materials exist in various forms like single walled carbon nanotubes (SWCNTs), multi walled carbon nanotubes (MWCNTs), carbon beads, carbon fibers and nanoporous carbon.

CNTs could be considered as cylindrical hollow micro-crystals of graphite with outer diameter range 4-30 nm and a length up to 1 μm. MWCNTs are made of concentric cylinders with spacing about 3.4 Å between the adjacent layers, as shown in Figure 4. Carbon nanotubes have high thermal and electrical conductivities, high strength, high stiffness, and special adsorption properties.

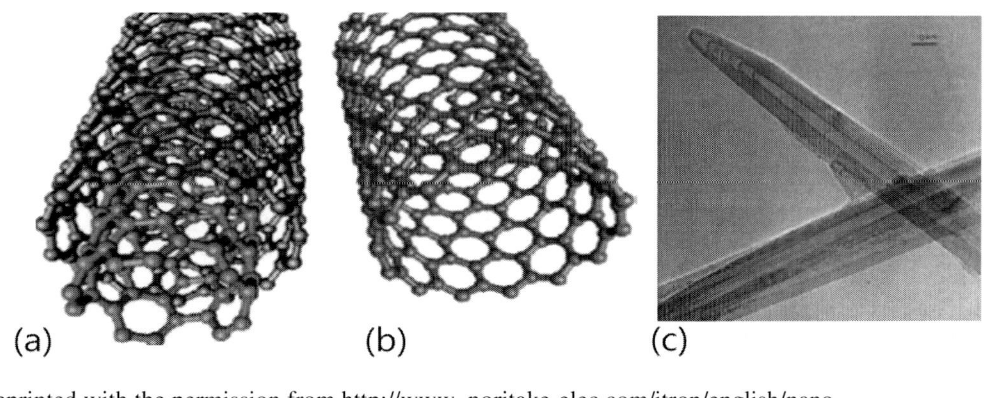

Reprinted with the permission from http://www. noritake-elec.com/itron/english/nano.

Figure 5. Structure of a) Multiwall Carbon Nanotube (2-layer) MWCNT b) Single wall Carbon Nanotube SWCNT c) TEM Image of MWNTs.

Table 4. Removal of pollutants using carbon nanotubes (CNTs)

Nano material	Target material	Remediation process
CNTs	Pb(II)	Metal remediation
CNTs oxidized by $KMnO_4$	Cd(II)	
SWCNTs	Zn(II)	
MWCNTs		
Al_2O_3/CNTs/ACNTs	Fluoride	-
CNT with Ag+/Cu nanoparticles	*E. coli*and *S. aureus*from contaminated water	Biological
CNTs(graphitized)	1,2- dichlorobenzene	Organic remediation
Nanocarbon Colloids	$Zn^{2+,}$ Cd^{2+}, Cu^{2+}, Hg^{2+}, $Ni^{2+,}$ Cr^{6+} from waste water	Metal removal (Sorption of contaminant metal ion and simultaneous formation of composite, coagulation and filtration)
Polyethylenimine		
CNT with Fe_2O_3 composites	Removal of $Co^{2+,}$ Sr^{2+} and Ni^{2+} from the aqueous solution	Metal removal (Enhanced surface area and adsorption capacity)
Hybrid diatomite/carbon composites	Polar aromatic compounds (*p*-cresol) removal from aqueous solutions	Organic remediation (Offer adequate open porous network to enhance removal rate)
CNT with TiO_2 nanoparticles	Remediate organic dyes, phenol and its derivatives, humic substances, and metallic ions from contaminated water	Organic and metal remediation
CNT with P-25 TiO_2 nanoparticles		

CNTs have exceptionally high adsorption efficiencies for various organic pollutants like benzene, 1,2-dichlorobenzene and ethyl benzene, trihalomethanes, n-nonane, CCl_4 [52], and the removal of heavy metals like lead [53] (Table 4). The adsorption capacities of all adsorbents slowly increase with CNT dosage. Nanocomposite of the carbon materials are shown in Table 4 with their advantages for environmental remediation. Depending on the pore size of CNTs and geometry of pores, adsorbent molecules interact with their carbon atoms on the surrounding walls. Due to highly graphitic nature of CNTs (much more than the activated carbons), the carbon nanotubes could adsorb molecules much stronger than the activated carbons that have slit-shaped or wedge-shaped pores [54].

4.4. Environmental Remediation by Dendrimers

Dendrimers are monodisperse and highly branched nanoparticles with size in the range of 1 – 100 nm, controllable composition and the architecture. In other words, dendrimers or cascade molecules have branching construction similar to a tree in which one trunk forms several large branches; each forming smaller branches, and so on. The roots of the tree have the same branching mode of growth. Dendrimers could be designed to encapsulate metal ions and zero-valent metals, enabling them to dissolve in suitable media or bind to appropriate surfaces. Modification of any compound in these branched polymeric structures give variety of means for controlling the critical macromolecular parameters such as internal and external rigidity, hydrophilicity and hydrophobicity, degrees of void, excluded volumes, and response to stimuli such as changes in solvent polarity and temperature [55]. Dendrimers consist of three components: (1) core, (2) interior branched cells and (3) terminal branched cells. A core atom is repeatedly added by carbon and other elements in a series of chemical reactions. This produces a spherical branching structure (Figure 5) in which divergent or convergent hierarchical assembly strategies are involved. During repetition process, successive layers are added that could be engineered to the required size. For example, ammonia (core molecule) reacts with methylacrylate in the presence of methanol and then ethylenediamine is added.

$$NH_3 + 3CH_2CHCOOCH_3 \rightarrow N(CH_2CH_2COOCH_3)_3$$
$$N(CH_2CH_2COOCH_3)_3 + 3NH_2CH_2CH_2NH_2 \rightarrow N(CH_2CH_2CONHCH_2CH_2NH_2)_3 + 3CH_3OH \quad (4.11)$$

Each branch of a dendrimer ends with a free amino group that could react with 2 methylacrylate monomers and 2 ethylenediamine molecules (Figure 6). The number of reactive groups of the core, the branch lengths and surface group dimensions determines the structure of the polymer. When molecules of different sizes are trapped inside the dendrimers, they could be selectively released by gradual hydrolysis (reaction with water) of the outer and middle layers. Dendrimers are a novel class of the polymers with a compact spherical structure, unique behavior and a narrow size distribution that could be used as templates or stabilizers to form relatively monodispersed organic/inorganic hybrid NPs.

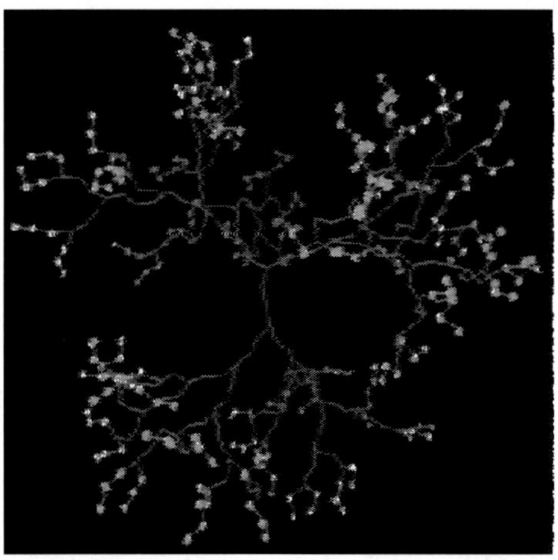

Figure 6. Schematic of dendrimer (rati.pse.umass.edu/ usim/gallery.html).

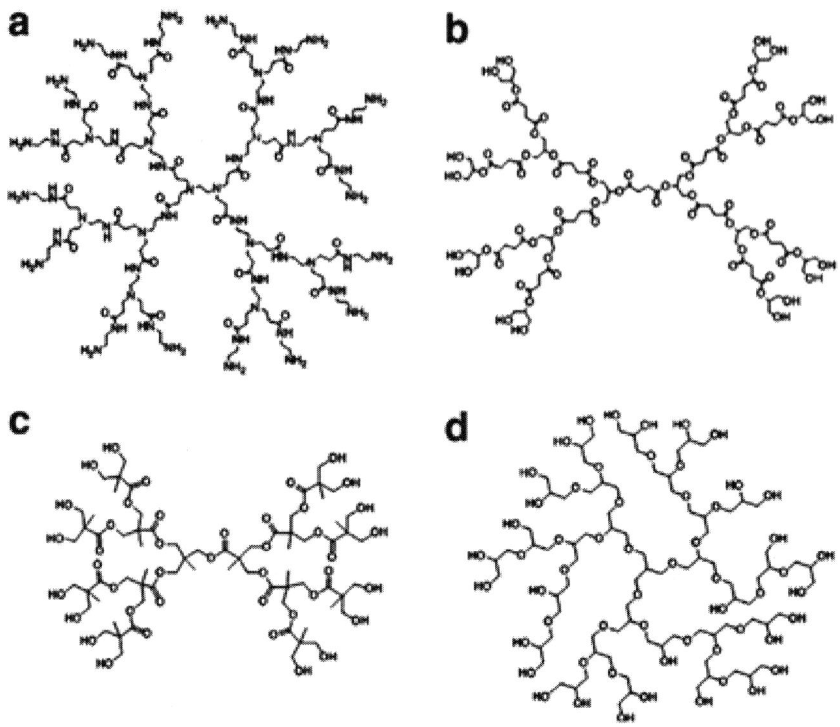

Reprinted with the permission from M. A. Quadir, R. Haag, J. Controlled Release 2012.

Figure 7. PAMAM dendrimers (a) PAMAM, (b) poly (glycerol-succinic acid) dendrimer, (c) Boltorn® and (d) hyperbranchedpolyglycerol.

The maximum size depends on how tightly dendrimer is packed. Poly (amidoamine) or PAMAM dendrimers are capable of removing variety of transition metal ions, such as copper (Cu (II)) from contaminated water. Dendritic nanopolymers encapsulate cations (e.g., copper, silver, gold, iron, nickel, zinc and uranium) through attachment to the functional groups of dendrimers (primary amines, carboxylates and hydroxymates) [56]. They are capable of deactivating activities of the bacteria and viruses. Due to smaller polydispersity and globular shape, dendritic nanopolymers have fewer tendencies to pass through the pores of ultrafiltration membranes than linear polymers of a similar molar mass. Thus, dendritic nanopolymers enhance the performance of UF and MF processes. The novelty of the dendritic polymer filtration process is the combination of dendritic polymers with multiple chemical functionalities with UF and MF that enable a new generation, flexible, reconfigurable and scalable water treatment processes.

(In this process: (step-1) contaminated water is mixed with a solution of functionalized dendritic nanopolymers; (step-2) the mixtures of nanopolymers and bound contaminants is transferred to UF or MF units to recover the clean water and then bound target substance is separated from the nanopolymers by changing the acidity (i.e., the pH) of the solution; (step-3) the recovered concentrated solution of the contaminants is collected for disposal; step-4) the nanopolymers recycled)

Dendritic polymer filtration processes are very useful to develop both small and mobile water treatment systems and large and fixed treatment systems. Intrinsic viscosities of dendritic nanopolymers are much smaller than linear polymers with the same molar mass. Another advantage of dendritic polymers is their comparatively lower operating pressures, energy consumption and the loss of ligands by shear force induced during the filtration that could be achieved with dendritic polymers in cross-flow UF systems [56]. Due to highly regular nature, low toxicity, branched and 3D structure, PAMAM dendrimers could host inorganic nanoclusters and form stable dendrimer complexes and nanocomposites [57]. During the reduction of silver cations with $NaBH_4$, the template role of the dendrimers ensures the dispersity of silver nanoparticles with a relatively small size distribution [58]. Removal of different pollutants using dendrimers is given in Table 5.

Table 5. Removal of pollutants using dendrimers

Type of composite dendrimers	Target contaminants	Remediation process
Octylurea DAB-32	Pyrene	Organic nano remediation
Octylurea DAB-64		
Poly(amidoamine)dendrimer PAMAM(G3)	Cu(II)	Metal nano remediation
Poly(amidoamine)dendrimer PAMAM(G5)		
Poly(amidoamine)dendrimer PAMAM(G8)		
Silica nanoparticles prepared by mixing salicylic acid and hyper-branchedpoly(propylene imine)	Removal of polycyclic aromatic hydrocarbons (PAH), like pyrene and phenanthrene, and Pb^{2+}, Cd^{2+}, Hg^{2+}, $Cr_2O_7^{2-}$ from contaminated aqueous solutions	Organic and metal nano remediation
PAMAM dendrimer composite membrane with hyaluronic acid in a chitosan gutter layer	Separation of CO_2 from a feed gas mixture of CO_2 and N_2 on porous substrates	Air nano remediation
PAMAM dendrimer composite membrane consisting of chitosan and a dendrimer	Separation of CO_2 from fossil fuel emission on porous substrates	
PAMAM dendrimer composite membrane consisting of chitosan and a dendrimer	Separation of CO_2 from a feed gas mixture of CO_2 and N_2 on porous substrates	Air nano remediation
Ni loaded hydrogel PAMAM	Separation of Cu^{2+}, Co^{2+} and Cr^{3+} from aqueous solutions	Metal nano remediation

4.5. Environment Remediation by Magnetic Nanoparticles

Materials usually react quite differently in the presence of an external magnetic field. Their reaction depends on their atomic and molecular structure, and the net magnetic field associated with the atoms. The magnetic moments associated with atoms might be due to electron orbital motion, change in the orbital motion due to presence of external magnetic field and electrons spin. Electrons could be in pairs and/or unpaired. Materials with paired electrons spin in the opposite directions and hence, their magnetic fields cancel each other. On other hand, materials with unpaired electrons have a net magnetic field and hence, show greater response to an external magnetic field. On the basis of their magnetic properties, materials could be classified as diamagnetic, paramagnetic or ferromagnetic. Diamagnetic materials have a very weak and negative susceptibility and have repulsive nature, while paramagnetic metals have small positive susceptibility and have attractive nature but both of these characteristics are temporary. Ferromagnetic materials exhibit a strong attraction to the magnetic fields and are able to retain their magnetic properties after the removal of external field. This nature of ferromagnetic materials is due to the presence of magnetic domains. In the presence of external magnetic field, atomic moments (10^{12} - 10^{15}) aligned parallel to the applied field that result in a strong magnetic force within the domain. Some transition ion atoms such as iron, manganese, nickel, and cobalt are the examples of ferromagnetic materials (Figure 6).

Applications of magnetic nanoparticles depend on their size and subsequent change in the magnetic property since the relaxation time of magnetic nanoparticle changes with the size of the nanoparticles. On the bases of their magnetic properties, different kinds of materials are useful tool in various kinds of applications like biomedical and data storage systems. Magnetic particles are useful for magnetic separation. These nanoparticles could separate a specific substance from the mixture. The separation time is very important parameters for magnetic separation method. Magnetic gradients separation techniques (especially high magnetic gradient separation (HGMS)) are very effective in the fields of medicine, diagnostics, catalysis, etc. In HGMS technique, a liquid phase containing magnetic particles is passed through a matrix of wires that are magnetized by applying a magnetic field. The particles are held onto the wires and could be released when magnetic field is cut off. The latter could be isolated from wastewater or slurries. The materials could be recycled and does not generate secondary waste. These processes sometime called "magnetically assisted chemical separation (MACS)". MACS processes contain micrometric particles that are made of magnetite nanoparticles embedded into a polymer microsphere with a diameter range between 0.1 and 25 μm. The magnetic component of micro and nanoparticles in MACS process is magnetite (Fe_3O_4) or its products of oxidation γ-Fe_2O_3. Polymer poly (1-vinylimidazole)-grafted nanosized magnetic particles (organic-inorganic hybrid magnetic materials) are being used in sorbent-based separation technology to a multiphase complex system. Surface-modified $MnFe_2O_4$ nanoparticles are very efficient adsorbents for the rapid removal of Cr(VI) from aqueous solutions. The disadvantages of magnetic particles are small adsorption capacity and slow adsorption rates due to their small surface area or their porous properties. Magnetic nanoparticles, contaminants and type of removal process are shown in the Table 6.

5. NANO SENSORS FOR ENVIRONMENTAL REMEDIATION

Nanoparticle based environmental sensors are rapid and reliable tools to perform the sample analysis. These sensors are very powerful to detect toxins, heavy metals and organic pollutants in air, water and soil. They could play an important role in environmental monitoring. NP sensors might have higher selectivity, sensitivity and stability and a lower cost in comparison to the traditional detection methods [59]. Nano sensor comprises nanoarrays or single nanostructures. Due to a large surface area and single crystalline surface, response of the sensors is much better than the conventional sensors. 1D ZnO nanostructure could detect NH_3, CO (charge donors), and NO_2, O_2 and dioxin (charge accepter). Co_3O_4, Fe_2O_3, SnO_2, TiO_2 and Pt metal tubes are some of the nanotube based nanosensors. Poly (3,4-ethylene-dioxythiophene) (PEDOT) are nanorods based nanosensors that have exciting response to NH_3 and HCl even at very low concentration. V_2O_5 nanobelts showed greater sensitivity to ethanol at either low (< 10 ppm) or high (1000 ppm) concentrations. The response and recovery times were 30 - 50 s.

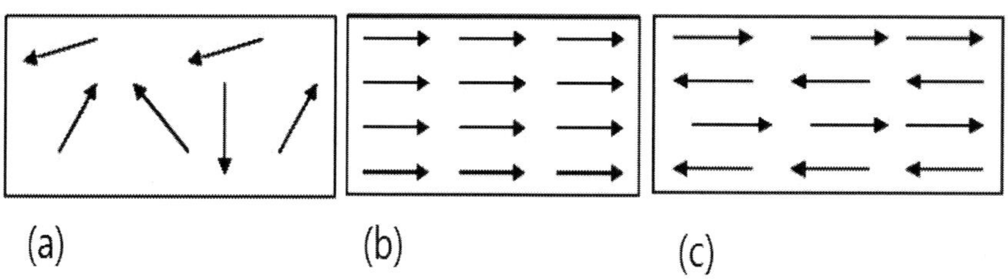

Reprinted with the permission from L. J. Zeman, A. L. Zydney, Microfiltration and Ultrafiltration, Marcel Dekker, (1996).

Figure 8. Various arrangements of individual atomic magnetic moments that constitute (a) paramagnetic, (b) ferromagnetic, (c) antiferromagnetic material.

In_2O_3 nanowires based sensors with an appropriate doping level have two opposite sensing responses which canceled each other and resulted in the immunity to NH_3. This unique property of In_2O_3 nanosensor offers a new way to achieve the selectivity. These sensors exhibited a higher response and a good selectivity towards C_2H_5OH at 370°C. Titanate nanotubes (TNT) showed a very good response to methylene blue (MB) for the detection of dopamine [60]. Porous TiO_2 sol–gel matrix are being used to construct nitrite sensors by immobilizing partially quaternized poly (4-vinylpyridine) complexed with PVP-Os as an electrode material [61].

Table 6. Removal of pollutants using magnetic nanoparticles

Removal Target	Types of nanoparticles	Remediation process
Cu(II)	Mag-Im_{18}	Metal remediation
Cr(II)	Modified jacobsite ($MnFe_2O_4$)	Metal remediation
	Maghemite nanoparticle	
Ni (II)	Magnetic alginate microcapsules	Metal remediation
Co (II)	Magnetic chitosan nanoparticle	Metal remediation
Methylene blue	PAA-bound iron oxide	Dye remediation
Bromelain	PAA-bound iron oxide	-

The limitation of nanoparticles in the practical applications, such as slow diffusion and aggregation, still exists. Immobilization of the nanoparticles by the polymer matrix is one of the most efficient approaches to overcome such limitations. Since the chemical and physical properties of the polymers might be tailored, they gained importance in the construction of sensor devices [62]. Conductive polymer nanomaterials have attracted particular interests as sensors for air-borne volatiles [63] (alcohols, NH_3, NO_2, CO) because of large surface area, adjustable transport properties and the chemical specificities, easy processing and scalable productions. Polyaniline–SnO_2/TiO_2 nanocomposite ultra-thin films have also been fabricated for CO gas sensing [64].

A Pd–polyaniline nanocomposite has been developed as a selective methanol sensor [65]. The synthesized nanocomposite sensor showed high selectivity and sensitivity to methanol vapors with rapid and reverse response. Some applications of polymer based nanocomposite sensors are shown in Table 7.

Table 7. List of nanosensors and their target sensing material

Type of nanosensor	Target material
SnO_2 on polystyrene/polyaniline	CO
SnO_2 on Polyaniline	Ethanol, Acetone
TiO_2 on Polyaniline	Trimethylamine
Iron on Oxide Polypyrrole	Co_2, N_2, CH_4
Pd on Polyaniline	Methanol
Au on Chitosan	Zn^{2+}, Cu^{2+}

CONCLUSION

The environmental remediation encompasses all the remediation methods employed to reduce the environmental and/or human health risks due to environmental pollution. The choice of remediation technique depends on the complexity, nature of the contaminated media and economic costs of the treatment. The environmental nanotechnology is playing a key role in shaping the current environmental engineering and science. It has stimulated the development and use of novel and cost effective technologies for the pollution detection, monitoring and remediation. Due to their extremely small size (1 - 10 nm) with large surface-to-volume ratios and various shapes and morphologies, nanoparticles offer excellent chemical reactivity and unique functionalities to treat the pollutants compared to the bulk materials. These materials are excellent adsorbents, catalysts and sensors due to their large specific surface areas and the high reactivity.

- Ag^+ ions are photoactive in the presence of UV irradiation and cause an improvement in the UV inactivation of bacteria and viruses. They could prevent DNA replication. Ag^+ ions (< 10 nm) are very effective in the disinfection of bacterias like *Escherichia coli* and *Pseudomonas aeruginosa.*
- Nanoparticles based iron are capable for reduction/adsorption of heavy metals, anions and organic pollutants. The disadvantages of iron nanoparticle based remediation are their hard recovery, sludge generation and high cost of sludge management.
- TiO_2 nanoparticles are very useful for heterogeneous photocatalysis to decompose a host of organic pollutants, such as phenolic compounds, metal ethylene diamine tetra acetate complexes, airborne microbes and odorous chemicals. These nanoparticles are non-toxic, water insoluble under most conditions and photo-stable. These particles are hard to recover, generates sludge and of high operation cost.
- A bimetallic nanoparticle enhances the reaction/adsorption rate than nanoparticles. Their reaction rate is much higher than iron nanoparticles.
- Nanofiltration (NF) removes low molecular weight solutes like salts, glucose, lactose and micro pollutants, hardness (i.e., Ca(II)) and natural organic matter from the contaminated water. However, the limitation of both RO and NF processes require high pressures (100 – 1000 psi). Combination filters have better capability of performing two separation tasks at once like solid-gas separation and adsorption of a gaseous impunity.
- Carbon nanotube clusters are unique due to their ability to remove the bacteria from water by an adsorption method. It is one of the preferred methods to remove the toxic contaminants from contaminated water and natural gas contaminated air. It has high thermal and electrical conductivities, high strength, high stiffness, and special adsorption properties. CNTs show exceptionally high adsorption efficiencies for various organic pollutants like benzene, 1,2-dichlorobenzene and ethyl benzene, trihalomethanes, n-nonane, CCl_4, and heavy metals like lead.

- Dendrimers are a novel class of polymers with a compact spherical structure, a unique behavior and a narrow size distribution that could be used as templates or stabilizers to form relatively monodispersed organic/inorganic hybrid NPs. They are capable of removing the varieties of transition metal ions, such as copper (Cu(II)) from the contaminated water. They are also capable of deactivating the activities of bacteria and viruses. Due to smaller polydispersity and globular shape, dendritic nanopolymers have fewer tendencies to pass through the pores of ultrafiltration membranes than linear polymers of a similar molar mass. Thus, dendritic nanopolymers enhance the performance of UF and MF processes. The novelty of the dendritic polymer filtration process is the combination of dendritic polymers with multiple chemical functionalities with UF and MF that enable a new generation, flexible, reconfigurable and scalable water treatment processes.
- Magnetic nanoparticles could separate a specific substance from the mixture. Magnetic gradients separation techniques (especially high magnetic gradient separation (HGMS)) are very effective in the fields of medicine, diagnostics, catalysis, etc. In HGMS technique, a liquid phase containing magnetic particles is passed through the matrix of wires that are magnetized by applying a magnetic field. The materials could be recycled and does not generate secondary waste.
- Nanoparticle based environmental sensors are very powerful to detect toxins, heavy metals and organic pollutants in air, water and soil. They could play an important role in an environmental monitoring. NP sensors have higher selectivity, sensitivity and stability and a lower cost in comparison to the traditional detection methods. The limitations of nanoparticles in practical applications are such as slow diffusion and aggregation. Immobilization of nanoparticles by polymer matrix could overcome such limitations.

REFERENCES

[1] (a) O. Carp, C. L. Huisman, A. Reller, Photoinduced reactivity of titanium dioxide, *Progress in Solid State Chemistry*, 32 (2004) 33, (b) N. Serpone , R. F. Khairutdinov (Eds. P.V. Kamat , D. Meisel). *Application of nanoparticles in the photocatalytic degradation of water pollutants*, Elsevier Science. B.V, 417 (1996).

[2] M. H. Fulekar, *Nanotechnology- Its Importance & Applications*, IK International (2010).

[3] Environmental Protection Agency, *US Environmental Protection Agency Report EPA 100/B-07/001* EPA Washington DC (2007).

[4] L. Zhang, M. Fang, Nanomaterials in pollution trace detection and environmental improvement, *Nano Today*, 5 (2010) 128.

[5] (a) L. Qi, Z. Xu, X. Jiang, C. Hu , X. Zou, Preparation and antibacterial activity of chitosan nanoparticles, *Carbohydr. Res.* 339 (2004) 2693, (b) J. R. Morones, J. L. Elechiguerra, A. Camacho, K. Holt, J. B. Kouri, J.T. Ramirez, M. J. Yacaman, The bactericidal effect of silver nanoparticles, *Nanotechnology* 16 (2005) 2346, (c) M. Cho, H. Chung, W. Choi , J. Yoon, Different inactivation behaviors of MS-2 phage and Escherichia coli in TiO_2 photocatalytic disinfection, *Appl. Environ. Microbiol.* 71 (2005) 270, (d) C. Wei, W. Y. Lin, Z. Zainal, N. E. Williams, K. Zhu, A. P. Kruzic, R. L. Smith , K. Rajeshwar, Bactericidal activity of TiO_2 photocatalyst in aqueous media: toward a solar-assisted water disinfection system, *Environ. Sci. Technol.* 28 (1994) 934, (e) S. Kang, M. Pinault, L. D. Pfefferle, M. Elimelech, Single walled carbon nanotubes exhibit strong antimicrobial activity, *Langmuir* 23 (2007) 8670.

[6] Karn et al. 2009; Bardos et al. 2014; B. Karn, T. Kuiken, M. Otto 2009. *Environmental Health Perspectives*. Bardos, B. Bone, P. Daly, D. Elliott, S. Jones, G. Lowry, C. Merly, 117(2014)1813.

[7] D. J. Wilson, A.N. Clarke, Soil vapour stripping. In: *Hazardous Waste Site Soil Remediation: Theory and Application of Innovative Technologies*. D.J. Wilson, A.N. Clarke (Eds). New York, NY: M. Dekker; 1994.

[8] (a) I. Jayaweera, Ed. M. A. Tarr, Marcel Dekker Inc., New York, (2003)129, (b) R. W. Shaw , N. Dahmen, Destruction of toxic organic materials using super-critical water oxidation: current state of the technology, *J. Supercrit. Fluids* 17 (2000) 425, (c) T. Adschiri, Y. W. Lee, M. Goto, S. Takami, Green materials synthesis supercritical water, *Green Chem.* 13 (2011) 1380.

[9] M. Hayyan, M. A. Hashim, I. M. AlNashef, Superoxide Ion: Generation and Chemical Implications. *Chem. Rev.* 5, (2016) 3029.

[10] http://www.geocleanse.com/chemical-remediation/reagents.

[11] Y. Matsumura, K. Yoshikata, S. Kunisaki and T. Tsuchido, Mode of Bactericidal Action of Silver Zeolite and Its Comparison with That of Silver Nitrate, *Appl. Environ. Microbiol.* 69 (2003) 4278.

[12] Q. L. Feng, J. Wu, G. Q. Chen, F. Z. Cui, T. N. Kim , J. O. Kim, A mechanistic study of the antibacterial effect of silver ions on Escherichia coli and staphylococcus aureus, *J. Biomed. Mater. Res.* 52 (2000) 662.

[13] (a) J. S. Kim, E. Kuk, K. N. Yu, J. H. Kim, S. J. Park, H. J. Lee, S. H. Kim, Y. K. Park, Y. H. Park, C. Y. Hwang, Y. K. Kim, Y. S. Lee, D. H. Jeong , M. H. Cho, Antimicrobial effects of silver nanoparticles, Nanomed.:Nanotechnol., Biol. Med. 3 (2007) 95, (b) R. O. Rahn , L. C. Landry, Ultraviolet irradiation of nucleic acids complexed with heavy atoms-II phosphorescence and photodimerization of DNA complexed with Ag*, *Photochem. Photobiol.* 18 (1973) 29.

[14] (a) X. Xu, W. J. Brownlow, S. V. Kyriacou, Q. Wan, J. J. Viola, Real-time probing of membrane transport in living microbial cells using single nanoparticle optics and living cell imaging, *Biochemistry* 43 (2004) 10400, (b) K. S. Gogoi, P. Gopina, A.

Paul, A. Ramesh, S. S. Ghosh, A. Chattopadhyay, Green fluorescent protein-expressing Escherichia coli as a model system for investigating the antimicrobial activities of silver nanoparticles, *Langmuir* 22 (2006) 9322.

[15] S. M. Ponder, J. G. Darab, T. E. Mallouk, Remediation of Cr(IV) and Pb(II) aqueous solutions using supported, nanoscale zero-valent iron, *Environ. Sci. Technol.* 34 (2000) 2564.

[16] S. R. Kanel, J. M. Greneche, H. Choi, Arsenic(V) removal from groundwater using nano scale zero-valent iron as a colloidal reactive barrier material, *Environ. Sci. Technol.* 40 (2006) 2045.

[17] K. M. Paknikar, V. Nagpal, A.V. Pethkar, J.M. Rajwade, Degradation of lindane from aqueous solutions using iron sulfide nanoparticles stabilized by biopolymer, *Sci. Technol. Adv. Mater.* 6 (2005) 370.

[18] L. Cumbal, A. K. Sengupta, Arsenic removal using polymer-supported hydrated iron(III) oxide nanoparticles: role of donnan membrane effect, *Environ. Sci. Technol.* 39, (2005) 6508.

[19] K. Hristovski, P. Westerhoff, T. Moller, P. Sylvester, W. Condit, H. Mashe, Simultaneous removal of perchlorate and arsenate by ion-exchange media modified with nanostructured iron (hydr)oxide. *J. Hazard. Mater.* 152 (2008) 397.

[20] H. L. Lien, W. X. Zhang, Nanoscale iron particles for complete reduction of chlorinated ethenes, *Colloids Surf. A* 191 (2001) 97.

[21] Y. H. Liou, S. L. Lo, C. J. Lin, C. Y. Hu, W. H. Kuan, S. C. Weng, Method for accelerating nitrate reduction using zerovalent iron at near neutral pH: effect of H_2-reducing pretreatment and copper deposition, *Environ. Sci. Technol.* 39 (2005) 9643.

[22] (a) L. J. Matheson and P. G. Tratnyek, Reductive Dehalogenation of Chlorinated Methanes by Iron Metal, *Environ. Sci. Technol.* 28 (1994) 2045 (b) H. L. Lien , W. X. Zhang, Enhanced dehalogenation of halogenated methanes by bimetallic Cu/Al, *Chemosphere*, 49 (2002) 371, (c) M. J. Alowitz , M. M. Scherer, Kinetics of nitrate, nitrite, and Cr(VI) reduction by iron metal, *Environ. Sci. Technol.* 36 (2002) 299, (d) F. Cheng, Q. Fernando, N. Korte, Electrochemical dechlorination of 4-chlorophenol to phenol, *Environ. Sci. Technol.* 31 (1997) 1074.

[23] D. E. Meyer, K. Wood, L. G. Bachas, D. Bhattacharyya, Degradation of chlorinated organics by membrane-immobilized nanosized metals, *Environ. Prog.* 23 (2004) 232.

[24] http://ruby.colorado.edu/~smyth/min/tio2.html.

[25] O. V. Makarova, T. Rajh, M. C. Thurnauer, A. Martin, P. A. Kemme, D. Cropek, Surface modification of TiO_2 nanoparticles for photochemical reduction of nitrobenzene, *Environ. Sci. Technol.* 34 (2000) 4797.

[26] M. Andersson, L. Osterlund, S. Ljungstrom, A. Palmqvist, Preparation of nanosize anatase and rutile by hydrothermal treatment of microemulsions and their activity for photocatalytic wet oxidation of phenol, *J. Phys. Chem. B* 106 (2002) 10674.

[27] T. Salthammer, F. Fuhrmann, Photocatalytic surface reactions on lndoor wall paint, *Environ. Sci. Technol.* 41 (2007) 6573.

[28] P. V. Kamat, D. Meisel, Nanoscience opportunities in environmental remediation, *C. R. Chim.* 6 (2003) 999.

[29] H. M. SungSuh, J. R. Choi, H. J. Hah, S. M. Khoo , Y. C. Bae, Comparison of Ag deposition effects on the photocatalytic activity of nanoparticulate TiO_2 under visible and UV light irradiation, *J. Photochem. Photobiol. A* 163 (2004) 37.

[30] C. C. Liu, Y. H. Hsieh, P. F. Lai, C. H. Li , C.bL. Kao, Photodegradation treatment of azo dye wastewater by UV/TiO_2 process, *Dyes Pigm.* 68 (2006) 191.

[31] K. Page, R. G. Palgrave, I. P. Parkin, M. Wilson, S. L. P. Savin A. V. Chadwick, Titania and silver-titania composite films on glass-potent antimicrobial coating, *J. Mater. Chem.* 17 (2007) 95.

[32] J. Thiel, L. Pakstis, S. Buzby, M. Raffi, C. Ni, D. J. Pochan , S. I. Shah, Antibacterial Properties of Silver-Doped Titania, *Small* 3 (2007) 799.

[33] Y. Kikuchi, K. Sunada, T. Iyoda, K. Hashimoto, A. Fujishima, Photocatalytic bactericidal effect of TiO_2 thin films: dynamic view of the active oxygen species responsible for the effect, *J. Photochem. Photobiol. A* 106, (1997) 51.

[34] A. Sclafani, J. M. Herrmann, Influence of metallic silver and of platinum-silver bimetallic deposits on the photocatalytic activity of titania (anatase and rutile) in organic and aqueous media, *J. Photochem. Photobiol. A* 113 (1998) 181.

[35] R. J. Watts, S. Kong, M. P. Orr, G. C. Miller , B. E. Henery, Photocatalytic inactivation of coliform bacteria and viruses in secondary wastewater effluent, *Water Res.* 29 (1995) 95.

[36] L. Zan, W. J. Fa, T. Y. Peng ,Z. K. Gong, Photocatalysis effect of nanometer TiO_2 and TiO_2 coated ceramic plate on hepatitis B virus, *J. Photochem. Photobiol. B* 86 (2007) 165.

[37] P. Hajkova, P. Spatenka, J. Horsky, I. Horska, A. Kolouch, Photocatalytic Effect of TiO_2 Films on Viruses and Bacteria, *Plasma Processes Polym.* 4 (2007) S397-S401.

[38] M. Cho, H. Chung, W. Choi , J. Yoon, Different inactivation behaviors of MS-2 phage and Escherichia coli in TiO_2 photocatalytic disinfection, *Appl. Environ. Microbiol.* 71 (2005) 270.

[39] S.E. Kentish, G.W. Stevens, Innovations in seperations technology for the recycling and re-use of liquid waste streams, *Chem. Eng. J.* 84 (2001) 149.

[40] J. Fu, M. Ji, Z. Wang, L. Jin, D. An, A new submered membrane photocatalysis reactor (SMPR) for fulvic acid removal using a nano-structured photocatalyst, *J. Hazard. Materi. B* 131 (2006) 238.

[41] Y. Kiso, Y. Sugiura, T. Kitao, K. Nishimura, Effect of hydrophobicity and molecular size on rejection of aromatic pesticides with nanofiltration membranes, *J. Membr. Sci.* 192 (2001) 1.

[42] M. Mulder, *Basic Principles of Membrane Technology.* Kluwer Academic Publishers

Group, London, 2nd Edn, (1996).

[43] K. Košutić, L. Furač, L. Sipos , B. Kunst, Removal of arsenic and pesticides from drinking water by nanofiltration membranes, *Sep. Purif. Technol.* 42 (2005) 137.

[44] A. W. Zularisam, A. F. Ismail ,R. Salim, Behaviours of natural organic matter in membrane filtration for surface water treatment-a review, *Desalination* 194 (2006) 211.

[45] J. Kim, B. Van der Bruggen, The use of nanoparticles in polymeric and ceramic membrane structures: review of manufacturing procedures and performance improvement for water treatment, *Environ. Pollut.* 158 (2010) 2335.

[46] A. Srivastava, O. N. Srivastava, S. Talapatra, R. Vajtai , P. M. Ajayan, Ajayan, Carbon nanotube filters, *Nat. Mater.* 3 (2004) 610.

[47] C. Lee, S. Baik, Vertically-aligned carbon nanotube membrane filters with superhydrophobicity and superoleophilicity, *Carbon* 48 (2010) 2192.

[48] C. Tang, V. Chen, *Nanofiltration: Principles and Applications*. Elsevier Advanced Technology, UK, (2005) 381.

[49] A. Boam, A. Nozari, Fine chemical: OSN-a lower energy alternative, *Filtr. Sep.* 43 (2006) 46.

[50] S. Li, X. Yue, Y. Jing, S. Bai, Z. Dai, Fabrication of zonal thiol-functionalized silica nanofibers for removal of heavy metal ions from wastewater, *Colloids Surf. A* 380 (2011) 229.

[51] L. Domingues, K. R. Benak, Economy J. Design of high efficiency polymeric cation exchange fibers, *Polym. Adv. Technol.* 12 (2001) 197.

[52] X. Peng, Y. Li, Z. Luan, Z. Di, Hu. Wang, B. Tian, Z. Jia, Adsorption of 1,2-dichlorobenzene from water to carbon nanotubes, Chem. Phy.Lett. 376 (2003) 154; C. Lu, Y. L. Chung, K. F. Chang, Adsorption of trihalomethanes from water with carbon nanotubes, *Water Research* 39 (2005) 1183.

[53] Y.H. Li, S. Wang, J. Wei, X. Zhang, C. Xu, Z. Luan, D. Wu , B. Wei, Lead adsorption on carbon nanotubes, *Chem. Phy. Lett.* 357 (2002) 263; Y. H. Li, Z. Di, J. Ding, D. Wu, Z. Luan , Y. Zhu, Adsorption thermodynamic, kinetic and desorption studies of Pb^{2+} on carbon nanotubes, *Water Research* 39 (2005) 605.

[54] Alph. T. Yang, *Adsorbents: Fundamentals and Applications*, John Wiley & Sons Inc (2003).

[55] M.S. Diallo, L. Balogh, A. Shafagati, J.H. Johnson Jr, W.A. Goddard, D.A. Tomalia, Poly(amidoamine) dendrimers: a new class of high capacity chelating agents for Cu(II) ions, *Environ. Sci. Technol.* 33 (1999) 820

[56] M. S. Diallo, S. Christie, P. Swaminathan, J. H. Johnson Jr and W. A. Goddard III, Dendrimer enhanced ultrafiltration. 1. Recovery of Cu(II) from aqueous solutions using PAMAM dendrimers with ethylene diamine core and terminal NH_2 groups, *Environ. Sci. Technol.* 39 (2005) 1366.

[57] M.S. Diallo, *Water treatment, by Dendrimer Enhanced Filtration*, US Patent

Application, US1006/0021938 Al (2006).
[58] A. Rether, M. Schuster, Selective separation and recovery of heavy metal ions using water-soluble N-benzoylthiourea modified PAMAM polymers, *React. Funct. Polym.* 57 (2003) 13.
[59] A. F. Ngomsik, A. Bee, M. Draye, G. Cote, V. Cabuil, Magnetic nano- and microparticles for metal removal and environmental applications: a review, *C. R. Chimie, Water Research* 8 (2005) 963.
[60] L. Wang, W. Ma, L. Xu, W. Chen, Y. Zhu, C. Xu , N. A. Kotov, Nanoparticle-based environmental sensors, *Mater. Sci. Eng.* 70 (2010) 265.
[61] T. Gao, T. H. Wang, Sonochemical synthesis of SnO_2 nanobelt/CdS nanoparticle core/shell heterostructures, *Chem. Commun.* (2004) 2558.
[62] M. W. Xiao, L. S. Wang, Y. D. Wu, X. J. Huang , Z. Dang, Electrochemical study of methylene blue/titanate nanotubes nanocomposite and its layer-by-layer assembly multilayer films, *J. Solid State Electrochem.* 12 (2008) 1159.
[63] B. Adhikari and S. Majumdar, Polymers insensor applications, *Prog. Polym. Sci.* 29 (2004) 699.
[64] M. K. Ram, Ö. Yavuz, V. Lahsangah, M. Aldissi, CO gas sensing from ultrathin nano-composite conducting polymer film, *Sens. Actuators B* 106 (2005) 750.
[65] A. A. Athawale, S. V. Bhagwata , P. P. Katre, Nanocomposite of Pd-polyaniline as a selective methanol sensor, *Sens. Actuators B* 114 (2006) 263.

In: Handbook of Remediation ...
Editors: Sadia Ameen, M. Shaheer Akhtar et al.
ISBN: 978-1-53615-118-3

Chapter 3

ENGINEERING OF NICKEL BASED CATALYST FOR DIRECT UREA FUEL CELL-ENERGY FROM MUNICIPAL LIQUID WASTE (MLW)

***Zafar Khan Ghouri*[1,*], *Saeed Al-Meer*[1,*,†] and *Nasser A. M. Barakat*[2,‡,*]**

[1]Chemical Engineering Program, Texas A&M University at Qatar, Doha, Qatar
[2]Central Laboratory Unit, Qatar University, Doha, Qatar
[3]Chemical Engineering Department, Faculty of Engineering, El-Minia University, El-Minia, Egypt
[4]Organic material and Fiber Engineering Department, Chonbuk National University, Jeonju, Republic of Korea

ABSTRACT

Alternatives to conventional fuels are required to meet the global energy demand, urea is a relatively non-toxic crystalline substance that is perhaps best known for its presence in municipal liquid waste (MLW) and human/animal urine. Urea has a high energy density compare to compressed/liquid hydrogen, hence is useful as an alternative energy vector for direct urea fuel cell. Luckily, urea-based fuel cells are *operated* in alkaline medium which offers good chance for the non-precious transition metals to be invoked as anode material due to the good chemical stability. Moreover, compared to noble metals, nickel catalysts showed higher current densities and lower oxidation potentials for the electrooxidation of urea. Consequently, this chapter details the study of physicochemical and electrochemical *properties* of two novel nickel based nanomaterials for direct urea fuel cells, together with their brief synthesis procedures.

* Corresponding Author's Email: zafar_khan.ghouri@qatar.tamu.edu; zafarkhanghouri@hotmail.com.
† Corresponding Author's Email: salmeer@qu.edu.qa.
‡ Corresponding Author's Email nasser@jbnu.ac.kr.

Keywords: noble metals, catalysts, electrochemical oxidation, carbon nanofibers, fuel cells

1. INTRODUCTION

Urea, commonly used as fertilizer and can be synthesized by using hydrogen, nitrogen and carbon dioxide [1, 2]. Urea also, is the main nitrogen-containing substance in the humans and animals urine, numerically the urine contains 2-2.5 wt% urea [3-5]. Therefore, urea-containing municipal liquid waste (MLW) are produced in relatively huge amounts in industrial and municipal levels. Moreover, 0.75 kg of urea-contaminated wastewater having around 1 wt% urea discharged from the industries during urea synthesis. Similarly, ~33 g of urea in urine is produced by an average adult per day [6]. Ecologically, presence of urea in the municipal liquid waste (MLW) does not only form water and soil pollution but also responsible for air pollution problem. In other words, the urea-rich municipal liquid waste (MLW) causes an unfavorable influence on entire ecosystem. Decomposition of urea containing municipal liquid waste (MLW) to dangerous species such as ammonia, nitrogen oxide and nitric acid can put some part to ecosystem by leading to acid rain which have negative effects on whole society [7]. Recently, it has been reported that energy can be generated from urea-rich municipal liquid waste (MLW). Up today, hydrogen or electricity can be directly generated through electrochemical oxidation of urea, according to the following reactions [8-10].

$$\text{Anode: } CO(NH_2)_2 + 6OH^- \rightarrow N_2 + 5H_2O + CO_2 + 6e \quad (1)$$

$$\text{Cathode: } 6H_2O + 6e \rightarrow 3H_2 + 6OH^- \quad (2)$$

$$\text{Overall: } CO(NH_2)_2 + H_2O \rightarrow N_2 + 3H_2 + CO_2 \quad (3)$$

The energy density of urea is higher than compressed or liquid hydrogen which makes it a potential energy carrier [11]. Unlike hydrogen, urea is a solid powder and is easily stored and transported. Urea effectively holds 10 wt% hydrogen and is a potential indirect hydrogen storage material. Further, urea-rich municipal liquid waste (MLW) can be used directly in fuel cell to generate electricity according to the following reactions.

$$\text{Anode: } CO(NH_2)_2 + 6OH^- \rightarrow N_2 + 5H_2O + CO_2 + 6e E^0 = +0.40\ V \quad (4)$$

$$\text{Cathode: } 2H_2O + O_2 + 4e \rightarrow 4OH^- E^0 = -0.746\ V \quad (5)$$

$$\text{Overall: } 2CO(NH_2)_2 + 3O_2 \rightarrow 2N_2 + 4H_2O + 2CO_2 E^0 = +1.146\ V \quad (6)$$

Noble metal based catalysts can be used for urea-based fuel cell [12-14]. But, unfortunately, the costly noble metals not only limit their potential application in large-scale removal of urea but also showed low current densities and higher oxidation potentials for the electro-oxidation of urea [5]. Therefore, many efforts have been made to replace the noble metal based catalyst with low-cost catalysts for urea electrolysis. It is well known, urea fuel cells operated only under alkaline conditions. Therefore, Ni-based catalyst possess worthy chance to be utilized as anode material, due to good chemical stability in alkaline environment. In addition, urea electrooxidation on Ni based catalyst was found to be very successful because it is anodically transformed to NiOOH (Ni^{3+}) according to the following reaction.

Electrochemical reaction: $E_{Ni}^{2+}/ E_{Ni}^{3+} \approx -0.45$ *V vs. SHE* (7)

$$Ni(OH)_2 + OH^- \rightleftharpoons NiOOH + H_2O + e^- \quad (8)$$

Catalytic reaction:

$$CO(NH_2)_2 + H_2O + 6\ NiOOH \rightarrow N_2 + CO_2 + 6Ni(OH)_2 \quad (9)$$

The onset potential is a significant indicator among the appealed parameters to demonstrate the electrocatalytic activity. The onset potential is the potential at which there exists a reaction pathway where all reaction steps have negative free energies. The onset potential of urea electro-oxidation depends on the redox potential of $\boldsymbol{E_{Ni}}^{2+}/\boldsymbol{E_{Ni}}^{3+}$. However, the translation of $\boldsymbol{E_{Ni}}^{2+}/\boldsymbol{E_{Ni}}^{3+}$ proceeds at a high potential leading to a large overpotential of more than 1 V for urea electro-oxidation on Ni. This large overpotential will affect the current density and energy efficiency of urea fuel cells [15-17]. Therefore, it is sensible to reduce the overpotential, for a better performance of urea fuel cells and it can be achieved through element doping, surface modification, design of morphology and advanced support materials. In this chapter, Ni-based electrocatalyst for urea oxidation based on morphology and composition modification are introduced.

2. Electrooxidation of Urea on Ni and Mn Nanoparticles Decorated Carbon Nanofibers

Among the various nanofibers making technique, electrospinning attracts the most attention due to simplicity, high yield, effectiveness and low cost. Figure 3 represents a typical electrospinning apparatus. Electrospinning is based on applying high DC electric voltage to stretch polymer solution or sol–gels. Accordingly, this technique was adequate to synthesis many functional polymeric nanofibers. Large axial ratio of the nanofibers delivers unique characteristic for this nano-morphology over the other structures [18-21]. Hence, the main aim of this section is to present Ni-based electrocatalyst for urea oxidation based on morphology and composition modification. Ni & Mn nanoparticles-decorated carbon nanofibers were synthesized using the electrospinning process (Figure 3) and investigated as anode in urea electrooxidation process. Typically, calcination of electrospun mats composed of nickel acetate, manganese acetate and poly (vinyl alcohol) in argon atmosphere resulted in abnormal decomposition of the acetate salts to form pristine metals according to the following equations, and good graphitization of the polymer to finally produce metal alloy nanoparticles-decorated carbon nanofibers.

$$Ni(CH_3CO_2)_2.4H_2O \rightarrow 0.86Ni(CH_3CO_2)_2{\cdot}0.14Ni(OH)_2 + 0.28CH_3CO_2H + 3.72H_2O \quad (10)$$

$$0.86Ni(CH_3COO)_2{\cdot}0.14Ni(OH)_2 \rightarrow NiCO_3 + NiO + CH_3COCH_3 + H_2O \quad (11)$$

$$NiCO_3 \rightarrow NiO + CO_2 \quad (12)$$

$$NiO + CO \rightarrow Ni + CO_2 \quad (13)$$

$$Mn(CH_3COO)_2{\cdot}4H_2O \rightarrow Mn(OH)(CH_3COO) + 3H_2O + CH_3COOH \quad (14)$$

$$Mn(OH)(CH_3COO) \rightarrow 0.5Mn + 0.5MnCO_3 + 0.5H_2O + 0.5CH_3COCH_3 \quad (15)$$

$$MnCO_3 \rightarrow MnO + CO_2 \quad (16)$$

$$MnO + CO \rightarrow Mn + CO_2 \quad (17)$$

Formation of CO and H_2 gases during the decomposition of nickel acetate, and their role in the abnormal decomposition of the nickel acetate was confirmed in details by De Jesus et al. [22].

2.1. Physicochemical Characterization

Figure 1A shows SEM image of the as-synthesized electrospun nanofibers, as shown in the figure good morphology, beads-free and smooth nanofibers were obtained. Due the polycondensation tendency of the utilized acetate precursors, calcination of the nanofiber mats did not strongly affect the nanofibrous morphology as shown in Figure 1B.

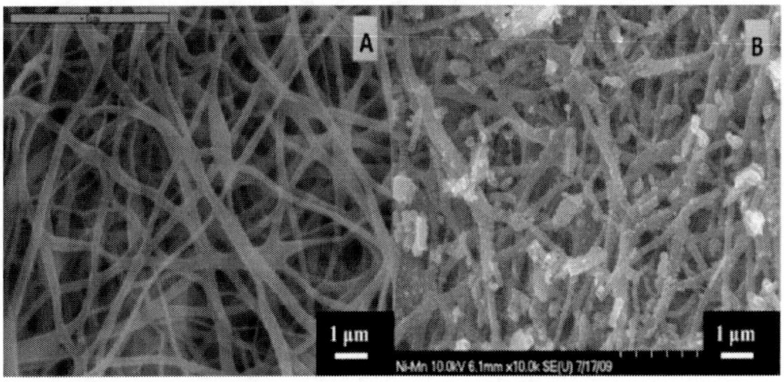

Figure 1. SEM image of the dry electrospun NiAc/MnAc/PVA nanofibers; (A), and FE SEM image of the calcined powder; (B). Reprinted with permission from, N.A.M. Barakat et al., Applied Catalysis A: General 510 (2016) 180–188. © 2016, Elsevier Ltd.

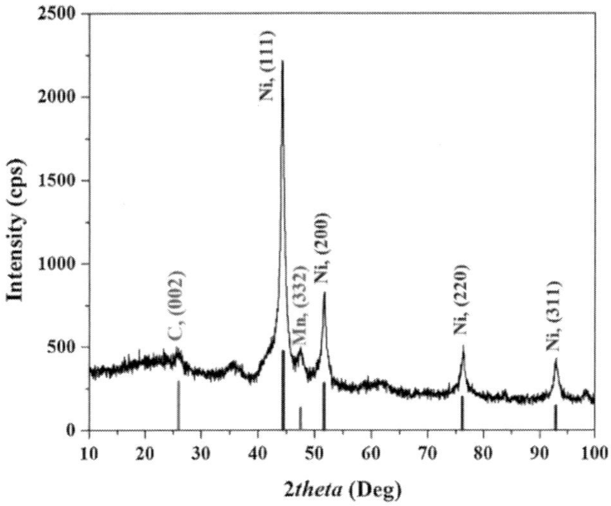

Figure 2. XRD pattern for the sintered nanofibers. Reprinted with permission from, N.A. Barakat et al., Applied Catalysis A: General 510 (2016) 180–188. © 2015, Elsevier Ltd.

However, the sintered nanofibers are decorated by some uniformly distributed nanoparticles. In contrast to other salts, calcination of the acetates in an inert atmosphere results in full decomposition to the corresponding metal rather than the expected metal oxides (eq. 8, 9,10,11,12, 13, 14 and 15).

Figure 2 shows the XRD results of the synthesized nanofibers, as shown no any peaks denoting to presence of any nickel or manganese oxides. Appearance of peaks at 2θ values of 44.30°, 51.55°, 76.05° and 92.55° corresponding to (1 1 1), (2 0 0), (2 2 0) and (3 1 1) crystal planes, respectively indicate the formation of pure nickel (JCDPS# 04-0850).

Additionally, the broad peak at 2θ of 26.3° corresponds to an experimental d spacing of 3.37 Å indicating presence of graphite-like carbon (d (0 0 2), JCPDS#; 41-1487). It is noteworthy mentioned that Mn can be represented in the pattern by the observed peak at 47.5° corresponding to crystal plane of (3 3 2) (JCDPS# 32-0637).

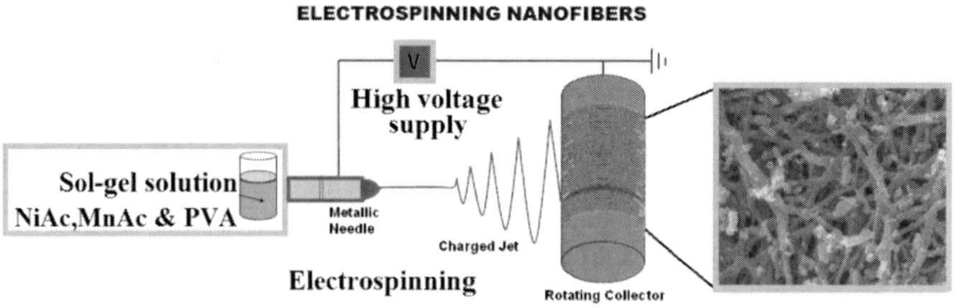

Figure 3. A typical electrospinning set up.

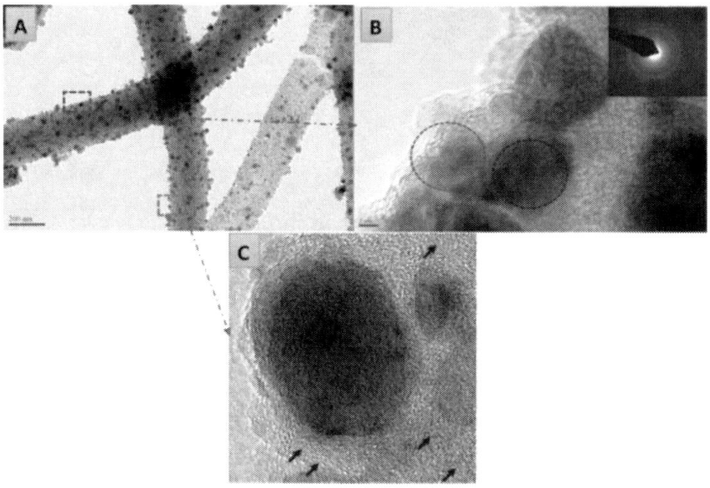

Figure 4. Normal TEM (A), and HR TEM images for the red marked area (B) the circles refers to two observed different crystal structures. The inset displays SAED pattern at the circles area. (C) HR TEM image at the green marked area. Reprinted with permission from, N.A.M. Barakat et al., Applied Catalysis A: General 510 (2016) 180–188. © 2016, Elsevier Ltd.

Figure 4B, HR TEM image, concludes that the nanoparticles are crystalline while the nanofiber body is amorphous. Moreover, as it is marked by the dashed circles, there are two different crystal structures for the attached nanoparticles which can translated as Ni and Mn metals. The SAED pattern, the inset, indicates that the investigated area composes of good crystalline material. Therefore, based on XRD and TEM results, it can be claimed that the nanofibers composed of amorphous carbon nanofibers decorated by crystalline metal (Ni of Mn) nanoparticles. The lattice spacing of the marked area (Figure 4b) indicated that there are two different kinds of crystalline nanoparticles and their planar distances are matching exactly with the standard cell parameters of Ni and Mn nanoparticles as the measuring planar spaces are 3.9 and 8.3 A°. The introduced SAED and HR-TEM images, shows polycrystalline nature of the introduced nanofibers (at the arrows).

2.2. Electrochemical Characterization

2.2.1. Activation of the Electrocatalyst Surface

The cyclic voltammetric behavior of the synthesized carbon nanofibers in 1.0 M KOH solution and absence of urea is shown in Figure 5. Polarization was started by a potential scanning at a scan rate of 100 mV/s from 800 to 0 mV (vs. Ag/AgCl) in the cathodic direction and then the scan was reversed in the back to 800 mV. As shown, redox peaks are observed in the positive potential side (~215 and 430 mV), this pair of peaks is related to the oxidation of $Ni(OH)_2$ to NiOOH [23-25].

Increasing the number of potential sweeps leads to an increase of the current density values of the cathodic peak because of the entry of OH^- into the $Ni(OH)_2$ surface layer, which leads to the progressive formation of a thicker NiOOH layer corresponding to the $Ni(OH)_2$/NiOOH transition [26]. Accordingly, the observed peaks in Figure 5 can be explained as $Ni(OH)_2$/NiOOH transformation reaction. Actually, activation of the Ni-based catalysts occurs in different steps. The oxidation of Ni to $Ni(OH)_2$ in the alkaline medium leads to form hydrous and anhydrous nickel hydroxide; α-$Ni(OH)_2$ and β-$Ni(OH)_2$, respectively [27]. However, with the time and at relatively high potential the unstable hydrous hydroxide (α-$Ni(OH)_2$) transforms to the more stable form; β-$Ni(OH)_2$ [28]. More applied potential leads to oxidize β-$Ni(OH)_2$ to β-Ni(OOH) which is partially converted to γ-Ni(OOH) and accumulates on the surface [29]. The appeared cathodic peak during the reverse scan indicates the presence of remaining different phases due to reduction of γ-Ni(OOH) to β or α-Ni(OOH) [29]. Increasing the current density at high potential can be attributed to the oxygen evolution reaction (OER).

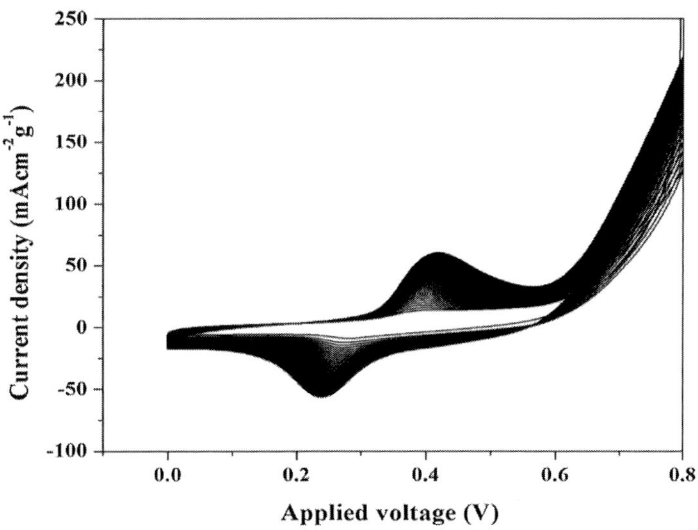

Figure 5. Cyclic voltamogramms of the NiMn-decorated carbon nanofibers in 1 M KOH at 25°C and 50 mV s−1scan rate. Reprinted with permission from, N.A.M. Barakat et al., Applied Catalysis A: General 510 (2016) 180–188. © 2016, Elsevier Ltd.

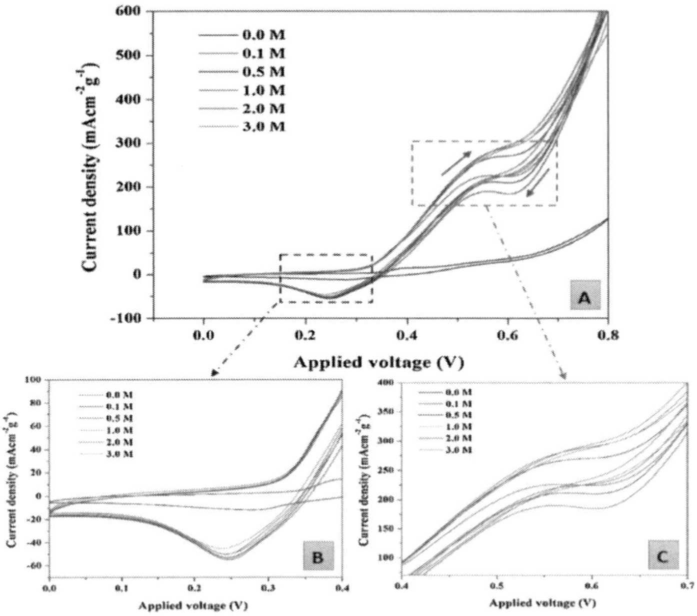

Figure 6. Cyclic voltammograms for the prepared NiMn-doped CNFs at different urea concentration in 1.0 M KOH at scan rate 50 mV $s^{-1}$and 25°C. Reprinted with permission from, N.A.M. Barakat et al., Applied Catalysis A: General 510 (2016) 180–188. © 2016, Elsevier Ltd.

2.2.2. Urea Oxidation

Presence of urea results in having a distinct increase in the anodic current density at potential higher than 280 mV (vs.Ag/AgCl), as shown in Figure 6A, which indicates strong electrocatalytic activity of the introduced nanofibers. Moreover, increasing the urea concentration in the alkaline solution leads to enhance the anodic current density. At low urea concentration (up to 0.5 M), it could be hypothesized that the increase in the anodic current density is due to a diffusion controlled process. However, as shown in Figure 6C, after urea concentration of 1.0 M, there is almost no change in the anodic current density as the current density was independent in the urea concentration that may be due to kinetics limitation control process. This finding explains that the surface coverage of urea molecules becomes high at high concentrations, which leads to fix the oxidation rate due to the local deprivation of OH^- for the excess urea molecules.

Another important finding can be observed in Figure 7C is that the urea concentration does not affect the anodic potential this can be explained by fast oxidation of urea and the intermediates. In other words, there are no adsorbed non-oxidized urea or intermediates at the surface even at high urea concentration need more high potential for oxidation [6].

Another anodic peak lying within the urea oxidation region can be observed during the reverse scan at ~520 mV. This finding can be assigned to further oxidation of urea molecules due to regeneration of the active (Ni^{+2}/N^{+3}) on the electrode surface which were blocked by urea molecules, intermediates, or reaction products during the forward scan. However, the decrease in the current density during the reverse scan compared to the forward one can be explain as incomplete regeneration of the surface. Moreover, a considerable decrease in the cathodic current density can be observed during the reverse scan (Figure 6B). The high increase in the anodic peak followed by considerable decrease in the cathodic peak in the reverse scan is consistent with a behavior typical to the catalytic regeneration effect. The principle of this mechanism involves the chemical oxidation of urea to its product along with the reduction of Ni^{+3} to Ni^{+2} species which is associated with the electrochemical regeneration of Ni^{+2} to Ni^{+3} [30, 31]. In other words, oxidation of urea strongly related to the regeneration of the catalyst surface as the require OH^- groups in reactions 1 and 4 to oxidize urea are obtained from NiOOH solid layer formed on the activated electrode surface. Compared to urea-free solution, as shown in the figure, the urea-containing solutions reveal low cathodic current density. This can be explained as the oxidation of urea is taking place with a rate higher than of oxidation $Ni(OH)_2$ which indicates good regeneration of the active surface at high potential and/or formation good active layer during the activation process (Figure 5). Figure 7 displays the influence of the scan rate on the electro-catalytic activity of the introduced nanofibers. As shown in the figure, the anodic current density increases with increasing the scan rate.

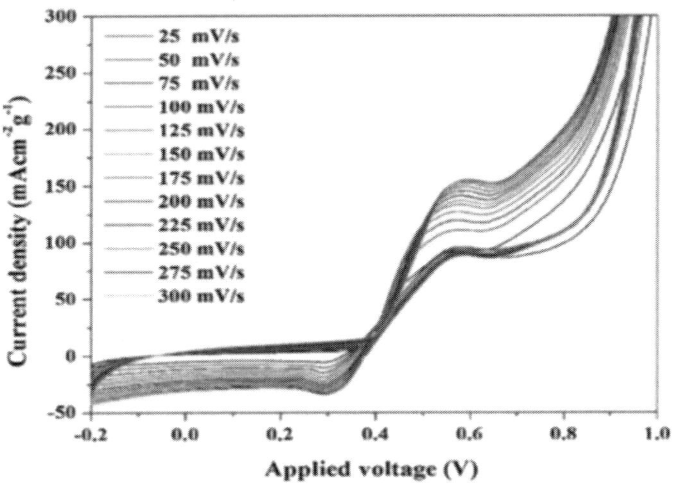

Figure 7. Cyclic voltagmmograms at different scan rates for the prepared nanofiber electrode in presence of 3.0 M urea solution. Reprinted with permission from, N.A.M. Barakat et al., Applied Catalysis A: General 510 (2016) 180–188. © 2016, Elsevier Ltd.

2.2.3. EIS Characterizations and Influence of Scan Rate

Nyquist plots for the investigated electrode at urea concentration of 3.0 M (in 1.0 M KOH) is displayed in Figure 8. In the Nyquist plot, the Faradaic reaction (urea oxidation) usually displayed by capacitive loop with a diameter almost matching the charge transfer resistance (R_{CT}). As shown, the capacitive loop appears clearly which c indicates the electrocatalytic activity of the investigated electrode.

3. Electrooxidation of Urea on Ni Nanoparticles Decorated Graphene

Various literatures have been presented work on a nickel-carbon sponge [32], nickel nanoribbons [33], nickel nanowires [17, 34], and nickel nanoparticles [6]. Due to superb electrical conductivity and particularly large surface area, graphene is highly considered support material for the electrocatalysis. Accordingly, in this section, Ni-decorated graphene was studied and investigated as an electrocatalyst for urea oxidation.

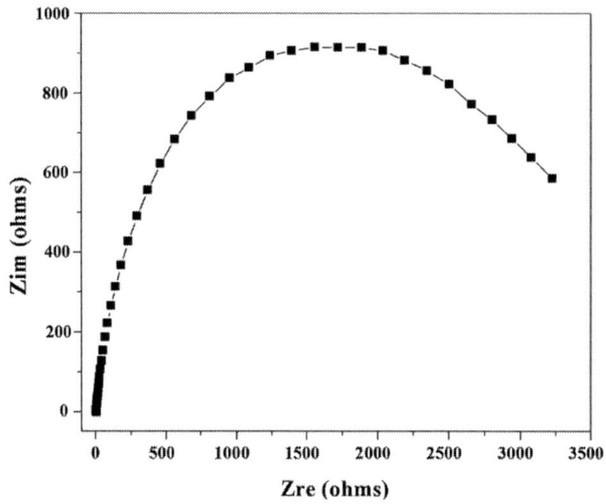

Figure 8. Nyquist plot for urea oxidation reaction at 0.60 V. Reprinted with permission from, N.A.M. Barakat et al., Applied Catalysis A: General 510 (2016) 180–188. © 2016, Elsevier Ltd.

Particularly, single- and few- layer graphene have gained much importance due to the heightened awareness of their potential applications in various fields. Ni-decorated graphene was prepared by reflux of graphene oxide with nickel acetate at 120°C for 10 h followed by calcination in argon atmosphere at 850°C for 2h. Figure 9 represents a typical heat reflux apparatus used for synthesis of Ni-decorated graphene.

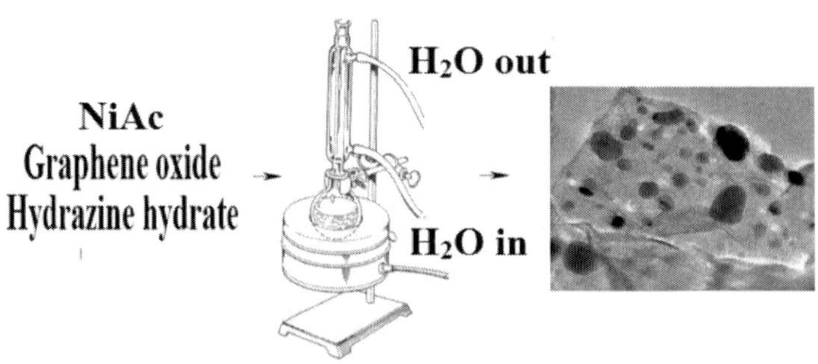

Figure 9. A typical heat reflux set up.

3.1. Physicochemical Characterization

Figure 10A TEM image for the produced graphene in absence of the metal precursor, as shown the exfoliation is clearly observed. As shown in Figure 10B, the formed graphene sheets, in presence of NiAc (100 wt% sample), are decorated by some nanoparticles. From Figure 10C which displays the HR TEM image, it can be concluded that the attached nanoparticles have good crystallinity which, based on XRD results (Figure 11), can be assigned to nickel.

Figure 11 shows the XRD results of the synthesized Ni-decorated graphene, as shown appearance of peaks at 2θ values of 44.30°, 51.55° and 76.05° corresponding to (1 1 1), (2 0 0) and (2 2 0) crystal planes, respectively indicate the formation of pure nickel (JCDPS# 04-0850). Additionally, the sharp peak at 2θ of 27.5° indicating the smaller crystalline size of graphene in a single or few layers structure[35, 36].

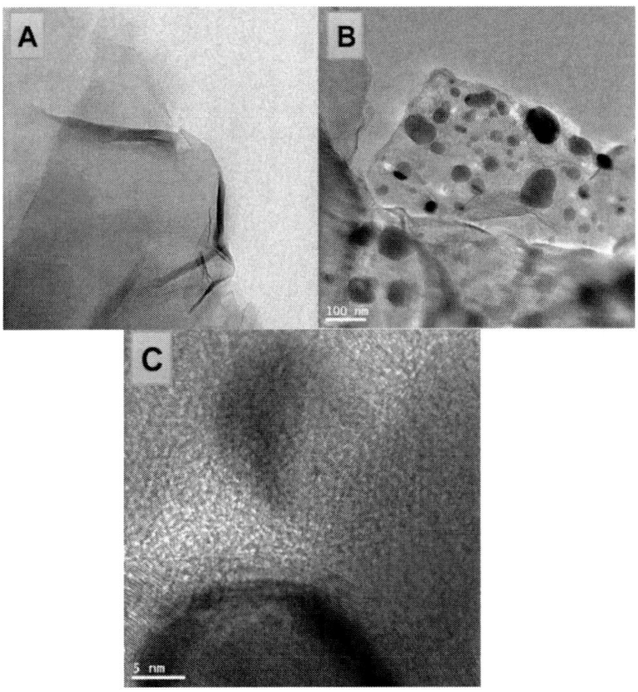

Figure 10. Normal TEM images for the pristine; (A) and Ni-decorated; (B) reduced graphene oxide. Panel C displays HR TEM image for the produced Ni-decorated graphene sheets (100 wt% NiAc sample). Reprinted with permission from, N. A. M. Barakat et al., Journal of molecular catalyst A: Chemical 421 (2016) 83–91. © 2016, Elsevier Ltd.

Besides XRD analysis, Raman spectroscopy can be used to distinguish graphite, GO and graphene. Figure 12 shows the Raman spectra of the prepared reduced graphene oxide with and without metal precursor. Addition. As shown in the figure, the patterns typically match with the standard graphene. However, there is an extra peak in Figure 12 which can be assigned to Ni.

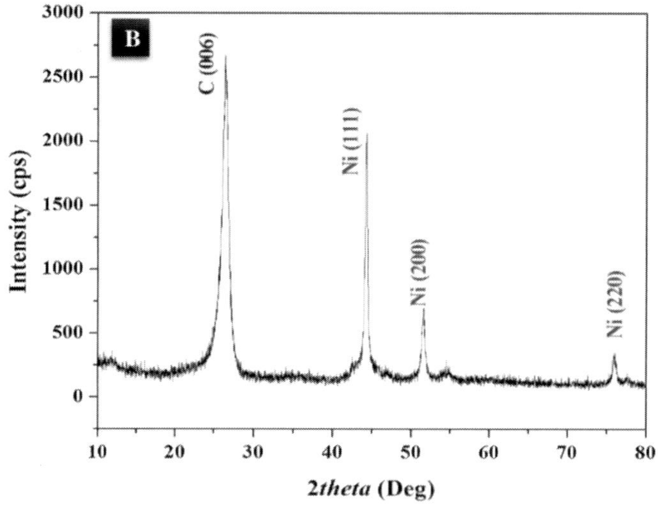

Figure 11. XRD pattern for the prepared electrocatalyst in presence of nickel acetate (100 wt% sample). Reprinted with permission from, N. A. M. Barakat et al., Journal of molecular catalyst A: Chemical 421 (2016) 83–91. © 2016, Elsevier Ltd.

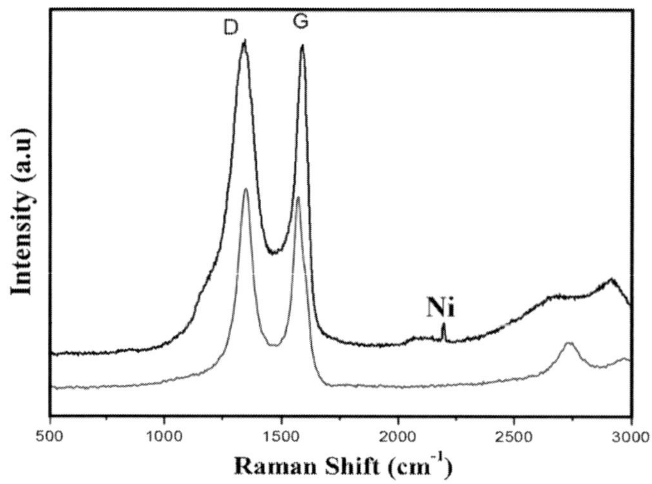

Figure 12. Raman spectroscopy for the prepared electrocatalyst in absence; (red) and presence of nickel acetate (100 wt.% sample); (black). Reprinted with permission from, N. A. M. Barakat et al., Journal of molecular catalyst A: Chemical 421 (2016) 83–91. © 2016, Elsevier Ltd.

3.2. Electrochemical Characterization

3.2.1. Activation of the Electrocatalyst Surface

Surface activation of the nickel-based materials is an important step to initiate the electrochemical activity through creation of NiOOH species [26]. As shown (Figure 13), polarization was initiated by a potential scanning at a scan rate of 50 mV/s from 1000 V to

−200 mV (vs. Ag/AgCl reference electrode) in the cathodic direction and the scan was then reversed in the anodic direction back to 1000 mV.

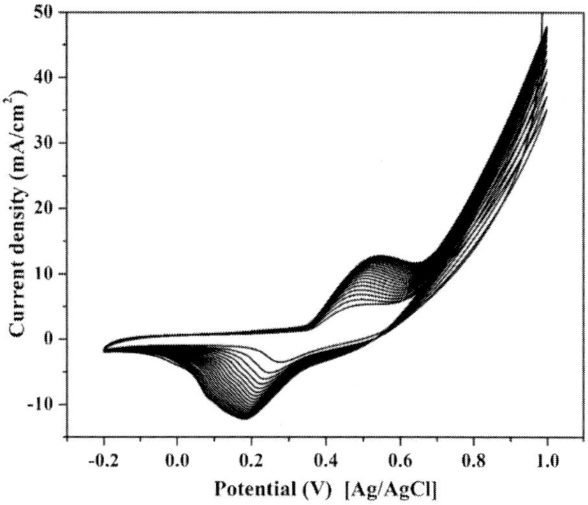

Figure 13. Cyclic voltammograms for the introduced Ni-loaded graphene (100 wt% NiAc sample) in 1 M KOH. Reprinted with permission from, N. A. M. Barakat et al., Journal of molecular catalyst A: Chemical 421 (2016) 83–91. © 2016, Elsevier Ltd.

From the obtained results, it can be shown that clear redox peaks are obtained in the positive potential. The observed peaks can be attributed to the oxidation of nickel hydroxide layer ($Ni(OH)_2$) to the required active NiOOH film according to the following equation 8. It is well known, during the activation step, more increase in the number of potential sweeps leads to an enhancement in the current density of the cathodic peak due to increase the entry of the hydroxyl group (OH^-) into the hydroxide film ($Ni(OH)_2$) which results in the progressive formation of a high thickness NiOOH layer based on $Ni(OH)_2$/NiOOH transformation [26]. Consequently, the obtained peaks in Figure 13 can be assigned to the conversion of $Ni(OH)_2$ to NiOOH.

3.2.2. Urea Oxidation

Figure 14 displays the CV behavior of Ni-decorated graphene (100 wt% NiAc sample) in presence and absence of urea (2 M) in 1.0 M KOH solution. Polarization was started by a potential scanning at a scan rate of 50 mVs^{-1} from 1000 mV to −200 mV (vs. Ag/AgCl reference electrode). As shown, presence of urea dramatically increases the anodic current density at potential higher than 320 mV (vs. Ag/AgCl) which indicates the strong electrocatalytic activity.

Another anodic peak lying with in the urea oxidation region can be observed. The secondary oxidation peak appears at a lower potential than that of the forward scan peak with the same current density. This can be attributed to oxidation of more urea molecules and/or the formed intermediates during the urea oxidation step. The further increase in the oxidation reactions is attributed to the regeneration of the active compounds (i.e., Ni^{+2} to Ni^{+3}) on the surface.

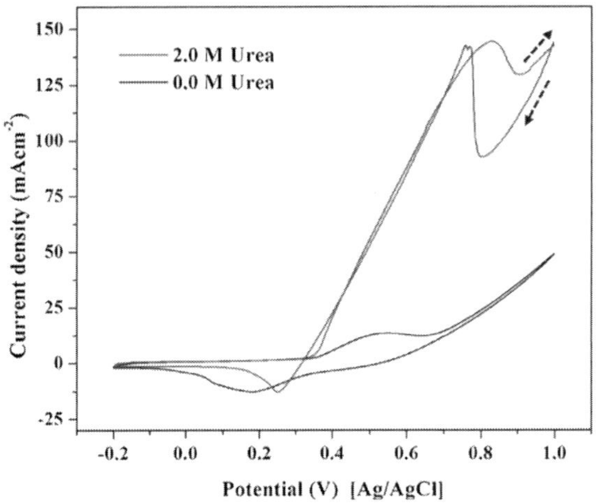

Figure 14. CV for the produced Ni-decorated graphene (100 wt% NiAc sample) in the presence and absence of urea: 1.0 M KOH, 50 mVs^{-1}scan rate. Reprinted with permission from, N. A. M. Barakat et al., Journal of molecular catalyst A: Chemical 421 (2016) 83–91. © 2016, Elsevier Ltd.

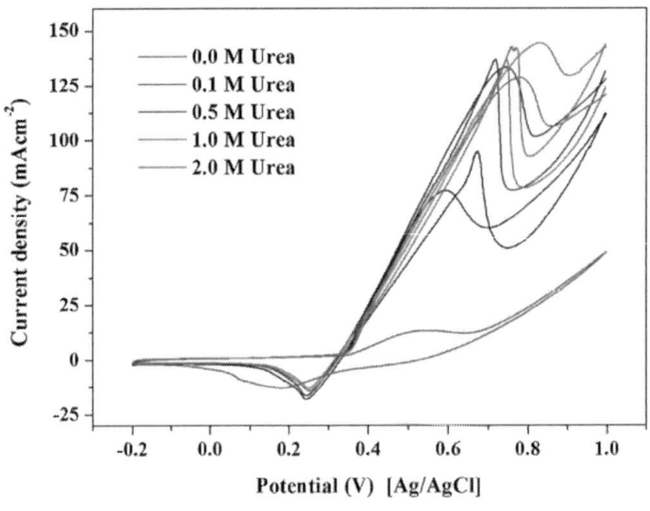

Figure 15. Influence of urea concentration in 1 M KOH on the electrocatalytic activity of the introduced Ni-decorated graphene (100 wt% NiAc sample): 50 mVs^{-1}scan rate and 25°C. Reprinted with permission from, N. A. M. Barakat et al., Journal of molecular catalyst A: Chemical 421 (2016) 83–91. © 2016, Elsevier Ltd.

3.2.3. Influence of Urea Concentration

Figure 15 shows the influence of urea concentration even at small urea concentration (up to 0.5 M). On the other hand, when the urea concentration was increased to 1.0 M, no change in the anodic current density could be detected even with more increase in the solution concentration. In other words, after 1.0 M, the current density is independent in the urea concentration. This finding can be hypothesized as the kinetics limitation is the main control process at high concentration of urea. This result indicates that the coverage surface by urea molecules is high at enhanced concentrations; accordingly, a fixed oxidation rate is observed because of the local deprivation of the hydroxyl groups (OH^-) for the extra molecules of urea.

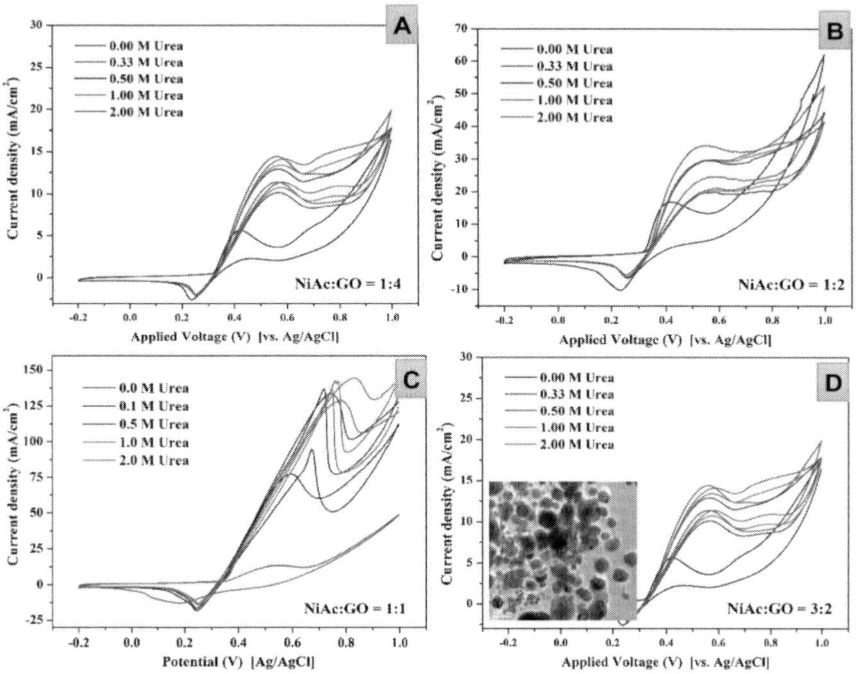

Figure 16. Study the influence of metal content on the electrocatalytic activity of Ni-loaded graphene at different urea solution concentrations (in 1 M KOH) at a scan rate of 50 mV/s and 25°C. Panels A, B, C and D represent 25, 50, 100 and 200 wt% NiAc, respectively. The inset in panel D displays TEM image for NiAc:GO 3:2 sample. Reprinted with permission from, N. A. M. Barakat et al., Journal of molecular catalyst A: Chemical 421 (2016) 83–91. © 2016, Elsevier Ltd.

A considerable additional finding is that an increase the urea concentration leads to a slight increase in the urea oxidation potential. Increasing the urea concentration results in an increase the surface coverage of urea and/or the intermediates, thus leading to a high current density by oxidizing the large amount of adsorbed molecules [6, 37].

3.2.4. Influence of Metal Loading

The catalyst composition is an important factor to design the proper electrocatalyst. Figure 16 displays the performance of the Ni-loaded graphene with different nickel acetate: graphene compositions. As shown, at the low metal content (25 wt% NiAc sample) the performance is weak which can be assigned to the small amount of the functional material (i.e., Ni NPs). However, increasing the nickel acetate percentage led to a distinct enhancement in the activity of toward urea oxidation as shown in Figure 16B.

However, as shown in Figure 16C, the best performance is attributed to the sample prepared when equal amounts from NiAc and GO were utilized. From Figure 16D, it can be concluded that more increase in the metal contents has a negative influence on the performance as the corresponding current density is low compared to the other formulations because at low nickel content (25 and 50 wt% NiAc samples; Figure 16A and B), the amount of the functional material (nickel nanoparticles) are small which results in low performance. However, increasing the metallic content positively enhances the performance as shown in Figure 16C.

But, more increase in the amount of the loaded metal nanoparticles leads to cover the graphene sheet surface which results in obstructing the graphene from achieving its main role; adsorption of urea molecules. The inset in Figure 16D displays the TEM image for the used sample, as shown the amount of the formed NiNPs are many with relatively big size compared to the best sample (Figure 10B).

3.2.5. Influence of Carbon Support and Its Morphology

The inset of Figure 17 shows a TEM image for the used Ni-incorporated CNFs. As shown, compared to the Ni-decorated graphene electrocatalyst, both materials have very low electrocatalytic activity toward urea oxidation. Moreover, the results confirm the importance of using carbon support as Ni-CNFs revealed higher current density compared to Ni NPs although the later possess higher amount from the functional material (Ni). Overall, it can be confidently claim that using carbon support is highly recommended to enhance the electrocatalytic activity of Ni toward urea oxidation. Moreover, due to its distinct properties, graphene is considered an optimum support.

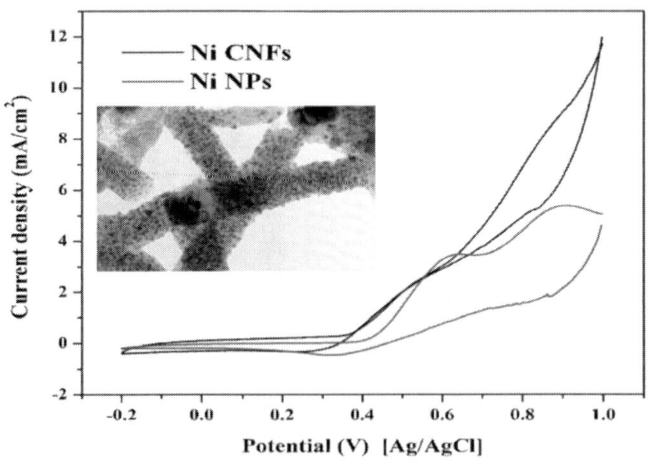

Figure 17. CV for Ni-incorporated carbon nanofibers (Ni-CNFs) and pristine nickel nanoparticles (Ni NPs) in presence of 1 M urea dissolved in 1 M KOH solution; scan rate 50 mV/s at room temperature. The inset displays the used Ni-incorporated carbon nanofibers. Reprinted with permission from, N. A. M. Barakat et al., Journal of molecular catalyst A: Chemical 421 (2016) 83–91. © 2016, Elsevier Ltd.

CONCLUSION AND FUTURE PERSPECTIVES

Direct urea fuel cells could be an effective method for generation of energy from municipal liquid waste containing urea. The kinetics of direct urea fuel cell is lethargic and hence the electrooxidation of urea deliberated to be an ineffective process. Nickel is widely used as an electrocatalyst in an attempt to increase electrochemical oxidation rate of urea in aqueous alkaline solution. Therefore, this chapter has unlocked a brief overview of two types of catalysts are taken into account for direct urea fuel cell: Ni and Mn decorated carbon nanofibers and Ni nanoparticles decorated graphene for anode catalyst. However state of the art nickel based catalyst delivers a low power output. Furthermore the catalytic activity of nickel based anode is found to decrease faster over time due to the possible poisoning of electrode during the oxidation of urea; it looks debatable that nickel based catalyst will substitute noble metal based catalysts. It is probable that in the future the performance of nickel based catalyst will increase and undoubtedly, this leads us to consider Ni based catalyst as potential option to replace Noble metal based catalysts.

REFERENCES

[1] J. H. Meessen, H. Petersen, Urea. *Ullmann's Encyclopedia of Industrial Chemistry*, 1996.

[2] P. Atkins, *Shriver Atkins' Inorganic Chemistry*, 2010.

[3] W.Yan, D. Wang, G. G. Botte, Electrochemical deposition of urea with Ni-based catalysts, *Appl. Cat. B: Environ.*, 127 (2012) 221.

[4] S. Kojima, A. Bohner, N. Von Wirén, Molecular mechanisms of urea transport in plants, *The J. Mem. Bio.*, 212 (2006) 83.

[5] B. K Boggs, R. L. King, G. G. Botte, Urea electrolysis: direct hydrogen production from urine, *Chem. Comm.*, 32 (2009) 4859.

[6] V. Vedharathinam, G. G. Botte, Understanding the electro-catalytic oxidation mechanism of urea on nickel electrodes in alkaline medium, *Electrochim. Acta*, 81 (2012) 292.

[7] M. Rahimpour, H. Mottaghi, M. Barmaki, Hydrogen production from urea wastewater using a combination of urea thermal hydrolyser-desorber loop and a hydrogen-permselective membrane reactor. *Fuel Proc. Tech.*, 91 (2010) 600.

[8] R. Lan, S. Tao, J. T. Irvine, A direct urea fuel cell-power from fertilizer and waste, *Ener. & Environ.Sci.*, 3 (2010) 438.

[9] R. L. King, G. G. Botte, Investigation of multi-metal catalysts for stable hydrogen production via urea electrolysis, *J. Pow.Sour.*, 196 (2011) 9579.

[10] D. Wang, et al., Electrochemically reduced graphene oxide-nickel nanocomposites for urea electrolysis, *Electrochim. Acta*, 89 (2013) 732.

[11] R. Graupner, *Guanidine–Safe, Clean, Flexible*. http://www.energy.iastate.edu/Renewable/ammonia/ammonia/2006/OregonSustainableEnergy.pdf

[12] W. Simka, J. Piotrowski, G. Influence of anode material on electrochemical decomposition of urea, Nawrat, *Electrochim. Acta*, 52 (2007) 5696.

[13] F. Manea, et al., Voltammetric detection of urea on an Ag-modified zeolite-expanded graphite-epoxy composite electrode, *Sensors*, 8 (2008) 5806.

[14] A. T. Miller, B. L. Hassler, G. G. Botte, Rhodium electrodeposition on nickel electrodes used for urea elctolysis, *J. App. Electrochem.*, 42 (2012) 925.

[15] W. Xu, Z. Wu, S. Tao, Urea-Based Fuel Cells andElectrocatalysts for Urea Oxidation. *Energy Tech.*, 4 (2016) 1329.

[16] J. D. Chen, et al., Easy-to-prepare electrochemical platform composed of ionic liquid-Ni(II)-graphite composites: laboratory study on electrochemical oxidation of urea, alcohols, and glucose, *Electrochim. Acta*, 182 (2015) 113.

[17] F. Guo, et al., Preparation of nickel nanowire arrays electrode for urea electro-oxidation in alkaline medium, *J. Pow. Sour.*, 278 (2015) 562.

[18] Z. K. Ghouri , et al., Facile synthesis of Fe/CeO_2-doped CNFs and their capacitance behavior, *Int. J. Electrochem. Sci*, 10 (2015) 2064.

[19] Z. K. Ghouri, et al., Ceram. Co/CeO_2-decorated carbon nanofibers as effective non-precious electro-catalyst for fuel cells application in alkaline medium, *Intern.*, 41 (2015) 2271.

[20] Z. K. Ghouri, et al., Synthesis and characterization of Co/$SrCO_3$ nanorods-decorated carbon nanofibers as novel electrocatalyst for methanol oxidation in alkaline medium, Ceram. *Intern*, 41 (2015) 6575.

[21] Z. K. Ghouri, et al., Nano-engineered ZnO/CeO_2 dots@CNFs for fuel cell application, *Arabian J.Chem.*, 9 (2016) 219.

[22] J. C De Jesus, et al., Thermal decomposition of nickel acetate tetrahydrate: an integrated study by TGA, QMS and XPS techniques, *J. Mol.Catal. A: Chem.*, 228 (2005) 283.

[23] M. Fleischmann, K. Korinek, D. Pletcher, The oxidation of organic compounds at a nickel anode in alkaline solution, *J. Electroan. Chem. Interf. Electrochem*, 31 (1971) 39.

[24] M. J.Vuković, Voltammetry and anodic stability of a hydrous oxide film on a nickel electrode in alkaline solution App. *Electrochem.*, 24(1994) 878.

[25] O. Enea, molecular structure effects in electrocatalysis-II. Oxidation of D-glucose and of linear polyols on Ni electrodes, *Electrochim. Acta*, 35 (1990) 375.

[26] M. A. Rahim, R. A. Hameed, M. Khalil, Nickel as a catalyst for the electro-oxidation of methanol in alkaline medium, *J. Pow.Sour.*, 134 (2004) 160.

[27] A. Van der Ven, J. et al., Phase stability of nickel hydroxides and oxyhydroxides, *J. Electrochem. Soc.*, 153 (2006) 210.

[28] J. Desilvestro, D. A. Corrigan, M.J. Weaver, Characterization of redox states of nickel hydroxide film electrodes by in situ surface raman spectroscopy, *J. Electrochem. Soc.*, 135 (1988) 885.

[29] A. N. Golikand, et al., Electrocatalytic oxidation of methanol on a nickel (II)-1-(2-pyridylazo)-2-naphthol complex modified glassy-carbon electrode in alkaline medium, *J. Russian Electrochem.* 42 (2006) 167.

[30] H. R. Zare, et al., Electrochemical evaluation of coumestan modified carbon paste electrode: study on its application as a NADH biosensor in presence of uric acid, *Sens. Act. B: Chem.*, 114 (2006) 610.

[31] U. Paulus, et al., Oxygen reduction on carbon-supported Pt-Ni and Pt-Co alloy catalysts, *The J. Phy. Chem. B*, 106 (2002) 4181.

[32] K. Ye, et al., Highly porous nickel@carbon sponge as a novel type of three-dimensional anode with low cost for high catalytic performance of urea electro-oxidation in alkaline medium, *J. Pow. Sour.*, 283 (2015) 408.

[33] D. Wang, et al., Enhanced electrocatalytic oxidation of urea based on nickel hydroxide nanoribbons, *J. Pow. Sour.*, 217 (2012) 498.

[34] W. Yan, et al., Nickel nanowires as effective catalysts for urea electro-oxidation. *Electrochim. Acta*, 134 (2014) 266.

[35] W. S. Hummers Jr, R.E. Offeman, Preparation of graphitic oxide, *J. Am. Chem.Soc.*, 80 (1958) 1339.

[36] S. Stankovich, et al., synthesis of graphene-based nanosheets via chemical reduction of exfoliated graphite oxide, *Carbon*, 45(2007) 1558.

[37] N. A. Barakat, et al., Ni&Mn nanoparticles-decorated carbon nanofibers as effective electrocatalyst for urea oxidation, *App. Cat. A: Gen.*, 510 (2016) 180.

In: Handbook of Remediation ...
Editors: Sadia Ameen, M. Shaheer Akhtar et al.
ISBN: 978-1-53615-118-3

Chapter 4

NATURAL PRODUCT (–) CATECHIN DERIVATIVES: SYNTHESIS, BIOLOGICAL EVALUATION, DOCKING STUDIES AND THEIR ENVIRONMENT IMPACTS

***Deepak Kumar*[1,2,*] *and Dong-Soo Shin*[1]**
[1]School of Natural Sciences, Changwon National University, Changwon, Republic of Korea
[2]Department of Medicinal and Pharmaceutical Chemistry, School of Pharmacy, Sharda University, Greater Noida, India

ABSTRACT

Catechin, a natural product obtained from the plant and its derivatives (-)-catechin were synthesized, characterized and screened for their cytotoxicity and antimicrobial activity. Most of the compounds showed promising activity and especially, compound **1d** exhibited promising activity against the cancer cell line against hepatocellular liver carcinoma (HepG2), lung adenocarcinoma (A549) and prostate (DU-145) cell lines. Compounds **1a, 1b,** and **1c** presented potent antimicrobial activity. The molecular docking studies of the potent compounds expressed good binding affinity and are in favor of the observed biological activity.

Keywords: plant extract, catechin, biological activity, cancel cell, molecular docking

* Corresponding Author's Email: guptadeepak002@gmail.com.

1. Introduction

The leaves of *Camellia sinensis*, are used to make green tea and it is one of the most popular and commonly consumed beverages next to water in the world [1]. The main ingredients of tea are catechins, which contain a benzopyran skeleton with a phenyl group substituted at the 2-position and a hydroxyl (or ester) function at the 3-position. Modification in the catechin structure involves the stereochemistry of the 2,3-substituents and the number of hydroxyl groups.

Green tea primarily comprise of catechins such as (−)-epigallocatechin-3-gallate (**I-1**), (+)-epicatechin (**I-2**), (−)-catechin (**I-3**), (−)-epicatechin (**I-4**), and gallocatechin gallate (**I-5**) (Figure 1). It also contains (−)-epigallocatechin and (−)-epicatechin-3-gallate [2]. (−)-EGCG is the major constituent and biologically promising polyphenol in green tea which has around 40% of the total polyphenolic mixture [3], and there has been extensive research into the ways of (−)-EGCG [4, 6].

It has been paid potential considerations over last decade with respect to the beneficial biological activities of its components along with catechins, which have been reported to have antimutagenic [7], antioxidant [8], antibacterial [9, 10, 11], antitumor and cancer preventive [12, 13]. Even though, they show stronger anticancer activities [7, 13], catechins possess the major limitations due to their low bioavailability, which could be related to their low stability in neutral or slightly alkaline solutions and their inability to cross cellular membranes [14]. Thus, the analogues of catechins are of great interest and became targets for many synthetic chemists.

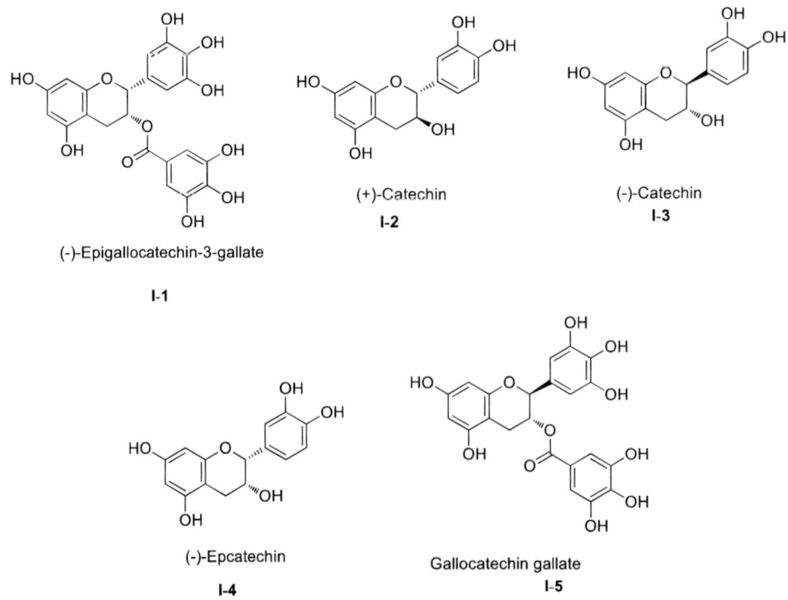

Figure 1. Major components of tea catechins.

Park and coworkers synthesized 3-*O*-alkyl and acyl analogues of (−)-epicatechin [15]. All alkylated ether analogues exerted better anticancer activities over acyl analogues against various cancer cell lines. The (−) - epicatechin analogues with moderate alkyl chains instead of longer ones, exhibited greater enhancement in their activity and also prevented from enzymatic or non-enzymatic cleavage. Introduction of moderate sized aliphatic chain at the C-3 hydroxyl may significantly improve inhibitory effect of cancer cell growth. This information was quite useful in making effective derivatives of catechins to study their activities of particular interest and in continuation of our studies towards biologically active chemical entities [16, 17]. Herein we report the synthesis of (−)-catechin analogues with an optimal alkyl chain length for their anticancer, antioxidant, antibacterial activities and docking studies.

1.1. Introduction to Synthesis of (−)-Catechin and Derivatives

Many reports on the synthesis of catechin have been reported [18] as first Nay et al. synthesized catechin. Catechin constituting the only flavan-3-ol is synthesized from the corresponding dihydroflavonol [19, 20]. ^{13}C-Labeled (±)-catechin is prepared by utilizing osmium-catalyzed dihydroxylation of a flav-3-ene intermediate as a main step to give the 2,3-*trans*-3,4-*cis*-isomer in high diastereoselectivity [21]. Later, a better approach was developed by the same group [22, 23] for the synthesis of ^{13}C-labeled (-)-procyanidin B-3. They obtained improved yields due to the number of steps and the reduction of intermediate flav-3-ene. The excellent results from this approach is related to the formation of AD-mix-*α* or AD-mix-*β* during asymmetric dihydroxylation of olefins. They also used this stereoselective catalysts for the synthesis of chirality at C-2 and C-3 of the flavan-3-ol framework [24, 28].

The diols was transformed more effectively into the corresponding flavan-3-ols. The utilization of methods have aimed to explore the selective removal of the 2'-*O*-methoxymethyl group and subsequent ring closure under mild acidic conditions. Deprotonation and cyclization of diols in the presence of 3M HCl and MeOH followed by acetylation, is yielded the **I-31** 2,3-*trans* and for the first time 2,3-*cis*-flavan-3-ols in excellent enantiomeric excess (>99%). This method is an extension of thc synthesis of flavan-3-ols, including different oxygenated phenolic substitutions as found in nature [29].

1.2. Design of Novel Synthesis of (−)-Catechin Derivatives

Park and coworkers synthesized 3-*O*-alkyl and acyl analogues of (−)-epicatechin [15]. All alkylated ether analogues are exerted the better biological activity over acyl analogues.

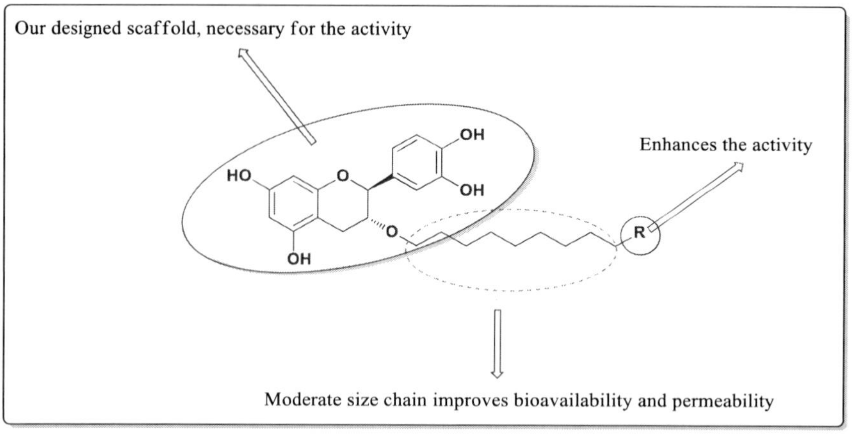

Figure 2. Our designed prototype for (-)-cathechin derivatives.

The (−)-epicatechin analogues with moderate alkyl chains instead of longer ones, exhibited greater enhancement in their activity and also prevented from enzymatic or non-enzymatic cleavage. Introduction of moderate sized aliphatic chain at the C-3 hydroxyl may significantly improve inhibitory effect of cancer cell growth. This information was quite useful in making effective derivatives of catechins to study their activities of particular interest.

1.3. Retrosynthetic Analysis

Retrosynthetic strategy of present work for the synthesis of (−)-catechin derivative is presented in scheme 1. (−)-Catechin derivatives **1** is obtained from **4** by nucleophilic substitution reaction followed by simple hydrogenation conditions in presence of Pd/C in ethylacetate in good yields.

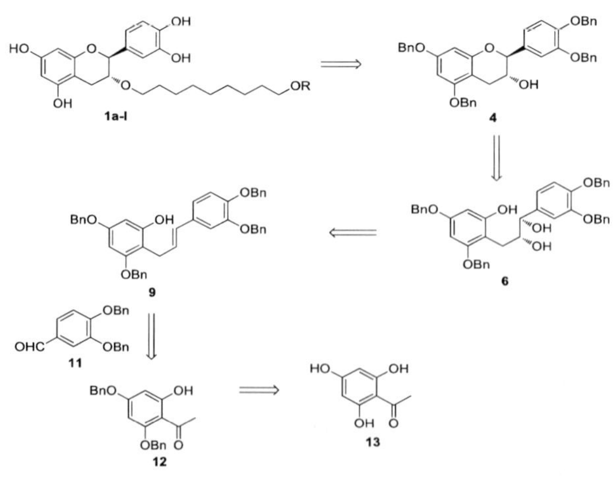

Scheme 1. Retrosynthetic analysis of (−)-catechin derivatives.

Tetraetrabenzylated (-)-catechin **4** was obtained from diol **6** by protecting with triethoxy orthoacetate, deprotonation followed by furan ring formation using K_2CO_3 as base. The triol **6** inturn obtained from alcohol **9** *via* Sharpless asymmetric dihydroxlation in presence of AD-mix-*β*. Compound **9** synthesized from alcohol condensation with aldehyde followed by reduction with $CeCl_3/NaBH_4$. Compound **12** was obtained by benzylation of commercially available trihydroxyacetophenone **13** in good yields.

2. Synthesis of Catechin

2.1. Synthesis of Tetrabenzylated (−)-Catechin

Tetrabenzylated (−)-catechin was synthesized according to the reported procedure [18].

2.2. Synthesis of 9-Bromononan-1-Ol

9-Bromononan-1-ol (**3**) was synthesized from 1,9-nonane diol (**18**) using hydrobromic acid at reflux (178–180°C) for 28 h. The water formed during the reaction was removed using a Dean–Stark trap. The compound **3** was obtained as white solid [30].

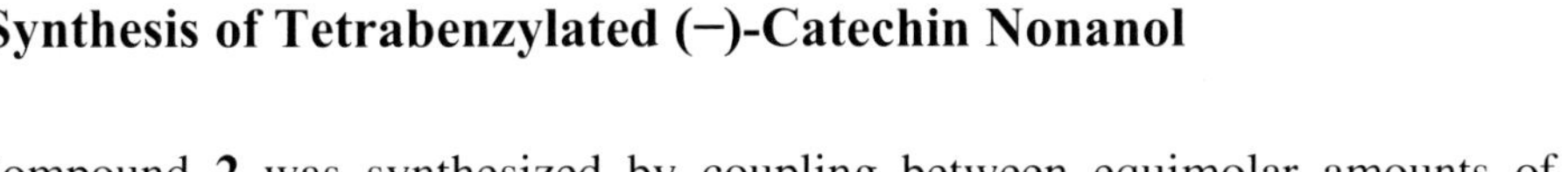

Scheme 2. Synthesis of 9-bromononan-1-ol.

2.3. Synthesis of Tetrabenzylated (−)-Catechin Nonanol

Compound **2** was synthesized by coupling between equimolar amounts of **4** and 9-Bromononan-1-ol (**3**) in dimethylforamide at room temperature using 60% of sodium hydride dispersed in mineral oil and the compound **2** was obtained in good yield.

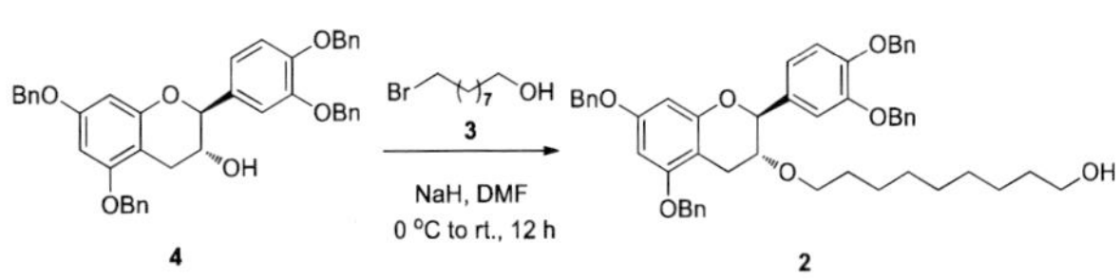

Scheme 3. Synthesis of tetrabenzylated (−)-catechin nonanol. Reprinted with permission from D. Kumar, 2015, Bull. Korean Chem. 36 (2015) 564© 2015 Korean Chemical Society, Seoul & Wiley-VCH Verlag GmbH & Co. KGaA, Weinheim.

2.4. Synthesis of (−)-Catechin Derivatives

The designed ether analogues at 3-hydroxy group **4** were prepared from **3**. The key synthetic intermediate **2** was substituted with different alkyl/aryl heterocyclic substitutes. The reductive debenzylation of **2, 2b-l** was carried in the presence of a 10% Pd/C in ethyl acetate as solvent to obtain the desired (−)-catechin analogues **1a-l** in 75-86% yields as sticky foam (Scheme 4). All the compounds were completely characterized by ^{1}H, ^{13}C NMR and mass spectrometry.

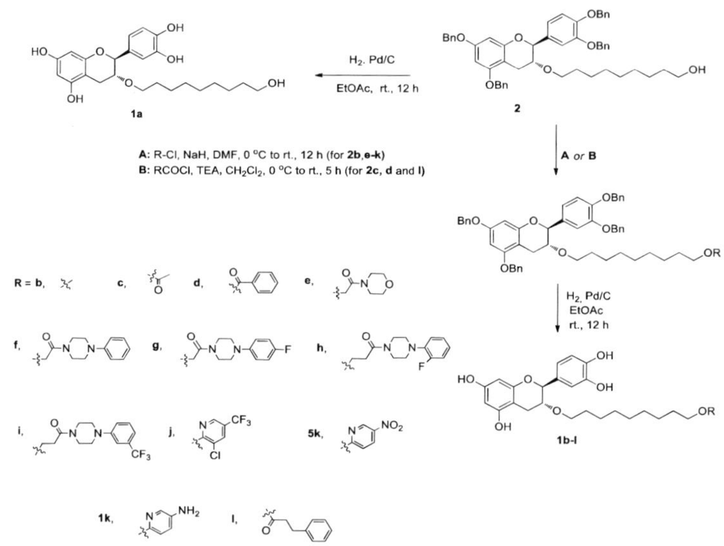

Scheme 4. Synthesis of (−)-catechin derivatives 1a-l. Reprinted with permission from D. Kumar, 2015, Bull. Korean Chem. 36 (2015) 564© 2015 Korean Chemical Society, Seoul & Wiley-VCH Verlag GmbH & Co. KGaA, Weinheim.

2.5. Anticancer Activity

The anticancer activity of compounds **1a-l** were screened for *in vitro* against four cancer cell lines such as human alveolar adenocarcinoma epithelial cell line (A549), human breast adenocarcinoma cell line (MDA-MB-231), human prostate cancer cell line (DU145) and human hepatocellular liver carcinoma cell line (HepG2) using MTT assay [31]. IC_{50} values of the test compounds for 24 h on each cell line was calculated Doxorubicin, which is a well-known anticancer drug and used as standard. The results are presented in Table 1.

Among the entire screened synthesized compounds against human alveolar adenocarcinoma epithelial cell line (A549), compound **1d** displayed very good activity with IC_{50} values of 4.8 μM. Compounds **1a**, **1e** and **1j** shows good activity (IC_{50} < 20 μM) while **1f**, **1h**, **1k** and **1l** showed moderate anticancer activity. The results are summarized in Table 1.

Table 1. *In vitro* anticancer effects against human alveolar adenocarcinoma epithelial cell line (A549), human breast adenocarcinoma cell line (MDA-MB-231), human prostate cancer cell line (DU145) and human hepatocellular liver carcinoma cell line (HepG2)

Test	IC_{50} values in (μM)				cLog *P*
compds	A549	MDA-MB-231	DU145	HepG2	
1a	9.4 ± 0.32	23.4 ± 0.21	22.4 ± 0.22	10.4 ± 0.34	3.39
1b	-	-	-	-	4.50
1c	-	-	-	5.8 ± 0.42	4.34
1d	4.8 ± 0.22	7.8 ± 0.12	5.4 ± 0.14	2.5 ± 0.22	6.27
1e	9.5 ± 0.24	8.9 ± 0.32	9.1 ± 0.16	7.4 ± 0.18	3.67
1f	21.9 ± 0.11	48.2 ± 0.37	51.2 ± 0.25	57.4 ± 0.46	5.66
1g	-	-	36.9 ± 0.26	48.2 ± 0.22	5.97
1h	49 ± 0.22	51.2 ± 0.25	-	-	6.18
1i	-	-	9.1 ± 0.22	7.0 ± 0.19	7.04
1j	7.9 ± 0.09	6.6 ± 0.11	8.2 ± 0.18	5.5 ± 0.08	7.10
1k	38.6 ± 0.42	-	46.8 ± 0.38	-	5.87
1l	28.0 ± 0.38	-	42.5 ± 0.42	-	6.43
Dox	0.7 ± 0.12	0.9 ± 0.11	0.9 ± 0.08	0.6 ± 0.18	-

IC_{50} = Compound concentration required to inhibit tumor cell proliferation by 50%. All the experiments were performed in triplicate and the results are reported as IC_{50} values (in μM) expressed as the mean ± S.D.

In case of human breast adenocarcinoma cell line (MDA-MB-231), compounds **1d**, **1e** and **1j** exhibited the highest growth inhibitory activity at concentrations of less than 8.9 μM, whereas compound **1a** shows good inhibitory activity. The compounds **1f** and **1h** also present the moderate anticancer activity when compared with all the synthesized compounds. The results are presented in Table 1.

Compound **1d** potent activity (IC_{50} values of 5.4 μM), while **1e**, **1i** and **1j** displayed good activity (IC_{50}< 9.1 μM) and **1a**, **1k** and **1l** expressed moderate cytotoxic activity when compared with other synthesized compounds.

The cytotoxic potential of compounds **1a-l** has evaluated against human hepatocellular liver carcinoma cell line (HepG2) which showed that compound **1d** to be the most active with an IC_{50} value of 2.5 μM. The compounds **1c** and **1j** exhibited an equipotent cytotoxic activity with IC_{50} values 5.8 and 5.5 μM respectively, whereas **1a**, **1e** and **1i** showed good activity while compounds **1f** and **1g** proved significant cytotoxic activity as compared with other synthesized compounds.

2.6. Structure Activity Relationship (SAR) Studies

Among all tested compounds, the compounds **1d**, **1e** and **1j** showed the broad spectrum anti-proliferative activity against the four cancer cell lines. The cLog *P* values of all catechins analogues were calculated using ACD/ChemSketch software version 12.01 (listed in Table 1) and were used for structure activity relationship (SAR) studies. SAR analyses indicated that the compound **1d** presented better anticancer activity with an IC_{50} < 7.8 μM and cLog *P* 6.27 as compared to other analogues. The increase in the activity may be attributed due to benzoyl ester group and also due to increase in hydrophilicity of the compound. The similar ester linked analogues **1c** and **1l** showed inferior activities to **1d** which may be due to aliphatic unit, even though **1l** showed good hydrophilicity.

While, compound **1e** containing a morpholine moiety also showed good anticancer activity (IC_{50} < 9.5 μM and cLog *P* 3.67), but to lesser extent than **1d**, which might be due to the decrease in hydrophilicity. When the morpholine group was replaced with substituted piperazine moieties **1f-i**, we observed decreased activities which may be due to the presence of aryl group on nitrogen. On the other hand, **1j** showed a good anticancer activity with IC_{50} values less than 8.2 μM and cLog *P* 7.10. The activity might be due to the presence of electron withdrawing functions like trifluoromethyl and chloro groups on pyridine ring. An electron donating group like a moiety in **1k** had drastically reduced the activities, which showed the importance of substituent on pyridine.

The primary compound **1a** (cLog *P* 3.67) has no ether or ester link to hydroxyl group also produced significant activity but lesser than **1d**, **1e** and **1j**. These results showed the scope of producing new analogues of (−)-catechin to study various activities.

2.7. Docking Studies

Most of the catechins showed the anticancer activity through Receptor Tyrosine Kinases (RTKs). All synthesized compounds are catechin derivatives. Compound **1d** showed the better anticancer activity in HepG2 cell line which was closely associated with Insulin-like Growth Factor-1 Receptor (IGF-1R) tyrosine kinase.

This kinase is also found to correlate with a variety of tumors such as colorectal cancer, breast cancer, prostate cancer, hematological malignancies etc [32]. Therefore, IGF-1R was chosen for the molecular docking studies in order to support the observed *in vitro* anticancer activity in hand and to add more insight to predict the possible mechanism of action. The docking studies were performed using the Surflex-dock program [33, 34] .The X-ray crystal structure of IGF-1R complexed with ligand was taken from RCSB Protein Data Bank (PDB code: 3F5P) [35].

All the poses with generated into the active site of IGF-1R were based on the scores (Table 2). The pose with the highest total score was chosen for analyzing the binding features of compound **1d** which represented good binding affinities between the IGF-1R and **1d** in terms of ionic, hydrogen bond, aromatic, and lipophilic interactions.

The docking results revealed that compound **1d** showed desired key interactions in the active site of IGF-1R [36]. The carbonyl group of the ester bond and oxygen of oxane group interacted with the side chains of Arg1003 and Lys1033 *via* O···H-N strong hydrogen bonds. The hydroxyl groups exhibited O···H-O binding with the side chains of Glu1050, Gly1152 and Asp1153 residues (Figure 3).

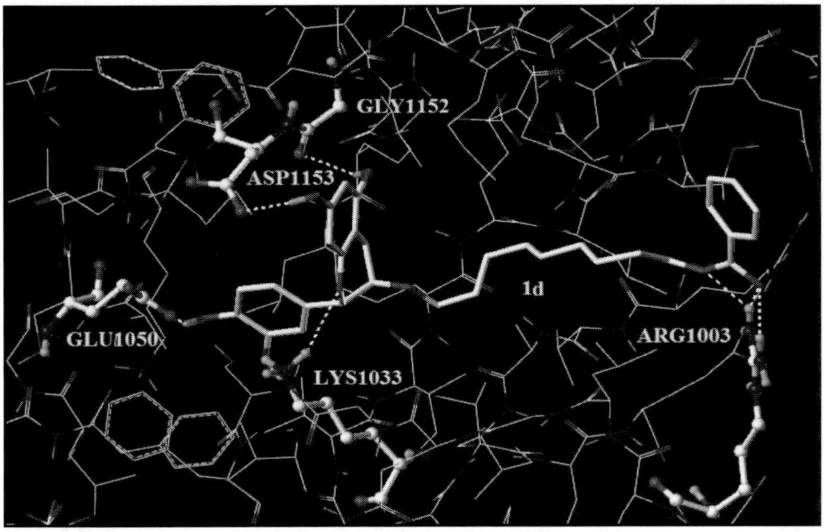

Figure 3. Key binding interactions of compound 1d with IGF-1R tyrosine kinase (H-bonds are depicted in dashed yellow lines). Reprinted with permission from D. Kumar, 2015, Bull. Korean Chem. 36 (2015) 564 © 2015 Korean Chemical Society, Seoul & Wiley-VCH Verlag GmbH & Co. KGaA, Weinheim.

Table 2. Docking scores of 1d and Doxorubicin

Compound	Total_Score	Crash	Polar	D_score	PMF_Score
1d	9.8417	-2.4396	4.8650	-162.6395	-20.5888
Doxorubicin	5.5002	-2.7699	2.4651	-145.5631	-4.9573

Compound	G_Score	CHEM_SCORE	CS Score	Global_Score	G_Score
1d	-271.1944	-35.1416	4	5	-271.1944
Doxorubicin	-242.0803	-27.1452	4	4	-242.0803

These critical interactions may be responsible for the promising anticancer activity and are in the favor of observed *in vitro* data. Thus, the result showed that the anticancer activity of synthesized compounds might be due to the above desired interactions with IGF-1RTK.

3. Antimicrobial Activity

The antimicrobial activity of the synthesized catechin derivatives was determined using well diffusion method [37] against different pathogenic bacterial and *Candida* reference strains procured from the Microbial Type Culture Collection and Gene Bank (MTCC), CSIR-Institute of Microbial Technology, Chandigarh, India.

3.1. Antibacterial Activity

All the synthesized compounds were screened for their antibacterial activity against *Staphylococcus aureus (S. aureus* MTCC 96, MLS16, MTCC 2940), *Micrococcus luteus (M. luteus* MTCC 2470), *Bacillus subtilis* (*B. subtilis* MTCC 121), *Klebsiella planticola (K. planticola* MTCC 530), *Escherichia coli* (*E. coli* MTCC 739) and *Pseudomonas aeruginosa* (*P. aeruginosa* MTCC 2453) bacterial species. Standard antibiotic solution of Neomycin at a dose range of 300-1.17 μg well^{-1} and the well containing methanol served as positive and negative controls, respectively.

The plates were incubated for 24 h at 37°C and 30°C for bacterial and the well containing the least concentration showing the inhibition zone was considered as the minimum inhibitory concentration. All experiments were carried out in duplicates.

The antimicrobial screening data showed that compounds **1a, 1b** and **1c** exhibited the excellent activity against *Bacillus subtilis* (*B. subtilis* MTCC 121), with *MIC* 2.34 μg/mL and showed moderate activity against *S. aureus* MTCC 96 2940 with MIC-75 μg/mL, showing a significant activity toward other Gram positive bacteria. Other compounds showed weak activity towards *Micrococcus luteus* MTCC 2470 (Table 4 and Figure 30).

Against Gram negative bacteria compounds **1a** showed excellent activity while **1b** showed good activity same with standard drug. While other compound viz. **1h**, **1c**, **1d** and **1e** possessed the moderate activity towards *Klebsiella planticola* MTCC 530 and *P. aeruginosa* MTCC 2453. Other compound showed weak activity towards *E. coli* MTCC 739 (Figure 4).

Among all tested compounds **1a**, **1b** and **1c** showed good antibacterial activity. The cLog *P* values of all catechins analogues were calculated using ACD/ChemSketch software version 12.01 (listed in Table 1) and were used for structure activity relationship (SAR) studies. SAR analyses indicated that the compound **1a, 1c** and **1c** were in excellent activities as compared to other analogues. The increase in the activity might be attributed due to ester linked analogues.

While, other derivatives are not shown good activity, it means, replacement of other substituent is not necessary. The ether or ester links to hydroxyl group or simply hydroxyl group is necessary for the activity.

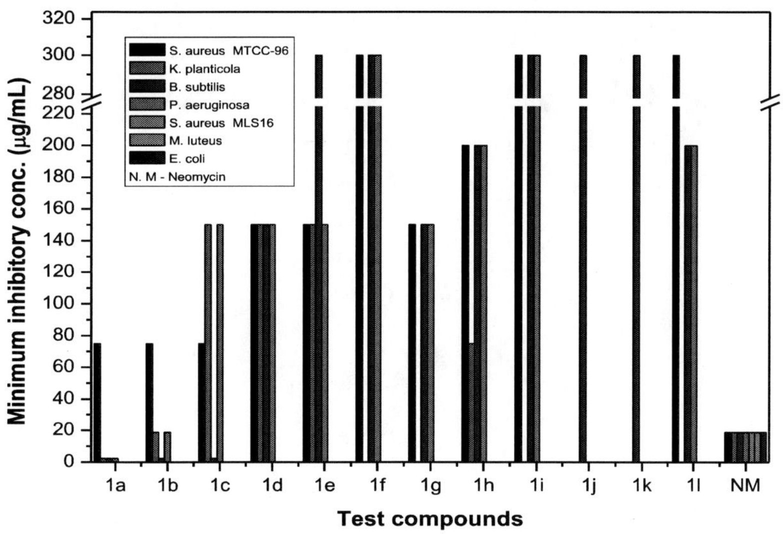

Figure 4. Antibacterial activities of compounds1a-l. Reprinted with permission from D. Kumar, 2015, J. Korean Soc. Appl. Biol. Chem. 58 (2015) 581© 2015 Springer.

3.2. Antifungal Activity

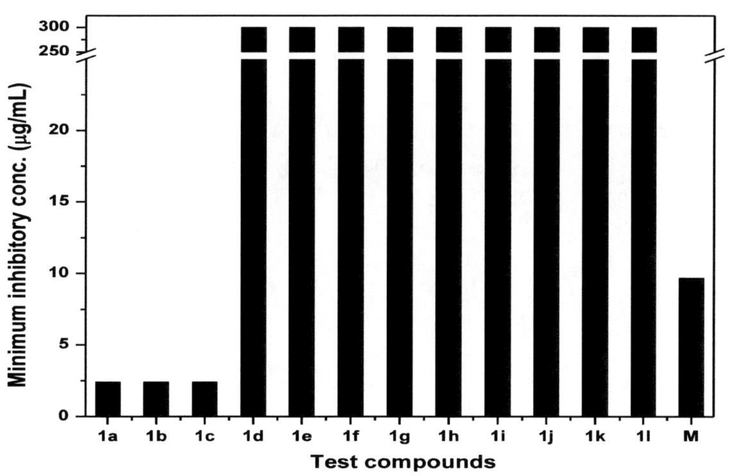

Figure 5. Antifungal activity of the compounds against *candida* strain. Reprinted with permission from D. Kumar, 2015, J. Korean Soc. Appl. Biol. Chem. 58 (2015) 581 © 2015 Springer.

The antifungal activity of all the compounds **1a-1l** were tested against *Candida albicans* (*C. albicans* MTCC 1962). Compounds **1a**, **1b** and **1c** exhibited excellent activity whereas other compounds gave significant antifungal activity the results, as shown in Figure 5. The compounds **1a**, **1b** and **1c** were evaluated for other strain of fungi such as (*C. albicans* MTCC 183), (*C. albicans* MTCC 854), (*C. albicans* MTCC 1637), (*C. albicans* MTCC 3018), *C. albicans* MTCC 395), (*C. albicans* MTCC 4748), (*C. albicans*

MTCC 7315), (*C. albicans* MTCC 227), *Candida aaseri (C. aaseri* MTCC 1962), *Issatchenkia hanoiensis* (*I. hanoiensis* MTCC 4755), *Candida glabrata* (*C. glabrata* MTCC 3019), *Issatchenkia orientalis* (*I. orientalis* MTCC 3020), *Candida parapsilosis (C. parapsilosis MTCC* 1744). These three compounds showed excellent activity against all type of fungi and the antifungal activity was found to be more than the standard drug.

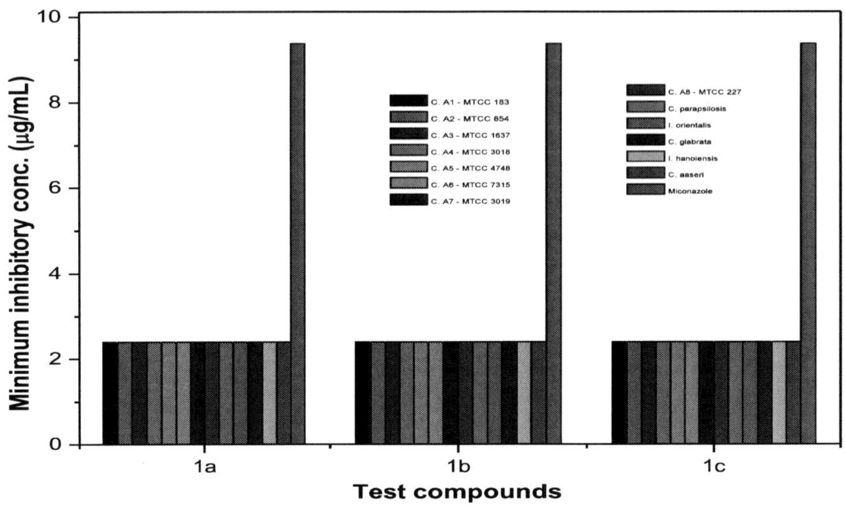

Figure 6. Antifungal activity of 1a, 1b and 1c against different strain of fungus. Reprinted with permission from D. Kumar, 2015, J. Korean Soc. Appl. Biol. Chem. 58 (2015) 581 © 2015 Springer.

The results have been represented in Figure 6 which presents very high inhibitory activity (MIC of 2.4 µg mL^{-1}) by compounds **1a**, **1b**, **1c** as compared with the standard drug. Introduction of lipophilic group at C-3 position is influenced the pharmacokinetic properties.

3.3. Docking of Catechins 1a and 1b on ATP Binding Pocket of DNA Gyrase B

Docking of catechins derivative on ATP binding pocket of DNA gyrase B was performed by Autodock4 [37, 38, 39] Two hydroxyl group of chroman-5,7-diol participates in hydrogen bond with main chain of Val97, Gly117 and side chain of Ser121.

The 9-hydroxynonyloxy group of compound **1a** makes hydrogen bond with Asp73, while methoxynonyloxy of **1b** is not involving in hydrogen but it can accommodate in ATP binding pocket (Figure 7a and b). Compound **1c** has nonyl acetate group which is difficult of accommodate at ATP binding pocket. Compounds **1d**, **1e**, **1f**, **1g**, **1h** and **1i** have alkyl chain containing bulky group that is difficult to fit in ATP binding pocket of DNA gyrase. The compounds **1a** and **1b** have hydroxyl and methoxy group, respectively, where both groups are small and easy to accommodate by ATP binding pocket of DNA gyrase. This may be a reason of a good activity against microbial strain for compound **1a** and **1b**.

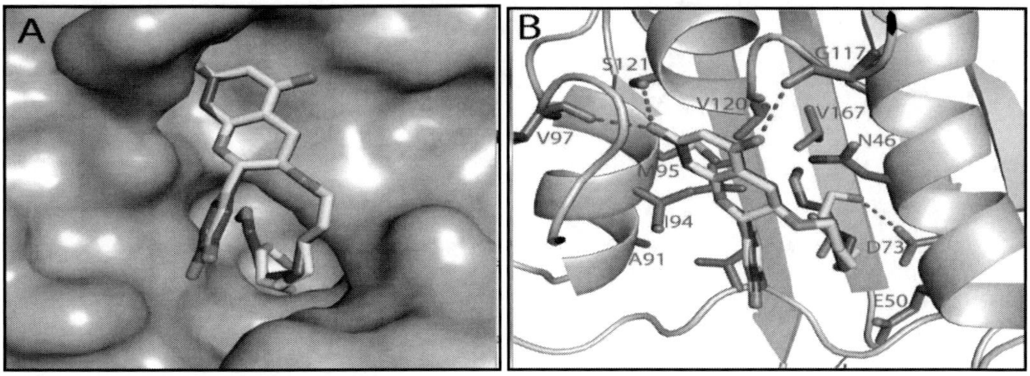

Figure 7. Docking pose of 1a and 1b on ATP binding pocket of DNA gyrase B (PDB code: 1AJ6). *Panel A* represents overall binding of compounds. Panel B, Stick representation of selected amino acids of DNA gyrase B interacting with compound 1a, and 1b. Red dot line represents hydrogen bonds. Reprinted with permission from D. Kumar, 2015, J. Korean Soc. Appl. Biol. Chem. 58 (2015) 581 © 2015 Springer.

4. Synthesis and Biological Activities

4.1. General

Nuclear magnetic resonance (^{1}H and ^{13}C NMR spectra) were recorded in $CDCl_3$ (400 MHz for ^{1}H and 100 MHz for ^{13}C, respectively) with TMS as the internal reference on Bruker Avance 400 FT spectrometer. Chemicals shifts were reported in parts per million. Mass spectra (MS) were measured by the EI method. Melting points are uncorrected. Silica gel (Merck D-6100, 70-230 mesh, ASTM) was used for flash column chromatography. All the reactions were monitored by TLC using 0.25 mm silica gel plates (Merck 60F-254) with or without UV indicator. *N*, *N*-Dimethyl-formamide was distilled over anhydrous magnesium sulfate and acetonnitrile was distilled over CaH_2 prior to use. All other reagents were commercially available (Acros, Aldrich) and were used without further purification.

4.2. Procedure for the Synthesis of (-)-Catechin Derivative

4.2.1. Synthesis of 9-Bromononan-1-ol (3)

9- Bomomonanol was synthesized by reported procedure [30]. 1,9-Nonane diol **18** (15 g, 93.6 mmol, 1 eq.) was dissolved in toluene (200 mL) by heating the solution at about 60°C, to this solution was added hydrobromic acid (48%, 15.6 mL), drop wise with stirring and mixture was reflux at (178–180°C) for 28 h. The water formed during the reaction was removed using a Dean–Stark trap. After cooling to room temperature, the mixture was washed with 6N sodium hydroxide solution (75 mL), 10% hydrochloric acid (75 mL), water (2 × 65 mL), and finally with brine (100 mL). The organic layer was dried over Na_2SO_4 and concentrated. On silica gel column chromatography of the resulting yellowish oil with 1:4 (ethyl acetate/hexane) gave compound **3** which was confirmed by ^{1}H NMR.

White solid; Yield 75% (15 g), mp 33–35°C. ^{1}H NMR (400 MHz, $CDCl_3$): ^{1}H NMR ($CDCl_3$) 3.63 (t, *J* = 6.6, 2H), 3.41 (t, *J* = 6.8, 2H), 1.85 (m, 2H), 1.74 (s, 1H), 1.56 (m, 2H), 1.42 (m, 2H), 1.31 (m, 8H).

4.2.2. General Procedure for the Synthesis of Intermediate 8-(2S,3R)-5,7-Bis (Benzyloxy)-2-(3,4-Bis(Benzyloxy)Phenyl)Chroman-3-yl)Oxy)Octan-1-ol (2)

To a solution of Tetra benzylated (□)-catechin **4** (4 g, 6.14 mmol, 1 eq.) in dry DMF (14 mL) was added NaH (60 percent dispersion in mineral oil) (0.281 g, 7.38 mmol) at 0°C. After 1 hr compound 9-Bromononan-1-ol (**3**) (1.53 g, 7.38 mmol, 1.2 eq.) was added. The reaction mixture was stirred overnight. The reaction was quenched with ice water and extracted with EtOAc. The organic phase was washed with water and brine respectively, which was dried over Na_2SO_4 and concentrated. The residue was purified by chromatography over silica gel using a gradient of 2:8 ethyl acetate/hexane to give compound **2**.

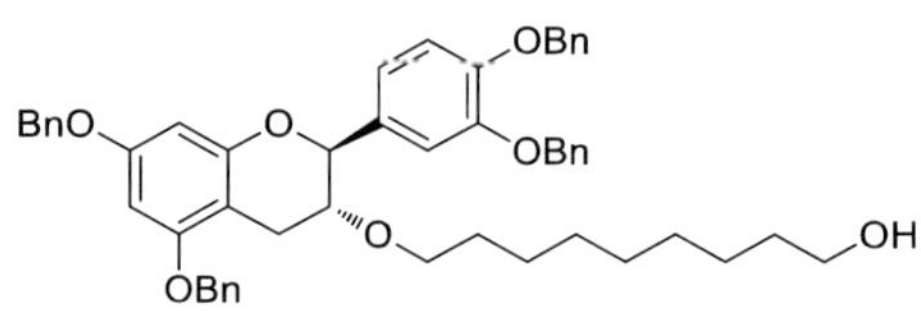

Semi solid; Yield 81% (4 g). ^{1}H NMR ($CDCl_3$, 400 MHz) δ 7.47-7.27 (m, 20H), 7.03 (d, 1H, *J* = 1.26 Hz), 6.94-6.92 (m, 2H), 6.25 (d, 1H, *J* = 2.272 Hz), 6.21 (d, 1H, *J* = 2.27 Hz), 5.21-5.11 (m, 4H), 5.03 (d, 2H, *J* = 2.77 Hz), 4.98 (s, 2H), 4.71 (d, 1H, *J* = 7.83 Hz), 3.62-3.55 (m, 3H), 3.34-3.27 (m, 1H), 3.06-2.91 (m, 2H), 2.62 (dd, 1H, *J* = 8.58, 16.42 Hz), 1.55-1.47 (m, 4H), 1.35-1.02 (m, 10H). ESI-MS (m/z): 815 $[M+Na]^+$

4.2.3. Procedure for the Synthesis of (2S,3R)-5,7-Bis(Benzyloxy)-2-(3,4-Bis(Benzyloxy)Phenyl)-3-(9 Methoxynonyl) Oxy) Chromone (2b)

To a solution of compound **2** (0.150 g, 0.189 mmol, 1 eq.) in dry DMF (4 mL) was added NaH (60 percent dispersion in mineral oil) (0.010 g, 0.283 mmol, 1.5 eq.) at 0°C. After 1 hr Methyl iodide (0.039 g, 0.283 mmol, 1.5 eq.) was added. The reaction mixture was stirred overnight and was quenched with ice water and extracted with EtOAc. Organic layer was washed with water and brine, dried over Na_2SO_4, and concentrated. The residue was purified by chromatography over silica gel using a gradient of 2:8 (ethyl acetate/hexane) gave compound **2b**.

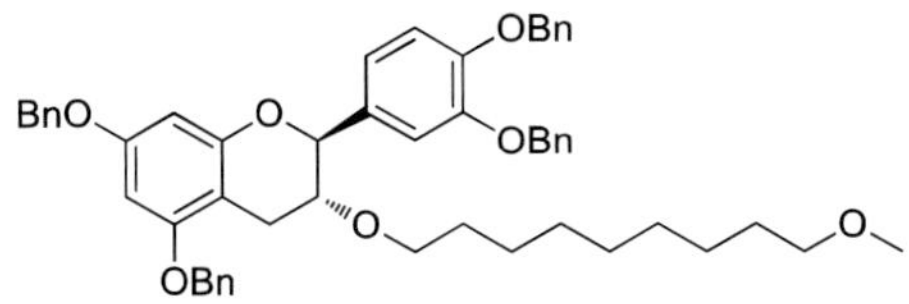

Yield 79 % (0.120 g). ^{1}H NMR ($CDCl_3$, 400 MHz) δ 7.48-7.25 (m, 20H), 7.02 (d, 1H, *J* = 1.51 Hz), 6.95-6.90 (m, 2H), 6.25 (d, 1H, *J* = 2.27 Hz), 6.21 (d, 1H, *J* = 2.27 Hz), 5.21-5.11 (m, 4H), 5.02 (d, 2H, *J* = 2.52 Hz), 4.97 (s, 2H), 4.71 (d, 1H, *J* = 8.08 Hz), 3.62-3.55 (m, 1H), 3.39-3.24 (m, 3H), 3.31 (s, 3H, OCH_3), 3.02 (dd, 1H, *J* = 5.55, 16.42 Hz), 2.97-2.90 (m, 1H), 2.62 (dd, 1H, *J* = 8.58, 16.42 Hz), 1.56-1.48 (m, 2H), 1.36-1.01 (m, 2H). ESI-MS (m/z): 829 $[M+Na]^+$.

4.2.4. Procedure for the Synthesis of 9-(2S,3R)-5,7-Bis(Benzyloxy)-2-(3,4-Bis(Benzyloxy Phenyl Chroman-3-Yl Oxy) Nonyl Acetate (2c)

A solution of compound **2** (0.150 g, 0.189 mmol, 1 eq.) dissolved in dry CH_2Cl_2 (5 mL) and treated with TEA (0.053 mL, 0.378 mmol, 2 eq.) under nitrogen. Acetic anhydride (0.037 mL, 0.378 mmol, 2 eq.) added at 0 °C. After completion of the reaction as monitored by TLC, work up was done and organic layer concentrated under reduced pressure, purified by chromatography over silica gel using a gradient of 2:8 (ethyl acetate/hexane) get the compound **2c**.

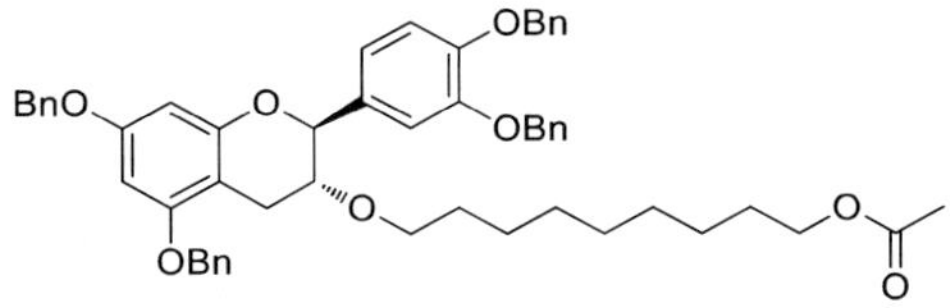

Yield 82 % (0.130 g). ^{1}H NMR ($CDCl_3$, 400 MHz) δ 7.46-7.24 (m, 20H), 7.03 (d, 1H, *J* = 1.51 Hz), 6.95-6.90 (m, 2H), 6.25 (d, 1H, *J* = 2.27 Hz), 6.21 (d, 1H, *J* = 2.27 Hz), 5.21-5.11 (m, 4H), 5.02 (d, 2H, *J* = 2.52 Hz), 4.98 (s, 2H), 4.71 (d, 1H, *J* = 7.83 Hz), 4.02 (t, 2H, *J* = 6.82), 3.62-3.55 (m, 1H), 3.32-3.28 (m, 1H), 3.02 (dd, 1H, *J* = 5.55, 16.42 Hz), 2.98-2.90 (m, 1H), 2.62 (dd, 1H, *J* = 8.58, 16.42 Hz), 2.02 (s, 3H), 1.62-1.52 (m, 2H), 1.36-1.0 (m, 12H). ESI -MS (m/z): 835 $[M+H]^+$.

4.2.5. Procedure for the Synthesis of 9-(2S,3R)-5,7-Bis(Benzyloxy)-2-(3,4-Bis(Benzyloxy Phenyl Chroman-3-Yl-Oxy) Nonyl Benzoate (2d)

A solution of compound **2** (0.150 g, 0.189 mmol, 1 eq.) was dissolved in dry CH_2Cl_2 (5 mL) and treated with TEA (0.053 mL, 0.378 mmol, 2 eq.) under nitrogen. Benzoylchloride (0.026 mL, 0.226 mmol, 1.2 eq.) was added at 0°C. After completion of the reaction as monitored by TLC, work up was done and organic layer concentrated under reduced pressure, purified by chromatography over silica gel using a gradient of 2:8 (ethyl acetate/hexane) gave compound **2d**.

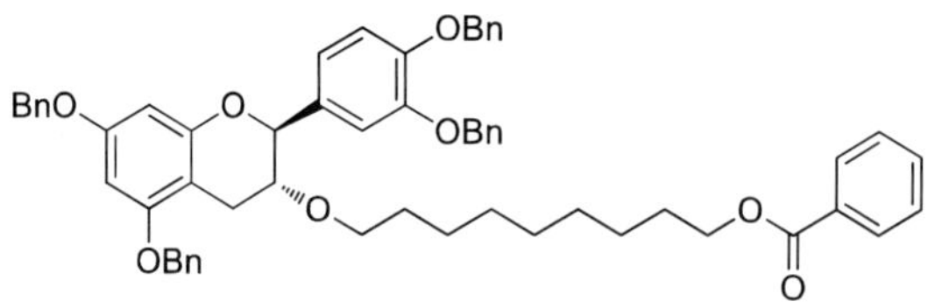

Yield 77 % (0.130 g). ^{1}H NMR ($CDCl_3$, 400 MHz) δ 8.18-8.13 (m, 2H), 7.69-7.63 (m, 1H), 7.55-7.49 (m, 2H), 7.46-7.24 (m, 20H), 7.04 (d, 1H, J = 1.26 Hz), 6.96-6.90 (m, 2H), 6.25 (d, 1H, J = 2.02 Hz), 6.21 (d, 1H, J = 2.27 Hz), 5.21-5.11 (m, 4H), 5.02 (d, 2H, J = 1.76 Hz), 4.98 (s, 2H), 4.72 (d, 1H, J = 8.08 Hz), 4.27 (t, 2H, J = 6.56 Hz), 3.63-3.56 (m, 1H), 3.32-3.27 (m, 1H), 3.06-2.91 (m, 2H), 2.63 (dd, 1H, J = 1.84, 16.67 Hz), 1.76-1.66 (m, 2H), 1.41-1.02 (m, 12H). ESI -MS (m/z): 910 $[M+Na]^+$

4.2.6. Procedure for the Synthesis of 9-(2S,3R)-5,7-Bis(Benzyloxy)-2-(3,4-Bis(Benzyloxy Phenyl Chroman-3-Yl-Oxy) Nonyl Oxy-1-Morpholinoethanone (2e)

To a solution of compound **2** (0.15 g, 0.19 mmol, 1 eq.) in dry DMF (4 mL) was added NaH (60 percent dispersion in mineral oil) (0.01 g, 0.28 mmol, 1.5 eq.) at 0°C. After 1 hr 2-chloro-1-morpholinoethanone (0.036 g, 0.226 mmol, 1.2 eq.) was added and reaction mixture stirred overnight. Workup was done with ice water and extracted with EtOAc. The organic phase was washed with water and brine, dried over Na_2SO_4, and concentrated. The residue was purified by column chromatography gave compound **2e**.

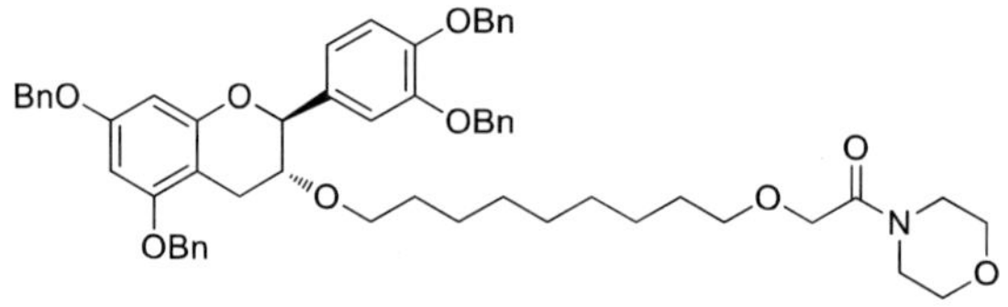

Yield 74 % (0.132 g). ^{1}H NMR ($CDCl_3$, 400 MHz) δ 7.46-7.24 (m, 20H), 7.02 (s, 1H), 6.95-6.89 (m, 2H), 6.24 (d, 1H, J = 2.02 Hz), 6.21 (d, 1H, J = 2.02 Hz), 5.20-5.11 (m, 4H), 5.04 (d, 1H, J = 1.76 Hz), 4.97 (s, 2H), 4.71 (d, 2H, J = 7.83 Hz), 4.09 (s, 2H), 3.68-3.61 (m, 4H), 3.61-3.56 (m, 3H), 3.53-3.48 (m, 2H), 3.44 (t, 2H, J = 6.56 Hz), 3.33-3.25 (m,

1H), 3.02 (dd, 1H, J = 5.30, 16.42 Hz), 2.98-2.91 (m, 1H), 2.62 (dd, 1H, J = 5.30, 16.42 Hz), 1.60-1.50 (m, 2H), 1.36-1.01 (m, 12H).ESI -MS (m/z): 942 $[M+Na]^+$.

4.2.7. Procedure for the Synthesis of 9-(2S,3R)-5,7-Bis(Benzyloxy)-2-(3,4-Bis(Benzyloxy Phenyl Chroman-3 Yl Oxy) Nonyl Oxy)-1-(4-Phenylpiperazin-1-Yl)Ethanone (2f)

To a solution of compound **2** (0.105 g, 0.132 mmol, 1 eq.) in dry DMF (4 mL) was added NaH (60 percent dispersion in mineral oil) (0.0060 g, 0.158 mmol, 1.2 eq.) at 0°C. After 1 hr 2-chloro-1-(4-phenylpiperazin-1-yl) ethanone (0.037 g, 0.158 mmol, 1.2 eq.) was added and reaction mixture stirred overnight. Workup was done with with ice water and extracted with EtOAc. The organic phase was washed with water and brine, dried over Na_2SO_4, and concentrated. The residue was purified by chromatography over silica gel using a gradient of 2:8 (ethyl acetate/hexane) gave compound **2f**.

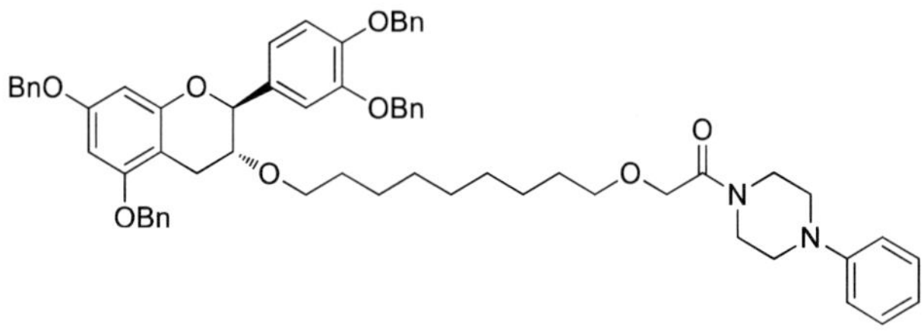

Yield 92% (0.129 g). ^1H NMR ($CDCl_3$, 400 MHz) δ 7.46-7.26 (m, 22H), 7.02 (d, 1H, J = 1.26 Hz), 6.95-6.87 (m, 5H), 6.24 (d, 1H, J = 2.27 Hz), 6.21 (d, 1H, J = 2.27 Hz), 5.20-5.10 (m, 4H), 5.02 (d, 2H, J = 2.77 Hz), 4.98 (s, 2H), 4.71 (d, 1H, J = 7.83 Hz), 4.15 (m, 2H), 3.76(t, 2H, J = 4.80 Hz), 3.68 (t, 2H, J = 4.54 Hz), 3.62-3.55 (m, 1H), 3.46 (t, 2H, J = 6.58 Hz), 3.32-3.25 (m, 1H), 3.19-3.12 (m, 4H), 3.02 (dd, 1H, J = 5.55, 16.42Hz), 2.98-2.90 (m, 1H), 2.62 (dd, 1H, J = 8.58, 16.42Hz), 1.63-1.52 (m, 2H) 1.37-0.98 (m, 12H). ESI-MS (m/z): 1017 $[M+Na]^+$.

4.2.8. Procedure for the Synthesis of 9-(2S,3R)-5,7-Bis(Benzyloxy)-2-(3,4-Bis (Benzyloxy)Phenyl)Chromanyl) Oxy) Nonyl) Oxy)-1-(4-(4-Fluorophenyl)Piperazin-1-Yl)Ethanone (2g)

A solution of compound **2** (0.100 g, 0.126 mmol, 1 eq.) in dry DMF (4 mL) was added NaH (60 percent dispersion in mineral oil) (0.005 g, 0.152 mmol, 1.2 eq.) at 0°C. After 1 hr 2-chloro-1-(4-(4-Fluorophenyl) piperazin-1-yl) ethanone (0.038 g, 0.152 mmol, 1.2 eq.) added and stirred overnight. The reaction was quenched with ice water and extracted with EtOAc. The organic phase was washed with water and brine, dried over Na_2SO_4, and concentrated. The residue was purified by chromatography over silica gel using a gradient of 2:8 (ethyl acetate/hexane) gave compound 2g.

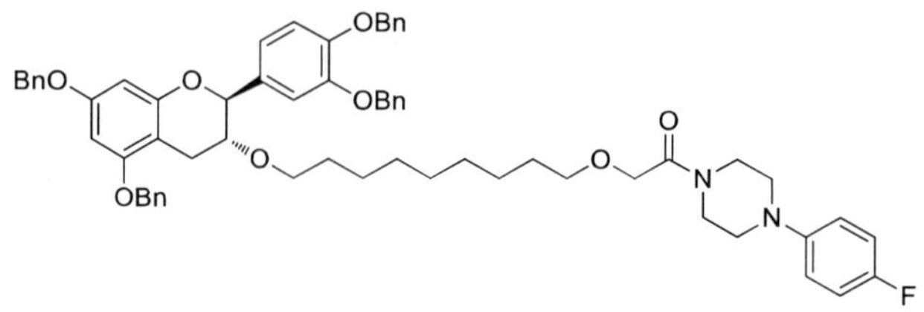

Yield 78% (0.100 g). ^{1}H NMR ($CDCl_3$, 400 MHz) δ 7.46-7.26 (m, 20H), 7.04-6.91 (m, 5H), 6.89-6.83 (m, 2H), 6.24 (d, 1H, *J* = 2.27 Hz), 6.21 (d, 1H, *J* = 2.27 Hz), 5.20-5.10 (m, 4H), 5.02 (d, 2H, *J* = 2.27 Hz), 4.98 (s, 2H), 4.71 (d, 1H, *J* = 8.08 Hz), 4.14 (s, 2H), 3.75 (t, 2H, *J* = 4.80 Hz), 3.67 (t, 2H, *J* = 4.80 Hz), 3.62-3.55 (m, 1H), 3.46 (t, 2H, *J* = 6.56 Hz), 3.32-3.26 (m, 1H), 3.08-3.02 (m, 4H), 3.01-2.85 (m, 2H), 2.25 (dd, 1H, *J* = 8.83, 16.42Hz), 1.61-1.51 (m, 2H) 1.42-0.95 (m, 12H). ESI-MS (m/z): 1035 $[M+Na]^+$.

4.2.9. Procedure for the Synthesis of 9-(2S,3R)-5,7-Bis(Benzyloxy)-2-(3,4-Bis (Benzyloxy Phenyl Chroman-3-Yl-Oxy) Nonyl Oxy-1-(4-2-Fluorophenyl)Piperazin-1-Yl)Propan-1-One (2h)

A solution of compound **2** (0.100 g, 0.126 mmol, 1 eq.) in dry DMF (4 mL) was added NaH (60 percent dispersion in mineral oil) (0.005 g, 0.152 mmol, 1.2 eq.) at 0°C. After 1 h 3-chloro-1-(4-(2-Fluorophenyl) piperazin-1-yl) propan-1-one (0.041 g, 0.152 mmol, 1.2 eq.) added slowly and reaction stirred overnight. The reaction was quenched with ice water and extracted with EtOAc. The organic phase was washed with water and brine, dried over Na_2SO_4, and concentrated. Residue was purified by chromatography over silica gel using a gradient of 2:8 (ethyl acetate/hexane) gave compound **2h.**

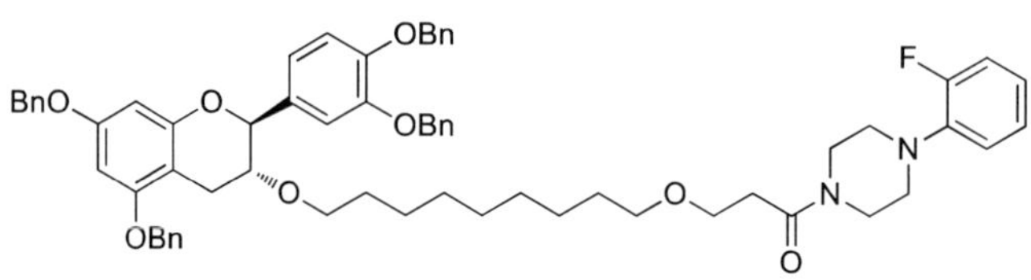

Yield 85%, (0.110 g). ^{1}H NMR ($CDCl_3$, 400 MHz) δ 7.45-7.22 (m, 20H), 7.05-6.85 (m, 7H), 6.24 (d, 1H, *J* = 2.02 Hz), 6.21 (d, 1H, *J* = 2.02 Hz), 5.20-5.10 (m, 4H), 5.02 (d, 2H, *J* = 2.02 Hz), 4.97 (s, 2H), 4.71 (d, 1H, *J* = 7.83 Hz), 3.76 (t, 2H, *J* = 5.05 Hz), 3.62 (t, 2H, *J* = 4.80 Hz), 3.60-3.55 (m, 1H), 3.40 (t, 2H, *J* = 6.56 Hz), 3.32-3.25 (m, 1H), 3.20-3.15 (m, 4H), 3.14-3.11 (m, 2H), 3.02 (dd, 1H, *J* = 5.55, 16.42Hz), 2.98-2.91 (m, 1H), 2.67-2.58 (m, 3H), 1.61-1.47 (m, 2H) 1.35-0.95 (m, 12H). ESI-MS (m/z): 1050 $[M+Na]^+$

4.2.10. Procedure for the Synthesis of 9-(2S,3R)-5,7-Bis(Benzyloxy)-2-(3,4-Bis(Benzyloxy Phenyl Chroman-3-Yl-Oxy) Nonyl Oxy-1-(4-(2-Trifluoromethyl Phenyl) Piperazin-1-Yl)Propan-1-One (2i)

A solution of compound **2** (0.100 g, 0.126 mmol, 1 eq.) in dry DMF (4 mL) was added NaH (60 percent dispersion in mineral oil) (0.005 g, 0.152 mmol, 1.2 eq.) at 0°C. After 1 h 3-chloro-1-(4-(3-(trifluoromethyl) phenyl) piperazin-1-yl) propan-1-one (0.048 g, 0.152 mmol, 1.2 eq.) added slowly, stirred overnight. Reaction mixture quenched with ice water and extracted with EtOAc. Organic phase washed with water and brine, dried over Na_2SO_4, and concentrated, purified by chromatography over silica gel using a gradient of 2:8 (ethyl acetate/hexane) to get compound **2i.**

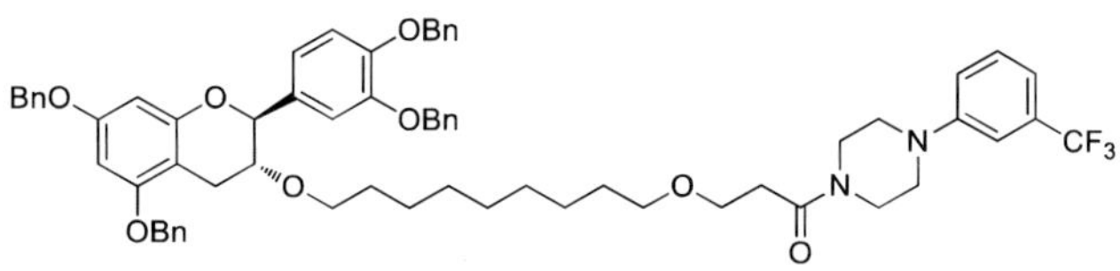

Yield 78% (0.100 g). ^{1}H NMR ($CDCl_3$, 400 MHz) δ 7.48-7.22 (m, 20H), 7.15-7.0 (m, 5H), 6.96-6.90 (m, 2H), 6.25 (d, 1H, *J* = 2.27 Hz), 6.21 (d, 1H, *J* = 2.27 Hz), 5.20-5.10 (m, 4H), 5.02 (d, 2H, *J* = 2.02 Hz), 4.98(s, 2H), 4.72 (d, 1H, *J* = 7.83 Hz), 3.78 (t, 2H, *J* = 5.05 Hz), 3.65 (t, 2H, *J* = 4.80 Hz), 3.61-3.56 (m, 1H), 3.40 (t, 2H, *J* = 6.56 Hz), 3.32-3.26 (m, 1H), 3.27-3.21 (m, 4H), 3.21-3.16 (m, 2H), 3.05-2.91 (m, 2H), 2.67-2.58 (m, 3H), 1.61-1.47 (m, 2H), 1.56-1.47 (m, 2H), 1.36-0.98 (m,12H). ESI-MS (m/z): 1099 $[M+Na]^+$.

4.2.11. Procedure for the Synthesis of (2S,3R)-5,7-Bis(Benzyloxy)-2-(3,4-Bis(Benzyloxy Phenyl Chroman-3-Yl-Oxy Nonyl) Oxy-3-Chloro-5-Trifluoromethylpyridine (2j)

A solution of compound **2** (0.200 g, 0.252 mmol, 1 eq.) in dry DMF (6 mL) NaH (60 percent dispersion in mineral oil) (0.0114 g, 0.302 mmo, 1.2 eq.) added at 0°C. After 1 h 2,3-dichloro-5-(trifluoromethyl)pyridine (0.064 g, 0. 302 mmol, 1.2 eq.) was added and reaction mixture was stirred overnight. The reaction was quenched with ice water and extracted with EtOAc. Organic phase was washed with water and brine, dried over Na_2SO_4, and concentrated, purified by chromatography over silica gel using a gradient of 2:8 (ethyl acetate/hexane).

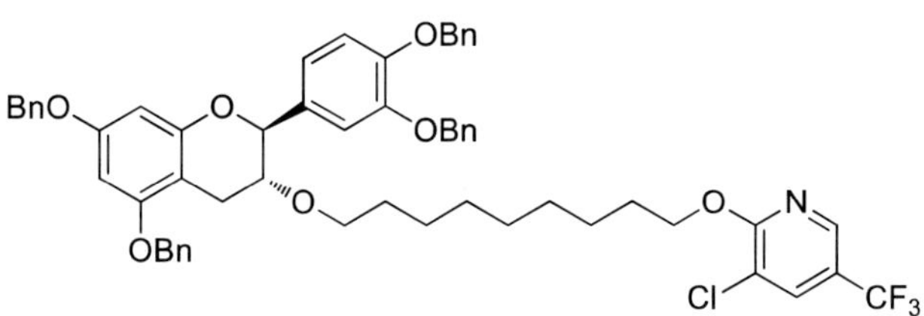

Yield 81% (0.200 g). ^{1}H NMR ($CDCl_3$, 400 MHz) δ 8.31-8.26 (m, 1H), 7.81-7.77 (m, 1H), 7.46-7.21 (m, 20H), 7.04 (d, 1H, J = 1.52 Hz), 6.96-6.89 (m, 2H), 6.25 (d, 1H, J = 2.27 Hz), 6.21 (d, 1H, J = 2.27 Hz), 5.20-5.10 (m, 4H), 5.01 (d, 2H, J = 2.27 Hz), 4.96 (s, 2H), 4.71 (d, 1H, J = 8.08 Hz), 4.37 (t, 2H, J = 6.82 Hz), 3.61-3.55 (m, 1H), 3.41-3.26 (m, 2H), 3.03 (dd, 1H, J = 5.55, 16.42 Hz), 2.62 (dd, 1H, J = 8.58, 16.42 Hz), 1.84-1.72 (m, 2H), 1.49-0.99 (m,12H). ESI -MS (m/z): 994 $[M+Na]^+$

4.2.12. Procedure for the Synthesis of (2S, 3R)-5,7-Bis(Benzyloxy)-2-(3,4-Bis(Benzyloxy Phenyl Chroman-3-Yl-Oxy) Nonyl Oxy-5-Nitropyridine (2k)

A solution of compound **2** (0.20 g, 0.252 mmol, 1 eq.) in dry DMF (6 mL) was added NaH (60 percent dispersion in mineral oil) (0.011 g, 0.302 mmol, 1.2 eq.) at 0°C. After 1 hr 2,3-Dichloro-5-nitropyridine (0.064 g, 0.302 mmol, 1.2 eq.) added slowly and reaction mixture stirred overnight. The reaction was quenched with ice water and extracted with EtOAc. Organic phase washed with water and brine, dried over Na_2SO_4, and concentrated, purified by chromatography over silica gel to get compound 2k.

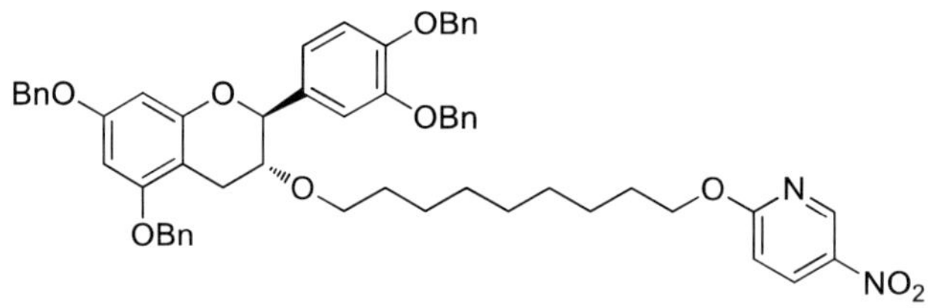

Yield 82% (0.190 g). ^{1}H NMR ($CDCl_3$, 400 MHz) δ 8.05 (d, 1H, J = 2.27 Hz), 8.31 (dd, 1H, J = 2.27, 9.09 Hz), 7.49-7.24 (m, 20H), 6.96-6.90 (m, 2H),6.76 (d, 1H, J = 9.09 Hz), 6.25 (d, 1H, J = 2.27 Hz), 6.21 (d, 1H, J = 2.27 Hz), 5.21-5.11 (m, 4H), 5.03 (d, 2H, J = 2.52 Hz), 4.98 (s, 2H), 4.71 (d, 1H, J = 7.83Hz),), 4.36 (t, 2H, J = 6.81 Hz), 3.66-3.55 (m, 1H), 3.41-3.32 (m, 1H), 3.32-3.25 (m, 1H), 3.37-2.90 (m, 2H), 2.62 (dd, 1H, J = 8.58, 16.42 Hz), 1.89-1.69 (m, 2H), 1.45-0.96 (m,12H). ESI-MS (m/z): 937 $[M+Na]^+$

4.2.13. Procedure for the Synthesis of Compound 2l

A solution of compound **2** (0.200 g, 0.252 mmol, 1 eq.) was dissolved in dry CH_2Cl_2 (5 mL) and treated with TEA (0.042mL, 0.302 mmol, 1.2 eq.) under nitrogen. Cinnamoyl chloride (0.050 mL, 0.302 mmol, 1.2 eq.) was added at 0°C. After completion of the reaction as monitored by TLC, the mixture was concentrated under reduced pressure and it was proceeded for next step.

4.3. General Procedure for the Synthesis of 1a-1l

Compound **2a-l** (1 eq.) was taken in dry ethyl acetate and 10% Pd/C (0.02 eq.) and stirred at room temperature under H_2 atmosphere for 12 h. TLC of the reaction mixture showed complete consumption of starting material. Then reaction mixture was filtered through celite with excess washing with ethyl acetate and flash chromatography and concentrated to get target compounds **1a-l**.

(2*S*,3*R*)-2-(3,4-Dihydroxyphenyl)-3-(9-hydroxynonyl-oxy)chroman-5,7-diol (1a):

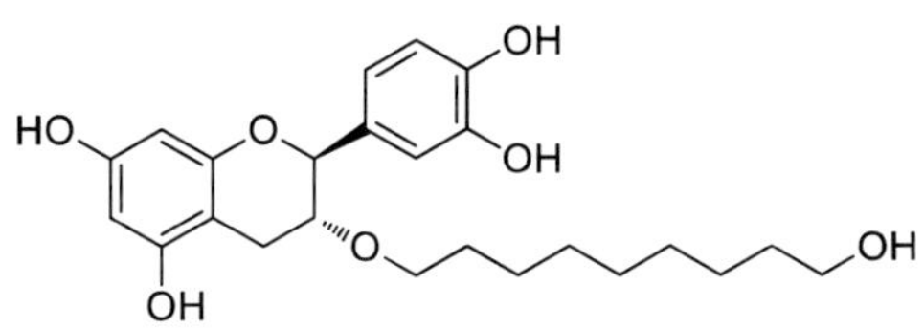

Yield 83% (0.093 g), $[\alpha]_D^{25}$ –52.4 (*c* 1.0, MeOH). ^{1}H NMR (CD_3OD, 400 MHz): δ 6.88 (s, 1H), 6.84-6.74 (m, 2H), 5.96 (d, 2H, *J* = 3.28 Hz), 4.68 (d, 1H, *J* = 7.83 Hz), 3.73-3.68 (m, 1H), 3.58 (t, 2H, *J* = 6.56 Hz), 3.47-3.39 (m, 1H), 3.21-3.13 (m, 1H), 2.98(dd, 1H, *J* = 5.30, 16.16Hz), 2.57 (dd, 1H, *J* = 8.58, 16.16Hz), 1.58-1.49 (m, 2H), 1.43-1.03(m, 12H). ^{13}C NMR (CD_3OD, 100 MHz): δ 160.0, 159.8, 159.4, 148.3, 148.3, 135.1, 123.0, 118.7, 117.9, 103.7, 99.2, 98.6, 83.92, 79.5, 73.6, 66.2, 36.2, 33.4, 33.2, 33.0, 32.6, 29.4, 29.2. ESI-MS (m/z): 455 $[M+Na]^+$. IR (KBr) ν_{max}: 3428, 2926, 2853, 1630, 1521, 1466, 1382, 1283, 1168, 1141, 1104, 1059, 818, 762, 627cm^{-1}

(2*S*,3*R*)-2-(3,4-Dihydroxyphenyl)-3-(9-methoxy nonyl-oxy)chroman-5,7-diol (1b):

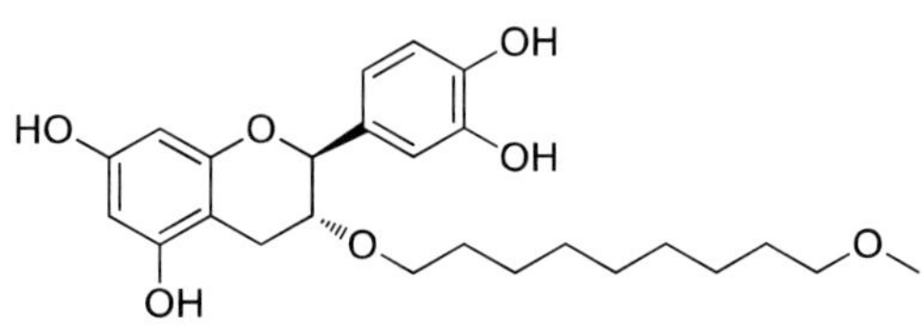

Yield 83% (0.060 g), $[\alpha]_D^{25}$ –59.9 (*c* 1.0, MeOH). ^{1}H NMR (CD_3OD, 400 MHz): δ 6.82 (d, 1H, *J* = 1.76 Hz), 6.76-6.68 (m, 2H), 5.92 (d, 1H, *J* = 2.27Hz), 5.85 (d, 1H, *J* = 2.27 Hz), 4.62 (d, 1H, *J* = 7.32 Hz), 3.69-3.63 (m, 1H), 3.48-3.40 (m, 1H), 3.37 (t, 2H, *J* = 6.56 Hz), 3.31 (s, 3H), 3.21-3.15 (m, 1H), 2.86 (dd, 1H, *J* = 5.33, 15.91 Hz), 2.50 (dd, 1H, *J* = 1.80, 16.16Hz), 1.58-1.49 (m, 2H), 1.42-1.10 (m, 12H).^{13}C NMR (CD_3OD, 100 MHz) δ 157.8, 157.5, 157.0, 146.1, 146.1, 132.6, 119.9, 116.1, 115.2, 100.8, 96.4, 95.6, 81.2, 77.1, 74.0, 70.7,58.6, 30.8, 30.6, 30.5, 30.4, 30.3, 27.1, 27.0, 26.2. ESI-MS (m/z): 469 $[M+Na]^+$. IR (KBr) ν_{max}: 3393, 2927, 2854, 1702, 1607, 1520, 1485, 1371, 1283, 1282, 1168, 1191, 1140, 1101, 867, 817, 780, 719, 629, cm^{-1}

(2*S*,3*R*)-2-(3,4-Dihydroxyphenyl)-5,7-dihydroxychroman-3-yl-oxy)nonylacetate (1c)

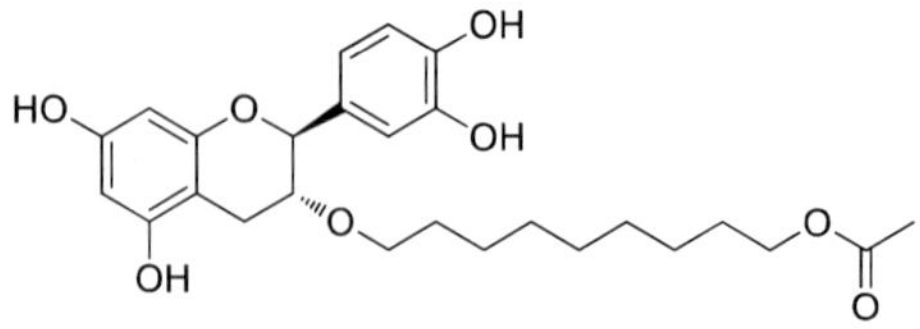

Yield 82% (0.055 g), $[\alpha]_D^{25}$ –46.9 (*c* 1.0, MeOH). ^{1}H NMR (CD_3OD, 400 MHz): δ 6.82 (d, 1H, *J* = 1.76 Hz), 6.77-6.68 (m, 2H), 5.92 (d, 1H, *J* = 2.27Hz), 5.85 (d, 1H, *J* = 2.27 Hz), 4.62 (d, 1H, *J* = 7.32 Hz), 4.04 (t, 2H, *J* = 6.56 Hz), 3.70-3.63 (m, 1H), 3.49-3.38 (m, 1H), 3.23-3.14 (m, 1H), 2.87 (dd, 1H, *J* = 5.30, 16.16 Hz), 2.51 (dd, 1H, *J* = 7.83, 15.91Hz), 2.01 (s, 3H), 1.65-1.55 (m, 2H), 1.43-1.10 (m, 12H). ^{13}C NMR (CD_3OD, 100 MHz): δ 173.2, 157.8, 157.6, 156.9, 146.1, 132.6, 119.9, 116.0, 115.3, 100.8, 96.4, 95.6, 81.3, 77.2, 70.6, 65.8, 30.8, 30.5, 30.4, 30.2, 29.7, 27.2, 26.9, 26.2, 20.8. ESI-MS (m/z): 475 $[M+H]^+$. IR (KBr) ν_{max}: 3382, 2928, 2854, 1706, 1607, 1520, 1466, 1368, 1278, 1194, 1140, 1042, 970, 668, 818, 764, 722, 631, 611 cm^{-1}.

(2*S*,3*R*)-2-(3,4- Dihydroxyphenyl)- 5,7- dihydroxychroman- 3- yl) oxy) nonylbenzoate (1d):

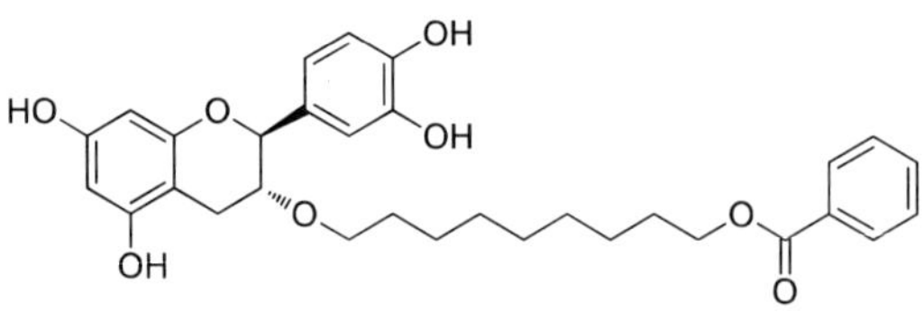

Yield 84% (0.061 g), $[\alpha]_D^{25}$ –24.4 (*c* 1.0, MeOH). ^{1}H NMR (CD_3OD, 400 MHz): δ 7.93-7.83 (m, 2H), 7.51-7.46 (m, 1H), 7.39-7.34 (m, 2H), 6.72 (d, 1H, *J* = 1.76 Hz), 6.66-6.58 (m, 2H), 5.83 (d, 1H, *J* = 2.27 Hz), 5.75 (d, 1H, *J* = 2.27 Hz),4.51 (d, 1H, *J* = 7.57 Hz), 4.21 (t, 2H, *J* = 6.56 Hz), 3.59-3.51 (m, 1H), 3.38-3.30 (m, 1H), 3.11-3.04 (m, 1H), 2.76 (dd, 1H, *J* = 5.30, 16.16 Hz), 2.40 (dd, 1H, *J* = 8.08, 16.16 Hz), 1.72-1.62 (m, 2H), 1.38-0.98 (m, 12H). ^{13}C NMR (CD_3OD, 100 MHz): δ 168.2, 157.9, 157.5, 157.0, 146.1, 134.1, 132.5, 131.7, 119.9, 116.0, 115.2, 100.7, 96.4, 95.6, 81.2, 77.1, 70.7, 66.3, 30.8, 30.4, 30.2, 30.2, 29.7, 27.1, 26.9, 26.2. ESI-MS (m/z): 560 $[M+Na]^+$. IR (KBr) ν_{max}: 3382, 2926, 2853, 1695, 1605, 1519.7, 1467, 1368, 1316, 1284, 1178, 1139, 1060, 1025, 960, 669, 818, 780, 713, 673, 630, 4317 cm^{-1}.

(2*S*,3*R*)-2-(3,4-Dihydroxyphenyl)-5,7-dihydroxychroman-3-yl-oxy nonyl oxy)-1-morpholinoethanone (1e)

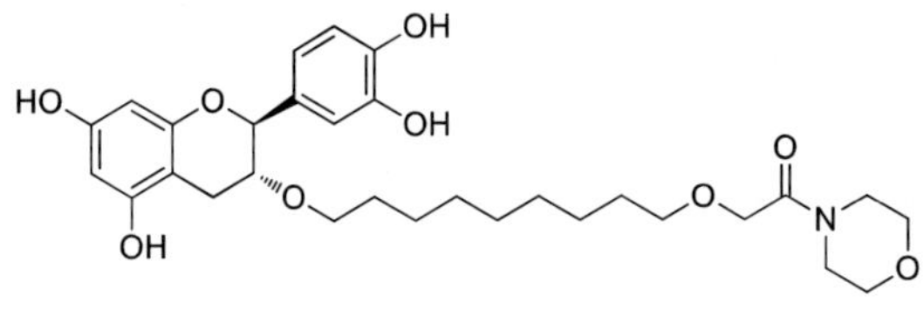

Yield 75% (0.046 g), $[\alpha]_D^{25}$ –27.5 (*c* 1.0, MeOH). ^{1}H NMR (CD_3OD, 400 MHz): δ 6.85 (d, 1H, *J* = 2.02 Hz), 6.79-6.70 (m, 2H), 5.96 (d, 1H, *J* = 2.27 Hz), 5.90 (d, 1H, *J* = 2.27 Hz), 4.65 (d, 1H, *J* = 7.57 Hz), 4.15 (s, 2H), 3.74-3.64 (m, 5H), 3.61-3.56 (m, 2H), 3.53-3.41 (m, 5H), 3.23-3.16 (m, 1H), 2.91 (dd, 1H, *J* = 5.30, 16.16 Hz), 2.54 (dd, 1H, *J* = 8.08, 16.16 Hz), 1.65-1.55 (m, 2H), 1.36-1.08 (m, 12H). ^{13}C NMR (CD_3OD, 100 MHz): δ 170.0, 157.3, 157.1, 156.5, 145.7, 132.1, 119.6, 115.7, 114.9, 100.5, 96.2, 80.9, 76.7, 76.7, 72.4, 70.5, 67.5, 61.3, 46.4, 43.3, 30.4, 30.2, 30.0, 29.9, 26.7, 26.6, 26.0, 20.9. ESI-MS (m/z): 582 $[M+Na]^+$. IR (KBr) ν_{max}: 3393, 2925, 2854, 1622, 1520, 1445, 1362, 1276, 1245, 1101, 1068, 985, 815, 782, 587 cm^{-1}

(2*S*,3*R*)-2-(3,4-Dihydroxyphenyl)-5,7dihydroxychroman-3-yl)oxy-nonyl-oxy-1-(4-phenylpiperazin-1(-yl)ethanone (1f):

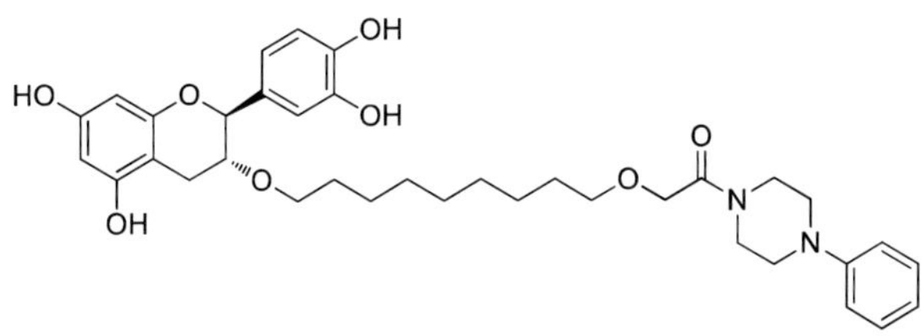

Yield 79%, (0.046 g), $[\alpha]_D^{25}$ –26.4 (*c* 1.0, MeOH). ^{1}H NMR (CD_3OD, 400 MHz): δ 7.25-7.20 (m, 2H), 6.98-6.93 (m, 2H), 6.88-6.83 (m, 1H) 6.70 (t, 3H, *J* = 2.01 Hz), 5.96 (d, 1H, *J* = 2.27 Hz), 5.85 (d, 1H, *J* = 2.27 Hz), 4.62 (d, 1H, *J* = 2.02 Hz), 4.20 (s, 2H), 3.74-3.69 (m, 1H), 3.67-3.61 (m, 2H), 3.61-3.55 (m, 1H), 3.54-3.39 (m, 4H), 3.22-3.10 (m, 3H), 2.85 (dd, 1H, *J* = 2.52, 16.16 Hz), 2.65-2.58 (m, 2H), 2.50 (dd, 1H, *J* = 8.58, 16.16 Hz), 1.67-1.55 (m, 2H), 1.42-1.09 (m, 12H).^{13}C NMR (CD_3OD, 100 MHz) δ 168.0, 158.7, 157.8, 149.0, 131.7, 129.2, 129.5, 128.5, 120.6, 120.5, 116.6, 115.1, 114.2, 102.5, 94.5, 93.7, 80.1, 75.5, 70.8, 70.2, 70.1, 49.8, 45.1, 29.8, 29.7, 29.6, 29.5, 26.1, 25.9.ESI-MS (m/z): 635 $[M+H]^+$. IR (KBr) ν_{max}: 3401, 2928, 2854, 1627, 1602, 1517, 1451, 1369, 1280, 1234, 1185, 1141, 1105, 1064, 815, 764, 428 cm^{-1}

(2*S*,3*R*)-2-(3,4-Dihydroxyphenyl)-5,7-dihydroxychroman-3-yl)oxy)nonyl)oxy)-1-(4-(4-fluorophenyl)piperazin-1-yl)ethanone (1g)

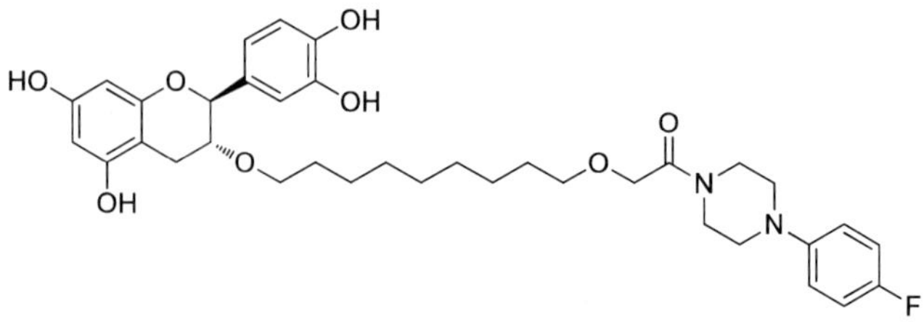

Yield 77% (0.046 g), $[\alpha]_D^{25}$ –25.5 (*c* 1.0, MeOH). ^{1}H NMR (CD_3OD, 400 MHz): δ 6.89-6.88 (m, 2H), 6.87-6.86 (m, 2H), 6.72 (d, 1H, *J* = 2.02 Hz), 6.66-6.68 (m, 2H), 5.92 (d, 1H, *J* = 2.27 Hz), 5.82 (d, 1H, *J* = 2.27 Hz), 4.61 (d, 1H, *J* = 7.57 Hz), 4.20 (s, 2H), 3.74-3.69 (m, 1H), 3.67-3.61 (m, 2H), 3.61-3.55 (m, 1H), 3.54-3.39 (m, 4H), 3.22-3.10 (m, 3H), 2.85 (dd, 1H, *J* = 2.52, 16.16 Hz), 2.65-2.58 (m, 2H), 2.50 (dd, 1H, *J* = 8.58, 16.16 Hz), 1.67-1.55 (m, 2H), 1.42-1.09 (m, 12H).^{13}C NMR (CD_3OD, 100 MHz) δ 168.9, 158.7, 157.9, 154.0, 152.6, 152.0, 148.2, 148.1, 148.1, 135.6, 134.9, 132.5, 130.5, 130.2, 125.6, 117.6, 116.3, 114.8, 102.6, 94.6, 78.7, 77.8, 75.8, 71.4, 59.4, 52.9, 52.1, 36.7, 36.4, 36.3, 34.6, 30.1, 25.1.ESI-MS (m/z): 653 $[M+H]^+$. IR (KBr) ν: 3438, 2925, 2854, 2345, 1734, 1630, 1509, 1459, 1383, 1280, 1280, 1139, 816, 619, 484 cm^{-1}

(2*S*,3*R*)-2-(3,4-Dihydroxyphenyl)-5,7dihydroxychroman-3 yl-oxy nonyl oxy)-1-(4-(2-fluorophenyl piperazin-1 yl) propan-1-one (1h)

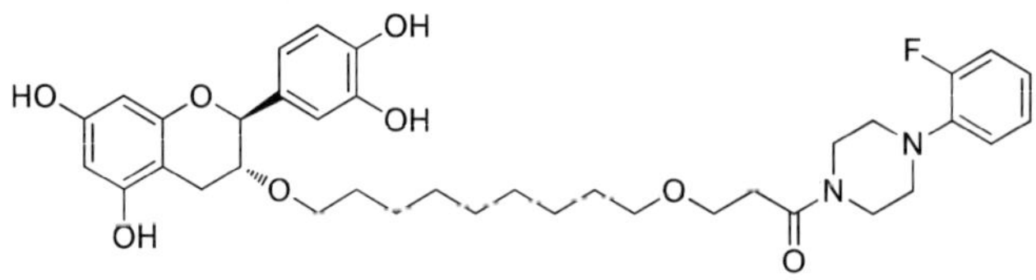

Yield 77% (0.054 g), $[\alpha]_D^{25}$ –22.4 (*c* 1.0, MeOH). ^{1}H NMR (CD_3OD, 400 MHz) δ 7.26-7.20 (m, 2H), 7.00-6.93 (m, 1H), 6.87-6.82 (m, 1H), 6.81 (d, 1H, *J* = 2.02 Hz), 6.76-6.67 (m, 2H), 5.92 (d, 1H, *J* = 2.27 Hz), 5.84 (d, 1H, *J* = 2.27 Hz), 4.61 (d, 1H, *J* = 7.32 Hz), 3.77-3.68 (m, 6H), 3.67-3.61 (m, 2H), 3.50-3.38 (m, 3H), 3.20-3.09 (m, 4H), 2.86 (dd, 1H, *J* = 5.30, 16.16 Hz), 2.69 (t, 2H, *J* = 6.06 Hz), 2.49 (dd, 1H, *J* = 5.30, 16.16 Hz), 1.57-1.49 (m,2H) 1.38-1.08 (m, 12H).^{13}C NMR (CD_3OD, 100 MHz) δ 168.9, 158.7, 157.9, 154.0, 152.6, 152.0, 148.2, 148.1, 148.1, 135.6, 134.9, 132.5, 130.5, 130.2, 125.6, 117.6, 116.3, 114.8, 102.6, 94.6, 78.7, 77.8, 75.8, 71.4, 59.4, 52.9, 52.1, 36.7, 36.4, 36.3, 34.6, 30.1, 25.1. ESI-MS (m/z): 653 $[M+H]^+$.IR (KBr) ν_{max}: 3677, 3429, 2925, 2854, 1623, 1459, 1383, 1233, 1142, 1106, 760, 505 cm^{-1}.

(2*S*,3*R*)-2-(3,4-Dihydroxyphenyl)-5,7dihydroxychroman-3yl-nonyl-oxy-1-(2-trifluoromethyl phenyl)piperazin-1-yl)propanone (1i):

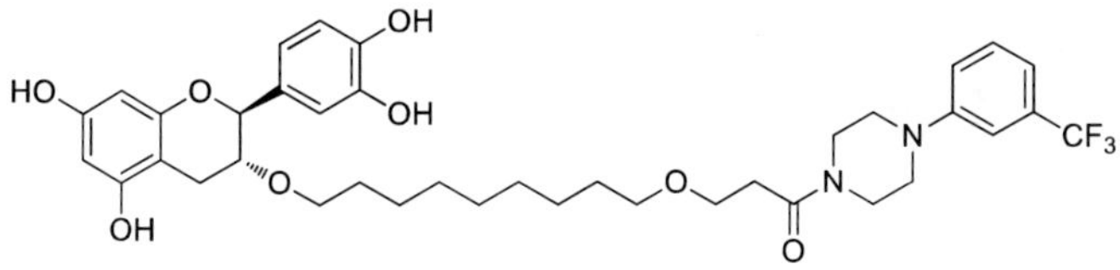

Yield 83% (0.093 g), $[\alpha]_D^{25}$ –14.9 (*c* 1.0, MeOH). ^{1}H NMR (CD_3OD, 400 MHz): δ 7.45-7.36 (m, 1H), 7.23-7.16 (m, 2H), 7.13-7.07 (m, 1H), 6.81 (d, 1H, *J* = 2.02 Hz), 6.76-6.67 (m, 2H), 5.92 (d, 1H, J = 2.27 Hz), 5.84 (d, 1H, *J* = 2.27 Hz), 5.85 (d, 1H, *J* = 2.27 Hz), 4.61 (d, 1H, *J* = 7.32 Hz),, 3.79-3.61 (m, 7H), 3.49-3.39 (m, 3H), 3.28-3.12 (m, 5H), 2.86 (dd, 1H, *J* = 5.30, 16.16 Hz), 2.69 (t, 2H, J = 6.06 Hz), 2.48 (dd, 1H, *J* = 7.83, 16.16 Hz), 1.58-1.48(m,2H) 1.39-1.02 (m, 12H).^{13}C NMR (CD_3OD, 100 MHz) δ 171.0, 157.8, 157.5, 152.6, 146.2, 145.4, 131.4, 130.9, 127.8, 126.6, 123.9, 121.9, 120.6, 117.1, 116.4, 116.0, 112.0, 109.4, 104.7, 95.5, 79.5, 79.2, 78.9, 72.3, 69.2, 59.0, 50.1, 45.3, 34.4, 30.7, 30.5, 30.3, 30.2, 27.2, 26.9, 26.2. ESI-MS (m/z): 740 $[M+Na]^+$. IR (KBr) ν_{max}: 3401, 2929, 2854, 1607, 1518, 1448, 1353, 1308, 1282, 1234, 1162, 1119, 1075, 993, 947, 868, 819, 789, 728, 696, 658 cm^{-1}

(2S,3R)-3-9-(3-Chloro-5-(trifluoromethyl)pyridin-2-yl)oxy)nonyl-oxy)-2-(3,4 dihydroxyphenyl)-chroman-5,7-diol (1j)

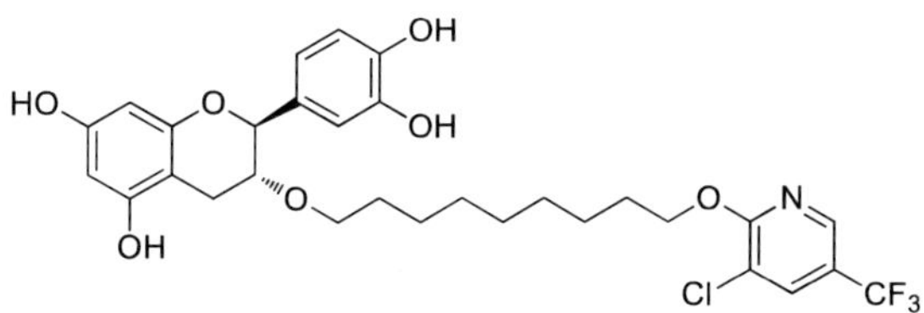

Yield 80% (0.085 g), $[\alpha]_D^{25}$ –29.9 (*c* 1.0, MeOH). ^{1}H NMR (CD_3OD, 400 MHz): δ 8.25-8.23 (m, 1H), 7.94-7.92 (m, 1H), 6.73 (d, 1H, *J* = 2.27 Hz), 6.66-6.58 (m, 2 H), 5.83 (d, 1H, *J* = 2.27 Hz), 5.75 (d, 1H, *J* = 2.27 Hz), 4.52 (d, 1H, *J* = 7.32 Hz), 4.33 (d, 2H, *J* = 7.32 Hz), 3.60-3.51 (m, 1H) 3.37-3.28 (m, 1H), 3.12-3.03 (m, 1H), 2.77 (dd, 1H, *J* = 5.30, 16.16 Hz), 2.40 (dd, 1H, *J* = 8.08, 16.16 Hz), 1.75-1.65 (m,2H) 1.46-0.99 (m, 12H). ^{13}C NMR (CD_3OD, 100 MHz): δ 162.8, 157.8, 157.5, 156.9, 146.1, 146.1, 143.7, 136.4, 132.5, 126.1, 123.4, 121.8, 119.9, 116.0, 115.2, 100.7, 96.4, 81.2, 77.1, 70.7, 69.2, 30.8, 30.7, 30.5, 30.2, 29.7, 27.2, 26.3, 20.9. ESI-MS (m/z): 634 $[M+Na]^+$. IR (KBr) ν_{max}: 3369, 2929, 2855, 1606, 1521, 1487, 1466, 1410, 1323, 1266, 1249, 1137, 1071, 913, 879, 818, 819, 781, 761, 712, 643 cm^{-1}

(2S,3R)-3-(9-(5-Aminopyridin-2-yl)oxy-nonyl-oxy)-2-(3,4dihydroxyphenyl) chroman-5,7-diol (1k)

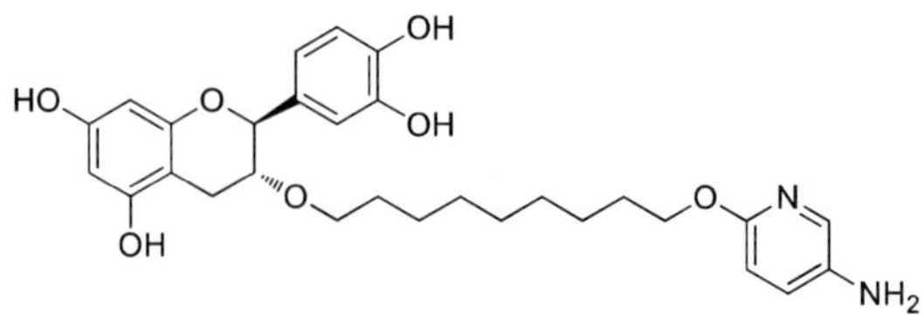

Yield 73% (0.085 g), $[\alpha]_D^{25}$ –44.4 (*c* 1.0, MeOH). ^{1}H NMR (CD_3OD, 400 MHz) δ 8.30-8.28 (m, 1H), 7.96 (dd, 1H, *J* = 2.27,9.09 Hz), 6.72 (d, 1H, *J* = 2.02 Hz), 6.66-6.58 (m, 3H), 5.82 (d, 1H, *J* = 2.27 Hz), 5.74(d, 1H, *J* = 7.33 Hz), 4.54 (d, 1H, *J* = 7.32 Hz), 4.36 (t 1H, *J* = 6.56 Hz) 3.60-3.53 (m, 1H), 3.46-3.41 (m, 1H) 3.39-3.32 (m, 1H), 3.12-3.05 (m, 1H), 2.76 (dd, 1H, *J* =5.05, 15.95 Hz), 2.40 (dd, 1H, *J* =7.83, 15.91 Hz), 1.76-1.16 (m, 2H), 1.48-0.97 (m, 12H).^{13}C NMR (CD_3OD, 100 MHz) δ 157.7, 156.9, 155.9, 154.0, 145.9, 143.0, 137.0, 131.9, 131.9, 126.8, 120.8, 119.5, 115.0, 111.1, 101.7, 95.8, 81.0, 79.8, 70.6, 69.8, 30.8, 30.7, 30.5, 30.1, 29.6, 27.2, 26.2, 20.7. ESI-MS (m/z): 547 $[M+Na]^+$. IR (KBr) ν_{max}: 3444, 2925, 2854, 1631, 1523, 1464, 1382, 1261, 1115, 803, 637, 615 cm^{-1}

(2*S*,3*R*)-5-(Benzyloxy)-2-(3,4-dihydroxyphenyl)-7 hydroxy chroman-3-yl)oxy)nonyl 3-phenylpropanoate (1l)

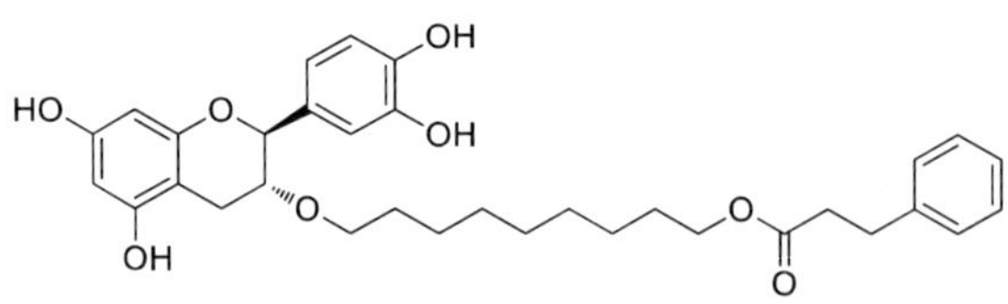

Yield 46% (0.080 g), $[\alpha]_D^{25}$ –20.9 (*c* 1.0, MeOH). ^{1}H NMR (CD_3OD, 400 MHz) δ 7.30-7.12 (m, 5H), 6.82 (d, 1H, *J* = 2.02 Hz), 6.76-6.67 (m, 2H), 5.92 (d, 1H, *J* = 2.27 Hz), 5.84 (d, 1H, J = 2.27 Hz), 4.61 (d, 1H, *J* = 7.57 Hz), 4.03 (t, 2H, J = 6.31 Hz), 3.69-3.62 (m, 1H), 3.41 (t, 2H, *J* = 6.31), 2.9 (m, 3H), 2.60 (m, 2H), 2.50 (dd, 1H, *J* = 8.08, 16.16 Hz), 1.60-1.48 (m, 2H) 1.42-1.06 (m, 12H).^{13}C NMR (CD_3OD, 100 MHz) δ 175.3, 157.8, 157.6, 146.2, 146.1, 141.9, 129.4, 129.3, 127.2, 119.9, 116.0, 115.2, 107.6, 96.4, 95.6, 81.3, 73.8, 71.9, 65.7, 36.8, 32.0, 30.7, 30.5, 30.4, 29.7, 27.1, 26.9, 26.2, 23.7. ESI-MS (m/z): 587 $[M+Na]^+$. IR (KBr) ν_{max}: 3382, 2926, 2853, 1695, 1605, 1519.7, 1467, 1368, 1316, 1284, 1178, 1139, 1060, 1025, 960, 669, 818, 780, 713, 673, 630, 4317 cm^{-1}

4.4. Anticancer Activity

Cytotoxicity of the synthesized (−)-catechin derivatives **1a-l** were determined on the basis of measurement of *in vitro* growth inhibition of tumor cell lines in 96 well plates by cell-mediated reduction of tetrazolium salt to water insoluble formazan crystals using doxorubicin as a standard. The cytotoxicity was assessed against a panel of four different human tumor cell lines: A549 derived from human alveolar adenocarcinoma epithelial cells (ATCC No. CCL-185), MDA-MB-231 derived from human breast adenocarcinoma cells (ATCC No. HTB-26), DU145 derived from human prostate cancer cells (ATCC No. HTB-81) and HepG2 derived from human hepatocellular liver carcinoma cells (ATCC No. HB-8065) using the MTT colorimetric assay [31]. The effect of the different synthesized (−)-catechin derivatives on the viability of tumor cell lines was measured at the wavelength of 540 nm on a multimode reader (Infinite® M200, Tecan, Switzerland). Dose–response curves were plotted for the test compounds and controls after correction by subtracting the background absorbance from that of the blanks. The IC_{50} values (50% inhibitory concentration) were calculated from the plotted absorbance data for the dose-response curves. All the experiments were performed in triplicate and the results are reported as IC_{50} values (in μM) expressed as the mean ± S.D.

4.5. Molecular Docking Method

The molecular docking study [32, 36] was carried out using the Surflex-Dock module with standard protocols in SYBYL-X 1.3 (Tripos Inc, St. Louis, MO, USA) modeling package which was installed on an HP Z400 desktop workstation equipped with a dual 3.20 GHz Intel® Xeon® processor running Windows 7 Professional (Service Pack 1) operating system. The X-ray crystal structure of IGF-1R complexed with ligand was retrieved from the RCSB Protein Data Bank (PDB code: 3F5P). During the protein preparation in the Surflex-Dock protocol, the ligand and crystallographic water molecules were removed from the crystal structure. Hydrogen atoms were added to the protein structure and ligand to define the correct configuration and tautomeric states. Charges were also added using Gasteiger-Huckel charge for both protein and ligand. Energy-minimizations were performed with standard protocol using the Tripos force field with a Gasteiger-Huckel charge, distance-dependent dielectric and the Powell conjugate gradient algorithms with a convergence criterion of 0.01 kcal/mol. The extracted ligand was used for a protomol (an idealized active site ligand to which putative ligands can be aligned) generation. Molecular docking simulation was carried out by binding to the binding site of IGF-1R by an empirical scoring function to score the ligand and protomol guided docking. Thus, this protomol-based method and empirically derived scoring function (e.g., Total score, crash score, polar etc.) were used to calculate the binding affinities i.e., interaction of ligand and IGF-1R. The potential hydrogen bondings were depicted by dashed lines.

4.6. Antimicrobial Activity

The antimicrobial activity of the synthesized catechin derivatives was determined using well diffusion method [37] against different pathogenic bacterial and *Candida* reference strains procured from the Microbial Type Culture Collection and Gene Bank (MTCC), CSIR-Institute of Microbial Technology, Chandigarh, India. The pathogenic bacterial and *Candida* reference strains were seeded on the surface of the media Petri plates, containing Muller-Hinton agar with 0.1 mL of previously prepared microbial suspensions individually containing 1.5×10^8 cfu mL^{-1} (equal to 0.5 McFarland). Wells of 6.0 mm diameter were prepared in the media Petri plates using a cork borer and the synthesized compounds at a dose range of 300-1.17 μg $well^{-1}$ was added in each well under sterile conditions in a laminar air flow chamber. Standard antibiotic solution of Neomycin and Miconazole at a dose range of 300-1.17 μg $well^{-1}$ and the well containing methanol served as positive and negative controls, respectively. The plates were incubated for 24 h at 37°C and 30°C for bacterial and *Candida* strains, respectively, and the well containing the least concentration showing the inhibition zone was considered as the minimum inhibitory concentration. All experiments were carried out in duplicates and mean values are represented.

4.7. Docking Studies of Compound 1b

DNA gyrase (PDB code: 1AJ6) was selected as the receptor for docking simulation in ATP binding pocket [37, 39]. After removing the ligand and solvent molecules, hydrogen atoms and Kollman charges were added to each protein atom. Coordinates of each compound were generated using Chemdraw11 followed by MM2 energy minimization. Docking was carried out by AutoDock4 in ATP binding pocket. Grid map in Autodock that defines the interaction of protein and ligands in binding pocket was defined. The grid map was used with 60 points in each x, y, and z direction, equally spaced at 0.375 Å. Docking was performed using the Lamarckian genetic algorithm. Each docking experiment was performed 100 times, yielding 100 docked conformations. Parameters used for the docking were as follows: population size of 150; random starting position and conformation; maximal mutation of 2 Å in translation and 50° in rotations; elitism of 1; mutation rate of 0.02 and crossover rate of 0.8; and local search rate of 0.06. Simulations were performed with a maximum of 1.5 million energy evaluations and a maximum of 50,000 generations. Final docked conformations were clustered using a tolerance of 1.0 Å root mean square deviation. The best model was picked based on the best stabilization energy.

Conclusion

In conclusion, novel (−)-catechin derivatives were designed, synthesized and fully characterized by analytical data. Most of the synthesized compounds exhibited significant activities, in which, compounds **1d**, **1**e and **1j** showed good antioxidant and broad spectrum anti-proliferative effects. Whereas, compound **1a**, **1b** and **1c** displayed excellent antibacterial activity for Gram negative and Gram positive bacteria, and these compounds also exhibited excellent antifungal activity more than the standard drug neomycin and miconazole respectively. Thus, (−)-catechin derivatives can be considered as potential antioxidant, antimicrobial and anticancer agents. Molecular docking studies were also exhibited good binding affinity of compound **1d** at the active site of IGF-1R complementing the biological activity for anticancer activity while docking studies of **1a** and **1b** which has hydroxyl and methoxy group respectively, both groups are small and easy to accommodate by ATP binding pocket of DNA gyrase. Finally, it is conceivable that further derivatization of such compounds can serve as new templates for anticancer agents, which might lead to more active molecules in the field of antimicrobial and cancer chemotherapy.

References

[1] H. N. Graham, Green tea composition, consumption, and polyphenol chemistry, *Prev. Med.* 21(1992) 334.

[2] Y. Kuroda, Y. Hara, Antimutagenic and anticarcinogenic activity of tea polyphenols, *Mutat. Res.* 436 (1999) 69.

[3] Y. D. Jung, M. S. Kim, B. A. Shin, K. O. Chay, B. Ahn, W. Liu, C.D. Bucana, G. E Gallick, L. M. Ellis, EGCG, a major component of green tea, inhibits tumour growth by inhibiting VEGF induction in human colon carcinoma cells, *Br. J Cancer* 84 (2001) 884.

[4] J. R. Carlson, B. A. Bauer, A. Vincente, P. J. Limburg, T. Wilson, Reading the tea leaves: anticarcinogenic properties of (-)-epigallocatechin-3-gallate, *Mayo. Clin. Proc.* 82 (2007) 725.

[5] D.G Nagle, D. Ferreira, Y. D Zhou, Epigallocatechin-3-gallate (EGCG): chemical and biomedical perspectives, *Phytochemistry* 67 (2006) 1849.

[6] L. Sanchez-del-Campo, F. A. Oton, J. Tarraga, H. Cabezas, S. Chazarra, L. J. N. Rodrıguez , Synthesis and Biological Activity of a 3,4,5-Trimethoxybenzoyl Ester Analogue of Epicatechin-3-gallate, *J. Med. Chem.* 51 (2008) 2018.

[7] A. G. Paschka, R. Butler, C. Y. Young, Induction of apoptosis in prostate cancer cell lines by the green tea component, (-)-epigallocatechin-3-gallate, *Cancer Lett.* 130 (1998) 1.

[8] Von Gadow, E Joubert, C. F. Hansmann, Comparison of the antioxidant activity of rooibos tea (*Aspalathus linearis*) with green, oolong and black tea, *Food Chem.* 60 (1997) 73.

[9] P. D. Stapleton, S. Shah, J. C. Anderson, Y. Hara, J. M. Hamilton-Miller, P. W. Taylor, Modulation of beta-lactam resistance in Staphylococcus aureus by catechins and gallates, *Int. J. Antimicrob. Agents* 23 (2004) 462.

[10] Z. Q. Hu, W. H. Zhao, Y.; Hara, T. Shimamura, Epigallocatechin gallate synergy with ampicillin/sulbactam against 28 clinical isolates of methicillin-resistant Staphylococcus aureus, *J. Antimicrob. Chemother.* 48 (2001) 361.

[11] D. Kumar, M. Poornima, R. N. Kushwaha, T. J. Won, C. Ahn, C. G. Kumar, K. Jang, D.-S.Shin, Antimicrobial and docking studies of (−)-catechin derivatives, *J. Korean Soc. Appl. Biol. Chem.* 58 (2015) 581.

[12] D. Kumar, S. J Harshavardhan, S. Chirumarry, Y. Poornachandra, K. Jang, C. G. Kumar, Y.-J. Yoon, B.-X. Zhao, J.-Y. Miao, D.-S. Shin, Design, Synthesis *In Vitro* Anticancer Activity and Docking Studies of (−)-Catechin Derivatives, *Bull. Kor. Chem. Soc.* 36 (2015) 564.

[13] S. Uesato, Y. Kitagawa, Y. Hara, H. Tokuda, M. Okuda, X. Y Mou, T. H Nishino, Antitumor promoting activities of 3-O-acyl-(-)epigallocatechins, *Bioorg. Med. Chem. Lett.* 10 (2000) 1673.

[14] J. Hong, H. Lu, X. Meng, J. H. Ryu, Y. Hara, C. S. Yang, Stability, cellular uptake, biotransformation, and efflux of tea polyphenol (-)-epigallocatechin-3-gallate in HT-29 human colon adenocarcinoma cells, *Cancer Res.* 62 (2002) 7241.

[15] K. D. Park, S. G. Lee, S. U. Kim, S. H. Kim, W. S. Sun, S. J. Cho, D. H. Jeong, Anticancer activity of 3-O-acyl and alkyl-(-)-epicatechin derivatives, *Bioorg. Med. Chem. Lett.* 14 (2004) 5189.

[16] A. Husain, R. Mohd, M. Ravinesh, P. Shama, D.-S. Shin, D. Kumar, Benzimidazole bearing oxadiazole and triazolo-thiadiazoles nucleus: Design and synthesis as anticancer agents, *Bioorg. Med. Chem. Lett.* 22 (2012) 5438.

[17] S. X. Yong, D. Kumar, B.V. D. Vijaykumar, T. Tarani, X. B. Zhao; Y. M. Jun, J. Kiwan, D.-S. Shin, Microwave-assisted parallel synthesis of benzofuran-2-carboxamide derivatives bearing anti-inflammatory, analgesic and antipyretic agents, *Tetrahedron Lett.* 55 (2014) 2796.

[18] Jr. L. J. Romanczyk, P. K. Sharma, D. Gou, Y. Gou, *US2010/0048920 A1*.

[19] K. L. Weinges, Über catechine und ihre herstellung aus leuko-anthocyanidin-hydraten [About catechins and their production from leuko-anthocyanidin-hydrates], *Liebigs. Chem. Ann.* 615 (1958) 203.

[20] Freudenberg, K.; Weinges, K., Leuko- und Pseudoverbindungen der Anthocyanidine [Leuko and pseudo compounds of anthocyanidins], *Liebigs Chem. Ann.* 61(1958) 613.

[21] Nay, V. Arnaudinaud, J. F. Peyrat, A. Nuhrich, G. Deffieux, J-M. Mérillon, J. Vercauteren, Total synthesis of labeled flavanoids. 2. 13C-labeled (±) -catechin from potassium [13C]cyanide, *Eur. J. Org. Chem.* 7 (2000) 1279.

[22] V. Arnaudinaud, B. Nay, A. Nuhrich, G. Deffieux, J. M. Merillon, J. P. Monti, J. Vercauteren, Total synthesis of isotopically labeled flavonoids. Part 3: 13C-labeled (−)-procyanidin B3 from 1[13C]-acetic acid, *Tetrahedron Lett.* 42 (2001) 1279.

[23] V. Arnaudinaud, B. Nay, S. Verge, A. Nuhrich, G. Deffieux, J. M. Merillon, J. P. Monti, J. Vercauteren, Total synthesis of isotopically labeled flavonoids. Part 5: Gram-scale production of 13C-labeled (−)-procyanidin B3, *Tetrahedron Lett.* 42 (2001) 5669.

[24] T. Gobel, K. B. Sharpless, Temperature Effects in Asymmetric Dihydroxylation: Evidence for a Stepwise Mechanism, *Angew. Chem.* 32 (1993) 1329.

[25] Z. M. Wang, X. L. Zhang, K. B. Sharpless, Asymmentric dihydroxylation of aryl allyl ethers, *Tetrahedron Lett.* 34 (1993) 2267.

[26] H. C. Kolb, M. S. Nieuwenhze, K. B. Sharpless, Catalytic Asymmetric Dihydroxylation, *Chem. Rev.* 94 (1994) 2483.

[27] H. C. Kolb, P. G. Andersson, K. B. Sharpless, Toward an Understanding of the High Enantioselectivity in the Osmium-Catalyzed Asymmetric Dihydroxylation (AD). 1. Kinetics, *J. Am. Chem. Soc.* 116 (1994)1278.

[28] P. O. Norrby, H. C. Kolb, K. B. Sharpless, Calculations on the reaction of ruthenium tetroxide with olefins using density functional theory (DFT). Implications for the possibility of intermediates in osmium-catalyzed asymmetric dihydroxylation, *Organometallics* 13 (1994) 344.

[29] R. J. J. Nel, M. Mthembu, J. Coetzee, H. V. Rensburg, E. Malan, D. Ferreira, Stereoselective synthesis of flavonoids. The novel flavan-3-ol, (2R, 3S)-guibourtinidol and its Diastereomers, *Phytochem.,* 52 (1999) 1153.

[30] G. W. Buchanan, R. Smits, E. Munteanu, Synthesis and 19F NMR study of RF-oleic acid-F13, *J. Fluorine Chem.* 123 (2003) 255.

[31] T. J. Mosmann, Rapid colorimetric assay for cellular growth and survival: application to proliferation and cytotoxicity assays, *J. Immunol. Methods* 65 (1983) 55.

[32] L. Steiner, G. Blum, Y. Friedmann, A. Levitzki, ATP non-competitive IGF-1 receptor kinase inhibitors as lead antineoplastic and anti-papilloma agents, *Eur. J. Pharmaco*. 562 (2007) 1.

[33] A. N. Jain, Surflex: fully automatic flexible molecular docking using a molecular similarity-based search engine, *J. Med. Chem.* 46 (2003) 499.

[34] Tripos Associates: *SYBYL Molecular Modelling Software, version X*. St. Louis, MO: Tripos Associates; 2009. Technical tips online. Available at http://www. tripos.com.

[35] http://www.rcsb.org/pdb/explore.do?structureId=3f5p

[36] L. M. Miller, S. C. Mayer, D. M. Berger, D. H. Boschelli, F. Boschelli, L. Di, F. Du, S. Petusky, D. Tkach, N. Torres, B. Wu, W. Xu, Lead identification to generate 3-cyanoquinoline inhibitors of insulin-like growth factor receptor (IGF-1R) for potential use in cancer treatment, *Bioorg. Med. Chem. Lett.* 19 (2009) 62.

[37] Amsterdam. Susceptibility testing of antimicrobials in liquid media. In: Loman, V. (Ed.) *Antibiotics in Laboratory Medicine*, 4th Edition, Williams and Wilkins, Baltimore, MD. 52(1996)111.

[38] G. A. Holdgate, G. A. Holdgate, A. Tunnicliffe, W. H Ward, S. A. Weston, G. Rosenbrock, P. T. Barth, I. W Taylor, R. A. Pauptit, D. Timms, The entropic penalty of ordered water accounts for weaker binding of the antibiotic novobiocin to a resistant mutant of DNA gyrase: a thermodynamic and crystallographic study, *Biochemistry* 36 (1997) 9663.

[39] H. Gradisar, P. Pristovsek, A. Plaper, R. Jerala, Green tea catechins inhibit bacterial DNA gyrase by interaction with its ATP binding site, *J. Med. Chem.* 50 (2007) 264.

In: Handbook of Remediation ...
Editors: Sadia Ameen, M. Shaheer Akhtar et al.
ISBN: 978-1-53615-118-3

Chapter 5

MODIFIED ZNO NANOPARTICLES USING IRON IONS (FE^{3+}) FOR PHOTO-CATALYTIC DEGRADATION OF TOXIC POLLUTANTS UNDER SUNLIGHT

Muneer M. Ba-Abbad[1,2,4,*], Mohd S. Takriff[1,2], Abdul Amir H. Kadhum[1], Abu Bakar Mohamad [1], Abdelbaki Benamor[3] and Abdul Wahab Mohammad[1,2]

[1]Department of Chemical and Process Engineering, Faculty of Engineering and Built Environment, Universiti Kebangsaan Malaysia, Bangi, Selangor, Malaysia
[2]Research Centre for Sustainable Process Technology (CESPRO), Faculty of Engineering and Built Environment, Universiti Kebangsaan Malaysia, Bangi, Selangor, Malaysia
[3]Gas Processing Centre, Qatar University, Doha, Qatar
[4]Department of Chemical Engineering, Faculty of Engineering and Petroleum, Hadhramout University of Science &Technology, Mukalla, Hadhramout, Yemen

ABSTRACT

The optical properties of ZnO nanoparticles were modified using various dopant concentrations of iron (Fe^{3+}) via a sol-gel method. To investigate of ZnO nanoparticles, the X-ray diffraction (XRD), scanning electron microscopy (SEM), transmission electron microscopy (TEM), ultraviolet–visible (UV-Vis) spectroscopy and photoluminscence (PL) were applied. The results showed that ZnO has a wurtzite, hexagonal structure and that the Fe^{3+} ions were well incorporated into the ZnO crystal lattice. As the Fe^{3+} concentration increased from 0.25 to 1.00 wt. %, the crystal size decreased in comparison with the undoped ZnO. The spectral absorption shifts of the visible light region (red shift) and the

[*] Corresponding Author's Email: muneer711@gmail.com.

band gap decreased for each Fe-ZnO sample. The photoluminescence results showed a clearly deep level emission in the visible region. The photocatalytic activities of pure ZnO and Fe-ZnO samples were evaluated based on the degradation of 2-chlorophenol in an aqueous solution under the sunlight. The samples with a small concentration of Fe^{3+} ions showed enhanced photocatalytic activity with an optimal maximum performance at 0.5 wt. %. The results indicated that toxicity removal of 2-chlorophenol at same line of the degradation efficiency. Small crystallite size and low band gap were attributed to high activities of Fe-ZnO samples under various concentrations of Fe^{3+} ions compared to undoped ZnO.

Keywords: ZnO, nanoparticles, iron doping, hazardous chemical, degradation efficiency

1. INTRODUCTION

ZnO has a wide direct band gap of ~3.37 eV and a large excitation binding energy of ~60 meV [1]. ZnO of a hexagonal wurtzite crystalline structure is commonly used as the semiconductors [2]. Nanoparticles of ZnO are used in the photo-catalysis and as sensors and dye sensitized solar cells because they have good optical, electrical and the chemical properties [3]. ZnO nanoparticles have been successfully used to convert toxic chemicals (organic pollutants) in water into innocuous materials via solar irradiation or UV light [4]. ZnO nanoparticles are excited by UV light, causing them to generate electron–hole pairs which then form $^{\bullet}OH$ and $^{\bullet}O^{2-}$ radicals which indicates that the absorption of a ZnO photocatalyst in the UV is disadvantageous because UV light makes up approximately 4% of the solar spectrum. To improve the activity of ZnO for environmental the applications, the absorption of light in the visible light spectrum must be enhanced since visible light makes up approximately 43% of the solar spectrum. If these enhancements are accomplished then ZnO nanoparticles could offer an attractive option for the environmental applications due to its efficient use of sunlight as the source of energy [5]. Recently, transition metal ions such as Co^{2+} [6], Ni^{2+} [7] and Fe^{3+} [8] have been used as dopants of ZnO to improve the optical properties. Doping with various concentrations of Fe^{3+} not only enhances the electron-hole pair separation by decreasing the band gap which advantages the shifting of absorption to the visible light spectrum but also decreases the size [9]. Doped ZnO has been prepared though a sol gel method [10] which has several advantages over other production techniques, including the ease of processing, control of composition, purity and homogeneity of the produced materials [11]. In this study, ZnO nanoparticles were doped with various wt. % of Fe^{3+} ions via the sol-gel synthesis method. Samples were characterized using XRD, TGA-DSC, FESEM, TEM, UV-Vis spectrophotometer and PL. The influence of Fe^{3+} concentration in ZnO nanoparticles under the visible light absorbance was investigated. The activity of ZnO and Fe-ZnO nanoparticles was tested using the photocatalytic degradation of 2-chlorophenol under sunlight.

2. Synthesis of Fe-ZnO Nanoparticles

Fe^{3+}-doped ZnO nanoparticles (Fe -ZnO) were synthesized via a sol gel method using zinc acetate, oxalic acid and iron nitrate as starting materials. A solution of zinc acetate was prepared by dissolving zinc acetate in 100 ml of alcohol and then heated the mixture in the water bath at 65 ± 5°C under reflux for 30 min (solution A). Oxalic acid solution was prepared by dissolving the oxalic acid in 50 ml of alcohol while stirring on a magnetic stirrer for 30 min (solution B). The optimization of all the parameters such as the molar ratio, pH and calcination temperature was reported elsewhere [12]. The required amount of iron nitrate (0.0, 0.25, 0.5, 0.75 and 1.0 wt %) was dissolved in 20 ml of ethanol under ultra-sonication and then added to solution (A) under continuous stirring. Solution (B) was added drop-wise to solution (A) under vigorous magnetic stirring until a gel was formed. The gel was continuously stirred for another 60 min. The resulting gel was dried in an oven at 80°C for 5 h and then calcined at 400°C for 2 h.

2.1. Characterization of Fe-ZnO Nanoparticles

The phase composition and the crystallite size of the Fe-ZnO samples were determined using an X-ray diffractometer (XRD) model D8 Advance Bruker AXS X-ray with CuKα radiation (1.5406 Å) in a 2θ scan range of 20-80°. The size of the particles was investigated using a transmission electron microscope (TEM), model JEOLJEM 2100 at 200 kV. The morphology of the particles was investigated using a field emission scanning electron microscope, model SUPRA 55VP under high resolution mode, along with an the EDX-System (Energy Dispersive X-ray analysis) model OXFORD INCA Penta FETx3. The thermal decomposition and the crystallization behaviors of the nanoparticles were determined using thermogravimetric-differential scanning calorimetry, model STA 449F3, Jupiter, Netzsch, Germany. A heating rate of 10°C/min from 25 to 800°C was used. The zeta potential of Fe-ZnO nanoparticles was measured using Malvern, Zetasizer, model Nano-ZS. A Perkin Elmer (lambda 35) UV/Vis spectrophotometer equipped with an integrating sphere was used to record the UV-Vis absorption spectra of the Fe-ZnO samples. The spectra were recorded at the room temperature with the wavelength ranging from 350-800 nm.

2.2. Photocatalytic Reactor and Procedures

The photocatalytic activity of ZnO and Fe-ZnO was evaluated using the degradation of an aqueous solution of 2-chlorophenol in a cylindrical slurry batch reactor. All the photocatalytic experiments were carried out under similar conditions, with ambient temperatures between 28-38°C. 100 ml aliquot of a 50 mgl^{-1} 2-chlorophenol solution was mixed with 1 gl^{-1} of photocatalyst for approximately 60 min to achieve the adsorption-desorption equilibrium which was later exposed to the solar radiation. A sample was withdrawn after every 15 min, centrifuged and then filtered.

2.3. Analytical Methods

High-performance liquid chromatography (Agilent 1200 HPLC) equipped with a Jones LC-18 column (250 mm x 4.6 mm x 4 μm) and a variable wavelength UV detector was used. An acetonitrile: water solution (20:80 v/v) with 0.01 M phosphoric acid (H_3PO_4) was used as the mobile phase with a flow rate of 1.0 ml min^{-1}. The injection volume of the sample used was 20 μl and 2-chlorophenol was detected at a wavelength of 254 nm. The photocatalytic degradation efficiency was calculated using Eq. (1):

$$\% \text{ Degradation} = (C_o - C/C_o) \times 100 \quad (1)$$

Where, C_o is the initial concentration and C is the concentration of 2-chlorophenol at different intervals of the irradiation time. The toxicity of 2-chlorophenol before and after the degradation procedure was evaluated using *Escherichia coli* bacteria. The clear inhibition zone method of bacteria growth was used for the toxicity detections. Media, culture conditions and toxicity testing were reported in our pervious study [13].

3. THERMAL STABILITY ANALYSIS OF FE-ZNO NANOPARTICLES

The thermal stability of ZnO and Fe-ZnO (0.5 wt. %) nanoparticles were investigated using the thermogravimetry-differential scanning calorimetry (TGA-DSC) analysis. Figure 1 shows the resultant TGA- DSC curves of both ZnO and Fe-ZnO precursors prepared using the same method. As could be observed from the DSC curves of the samples, two endothermic peaks are observed in the heating profile, with a corresponding loss of weight observed on the TGA curves. The first endothermic peak is observed below 190°C, this corresponds to the evaporation of ethanol and the remaining water. The second peak could be attributed to the thermal decomposition of oxalate for ZnO and Fe-ZnO. The TGA curves of the samples show that there are two different stages of weight loss for the complete decomposition of zinc oxalate. The first weight loss of approximately ~17.9% and ~19.4% for ZnO and Fe-ZnO respectively, occurs between 110 to 190°C. This weight loss is attributed to the removal of water from zinc oxalate dihydrate which has a reported theoretical water content of approximately ~19.05% ($ZnC_2O_4.2H_2O$ ZnC_2O_4 +

$2H_2O$). The second stage of heating from 360 to 420°C, results in a larger weight loss approximately 36.1 and 36.6% for ZnO and Fe-ZnO, respectively.

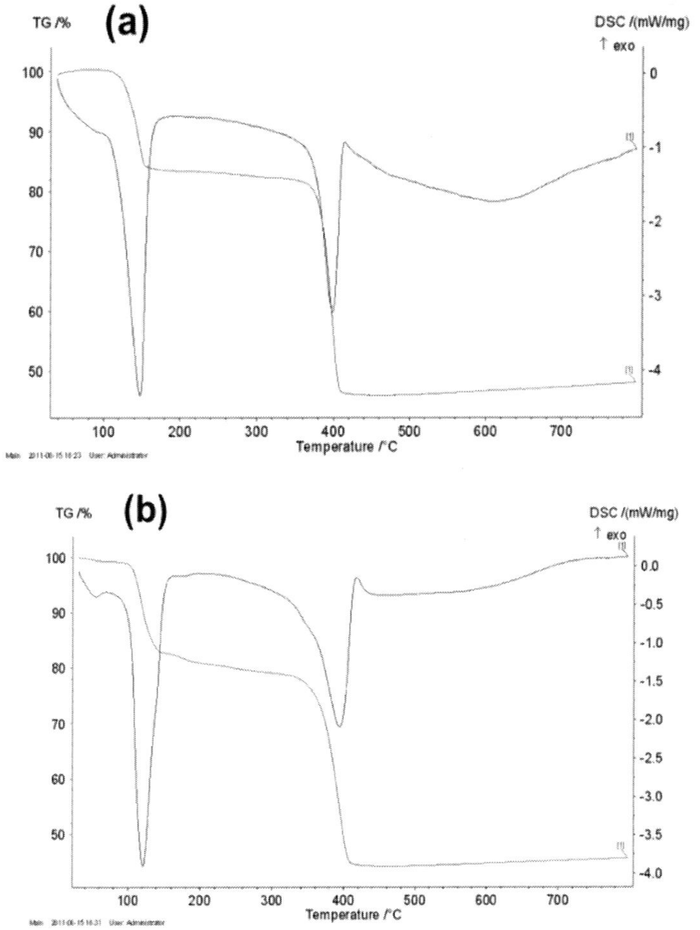

Figure 1. TGA-DSC analysis for thermal decomposition of (a) ZnO undoped (0.00 wt %) (b) 0.5 wt % Fe^{3+} doped ZnO NP samples. Reprinted with permission from M. M. Ba-Abbad, 2013, Chemosphere, 91 (2013) 1604 © 2013 Elsevier Ltd.

The weight loss ccould be attributed to the complete decomposition of zinc oxalate dihydrate into Zinc metal oxide ($ZnC_2O_4 \longrightarrow ZnO + CO + CO_2$), which is close to the theoretical value [14]. The difference in the weight loss of ZnO and Fe-ZnO samples might due to the thermal decomposition of the additional iron nitrate in the doped ZnO [15]. No further weight loss is observed by TGA at the temperatures higher than 410°C. This temperature is the point at which the thermal analysis of ZnO and Fe- ZnO oxalate dihydrates is completed, as reported by an earlier study [16].

3.1. X-Ray Diffraction of Fe-ZnO Nanoparticles

Figure 2 shows the XRD patterns of the doped (0.0, 0.25, 0.5, 0.75 and 1.0 wt. % Fe^{3+}) and undoped ZnO nanoparticles which are calcined at 400°C for 2 hr.

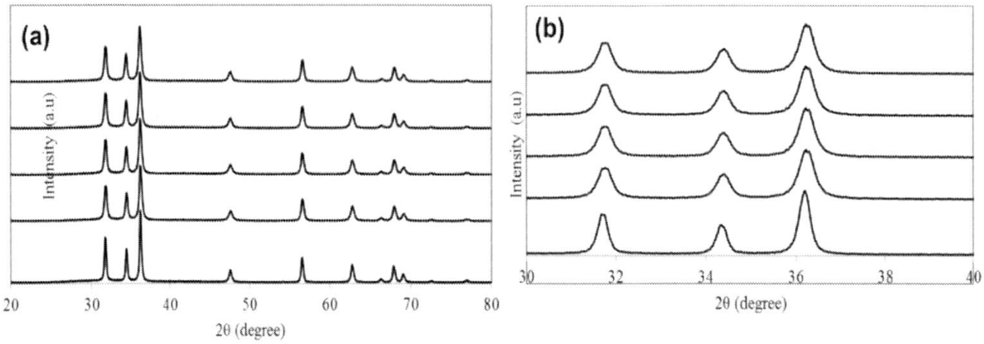

Figure 2. XRD patterns spectrum of (a) Fe^{3+}-doped ZnO NP with different wt. % (b) shifted of peak (101) position to right for Fe^{3+}-doped ZnO NP samples. Reprinted with permission from M. M. Ba-Abbad, 2013, Chemosphere, 91 (2013) 1604 © 2013 Elsevier Ltd.

The indexed diffraction peaks indicate a hexagonal wurtzite structure with a space group (*P63mc*) for all the samples, the diffraction data is in good agreement with the JCPDS card for ZnO (JCPDS 36-1451). The diffraction spectrums of the doped Fe-ZnO samples do not indicate the presence of pure Fe or Fe_2O_3. This could probably be attributed to the good dispersion of Fe^{3+} into the structure lattice of the ZnO or to the minute amount of Fe^{3+} dopant that was used [17].

On the other hand, the XRD peaks of Fe-ZnO are broad in comparison with the undoped ZnO, indicating that the Fe doping into the lattice of ZnO inhibits the crystallization of the samples [18]. These results show that smaller particles grow more easily when incorporated with ZnO as earlier shown in other studies [19]. This phenomenon could be attributed to the differences in size of the ionic radii of Fe^{3+} (0.67 Å) and Zn^{2+} (0.74 Å) ions. The lattice parameters of Fe-ZnO (0.5 wt. %) and ZnO are calculated from the XRD as a = b = 3.2488 Å, c = 5.2054 Å and a = b = 3.2498 Å, c = 5.2066 Å, respectively. Lattice parameters of Fe-ZnO are slightly less than those of ZnO, confirming that the Fe ions doped into the ZnO crystal lattice has not changed the wurtzite structure [20]. On the other hand, as compared to undoped ZnO, the increasing of the lattice parameters of Fe-ZnO sample when using Fe^{2+} are reported by other researchers [21]. The effect of Fe2+ ionic radii (0.77 Å) shows the significant enlargement of the interatomic spacing due to increase of the lattice parameters [22]. The average crystallite size D, of ZnO and Fe^{3+} doped ZnO are calculated by using the Debye-Scherer equation Eq. (2), and the data are presented in Table 1.

Table 1. Crystallite size and lattice parameters of Fe^{3+}-doped ZnO in comparison to undoped ZnO.

Doping by Wt. %	[a]Crystallite size (nm) ±2	[b]Dopant w.t%	Lattice parameters a = b	Lattice parameters c
0.00	18.00	0.00	0.32490	0.52060
0.25	16.10	0.235	0.32490	0.52060
0.50	15.00	0.473	0.32488	0.52056
0.75	13.20	0.732	0.32480	0.52054
1.00	11.82	0.965	0.32488	0.52054

[a]measured by XRD, [b]measured wt. % Fe^{3+} by EDX.

Reprinted with permission from M. M. Ba-Abbad, 2013, *Chemosphere*, 91 (2013) 1604 © 2013 Elsevier Ltd

$$D = K\lambda/\beta\cos\theta \quad (2)$$

Where, K is the Scherer constant (K = 0.89), λ is the incident X-ray wavelength, β is the peak width at half maximum, and θ is the Bragg diffraction angle.

At the same time, to confirm that doping of Fe^{3+} in the lattice of ZnO by the replacement of Zn in the matrix is shown by the obvious position of XRD pattern peak of (002) shifting [23]. The positions of peaks shift towards high value of 2θ with increasing size of Fe^{3+} are observed. These results of shifted position peaks reflect the doping of Fe^{3+} in Zn^{2+} in the cell of ZnO and changes in the average crystal size. Presence of Fe^{3+} as dopant is suitable because of easy and possible Fe^{3+} doping of ZnO cell by cation (Zn^{2+}) vacancies, as reported earlier. At same time, the shifting of peaks position to lower value of 2θ when ZnO is doped by Fe^{2+} is reported due to the ionic radii of Fe^{2+} (0.77 Å) which causes the increases of lattice parameters [24].

3.2. Morphology of Fe-ZnO Nanoparticles

The morphologies of ZnO and Fe-ZnO nanoparticles were investigated by using the FESEM and TEM analysis, as shown in Figure 3. The images reveal that for both ZnO and Fe-ZnO, the particles are mostly spherical in shape, although a less agglomeration is observed for the doped ZnO, as shown in Figure 3 (b, c).

As shown in Figure 3 (b_1, c_1), the particle size of the Fe-ZnO sample decreases with an increase in the concentration of Fe^{3+} ions. Upon further increase of the Fe^{3+} contents from 0.5 to 1 wt. %, the particle size decreases from 16 to 14 nm, respectively, this is consistent with the results obtained from XRD data. Energy dispersive X-ray (EDX) analysis is conducted for Fe-ZnO samples to determine the wt. % of the Fe dopants. It clearly indicates the doping of Fe into ZnO and the results are represented in the Table 1.

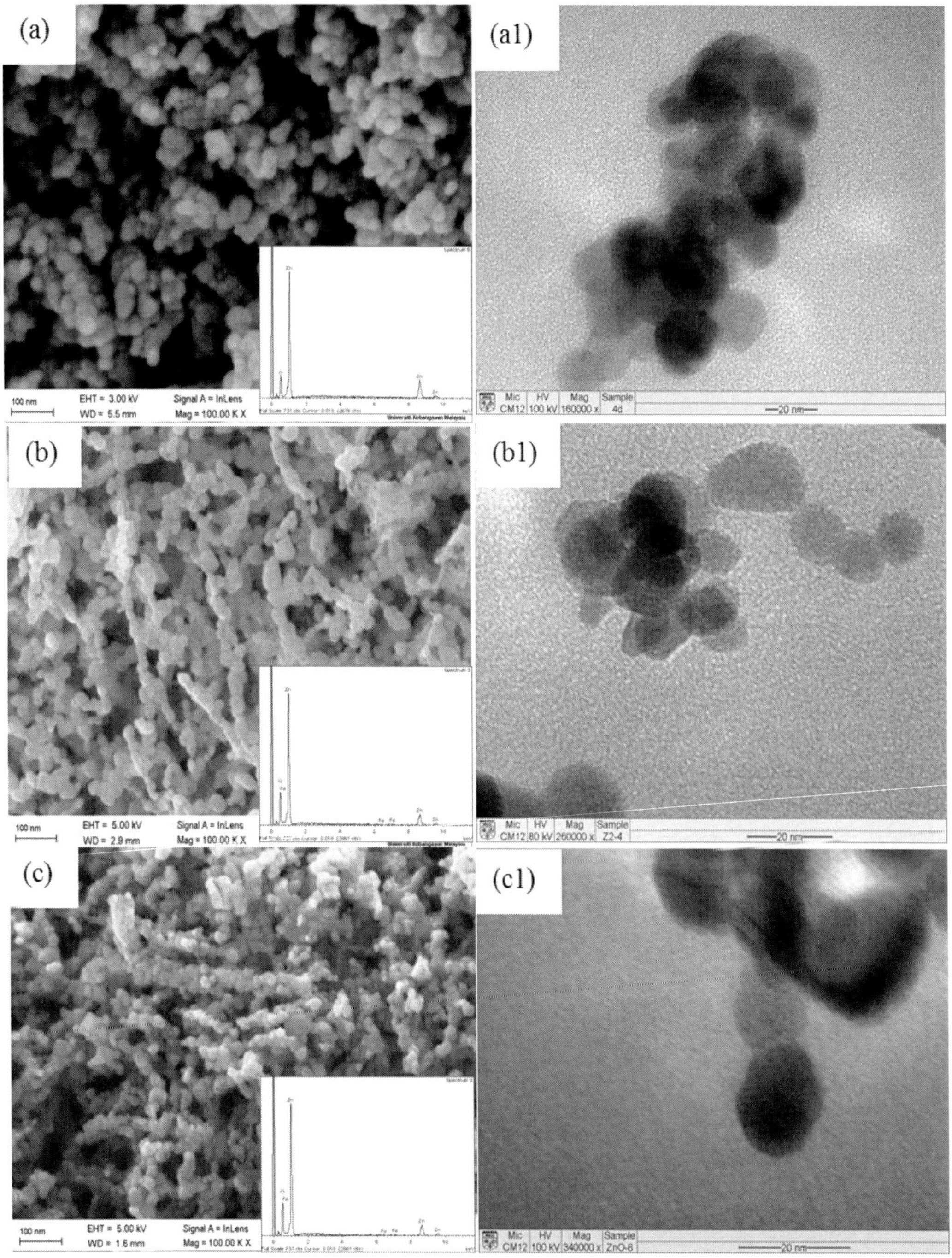

Figure 3. FESEM,TEM and EDX images of (a, a_1) ZnO undoped (b, b_1) 0.50 wt. % Fe^{3+} doped ZnO (c, c_1) 1.0 wt % Fe^{3+} -doped ZnO NP. Reprinted with permission from M. M. Ba-Abbad, 2013, Chemosphere, 91 (2013) 1604 © 2013 Elsevier Ltd.

3.3. Surface Area of Fe-ZnO Nanoparticles

Effects of the doping of Fe^{3+} ions with the concentration between 0.00 to 1.0 wt.% on the surface of ZnO NP was investigated.

Figure 4 shows the nitrogen adsorption-desorption isotherms graph of all doped ZnO NP samples, calcined at 400°C. All doped ZnO NP sample exhibits the V type of adsorption-desorption isotherms based on the IUPAC calcification for the sorption isotherms. Additionally, the hysteresis loops as H1based on the IUPAC, the calcification of hysteresis loops was reported in the range of 0.8 to 0.99 [25].

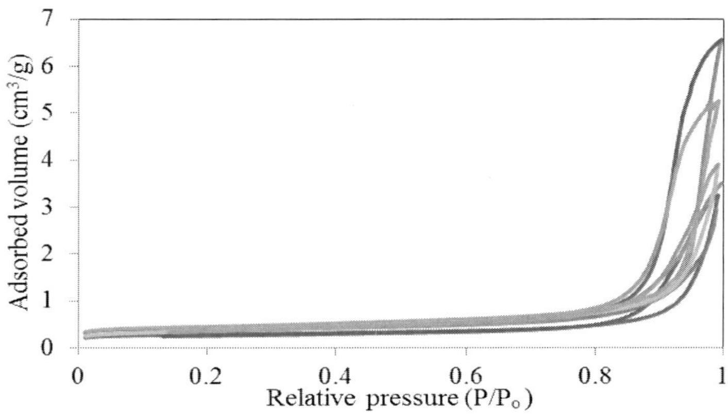

Figure 4. Nitrogen adsorptions-desorption of undoped ZnO and Fe^{3+}-doped ZnO NP with different wt %.

All the surface properties such as surface area, pore volume and average pore size of doped ZnO NP samples are given in Table 2 and are compared with ZnO undoped NP. With increasing the concentrations of all dopants, the volume of the adsorbed gas increased with the hysteresis loops became boarded under range of relative pressure of 0.8 to 0.99.

Table 2. Surface properties of undoped and Fe^{3+} -doped ZnO NP with different wt. %

Doping by Wt. %	Crystal size by XRD (nm) ±2	Surface area (m^2/g)	Pore volume (cm^3/g)	Average pore size (nm)
0.00	18.00	26.2	0.1123	25.20
0.25	16.10	26.8	0.1663	24.19
0.50	15.00	27.8	0.1649	21.51
0.75	13.20	30.10	0.1761	22.77
1.00	11.82	31.2	0.2274	24.40

These results were attributed to the increase of the surface area with the concentration increase of dopant as compared with undoped sample which was reported earlier [26]. At low concentration (0.25 wt. %) of dopant, the surface areas was almost close to 26.8 ± 0.2 m^2/g as compared to undoped ZnO of 26.2 m^2/g. This result confirmed that doping enhanced the surface area of ZnO NP even at a low concentration. However, the concentration of Fe^3 ions at 0.5 wt. % and above, significant increased the surface area.

The surface area of doped ZnO by Fe^{3+} was found for concentration more than 0.25 wt %. The highest surface area was obtained due smaller size of particle which confirmed by XRD and TEM analysis. Generally, with increasing doping concentration, the average pore size decreases, which could be attributed to the aggregation of smaller crystallites [27].

3.4. Zero- Point Charge of Fe-ZnO Nanoparticles

The surface charge of doped ZnO NP using Fe^{3+} ions was determined under similar experimental procedures of undoped ZnO NP. Effects of Fe^{3+} ions on the pH zpc of ZnO NP were evaluated for middle and higher concentrations. A pHzpc value of doped ZnO NP indicated a slight decrease from 9.1 to 8.85 for Fe^{3+} ions as given in Figure 5. The slight change in pHzpc of all dopants might attributed to the small difference on the zeta potential as ±5.0 mV due to the replacement of the dopant ions on ZnO matrix.

However, the surface of all doped ZnO samples were positively charged in a solution with a pH of less than 9.0, and were negatively charged for a pH of more than 9.0. Furthermore, there is slightly effect on surface charge of doped ZnO using Fe^{3+} ions in the case of substitution in host of ZnO NP as reported earlier [28].

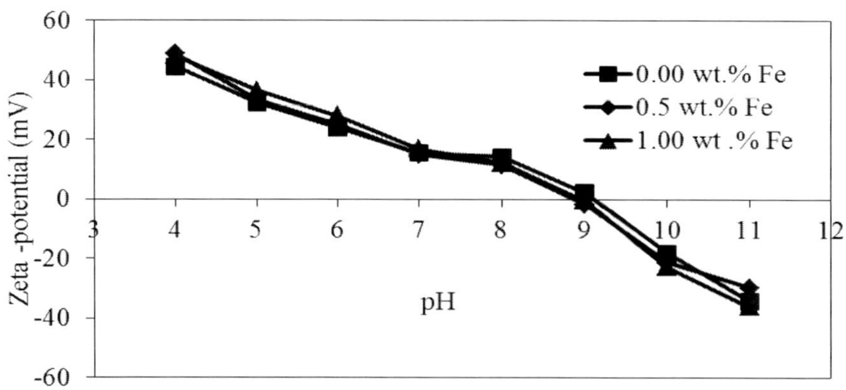

Figure 5. Zero point charge (pHzpc) of Fe^{3+}-doped ZnO NP with different wt. % compared to undoped ZnO. Reprinted with permission from M. M. Ba-Abbad, 2013, Chemosphere, 91 (2013) 1604 © 2013 Elsevier Ltd.

3.5. Absorbance and Band Gap of Fe-ZnO Nanoparticles

The absorbance spectra for all Fe-ZnO samples are measured at the room temperature and comparisons are made with the undoped ZnO, as shown in Figure 6(a). The light absorption of the ZnO indicates the effect of doing with Fe^{3+} ions [29]. The undoped ZnO spectrum shows an absorption onset at approximately 390 nm, which is in good agreement with the intrinsic energy band gap of ZnO at ~3.19 eV. The absorption spectra for all Fe-ZnO is shifted to a longer wavelength (red shift) which is also indicated by the change in the catalyst color from white to reddish yellow. A significant increase of Fe content is observed, the stronger absorption is assigned in the range of 400-600 nm (visible light region), which indicated a red shift in the band gap transition. The shift of the absorption towards the red part of the spectra upon Fe^{3+} doping was attributed to sp–d exchange interactions between the band electrons in ZnO and the localized d electrons of the Fe^{3+} [30]. However, absorbance shifting of Fe^{3+} doping samples to the visible region could be attributed to Burstein –Moss effect, which was combination of Fermi energy level to the conduction band because of increasing carrier concentration of doping, as reported earlier by Nair et al. [31]. Fe^{3+} ions in the valance band worked as defects to reduce the band gap and the absorbance of the light, resulted in the formation of electron–hole pair. Therefore, the energy needed for the transition from valance to the conduction was of low band gap of undoped ZnO [32].

The direct band gap (E_g) of ZnO and Fe-ZnO were determined by using the direct transition equation, Eq. (3):

$$\alpha h\nu = E_d(h\nu - E_g)^{1/2} \quad (3)$$

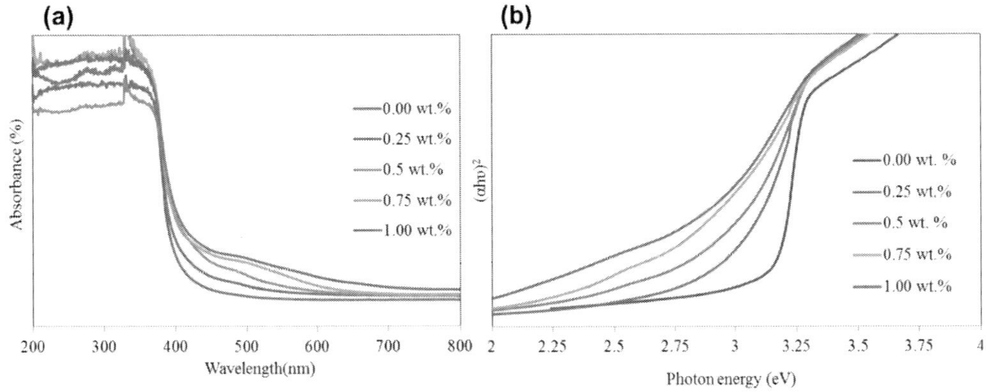

Figure 6. The optical properties of undoped ZnO and Fe^{3+}-doped ZnO NP (a) Absorption spectra (b) Band gap determination. Reprinted with permission from M. M. Ba-Abbad, 2013, Chemosphere, 91 (2013) 1604 © 2013 Elsevier Ltd.

where α is a the optical absorption coefficient, $h\nu$ is the photon energy, E_g is the direct band gap, and E_d is the constant. By plotting $(\alpha h\nu)^2$ as a function of photon energy and extrapolating the linear portion of the absorption curve equal to zero, the band gap could be calculated as shown in Figure 6 (b). The band gap (E_g) of Fe-ZnO samples are found to decrease (3.19, 3.0, 2.87, 2.72 and 2.65 eV) with an increasing Fe doapnt concentration (0, 0.25, 0.5, 0.75 and 1 wt %) as presented in Table 1. It could be observed, that the band gap (E_g) of ZnO might be tuned over a large energy range using the doping process, as reported in a the previous study [33].

3.6. Photoluminscence of Fe-ZnO Nanoparticles

The photoluminscence (PL) analysis of ZnO nanoparticles was measured at the room temperature with fluorescence spectrophotometer using a Xe lamp. A strong UV emission for ZnO nanoparticles, calcined at 400 °C, corresponded to an excitation recombination close to the band edge emission of ZnO nanoparticles with maxima at ~395 nm, as shown in Figure 7(a). The PL spectra of Fe^{3+}- ZnO samples are similar to that reported by other researchers, exhibiting the emission near band gap edge (NBE) region and the clearly deep level emission in the visible region [34]. High intensity of emission broad peaks at 430 nm (green band) is observed. The change in the amount of iron doping to ZnO nanoparticles, remarkably changes the emission intensity peak [35]. When Fe^{3+} content is increased, the emission intensity increases upto 0.5 wt %. Further with iron content doping, the emission intensity is negative affected, as shown in Figure 7 (b).

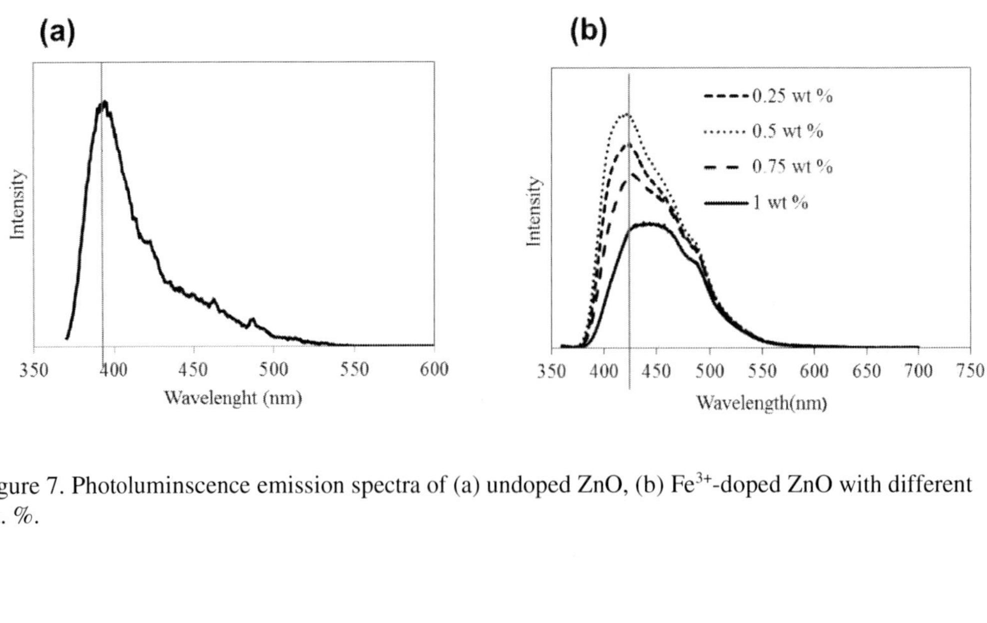

Figure 7. Photoluminscence emission spectra of (a) undoped ZnO, (b) Fe^{3+}-doped ZnO with different wt. %.

3.7. Photocatalytic Activity of Fe-ZnO Nanoparticles

The activities of ZnO and Fe-ZnO calcined at 400°C were evaluated using the degradation of 2-chlorophenol as a target compound. The effects of various Fe doping concentrations from 0.25 to 1.0 wt % on the photocatalytic activity (photocatalytic degradation efficiency, DE %) were investigated using the similar experimental conditions, as given in Figure 8. Effects of solar light intensity were evaluated under different conditions like cloudy, semi-cloudy and sunny day between the time period of 10:00 a.m. and 14:30 p.m. The results indicated that the sunny day was the best and the highest removal of 2-chlorophenol was found [35]. It could be observed that the photocatalytic activity increases as Fe concentration was increased.

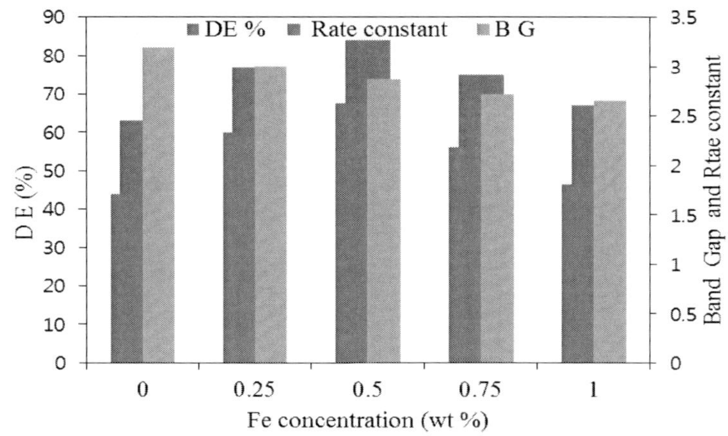

Figure 8. The photocatalytic activity (DE %), apparent kinetic constants (min^{-1}) and band gap (B G) vs. the different wt. % of Fe^{3+} -doped ZnO NP. Reprinted with permission from M. M. Ba-Abbad, 2013, Chemosphere, 91 (2013) 1604 © 2013 Elsevier Ltd.

When a small amount (0.25 wt. %) of Fe was doped into the ZnO, the photocatalytic activity increased in comparison with the undoped ZnO. The maximum value of the catalytic activity was achieved using 0.5 wt. % dopant, this produced the catalytic activity approximately one and a half times more than undoped ZnO. Further the increases of the dopant concentration resulted in a gradual decrease in the photocatalytic activity. Accordingly, the photodegradation efficiency, increases with the kinetic rate constant up to an optimal concentration of 0.5 wt. % as shown in Figure 8.

For Fe doped ZnO, the absorption was enhanced in the visible region through a red shift of ZnO band gap. These results show that the small concentrations of Fe ions from 0.25 to 0.5 wt. %, could be used to improve the photocatalytic activity to the maximum. At the same time, an excesses concentration of Fe ion (>0.5 wt. %) resulted in the decrease of the photocatalytic activity. This result was attributed to the fact that a small concentration of Fe ions could function as a photo-generated hole and trap, inhibiting the recombination of hole–electron [36], as described by the equations below:

$$ZnO + h\nu \rightarrow e^{-} + h^{+} \text{ (charge pair generation)} \quad (4)$$

$$Fe^{3+} + e_{cd}^{-} \rightarrow Fe^{2+} \quad (5)$$

$$Fe^{2+} + O_{2\,(ads)} \rightarrow Fe^{3+} + O_2^{\bullet-} \text{ (electron release)} \quad (6)$$

$$Fe^{3+} + h_{vb}^{+} \rightarrow Fe^{4+} \text{ (hole trap)} \quad (7)$$

$$Fe^{4+} + OH^{-} \rightarrow Fe^{3+} + OH^{\bullet} \text{ (hole release)} \quad (8)$$

Fe^{3+} ions react as electron holes and trapping materials which then form Fe^{2+} and Fe^{4+} (Eq. 5 and 7). Fe^{2+} and Fe^{4+} ions are relatively unstable as compared with Fe^{3+} ions [36]. To regain stability, these ions will react with oxygen and hydroxyl ions adsorbed onto the catalytic surface to produce hydroxyl ($OH^{\bullet}$) and superoxide ($O_2^{\bullet-}$) radicals (Eq. 6and 8). The recombination rate was found to be much faster at the optimum concentration of Fe^{3+} ions, which was expected to have the highest catalytic activity as reported by previous studies [37]. However, at higher concentration of Fe^{3+} (> 0.5 wt. %), the availability of traps is reduced because Fe^{4+} reacts with electrons ($Fe^{4+} + e^{-}$ Fe^{3+}) and Fe^{3+} acts as the recombination center ($Fe^{3+} + e^{-} \rightarrow Fe^{2+}$ and $Fe^{2+} + h_{vb}^{+}$, Fe^{3+}) thus reducing the activity of the Fe-ZnO photocatalyst [38]. Additionally, the narrower band gap of the photocatalyst indicates a lower oxidation-reduction potential. According to the results of our study, the photocatalytic process could be evaluated qualitatively as an electrochemical cell based on equation (Eq. 9) [39].

$$\Delta G = \quad n\,FE \quad (9)$$

Where, **ΔG** is the free energy change of the redox process occurring in the system, **n** the number of moles of electrons transferred involved in the process of redox, **F** the Faraday constant and **E** represents the band gap of the semiconductor. The equation indicates that the reducing band gap of Fe-ZnO enhances the photocatlytic activity. Therefore, the process needs to optimize the dopant concentration such as performed in our study, band gap at optimum (0.5 wt %) is 2.87 eV, which decreases to 2.65 eV at 1 wt %, and the lowest activity is obtained, as shown in Figure 5.

The kinetic of 2-chlorophenol degradation using ZnO and various % wt of Fe^{3+} doped ZnO was also investigated and the first order kinetic of photodegradation process was found with rate constants as summarized in Table 3. The rate constant value of 2-chlorophenol degradation showed the maximum value when Fe^{3+} concentration was at 0.5 wt %.

Table 3. Rate constant values of 2-chlorophenol degradation using Fe^{3+} -doped ZnO NP with different wt. %.

Doping by Wt. %	Crystal size by XRD (nm) ±2	Surface area (m^2/g)	Rate constant $\times10^{-2}(min^{-1})$	R^2
0.00	18.00	26.2	1.71	0.9928
0.25	16.10	26.8	2.33	0.9878
0.50	15.00	27.8	2.63	0.9927
0.75	13.20	30.10	2.19	0.9855
1.00	11.82	31.2	1.81	0.9943

3.8. Toxicity Evaluation of 2-Chlorophenol Degradation

The toxicity of 2-chlorophenol before and after the degradation using 0.5 wt % Fe-ZnO nanoparticles was investigated. *Escherichia coli* was used as an indicator strain, the growth inhibition was detected by agar disk diffusion assay reflecting the toxicity levels of 2-chlorophenol. Figure 9 shows the degree of toxicity according to the degradation percentage of 2-chlorophenol under direct sun radiation. It could be seen that clear inhibition zone was decreased, which attributed to non-toxicity effect of the final intermediate compounds.

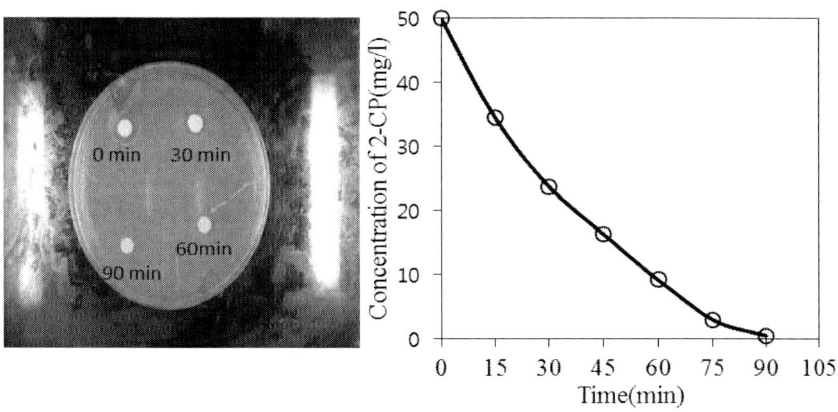

Figure 9. Toxicity evaluation of 2-chlorophenol before and after degradation using 0.5 wt. % Fe^{3+} doped ZnO NP. Reprinted with permission from M. M. Ba-Abbad, 2013, *Chemosphere*, 91 (2013) 1604 © 2013 Elsevier Ltd.

However, the higher clear inhibition around the paper disc at 0 min of irradiation time was obtained. After 30 min, gradual reduction of the inhibition was occurred, which illustrated that the concentration of toxicant chemical (2-chlorophenol) was decreased. There was no inhibition of *Escherichia coli* within 90 min, which indicated that the higher degradation of 2-chlorophenol with 99.0% was found. These results showed that the toxicity removal of 2-chlorophenol was at same line of degradation efficiency and in good agreement with earlier studies [40, 41].

Conclusion

ZnO nanoparticles were successfully doped with various concentrations of iron ions (Fe^{3+}) using a sol gel method. The Fe^{3+} doped ZnO samples showed that the size of final particles decreased with the increase of Fe^{3+} concentration. Longer wavelength absorption in the visible light region and a red shift of the band gap were observed for all Fe-ZnO samples. The photocatalytic activity of ZnO nanoparticles showed obvious enhancement with a small Fe^{3+} dopant concentration. At the optimal dopant concentration of 0.5 wt. % Fe^{3+}, the maximum photocatalytic activity was obtained which exceeded the undoped ZnO by one and a half time. In this study, the activity of ZnO nanoparticles indicated that Fe^{3+} ions dopant led to a more effective photodegradation of 2-chlorophenol in an aqueous solution under the visible light of the solar spectrum.

References

[1] W.K. Hong, J.I. Sohn, D.K. Hwang, S.S. Kwon, G. Jo, S. Song, S.M. Kim, H.J. Ko, S.J. Park, M.E. Welland, T. Lee, Tunable electronic transport characteristics of surface-architecture-controlled ZnO nanowire field effect transistors, *Nano Lett.* 8 (2008) 950.

[2] Y. T. Chung, M.M,.Ba-Abbad, A. W. Mohammad, N. H. Hairom, Benamor A., Synthesis of minimal-size ZnO nanoparticles through sol–gel method: Taguchi design optimization, *Mater. Des.* 87 (2015) 780.

[3] X. Sheng, Y. Zhao, J. Zhai, L. Jiang, D. Zhu, Electro-hydrodynamic fabrication of ZnO-based dye sensitized solar cells, *Appl. Phys. A* 87 (2007) 715.

[4] A.K. Srivastava, M. Deepa, N. Bahadur, M.S. Goyat, Influence of Fe doping on nanostructures and photoluminescence of sol–gel derived ZnO, *Mater. Chem. Phys.* 114 (2009) 194.

[5] C. R. Estrellan C. Salim, H. Hinode, Photocatalytic activity of sol–gel derived TiO_2 co-doped with iron and niobium, *Catal. Lett.* 98 (2009) 187.

[6] X.C. Liu, E.W. Shi, Z.Z. Chen, H.W. Zhang, L.X. Song, H. Wang, S. D. Yao, Structural, optical and magnetic properties of Co-doped ZnO films, *J. Cryst. Growth*, 296 (2006) 135.

[7] E. Liu, P. Xiao, J.S. Chen, B.C. Lim, L. Li, Ni doped ZnO thin films for diluted magnetic semiconductor materials, *Curr. Appl. Phys.* 8 (2008) 408.

[8] Y.Q. Wang, S.L. Yuan, L. Liu, P. Li, X.X. Lan, Z.M. Tian, J.H. He, S.Y. Yin, Ferromagnetism in Fe-doped ZnO bulk samples, *Magn. Magn. Mater.* 320 (2008) 1423.

[9] W. C. Hung, S. H. Fu, J. J. Tseng, H. Chu, T. H. Ko, Study on photocatalytic degradation of gaseous dichloromethane using pure and iron ion-doped TiO_2 prepared by the sol-gel method, *Chemosphere* 66 (2007) 2142.

[10] Y.W. Chen, Y.C. Liu, S.X. Lu, C.S. Xu, C.L. Shao, C. Wang, J.Y., Zhang, Y.M. Lu, D.Z. Shen, X.W. Fan, Optical properties of ZnO and ZnO:In nanorods assembled by sol-gel method, *J. Chem. Phys.* 123 (2005) 134701.

[11] M.M. Ba-Abbad, P.V. Chai, M. S. Takriff, A. Benamor, A.W. Mohammad, Optimization of nickel oxide nanoparticle synthesis through the sol–gel method using Box–Behnken design, *Mater. Des.* 86 (2015) 948.

[12] M. M. Ba-Abbad, H. A. A. Kadhum, A. B. Mohamad, M. S. Takriff, K. Sopian, The effect of process parameters on the size of ZnO nanoparticles synthesized via the sol–gel technique, *J. Alloys Compd.* 550 (2013) 63.

[13] M. M. Ba-Abbad, H. A. A. Kadhum, A. A. Al-Amiery, A. B. Mohamad, M. S. Takriff, Toxicity evaluation for low concentration of chlorophenols under solar radiation using zinc oxide (ZnO) nanoparticles, *Int. J. Phys. Sci.* 7 (2012) 48.

[14] Z. Jia, L. Yue, Y. Zheng, Z. Xu, Rod-like zinc oxide constructed by nanoparticles: synthesis, characterization and optical properties, *Mater. Chem. Phys*, 107 (2008) 137.

[15] J.G. Yu, J.C. Yu, W.K. Ho, M.K.P. Leung, B. Cheng, G.K. Zhang, X.J. Zhao, Effects of alcohol content and calcination temperature on the textural properties of bimodally mesoporous titania, *Appl. Catal. A* 255 (2003) 309.

[16] M.M. Ba-Abbad, A.H. Kadhum. A.B. Mohamad, M. S. Takriff, K. Sopian, Optimization of process parameters using D-optimal design for synthesis of ZnO nanoparticles via sol–gel technique, *J. Ind. Eng. Chem.*19 (2013) 99.

[17] P. Vijayana, C. Mahendirana, C. Suresha, K. Shanthi, Photocatalytic activity of iron doped nanocrystalline titania for the oxidative degradation of 2,4,6-trichlorophenol, *Catal. Today* 141 (2009) 220.

[18] H. Liu, J. Yang, Y. Zhang, L. Yang, M. Weil, X. Ding, Structure and magnetic properties of Fe-doped ZnO prepared by the sol-gel method, *J. Phys: Condens. Matter.* 21 (2009) 145803.

[19] K.J. Chen, T.H. Fang, F.Y. Hung, L.W. Ji, S.J. Chang, S.J. Young, Y.J. Hsiao, The crystallization and physical properties of Al-doped ZnO nanoparticles, *Appl. Surf. Sci.* 254 (2008) 5791.

[20] C.J. Cong, J.H. Hong, Q.Y. Liu, L. Liao, K.L. Zhang, Synthesis, structure and ferromagnetic properties of Ni-doped ZnO nanoparticles, *Solid State Commun.*138 (2006) 511.

[21] J. Anghel, A. Thurber, D.A. Tenne, C.B. Hanna, A. Punnoose, Correlation between saturation magnetization, bandgap, and lattice volume of transition metal (M=Cr M=Cr, Mn, Fe, Co, or Ni) doped $Zn_{1-x}M_xOZn_{1-x}M_xO$nanoparticles, *J. Appl. Phys.* 107 (2010) 09E314.

[22] J.T. Luo, Y.C. Yang, X.Y. Zhu, G. Chen, F. Zeng, F. Pan, Enhanced electromechanical response of Fe-doped ZnO films by modulating the chemical state and ionic size of the Fe dopant, *Phys. Rev. B* 82 (2010) 014116.

[23] F. Ahmed, S. Kumar, N. Arshi, M. S. Anwar, B. H. Koo, Morphological evolution between nanorods to nanosheets and room temperature ferromagnetism of Fe-doped ZnO nanostructures, *Cryst. Eng. Comm.* 14 (2012) 4016.

[24] B Panigrahy, M. Aslam, D. Bahadur, Effect of Fe doping concentration on optical and magnetic properties of ZnO nanorods, *Nanotechnology*, 23 (2012) 115601.

[25] S.W. Sing, Everett, D.H., A.W. Moscou, R.A. Pierotti, J. Rouquerol, T. Siemieniewska, *Pure Appl. Chem.* 57 (1985) 603.

[26] M. M. Ba-Abbad, M. S. Takriff, A. W. Mohammad, Enhancement of 2-chlorophenol photocatalytic degradation in the presence Co^{2+}-doped ZnO nanoparticles under direct solar radiation, *Res. Chem. Intermed.* 42 (2016) 5219.

[27] M. M. Ba-Abbad, H. A. A Kadhum, A. B. Mohamad, M. S. Takriff, K. Sopian, S. S. Al-Shamali, Kinetics Transformation of Anatase to Rutile Phase for Titanium Dioxide Nanoparticles Prepared by Sol-Gel Method, *Mater. Sci. Forum* 756 (2013) 11.

[28] F. Ahmed, S. Kumar, N. Arshi, M. S. Anwar, B. H. Koo, Morphological evolution between nanorods to nanosheets and room temperature ferromagnetism of Fe-doped ZnO nanostructures, *Cryst. Eng. Comm.* 14 (2012) 4016.

[29] K.T. Ranjit, B. Viswanathan, Synthesis, characterization and photocatalytic properties of iron-doped TiO_2 catalysts, *J. Photochem. Photobiol.* 108 (1997) 79.

[30] Y. Zhang, L. Wu, H. Li, J. Xu, L. Han, B. Wang, Z. Tuo, E. Xie, Influence of Fe doping on the optical property of ZnO films, *J. Alloys Compd.* 473 (2009) 319.

[31] M. G. Nair, M. Nirmala, K. Rekha, A. Anukaliani, Structural, optical, photo catalytic and antibacterial activity of ZnO and Co doped ZnO nanoparticles, *Mater. Lett.* 65 (2011) 1797.

[32] M. Zhou, J. Yu, B. Cheng, Effects of Fe-doping on the photocatalytic activity of mesoporous TiO_2 powders prepared by an ultrasonic method *J. Hazard. Mater.* 137 (2006) 1838.

[33] T. Makino, Y. Segawa, M. Kawasaki, A. Ohtomo, R. Shoroki, K. Tamura, T. Yasuda, H. Koinuma, Band gap engineering based on $Mg_xZn_{1-x}OMg_xZn_{1-x}O$ and $Cd_yZn_{1-y}OCd_yZn_{1-y}O$ ternary alloy films, *Appl. Phys. Lett.* 78 (2001) 1237.

[34] S.L. Cho, J. Ma, Y.K. Kim, Y.G. Sun, K.L. Wong, J.B. Ketterson, Photoluminescence and ultraviolet lasing of polycrystalline ZnO thin films prepared by the oxidation of the metallic Zn, *Appl. Phys. Lett.* 75 (1999) 2761.

[35] M.M. Ba-Abbad, A.H. Kadhum, A.B. Mohamad, M.S. Takriff, K. Sopian, Synthesis and Catalytic Activity of TiO_2 Nanoparticles for Photochemical Oxidation of Concentrated Chlorophenols under Direct Solar Radiation, *Int. J. Electrochem. Sci.* 7 (2012) 4871.

[36] M. M. Ba-Abbad, H. A. A Kadhum, A. B. Mohamad, M. S. Takriff, K. Sopian, Photocatalytic degradation of chlorophenols under direct solar radiation in the presence of ZnO catalyst, *Res. Chem. Intermed.* 39 (2013) 1981.

[37] M.H. Zhou, J.G. Yu, B. Cheng, H.G. Yu, Preparation and photocatalytic activity of Fe-doped mesoporous titanium dioxide nanocrystalline photocatalysts, *Mater. Chem. Phys.* 93 (2005) 159.

[38] W. Huang, X. Tang, I. Felner, Y. Koltypin, A. Gedanken, Preparation and characterization of Fe_xO_y–TiO_2 via sonochemical synthesis, *Mater. Res. Bull.* 37 (2002) 1721.

[39] J. Zhu, W. Zheng, B. He, J. Zhang, M. Anpo, Characterization of Fe–TiO_2 photocatalysts synthesized by hydrothermal method and their photocatalytic reactivity for photodegradation of XRG dye diluted in water, *J. Mol. Catal. A* 216 (2004) 35.

[40] J.C. Yu, L.Z. Zhang, J.G. Yu, Direct Sonochemical Preparation and Characterization of Highly Active Mesoporous TiO_2 with a Bicrystalline Framework, *Chem. Mater.* 14 (2002) 4647.

[41] T. Xia , Y. Zhao , T. Sager , S. George , S. Pokhrel , N. Li , D. Schoenfeld , H. Meng, S. Lin , X. Wang , M. Wang , Z. Ji , J.I. Zink , L. Mädler , V. Castranova , S. Lin , A.E. Nel, Decreased dissolution of ZnO by iron doping yields nanoparticles with reduced toxicity in the rodent lung and zebrafish embryos, *ACS Nano.* 5 (2011) 1223.

In: Handbook of Remediation ...
Editors: Sadia Ameen, M. Shaheer Akhtar et al.
ISBN: 978-1-53615-118-3

Chapter 6

ENVIRONMENTAL FRIENDLY THERMAL INTERFACE MATERIALS (TIMS) FOR ELECTRONIC CONTROL UNITS

M. Wasim Akhtar*

School of Semiconductor and Chemical Engineering,
Chonbuk National University, Jeonju, Republic of Korea
Department of Metalurgy & Materials Enginering,
Mehran University of Engineering & Technology, Jamshoro, Sindh, Pakistan

ABSTRACT

The burning of fossil fuel from vehicles and other engines is the main cause of environmental pollution. These causes can be reduced by utilizing alternative sources. One of the major sources of alternative energy is electricity, which is efficiently used for many applications. This lead to intensifying the demand of electronic components from our cars to handphones. As the devices are more compacting, they require more power, several issues related to their performance and reliability has been reported. One of the main issues is related to thermal management of these devices, which decreases their performance up to several extent. Thermal interface materials are a good choice to address these issues. Herein; we review the convention types of thermal interface materials (TIMs) and their potentiality for future perspective. Conventional TIMs are still ininborn stage and associated with different issues. Polymer-based highly thermally conductive filler might be future choice for TIMs. However; the interfacial contact of the fillers with the matrix are still major problem which affect the overall thermal performance of TIMs. In this chapters we suggest the surface treatment and development of highly packed composites, improv the performance of fillers. Furthermore, the use of eco-friendly filler has good impact on environment and reliability of electronic devices. The development of optimum

* Corresponding Author's Email: wasim.akhtar@jbnu.ac.kr.

and high performance TIMs is pivotal in sustainably and reliability of electronic control units (ECUs) which is based on main alternative source of fossil fuels. The reliability of ECUs based on the TIMs performance, better TIMs relates to better environment.

Keywords: thermal interface materials, carbon materials, thermal fillers, thermal conductivity, interfaces

1. Introduction

Alternative energy is critically important for the development of a greener society. The use of fossil fuels are the major source of pollution. According to reports, more than half of air pollution is contributed through by the good and services we use in our daily life Figure 1.

Focusing on the needs, it is evident that in the past few decades the uses of green energy are intensifying to minimize environmental pollution. The use of electronic devices is a good alternative resource of conventional devices that require fossil fuels and pollute the environment. These devices include more enhanced features and computation capabilities as well. The progress in their development is exponentially increasing as the demand is increasing. According to consumer technology reports (CTA), the demands in electronic devices are increasing rapidly and share a large proportion in world economics.

Although, this development led to several associated problems that need to take in consideration for a more reliable and better performance of these devices. One of the major issues with these devices is heat dissipation, which is generated during their application Figure 2.

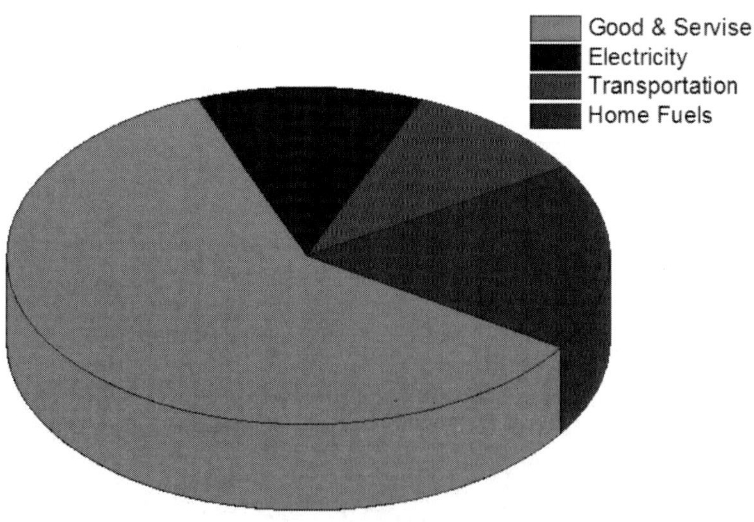

Figure 1. Air pollution source.

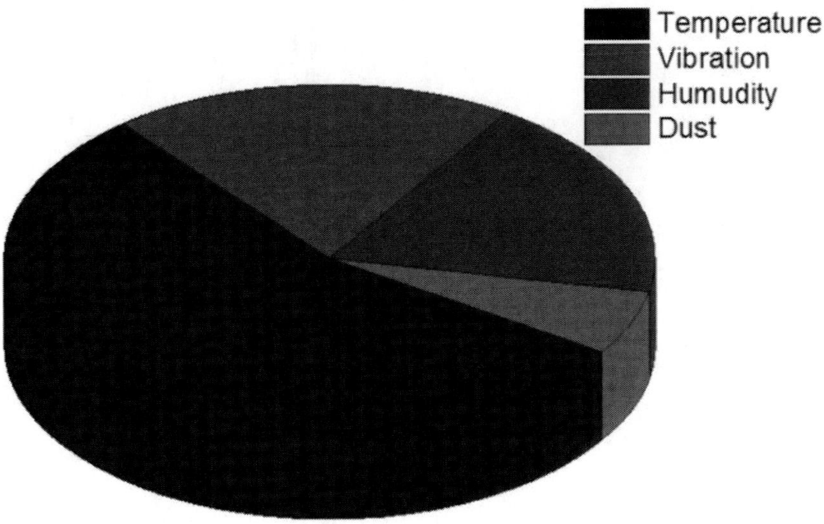

Figure 2. Major causes of deffect in electronic components.

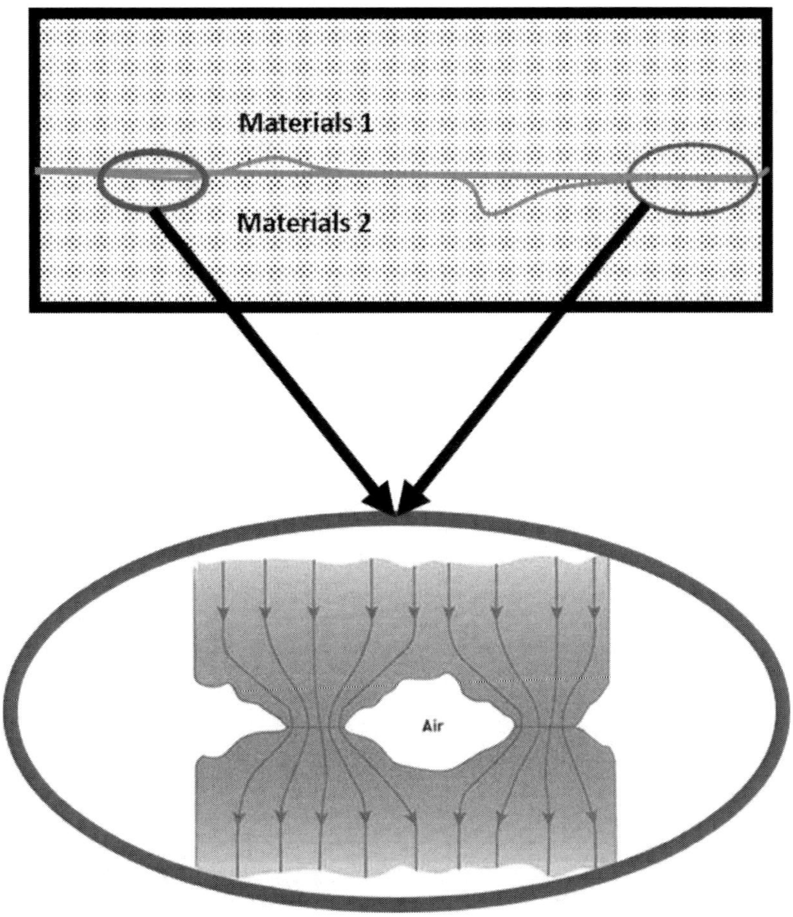

Figure 3. Roughness between two mating component of electronic devices.

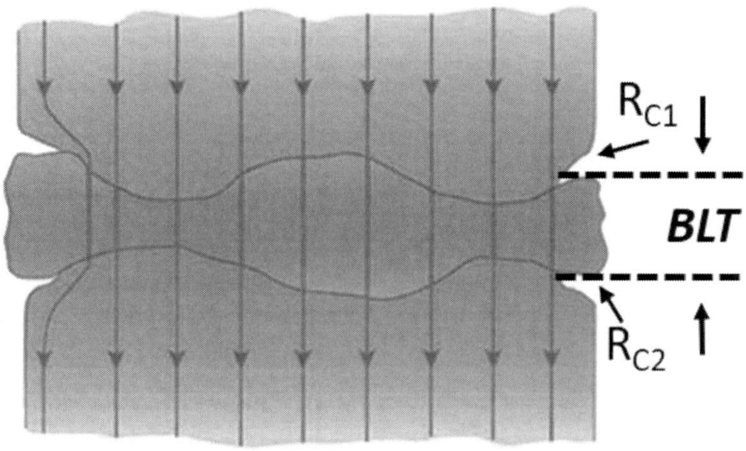

Figure 4. Thermal interface material between two mating bodies.

The removal of heat from these devices is important to maintain such devices within their optimum working conditions. To maintain the optimum performance and reliability of electronic devices, a properly designed thermal management system is required. The removal of heat is done through conduction from the package surface to heat spreader, which can efficiently transfer the heat to the ambient environment. When these surfaces brought in to intimate contact, due to micro scale roughness and non-planarity surfaces such as concave, convex or twisted shape, these surface are contact at only high points whereas the low point are air filled voids as shown in Figure 3 which is the significant cause of heat resistance due to lower thermal conductivity of the air.

High thermal conductive materials are requires to filled these interstitial air gap. Thermal interface materials have greater thermal conductivity as compares to air. TIMs are potential materials that can be used to decrease the resistance across the joined and component junction, result in better heat dissipation.

The total thermal resistance at an interface of materials is sum of bulk resistance and contact resistance as shown in Figure 4.

This is expressed by the following equation.

$$R_{eff} = \frac{BLT}{K_{TIM}A} + R_{c1} + R_{c2}$$

Where BLT is the bond line thickness and K_{TIM} is bulk thermal conductivity of the interface material. R_{c1} and R_{c2} is the contact resistance of TIM at the boundary with two surfaces and A is the area. The heat dissipation at the interface is increased by reducing the value of $R_{eff.}$ The total thermal resistance is decrease by reducing bond line thickness and introduction of highly thermal conductive thermal interface materials. Similarly; smoother surface has lower resistance; improving surface flatness and applying pressure decreases the contact resistance. Furthermore; for several more important factors are related with the reliable operation of thermal interface materials. Typical characteristics for thermal interface materials are discussed below.

1.1. Electrical Insulation

Electrically insulate thermal interface materials are more reliable as there is no effect on their operation while installed in the electronic field. The use of carbon filler to make highly thermal conductive TIMs increase its electric conductivity, but since these fillers are usually encapsulated in the polymer, where electrical conduction is not remained a major issue.

1.2. Thermal Stability

It is one of the most important characteristics for thermal interface materials. The TIMs should be thermally stable with the maximum operating temperature. It also reliable for the environmental application with a specific operating temperature ranges.

1.3. Phase Change Temperature

For TIMs it works with transition state, the phase change temperature is important. It is a temperature at which the transition from solid state to liquid state occurs in TIMs and the interface materials fill the voids and expelled out the air. The melting temperature of phase change materials should be lower than the operating temperature of the device.

1.4. Viscosity

For phase change materials, the viscosity is also an important thing to consider. The viscosity should be well enough to prevent the interface material from flowing and draining when handling in a different orientation.

1.5. Pressure

The pressure that exerts due to clamping and mounting of TIMs with devices also make a significant differencc. High pressure minimizes contact resistance.

1.6. High Surface Finish

Irregularities and waviness of the contact surfaces increase the contact resistance, which effects the thermal dissipation. High surface finishing is required for proper contacting of the mating surfaces.

1.7. Outgassing

During operation, some volatile gases are generated at elevated temperature especially with polymers and elastomers. This must me concern in an aerospace application, due to low pressure outgassing is increased.

1.8. Optimum Mechanical Properties

TIMs with high loading of filler fraction lead to poor mechanical properties. Proper modification of filler is required to obtain certain mechanical properties as well.

1.9. High Life Cycle

The life of TIMs should be higher than devices, the lower life span definitely reduces the overall life of the equipment.

1.10. Economical

Highly reliable with minimum cost is key factor for TIMs. Ease of fabrication and installation and use of commonly available filler make it feasible to produce low cost TIMs.

2. Types of Thermal Interface Materials

Keeping these thing in mind; several types of interface materials has been test until now. A comparison of typical types of TIMs is illustrated in Figure 5.

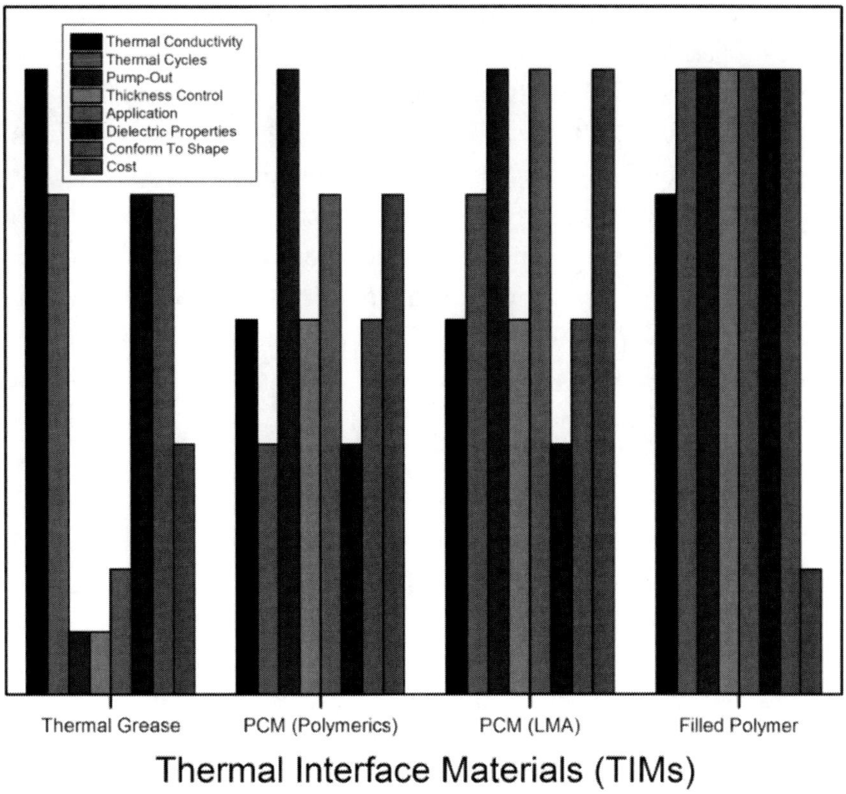

Figure 5. A comparison of typical thermal interface materials.

- Thermal grease
- Phase change materials
- Filled polymer matrices

2.1. Thermal Grease

Thermal grease is synthesized with silicon or hydrocarbon oils containing a high amount of thermally conductive fillers. They are in liquid states that are applied to heat spreader or heat sinks. It has good surface wetting characteristics and flowability to fill the voids at the interface, which result in low thermal impedance. The low viscosity of thermal grease enables matrix material to fill the interstitials air voids and thin boundary line thickness is also achievable. Filler morphology in also important phenomena in the processing of thermal grease [1]. The thermal conductivity of thermal grease is increasing by addition of filler content; however; it decreases the workability. Furthermore; optimal particle size and shape helps to synthesize highly thermal conductive grease with low viscosity.

2.1.1. Silicone Based Thermal Greases

This type of thermal grease has a tendency to separate from the filler and migrate to other parts, which cause contamination. Recently; CNT filled silicone grease showed low thermal impedance with 35% decrease compared with silicone greases by addition of 2. wt. % of CNTs-COOH [2]. Silicone based thermal grease is also related with "pump out" and dry-out problem [3]. Repeated on/off cycle may cause to dissipate the thermal grease due to pump-out [4, 5]. The reliability of thermal grease is also affected to the separation of the filler with a polymeric material at high temperature, known as dry-out, resulting in increase of thermal resistance [6]. Silicone fluids are volatilizes at high temperature and produce silicone dioxide, which cause defective contacts in motor, relays and switches because of its insulating properties.

2.1.2. Non-Silicone Based Thermal Greases

Non-silicone based compounds are synthesized with the help of binding agents to keep the filler and oil in contact for long time. High surface tension of the fluid keeps the oil and filler together. The use of sodium silicate instead of silicone results in a higher thermal conductive paste [7]. A polyethylene glycol based thermal paste has attractively incorporated as TIMs. The polymer based paste is superior to silicon based thermal grease and quite comparable with sodium-silicate based thermal grease.

A new type of dry to touch thermal grease was reported elsewhere, which easily applied to the component without any non-thermally conductive adhesive materials, which affect the thermal performance [8].

2.2. Thermal Pads

They are solid and stable at room temperature, which makes their handling more convenient and easy. However; it changes its state as the temperature of electronic component reaches its operating temperatures. The PCMs softens at that temperatures and starts to flow like conventional thermal grease, and homogenously acquires the empty voids at the joints. Due to this thermal resistance is minimized the thermal flow is increased at the interface between the heat sink and component. Extensive research has been conducting up to date to enhance the properties of thermal pads. The use of alumina or zinc oxide as a filler in silicon rubber improves its thermal conductivity up to sever extent [9]. Recently; functionalized graphene sheets and silver nanowires are incorporated in silicon rubber. The thermal conductivity is enhanced up to 1.289 $W.m^{-1}.K^{-1}$ by adding 10 phr of AA-RGO and 10-phr mAgNWs [10]. Zinc oxide filled silicon rubber showed better thermal performance in comparison to alumina filled silicon rubber, this is due to higher thermal conductivity of the zinc oxide. Thermal pads require high pressure in order to reduce contact resistance at the interface. Thermal pads should be compressible up to 25%

of their total thickness because of the tolerance variation of large gaps [6]. Increasing the amount of filler improves the thermal conductivity but limited to compressibility and increase thermal resistance. Thermal pad eventually is a better choice in comparison to thermal greases because they avoid the chances of pump-out effects and dry out problems. They also act as vibration damper and protect the components mounted on printed circuit boards (PCBs), but due to their high thermal resistance in the range of 1–3 $K.cm^2.W^{-1}$ limited its application to small chip sets.

2.3. Phase Change Materials

Due to the high heat of fusion; phase change materials (PCMs) are capable of storing and releasing high amount of heat energy by changing from one phase to another during melting and freezing at certain temperature. In general, the viscosity of these materials are reduced due to operating temperature, which allows them to flow, to conform the empty spaces in mating surface like grease. This unique property makes PCMs, as ideal materials for thermal management. These materials release large amount of heat energy during freezing due to crystallization, similarly they observed equal amount of energy for the environment during its melting. PCMs are made of organic binders, thermal conducive fillers and supporting substrate of aluminum or woven glass mesh [11]. Paraffin, polymer or copolymer can be used as a base materials [12]. The melting point of PCMs base materials should be below the operating temperature of the electronic component. In general, these materials have low melting temperature in a range of 50-90°C [11]. Recently; paraffin wax and microcrystalline wax composite is fabricated with low wt. % of α nano alumina for PCMs. The experimental data shows that introduction of alumina particle enhance the thermal conduction of wax filled alumina composite [13]. However; too low temperature of the PCMs may be affected with high environmental temperature.

2.3.1. Low Melting Alloys (LMAs)

Taking in consideration; low melting point of PCMs, phase change thermal interface materials are manufactured with low melting point alloys [14, 15]. These alloys possess liquidus point below the operating temperature of the component allow to flow homogenously throughout the surface asperities. Bismuth and Gallium are commonly used for low melting alloys for thermal interface materials because of their low melting temperature and are environmental friendly [16]. However; Tin, Indium, Cadmium, lead and Mercury can also be used as LMAs. Cadmium, Lead and Mercury is avoided due to their toxicity.

The concept of LMAs was first proposed by Cook and Token [17]. They developed low melting point solder alloy as a thermal interface material. These alloys may be used as a thin strip, or by applying the material to a substrate by a tinning process. Although, LMAs have some drawbacks, excessive intermetallic growth, increases the brittleness of the interface layer and may cause of the reliability of the joints [18]. Due to low viscosity at operating temperature; drip out is the common problem in vertically mounted interfaces[19]. At high operating temperature these metallic interface is corroded and oxidize, which affects their performance. The liquid state of these interfaces during operation facilitate the corrosion by accelerating the diffusion process [19, 20]. Several LMAs has been patent and marketed by this time. Substrate coated low melting alloys are also used to cope these issues up to some extent. But still these fabricated TIMs shows thermal degradation. Webb et al. [16] reported a high performance TIMs, in which copper is used as a substrate, with a 25 micron coating of LMAs at both side, however the thermal cycles test shows degradation after 25K above the melting point. Recently; Chandan et al. perform the evolution of LMAs placed at different surfaces like copper and nickel. Their reliability is checked on isothermal aging at 130°C and thermal cycling from −40 °C to 80 °C. it is observed that these materials shows remarkable reliability between dissimilar joints for up to 3000 hours of aging and approximately 1500 thermal cycles [21].

2.3.2. Smart and Multifunctional Materials

Smart materials provide new directions to thermal management solution because of their unique properties. But their implementation is still inborn stage in the management system. The use of emerging nano-materials; incredibly, improve thermal enhance and application as a thermal interface materials. However; only few reports have been published up to date. Shape memory alloys (SMAs) are also used as a thermal interface between the heat spreader and heat sink in electronic component. The high operating temperature of the working device causes the transformation of these shape memory alloys from low temperature martensitic phase to high temperature austenitic phase. These SMAs are consists of convention thermal grease and applied to interface at operating temperature of the device. These materials subsequently cooled and solidify to form an interface. A nano-NiTiCu alloy based SMAs in thermal grease was patent and reported to enhance thermal contact between the electronic device [22].

2.3.3. Gels

Thermal gels are high viscous pastelike materials similar to thermal grease, which are lightly cross-linked. They are consists of thermal conductive fillers to enhance their thermal conductivity. Silicon is used as base materials because of its high thermal stability, low modulus of elasticity and wetting characteristics. Gels are defined as a substantially dilute cross-linked polymers are similar to grease before cure, with high thermal conductivity 2~5 $W.m^{-1}.K^{-1}$. After curing the elasticity is reduced but however they maintain their shape, cohesive strength and has lateral stability to avoid problem of "pump-out," which is normally associated with liquid TIM materials [6]. Thermal gels are very much comparable with phase change materials. The degradation of thermal gels showed very much similar to PCMs when incorporated in Pentium M process as thermal interface materials [23]. These gels are very much similar to thermal grease with some differences. These gels are recycled but normally have lower thermal conductivity as compare to thermal grease.

2.3.4. Thermal Conductive Adhesive

Thermal conductive adhesives are filled with metallic fillers such as silvers or inorganic fillers like alumina or boron nitride, which are electrically insulating. These adhesive tapes are generally adapted for attaching metal heat sink to a plastic-encapsulated heat-generating electronic component. It eliminates the needs of physical attachments (screws, rivets, fasteners, etc.) due to its adhesiveness to the surface. These adhesives are available in single and double sided tapes. Conductive adhesives have several advantages over solder technology. Easy to fabricate, economical, low stress on the substrate and environmental friendly [24]. Thermal adhesives are free from pump-out and migration, however curing process is required during its application. Furthermore; these materials contains bonding agents that lower its thermal conductivity [11]. Bunyan et al. double sided thermal conductive adhesive tape for electronic components. He claims first pressure sensitive adhesive (PSA) surface which is bondable under pressure [25]. Multi layers anisotropic conductive films (ACFs) are developed for LCDs. The demand of low-temperature curable ACFs are increasing to produce low heat resistive display module and substrate [26].

3. THERMAL CONDUCTIVE FILLERS

One of the most important types of the thermal interface material (TIMs) is polymer filled composites. The traditional thermal conductive polymers are usually prepared by introduction of high thermal conductive filler in polymer matrix. Different types of fillers like metals [27], ceramics [28] and carbon based [29] and hybrid fillers already been introduced to enhance the thermal conductivity of the polymer composite. Typical thermal conductivity values for different filler is listed in Table 1.

Table 1. Thermal conductivity of different materials at room temperature

Materials	Thermal Conductivity @ 25°C ($W.m^{-1}.K^{-1}$)
Graphite	100~400 (on plane)
Carbon black	6~174
Carbon Nanotubes	2000~6000
Diamond	2000
PAN-based Carbon Fiber	8~70 (along the axis)
Pitch-based Carbon Fiber	530~1100 (along the axis)

Table 1. (Continued)

Materials	Thermal Conductivity @ 25°C ($W.m^{-1}.K^{-1}$)
Copper	483
Silver	450
Gold	345
Aluminum	204
Nickel	158
Boron Nitride	250~300
Aluminum nitride	200
Beryllium oxide	260
Aluminum oxide	20~29

Following section discussed in detail about the thermal conductive fillers.

3.1. Metallic Fillers

Metals filled polymers have outstanding potential as filler for polymer composites for a wide variety of applications. Metals have high electrical and thermal conductivity. Metals are widely used to enhance the electrical and thermal properties of the polymers composites. Conductive phase is formed between the polymer and metallic particles that help to enhance the conductivity of the composites. Many reports have been published that shows the metallic particles is commonly used to enhance the thermal conductivity of the polymer composite. The thermal conductivity of epoxy composite is increased up to several extend by simply adding copper and nickel powder [30]. It is observed that the thermal conductivity of copper filled composite is higher than as compare to nickel filled composite i.e.; that is due to morphology of the filler. It is concluded that packing factor is important phenomenon in thermal conductivity and dendritic shape copper particles has high thermal conductivity in compare to spherical alumina [30]. The thermal conductivity of the high density polyethylene (HDPE) filled with aluminum powder is increased up to ~3.8 $W.m^{-1}.K^{-1}$ by adding 33 Vol. % of the aluminum powder in the HDPE matrix [31]. The enhancement in thermal conductivity of the composite is relatively slow up to 12 Vol. % of the filler. The thermal conductivity is increased up to 1.8 times in compare to neat HDPE at 12 Vol. % of the filler. This is due to homogenous dispersion of the fillers which generate effective thermal conductive network. Zinc incorporated high density polyethylene showed significant enhancement in thermal diffusivity and thermal conductivity [32]. High aspect ratio metal fibers also used as a potential filler to improve thermal conductivity of the composites. The thermal properties of polymer filled with aluminum fiber are monotonically increasing with amount of filler. The results shows good agreement with Neilson Model for thermal conductivity [33]. Because of high thermal conductivity metal powder can be used as effective filler to enhance the overall thermal conductivity of polymer, however metal are also good electrical conductors, which limits their applications up to some extent as thermal interface materials. Furthermore the coefficients of thermal expansion of metal are quite different with the die and substrate that cause severe stress damage over a period of time.

3.2. Ceramic Fillers

Ceramics fillers are thermally conductive but electrically insulating as well as cost effective in compare to metallic fillers. Due to ease of synthesis and availability of ceramics filler, they are widely used as conductive filler for thermal interface materials [34-36]. Crystalline silica filled epoxy molding compounds (EMCs) with thermal conductivity of about 1.7 $W.m^{-1}.K^{-1}$, is already marketed. Alumina is also seems good as a fillers and many reports are published with alumina filled polymer composites. At 50 Vol.% of alumina, the thermal conductivity is enhanced to ~1.2 $W.m^{-1}.K^{-1}$; which is approximately six fold as compare to neat epoxy [37]. Aluminum Nitride (AlN) is found to be better thermal conductive filler in compare to crystalline silica. The introduction of 70 Vol. %, of filler in epoxy matrix, the thermal conductivity is improved to 2.2 times in compare to crystalline silica with similar volume fraction [34]. In another report, the addition of 57 Vol. % of AlN to epoxy matrix, the thermal conductivity is enhance to 15 time with the thermal conductivity of neat epoxy [38]. Silicon Nitride is also used as effective thermal conductive fillers. The addition of just 20 vol. % of the Si_3N_4 in UHMWPE/LLDPE blend, the thermal conductivity is enhanced up to several extent [39]. The smaller particle shows comparatively high thermal conductivity with large particles. The introduction of magnetite, barite and talc in polypropylene (PP) matrix, enhanced the thermal conductivity up to certain degree[40]. The thermal conductivity is improved up to nine fold by just adding 30 vol. % of the talc in polypropylene (PP) matrix. The higher thermal conductivity of talc with moderate value of thermal conductivity (λ_{talc}≈10.6 $W.m^{-1}.K^{-1}$) is because of crystal morphology of talc, that established more inter connected networks and firm structure that is more suitable for phonon transportation [40]. Magnetite is also used as thermal conductive filler for polypropylene and polyamide composites, Even the moderate thermal conductivity of magnetite ($\lambda_{magnetite}$ ≈9.7 $W.m^{-1}.K^{-1}$), the thermal conductivity of the composite filled with 40 vol.% of magnetite is improved up to four-folds in compare to neat polymer [41]. Recently; Boron nitride (BN) has caught the attention of many scientist due to its extra-ordinary characteristics, such as high thermal stability, good electrical insulation, excellent resistance to chemicals and intestinally; high coefficient of thermal conductivity [42]. These properties make boron nitride is promising candidate to use as an effective thermal conductive fillers [43]. A very high thermal conductivity of 32.5 $W.m^{-1}.K^{-1}$ boron nitride filled polybenzoxazine is achieved with 88 wt. % of BN loading. Very low viscosity bisphenol-A–methylamine-based polybenzoxazine, which helps in homogenous dispersion of the filler in the matrix as a result the composite are able to form highly thermal conductive networks, with low resistance to phonon transportation, due to heat conductive paths [44]. In another report; the effect of particle size, aspect ratio, surface area and concentration of BN on the thermal conductivity of boron nitride (BN)-filled poly(butylene terephthalate) (PBT) was investigated[45]. It is observed that larger surface area may result in lower thermal conductivity due to scattering of phonons at the interfaces. Boron nitride nano tubes (BNNTs) is also reported as thermal conductive fillers, BNNTs is incorporated in poly (methyl methacrylate), polystyrene, poly (vinyl butyral), or poly (ethylene vinyl alcohol) as the matrix and their thermal properties are measured. It is noted that the thermal conductivity is improved about 20 folds as compared to neat polymer. The thermal conductivity improved up to 3.61 $W.m^{-1}.K^{-1}$ and CTE values down to 16 X 10^{-6} K^{-1} were obtained for BNNT-loaded PS and PEVA composites [46]. Although; many research is still undergoing with ceramic filled polymers for their application as a thermal interface material (TIMs), due to their ease in availability and cheap in cost.

3.3. Carbon Based Fillers

The incredible property of carbon to combine it and other chemical elements is the basis of the organic chemistry. This adaptability of carbon makes it feasible to generate different structural forms. Carbon based materials due to its high thermal conductivity and light in weight are commonly used as an effective filler to improve the thermal properties of the interface materials. Carbon black is well known due to its reinforcing properties and good conductivity and low cost [47]. Due to permanent thermal and electrical conductivity, carbon black is widely used as thermal conductive filler. Carbon black shows good dispersibility in epoxy matrix, the thermal conductivity is linearly improving with addition of carbon black in epoxy matrix [48]. Graphite is an allotrope of carbon and well known due to its conductive properties. It has good thermally conductivity and dispersibility in polymer matrix [49]. The thermal conductivity is linearly increasing with the graphite content in polypropylene matrix, addition of 20 wt. % of graphite, may result ~200% enhancement in thermal conductivity with compare to neat polypropylene. In another report; the thermal conductivity of polystyrene is improved to 1.95 $W.m^{-1}.K^{-1}$, by addition of 34 vol.% of colloidal graphite [50]. Exfoliated graphite showed better thermal conductivity as compared to bulk graphite. The thermal conductivity is significantly enhanced up to 3 fold with addition 7 wt. % of exfoliated graphite (20~100 nm thickness), when incorporated in paraffin wax [51]. Exfoliated graphite is able to develop more interfaces between the matrix and the paraffin wax. Ref [52] suggest increasing the content of expanded graphite from 2% to 10% gradually increases the thermal conductivity of paraffin/EG composite. A correlation factor of (r=0.998) indicating strong interaction between the filler and matrix. Similarly; The EPON epoxy Resin containing 20 wt. % of exfoliated graphite shows 2000% enhancement in thermal conductivity. The thermal conductivity is improved from 0.219 $W.m^{-1}.K^{-1}$ to 4.3 $W.m^{-1}.K^{-1}$[53]. Intercalated graphite nano plates (GNPs) shows excellent thermal enhancement when embedded in epoxy matrix. In elsewhere [54], the improvement of thermal performance is related to the degree of exfoliation of graphite. The thermal conductivity is enhanced up to 6.44 $W.m^{-1}.K^{-1}$, i.e., ~3000% enhancement with compare to neat epoxy. In another report; the thermal conductivity of silicon rubber is improved to 1.909 W.m-1.K-1 by addition of 20. wt. % of the graphite nano platelets (GNPs) into a silicone matrix by mechanical mixing [55]. The mechanical mixing generates the thinner graphene plates as compared to speed mixing as a result, the thermal conductivity of the composites produced by mechanical mixing are higher than the composites produced by speed mixing at large loadings. Thinner graphene results in more surface area; which form more efficient thermal conductive networks with the matrix. Diamond is another allotropic form of carbon which is also used for thermal conductive filler. The micro scale diamond filler also used as thermal conductive filler in polysiloxane matrix. The diamond filler is homogenously dispersed with high speed mixing and the relocated in AC and DC electric fields before the mixture is cross linked. Composite with linearly aligned diamond fillers exhibits enhanced thermal conductivity. The enhancement of 155% in thermal conductivity with addition of 10 Vol.% of diamond in switching electric field was recorded [56]. Epoxy filled diamond showed four folds enhancement in thermal conductivity by addition of 60 wt. % of the filler in epoxy. At high filler loading the filler can easily cohere and formed heat conduction paths, thereby raising overall thermal conductivity of the composite [57].

Another important form of carbon-based filler is carbon fiber. Commercial carbon fiber has thermal conductivities similar to many organic and inorganic fillers. With the low density about 2260 kg/m^3 in compare to other organic fillers, carbon fiber is the important choice as a filler for thermal conductivity enhancement [58]. Agari et al; investigated thermal conductivity of randomly dispersed and disoriented carbon fiber filled with polyethylene; with different aspect ratios, it is observed that the thermal conductive chain formation is easier with long fiber, which enhanced the thermal conductivity of the composite in compare to short fibers [59]. In another report; the carbon fiber is embedded in paraffin matrix for PCM to enhance the thermal conductivity. PCM with an embedded light weight carbon fiber heat sink serve as an effective thermal management technique for high power transient applications [60]. Vapor grown carbon nanofibre (VGCNF) filled epoxy composites were synthesized by three-roll milling (RM) showed a 10 fold increment in thermal conductivity of composites at 40 wt.% VGCNFs i.e, 1.845 $W.m^{-1}.K^{-1}$, compared to Rubbery epoxy (RE) alone. This is attributed due to the intense shearing mixing of filler with matrix under shear stress of rolling mills, that significantly improves the dispersion of the fibers in the matrix [61].

Fundamental studies at nano scale showed that due to one dimensional (1D) structures carbon nano tubes (CNTs) have high thermal conductivity, large aspect ratio and good heat transport properties. Carbon nano tubes (CNTs) have an extremely high thermal conductivity at their axis direction[62]. Some reports predict the thermal conductivity value of CNTs is 1400 and 2400 $W.m^{-1}.K^{-1}$ for single-walled carbon nanotube (SWCNT) and multi-walled carbon nanotube (MWNT) respectively[63]. These prominent thermal conductive properties make the carbon nano tubes suitable candidate for thermal interface applications, with an effective implement to dissipate the heat from the electronic component to heat sink during operation [11]. Moisala et al. with 0.05 wt.% loading of MWCNT in the epoxy lead to increase in the thermal conductivity, however, the thermal conductivity enhancement is far below from the theoretical values [64]. Interestingly, in the same report the addition of SWCNT lead to the decrement in the thermal conductivity in a range of 0–0.5 wt. % of SWCNT. Gojny et al. the mechanism of thermal transport in CNT is related with thermal conduction mechanism. He found that phonons conduction is affected by several factors such as phonon active modes, the boundary surface scattering, the distance of free path for the electrons and phonons and inelastic Umklapp-scattering [65]. Therefore; the description of phonon conduction mechanism is non-trival [66]. The increment in thermal and electrical conductivity of the composite is supposed to phonon conduction due to percolated network of CNTs. The increment in conductivity is observed when the critical amount of filler form percolated network. The electrical conductivity increases with CNT content, however, the thermal conductivity remains unaffected. Therefore; correlation based on percolation mechanism and the thermal conductivity is not directly observed. CNTs orientation in the matrix has also significant effect, aligned CNTs have high phonon active modes and low boundary scattering. A great enhancement in thermal conductivity with low thermal measured for the CNT/Polymer composite. It is observed that by the only small fraction of aligned CNTs (0.3 %), the thermal conductivity of the polymer composite is improved to 0.65 $W.m^{-1}.K^{-1}$. This is fairly than loaded with dispersed CNTs. Only 4.6% of enhancement in thermal conductivity was recorded by adding the same amount of dispersed CNTs [67]. The low efficiency is due to random dispersion of CNTs, which increases the thermal resistance and heat conduction in only through the few portion of CNTs. In addition; phonons cannot transport from one side to another side through CNTs, as a result of the discontinuation of other lower thermal conductive filler of base fluid, which could affect the thermal performance of the CNT composites. The outcomes of CNTs based TIMs were argumentative. In few reports, the thermal conductivity has no effect on loading of CNTs or even decrement in the thermal conductivity values has been recorded [64, 68]. This is due to high thermal boundary between CNTs and polymer matrix i.e., ~ 10−7 m^2 K/W [69]. Even with high intrinsic thermal conductivity of CNTs, they do not interface with the matrix. These results suggest strong motivation towards graphene. With high intrinsic thermal conductivity graphene is most widely used filler to enhance the thermal conductivity of the polymer composite.

Small addition of the graphene enhances the thermal conductivity up to several extent. Yavari et al. [70] the thermal conductivity of graphene/1-octadecanol (stearyl alcohol) nanocomposite is increased by ~2.5 fold by introducing 4 wt.% of the graphene in 1-octadecanol matrix. The result distinctly support that addition of graphene markedly improve the thermal properties in compare to the effect of fillers such as silver nanowires and carbon nanotubes. This improvement in thermal conductivity is due to inter connective networks of graphene which lower resistance to phonons transportation.

Table 2. Chemical Classification of fillers used for thermal conductivity enhancement of polymer composites

Chemical Classifications	Examples	References
Organic Fillers		
Oxides	Alumina, Magnesia, Magnetite, Silica, Zinc Oxide	[9, 13, 36, 40, 41]
Nitrides	Aluminum Nitride, Boron Nitride, Silicon Nitride	[34, 39, 43-45]
Carbides	Silicon Carbide,	[114]
Silicates	Talc, Zeolite, Wollastonite	[114, 115]
Sulfates	Barite	[40]
Metals	Aluminum, Copper, Nickel, Zinc, Silver	[10, 21, 30, 31]
Inorganic Fillers		
Carbon	Carbon black, Carbon fibers, Carbon nano tubes	[48, 58-60]
Graphite & Diamonds	Graphite fiber, Graphene, Diamond	[49-51, 56, 57, 70, 71]

The large contact area and high aspect ratio of graphene helps thermal energy transport in graphene/PCM composite. In elsewhere; extra ordinary 2300% enhancement in thermal conductivity of has observed by addition of 10 vol. % loading of graphene in compare to neat epoxy. The thermal conductivity is improved to 5.1 $W.m^{-1}.K^{-1}$. This is quite comparable with the thermal conductivity of conventional TIMs. Multi-layer graphene (MLG) shows exceptionally high thermal conductivity when vertically aligned in epoxy composite. The exceptionally high in-plane thermal conductivity of 33.54 $W.m^{-1}.K^{-1}$ has been achieved in the aligned MLG/epoxy composite (AG/E. Efficient alignment of MLG enables ultrahigh thermal conductivity in the parallel direction. The densely packed MLG in the epoxy matrix enables resistance free transportation of phonons and inter-connective networks with epoxy matrix. Various reported fillers used to enhance the thermal conductivity of polymer composites are listed in Table 2.

3.4. Hybrid Fillers

Hybridization of two or more traditional fillers showed improved composite performance. The use of hybrid filler for polymer composite has been rapidly increases in last decades due to increasing in demand of high performance engineering materials; especially where low amount of filler is required to achieve desirable properties. The integration of different fillers into the single matrix, that leads the development of hybrid composites [72]. Commercially; available thermal grease and adhesive used several fillers to enhance their thermal and physical properties is already well documented [73]. Hybridization improves connectivity in hybrid filler with high aspect ratio, which would be effective in the formation of thermally conducting paths. Secondly, hybridization improves the packing density of the fillers in the matrix, which minimizes the phonon scattering between the filler and the matrix. The use of hybrid filler to enhance the thermal conductivity of the polymer composite is already reported in many reports, one can easily distinguishthese fillers on the nature of hybrid fillers. We found that there are three types of hybrid fillers are usually reported to enhance the thermal conductivity of the polymer composites i.e.,

- Non-Carbon-based hybrid fillers
- Carbon-based hybrid fillers
- Inorganic-organic hybrid fillers

3.4.1. Non-Carbon Based Hybrid Fillers

The main advantage of this hybridization is that the composite has high electrical resistivity and the packing density of these fillers is comparatively high in comparision to organic hybrid fillers. It is observed that usually high amount to inorganic filler is required to achieve the moderate thermal conductivity. Even highly thermally conductive filler is unable to enhance the thermal conductivity up to that extent. The main cause is thermal losses due to nano voids created during the synthesis of composites or in between the two particles. Based on the dispersion of different particles sizes of Boron nitride, a thermal conductive polyimide composite has been synthesized. The composite showed enhancement of ~6.5 folds in thermal conductivity with the addition of 30 wt. % of the mixture of micro and nano Boron nitride at the weight ratio of 7:3, when incorporated in polyimide matrix [74]. The nano filler occupied the empty void that found between to similar micro size Boron Nitride particles, which sufficiently enhance the phonon transportation across the matrix and overall thermal conductivity is improved. In another report; the mixture of small-sized aluminum oxide particles are mixed with large size aluminum nitride particles in a ratio of 3:7. The incorporation of this hybrid filler is found to be effective to enhance the thermal conductivity of the epoxy composite. The thermal conductivity is increased to 3.402 $W.m^{-1}.K^{-1}$, which is ~17 fold increment in compare to neat epoxy by addition of 58.4 vol. % of the filler [75]. The effect of different size and relative composition of on thermal conductivity of hybrid aluminum nitride and boron nitride is investigated. It is observed that in bimodal distribution of different shape morphology of particles helps to developed more conductive thermal networks. The packing factor and interfacial resistance is directly related to particle distribution. It is found that the thermal conductivity are significantly improved to 8.0 $W.m^{-1}.K^{-1}$, when the volume ratio of AlN:BN particles 1:1 and 80 vol. % of the filler is introduced in epoxy matrix [76]. Hybridization of sisal with glass fiber was also investigated. The thermal conductivity of sisal-reinforced polyethylene was noted to improve up to 0.45 $W.m^{-1}.K^{-1}$. This is due to the reduction in the mean free path of phonons that established due to sisal with glass fiber hybridization [77]. The use of hybrid filler is usually attempted to achieve high thermal conductivity. There is no authentic principle that determines the optimum ratio of the fillers in hybridization, however; for filler system, the maximum packing factor leads to established highly thermal conductive paths. The determination of maximum packing factor is rationally important to improve the thermal conductivity of the composite. Although; very high thermal conductivity is achieved with hybridization of inorganic filler in polymer matrix, but on the downside; they require high proportion of filler to achieve these values, which leads to poor mechanical properties of composite and certainly increases the problem during their synthesis.

3.4.2. Carbon Based Hybrid Fillers

The use of carbon materials shows extraordinary enhancement in thermal conductivity of polymer composites with low volume fraction of the filler. But still, some problem is associated with this filler. For instance; Carbon nano tubes (CNTs) are promising materials, but in many reports it is found out that the thermal conductivity of CNTs based polymer composites fails to reach the theoretical predictions. It is noted that CNTs induce a drastic reduction of thermal transport due to their defect scattering and interfacial boundaries [78]. Graphite and graphene have also shown good thermal characteristics due to their 2D nanostructures. They have shown even higher thermal conductivity enhancement than CNTs [79]. The use of different fillers comprised of carbon materials for hybridization has been explored and composite performance has been improved by combining the advantages of each type of filler. The hybrid filler comprised of two or more conventional filler materials has already been explored and well documented. By utilizing the advantages of both fillers, improved composite performance can be achieved. The thermal conductivity of graphene filled epoxy is ~ 23% enhanced by addition of just addition of 0.36 wt. % of MWCNT. This is due to the synergistic effect between GNPs and MWCNTs [80]. The enhancement in thermal conductivity by hybridization is quite obvious, in another report; the one dimensional single wall carbon nano tubes (1D-SWCNTs) are bridging adjacent with two dimensional Graphene nano plates (2D-GNP), which provide additional channels for the heat flow bypassing the polymer matrix. The thermal conductivity is improved to

~17 fold by the introduction of 30 wt. % of the hybrid GNP: SWCNT filler in 3:1 ratio. The hybridization activates heat transportation channels between randomly oriented GNPs and SWCNTs in the 3D polymer matrix and thermal enhancement is isotropic [73]. This improvement in thermal conductivity is due to formation of more efficient percolating networks, which significantly reduce the thermal interface and enhance the thermal conductivity of the composite. The through plane thermal conductivity is also recorded to increase up to 2.1 $W.m^{-1}.K^{-1}$, by CNTs directly grown on graphene nano plates [81]. The improvement in thermal conduction is due to direct hybridization of the filler, which cannot be achieved by simply mixing for this filler due to differences in their densities. However; the enhancement in thermal conductivity is far from the theoretical values limited by the interfacial thermal resistance between the matrix and fillers that is originated from the scattering of phonon due to interfacial resistance [82]. Secondly; the electrical insulation is lost due to interdiction carbon based fillers which form electrical conduction networks within the matrix [83, 84].

3.4.3. Organic-Inorganic Hybrid Fillers

The collegial effect of thermally conductive fillers to achieve hybridization of organic and inorganic fillers has been used to improve the thermal conductivity with lower the filler content while maintaining electrical insulation. The synergism created from different phenomena such as bridging, improved packing density and better interfacial contact between the fillers. Hybridization is effective for formation of conductive networks that facilitate phonon transportation result in improving the thermal conductivity of the composite. Choi et al. found a synergistic effect of combining MWCNTs and Aluminum Nitride (AlN)[85]. Thermal conductivity is clearly improved by addition of combined filler. As mentioned in another report, due to the excellent thermal conductivity of Aluminum Nitride, AlN shows enhancement in thermal conductivity of epoxy composites when hybridized with alumina in presence of CNT and graphene. However; the hybrid AlN/alumina with graphene showed high thermal conductivity in compare to AlN/alumina with CNT. The following factors might be reason of better thermal conductivity with graphene. The layer-structure of graphene are more favorable for thermal transport [86] and secondly; graphene rigidity allowed for better preservation of their high aspect ratio in comparison with the more flexible carbon nanotubes [73]. The hybrid MWCNTs/SiC 24.3 time's improvement in thermal conductivity in compare to neat epoxy as obtained with 5. wt% MWCNTs + 55 wt% micro-SiC [87]. MWCNTs with high aspect ratio effectively act as thermal conductive bridges among 3D micro-SiC particles. This facilitates the formation of the efficient 3D percolating network, resulting in the better thermal transportation of phonons. The hybrid filler of MWCNTs and Boron nitride also show good enhancement in thermal conductivity, adding small quantity of high aspect ratio or MWCNTs helps in enhancement of thermal conductivity. Due to 3D network of hybrid filler system heat conduction improved. Adding both 25 vol% BN and with small quantity (1 vol%)

MWCNTs together, thermal conductivity increased to 699.74% [88]. Similar phenomena are observed when epoxy composite containing MWNTs/Alumina hybrid filler. Due to high thermal conductivity and high aspect ratio of MWNTs, the MWCNTs create percolating network and high thermal conductive path leading to a significant increase in thermal conductivity of the composites. The difference between individual particles is clearly notable [89]. High aspect ratio of CNTs has great influence to create percolating networks even with small proportion while comprising with hybrid filler. One can clearly distinguish with low aspect ratio there is no improvement in thermal conductivity with the introduction of these high thermal conductive materials. The addition of carbon fiber with boron nitride in polybutylene terephthalate (PBT) matrix, did not show any kind of improvement in thermal conductivity [90]. Recently; alumina-coated graphene sheets showed significant enhancement in thermal conductivity which maintaining electrical resistivity. The hybrid structure of graphene with alumina while incorporated in polyvinyl diene fluoride (PVDF) showed ~2.4 time's higher thermal conductivity in respect to neat PVDF with 40. wt% of filler loading. The presence of 2D graphene sheets effectively enhances the interconnectivity between the nano alumina and overall high thermal conductivity of graphene facilitates the more phonon conduction within the matrix. The use of hybrid filler certainly improve and enhance the thermal conductivity and thermal properties of the polymer composites, however; still thermal conductivities values are far beyond with their theoretical predictions.

3.4.4. Surface Modification of Fillers

Polymer based composite for thermal interface materials (TIMs) has already been reported and showed improved thermal characteristics, however; even today we are far beyond with theoretical predictions and we stand with similar threshold in the realm of polymer nanocomposites high strength, reliable, and multifunctional materials with low volume fraction of filler content. One of the prime reasons of this cause is poor interfacial bonding between the filler and matrix that lead to non-homogenous dispersion of fillers. These properties can be further enhanced by modifying fillers, optimizing the dispersion of the filler and interface chemistry. Strong interfacial bonding between filler and matrix certainly improve the composite properties, which might be achieved through the introduction of chemical functionalities on the surface of the fillers. Even with low fraction of filler, with an optimal dispersion of nano particles has significant effect on the behavior of the polymers, which formed a continuous network of altered polymer chains and changing the thermal and mechanical properties of the composite [91-93]. Functionalization of filler can be achieved through chemical and physical bonding of chemical functionalities on the surface of the filler, here we review the covalent and non-covalent functionalization of fillers for enhancement in thermal properties of the composites.

3.5. Covalent Functionalization

The functionalization of fillers with organic functional groups has been used for different purposes. The main purpose is dispensability of filler in polymer matrix and strong interfacial contact between filler and matrix. When organic molecules are covalently attached on the filler surface, the offer new properties combine with the properties of the filler. Covalent functionalization reactions of filler include two general routes: (a) the formation of covalent bonds between free radicals or dienophiles and free radicals of the fillers (b) the formation of covalent bonds between organic functional groups and the oxygen groups that presence on the fillers. Grafting of silane molecules on the graphite flakes improved the dispersion of graphite in epoxy. Functionalization with silane reduces the thermal resistance by developing interconnected networks between epoxy and graphite platelets, as a result the acoustic mismatch of phonons are reduced and interfacial heat transfer between graphite and epoxy is improved. Silane treated graphite showed enhancement in thermal conductivity up to ~28 fold by addition of 20% of the filler [94]. Yuen et al. also found, the surface modification of MWCNTs with silane is effective to improve the thermal conductivity of epoxy composite. The interface between MWCNTs and epoxy is increased with surface modification of MWCNTs. The thermal conductivity is improved to 87% by addition of 0.99 wt. % MWCNT, which is quite better with untreated MWCNTs[95]. Amine modification is also useful to improve the interfacial contact between filler and the matrix. Dodecylamine (DDA) functionalized expanded graphite nano plates (EGNP) shows considerable improvement in thermal conductivity. Even small fraction of just 2 wt. % of ENGP loading, improved the thermal conductivity ~3.5 fold as compared to neat epoxy. In another report; Triethylenetetramine (TETA) grafted CNTs shows the improved interfacial interaction between MWCNT–epoxy. The acid treatment has adverse effect on CNTs aspect ratio, but however, the attachment of TETA on CNTs promotes the phonon transport due to interfacial connection between the epoxy matrix and the MWCNTs, and also help in homogenous dispersion of MWCNTs in the matrix, as a result thermal conductivity is improved to 1.3 $W.m^{-1}.K^{-1}$ with 10 vol.% of nanotubes [96].

The use of severe reaction condition damages the filler structure. Strong acid modification of CNTs damages the structures of CNT side walls and even shorten the CNTs is well documented. The effective chemical modification with less destructive in filler structure is more important. Recently; our group; modified the shear exfoliated graphene with O-phenylenediamine, the improvement in thermal conductivity in through and in-plane is significant at a low fraction of the filler [97]. The in-plane and through-plane thermal conductivity values of the composite filled with 6 Wt. % of OPD-*f*-Graphene were 2.671 and 0.976 $W.m^{-1}.K^{-1}$ respectively, the in–plane and through plane thermal conductivities were improved by an enhancement factor of ~12.35 and ~3.88 respectively as shown in Figure 6.

The prime reason for such significant enhancement in thermal conductivity is based on (a) Shear exfoliated graphene and mild oxidation helps to maintain the pristine properties of graphene [98] which possibly alters with strong oxidizing agents during graphene synthesis shown in Figure 7 the amine modified graphene help to develop better interface through interconnection of graphene with epoxy Figure 8.

The amino group in functionalized graphene reacts with epoxy matrix which developed firm interconnective networks between graphene and epoxy as shown in Figure 9. The direct attachment of carboxylic group on CNTs is achieved via Friedel–Crafts reaction. The functionalized MWCNTs were treated with benzene tricarboxylic acids (BTC) which create chemical bonding between the MWCNTs and the epoxy. The thermal conductivity is increased by increased ~684%, by adding 5 vol.% BTC-MWCNTs in epoxy matrix [99]. Covalent functionalization helps carbon based filler in two ways (a) generates strong linkage between the filler and matrix and (b) prevent the filler from agglomeration, which is more obvious due to high surface area of the carbon based filler. Covalent functionalization of inorganic filler is also beneficial as they create strong interface between rigid filler and weak polymer matrix. Even the organic molecules are less thermal conductive but there are many reports that proved that covalent functionalization of inorganic filled helps to enhance the thermal conductivity because of better linkage and firm network, which helps phonon transportation within the composite. Silane modified boron nitride showed considerable improvement in thermal conductivity in compare to unmodified BN. The thermal conductivity is increased to ~ 6 times and reached to 1.178 $W.m^{-1}.K^{-1}$ with loading of 30.wt. % modified BN particles [100]. The platelet like morphology of boron nitride with smooth basal plane has poor dispensability in epoxy. The poor filler matrix interaction and non-homogenous dispersion lead to create agglomeration and voids, which cause phonon scattering and increase thermal resistance. However; after modification of BN particles interfacial affinity is increased between filler and the matrix. The coupling agents serve as a connecting bridge between the BN and react to curing matrix. This interconnection improves phonon transportation, which results in improving the thermal conductivity of the composite. Functionalization of alumina also shows improvement of about 29% in thermal conductivity when modified with silane [101]. of Covalent functionalization is well effective for hybridization of fillers, alumina-coated graphene sheet (GS@Al_2O_3) hybrid fillers show high thermal conductivity of 0.586 $W.m^{-1}.K^{-1}$ with 40 wt. % GS@ Al_2O_3 (mass ratio of GS to Al_2O_3 of 1:20). The functionalization helps to couple graphene with alumina as well improve the compatibility of the hybrid filler with the matrix material [102]. Covalent functional prove to develop a good interface between filler and matrix that develop strong interconnective networks with low thermal resistance heat paths.

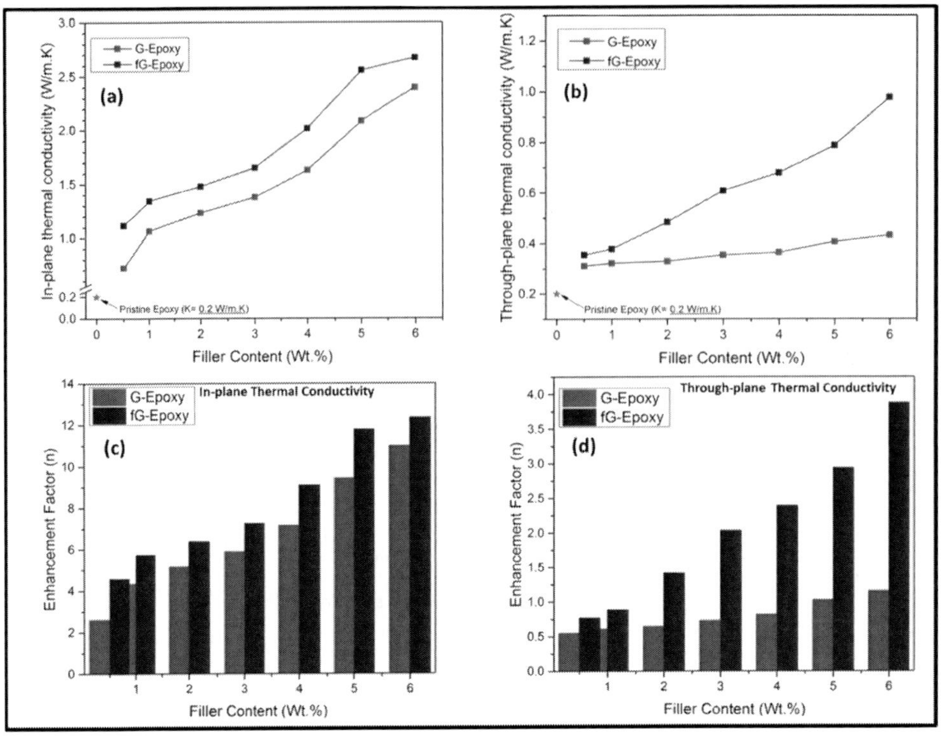

Figure 6. The effect of surface modification on thermal conductivity of TIMs; (a) in-plane and (b) through-plane; enhancement in thermal conductivity as a function of graphene content (c) in-plane and (d) through-plane. Reprinted with permission from [Akhtar M., 2016], *RSC Advances* 2016; 6©2016, Royal Society of Chemistry.

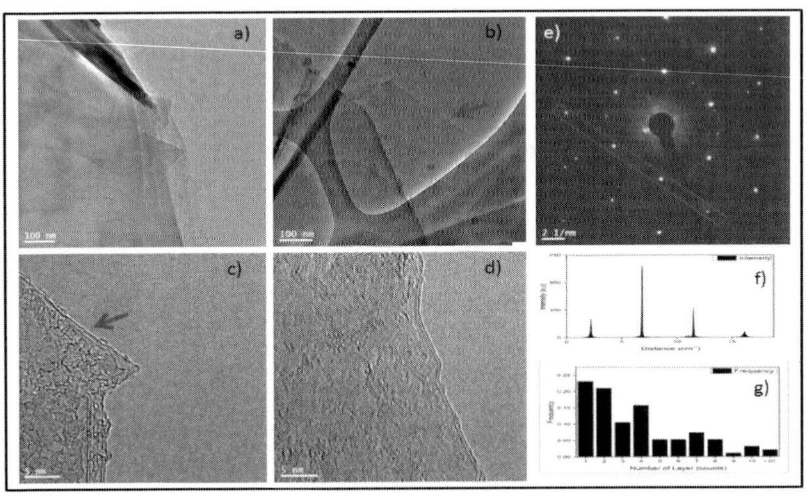

Figure 7. HR-TEM image of exfoliated defect free graphene (a–b) Bright-field Transmission electron microscope (TEM) images of graphene flakes (c–d) A HRTEM image of a section of a graphene: monolayer at an edge of (a), bilayers at an edge of (b). (e) The electron diffraction pattern corresponding to the arrow indicated regions in (c). (f) Intensity distribution of spots in (e) marked by red box, (g) histograms of the number of layers per flake of exfoliated graphene. Reprinted with permission from [Akhtar M., 2015], *Journal of Nanoscience and Nanotechnology* 2015; 15 ©2015, American Scientific publisher.

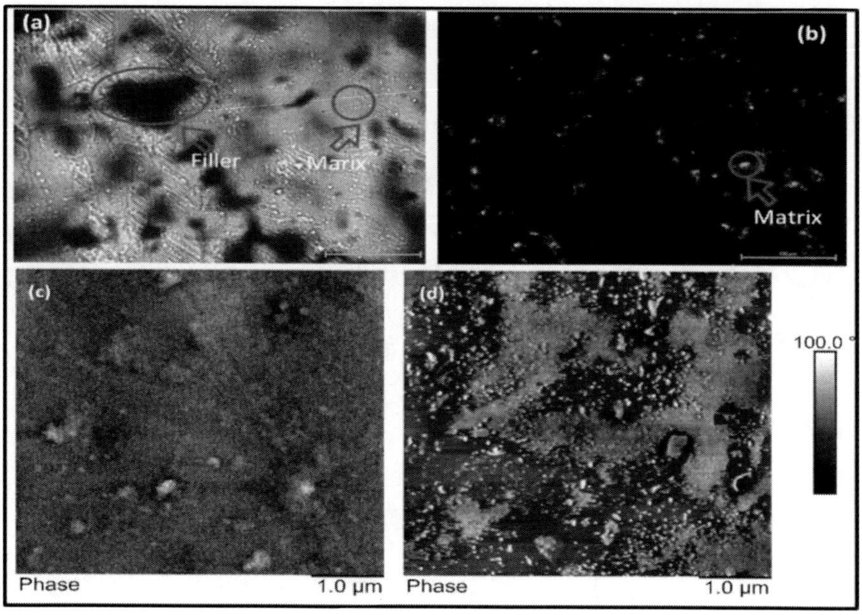

Figure 8. Improved dispersion of filler after functionalization with OPD; (a & b) Optical images at 20X of G–epoxy and fG–epoxy; EFM images of G–epoxy (c) and fG–epoxy (d). Reprinted with permission from [Akhtar M., 2016], *RSC Advances* 2016; 6©2016, Royal Society of chemistry.

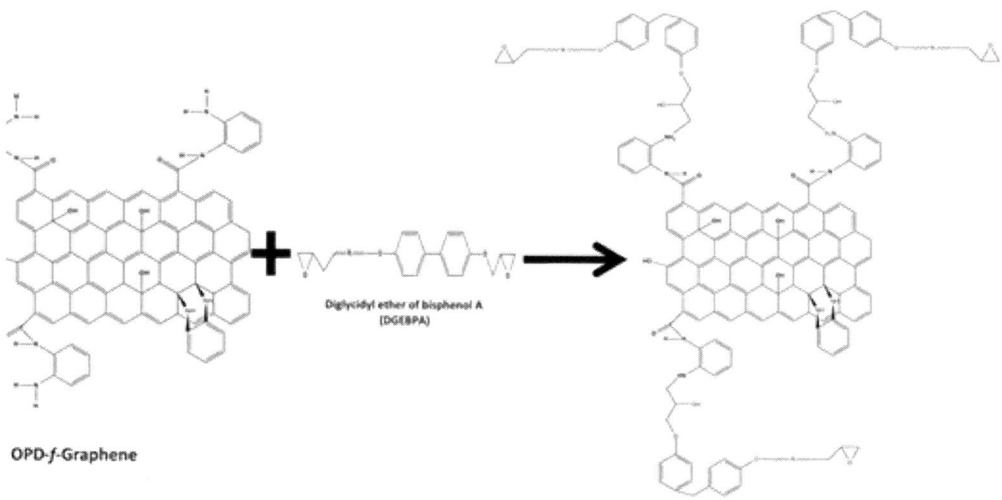

Figure 9. Interface mechanism of the OPD-f–epoxy nanocomposite Reprinted with permission from [Akhtar M., 2016], *RSC Advances* 2016; 6©2016, Royal Society of chemistry.

3.6. Non-Covalent Functionalization

Noncovalent functionalization by π-interactions is an alternate method to attach functional groups with fillers. The major advantage of this method is pristine properties of filler is maintain because functional groups are physically attached without distrusting the chemical structure of the filler [103-106]. Recently; non covalent functionalization of graphene is achieved through 1-pyrenebutyric acid. The non-oxidized, low defect graphene use as a filler in polymer that show improved thermal conductivity. The functionalized graphene composite at 10 wt% loading in epoxy has thermal conductivity value of 1.53 $W.m^{-1}.K^{-1}$. This is far better in compare to previous reported work. This improvement is attributed to high quality graphene structure and homogenous dispersion that achieved through non covalent functionalization [107]. In similar work graphene sheets are non-covalently functionalized with pyrene molecule with a functional segmented poly (glycidyl methacrylate) (Py-PGMA). The physical absorption such as π–π stacking help to attach Py-PGMA on the graphene surface. Py-PGMA–GNS/epoxy composites show extra ordinary enhancement in thermal conductivity, the thermal conductivity is improved to 1.91 $W.m^{-1}.K^{-1}$, which generally achieved with high loading of the graphite filler [29].

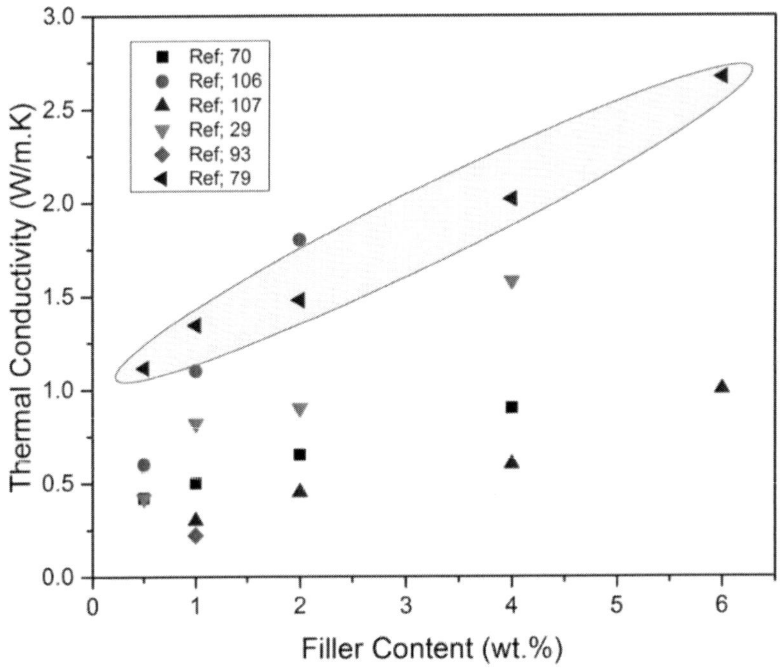

Figure 10. Comparison of improvement in thermal conductivity with other reported literature. Reprinted with permission from [Akhtar M., 2016], RSC Advances 2016; 6©2016, Royal Society of chemistry.

Wang et al, proved by molecular dynamics simulation, that non covalent functionalization is beneficial to improving the thermal transportation of phonon. He investigates by MD simulations that the attachment of non-covalent molecules on the surface of graphene develop better dispersion and interfacial contact between the filler and the matrix, which results in a reduced interfacial thermal resistance [108]. This improvement in thermal conductivity due to non-Covalent functionalization is based on the several reasons (a) physical attachment of function group with pristine structure of filler, maintain filler integrity, which preserved all basic properties of filler which normally alters during the attachment of functional group by strong chemical agents; (b) The attachment of functional moieties on the fillers supports better dispersion and reduce agglomeration, as a result homogenous dispersion of the filler in the matrix is possible, which poses a large contact area with polymer, promoting phonon diffusion with minimum resistance; (c) these functional group on the surface of filler; react with polymer moieties. This form interconnected and cross-linked structure between filler and matrix. The strong interaction between particle and matrix, reduce thermal interfacial resistance, and minimize the phonon scattering and usually occur between a rigid filler and weak polymer interface, as a result; significant improvement in thermal conductivity is recorded. Improvement in thermal conductivity after surface modification of filler is presented in Figure 10.

4. Thermal Conductivity Phenomenon in Polymer Composites

Heat can be transfer in there ways conduction, convention and radiation [109]. Thermal conduction is the main source of heat transfer in the solids. The heat is transfer in solids through vibrational energy from particle to particle without the motion of mater, due to the collision. The mechanism of heat transfer in solids varies from matter to matter, for instance; metals are good conductors of heats whereas polymers are a heat insulator. Crystalline polymers has better thermal conductivity in comparison to amorphous polymers similarly like others crystalline materials shows highest thermal conductivity such as graphene, CNT, diamonds, silicon and glass. Thermal conduction in polymers is usually occured when the heat reaches to the surface monomer of the polymers, the heat eventually transmitted to next monomer and this happen until it reaches to last monomer of the polymer. Propagation of heat in the polymer is not the same as in crystalline solids. The heat energy will diffuse slowly in the polymers chains, causing disorder vibration and rotates in all around the polymer matrix in the equilibrium position and scattered to adjacent chains.

Ordered structure of crystalline solids diffuse the initial vibration due to heat energy to adjacent faces more rapidly, however; the initials kinetic energy is usually spread in the disordered structure. Polymers possess low thermal conductivity due to an unideal morphology of polymer. Polymers are not well organized as crystals, phonon easily scattera due to acoustic mismatch and disordered structure, unable to transfer the heat quickly to the other side of the polymer. Furthermore; another factor such as different particle size, anharmonicity and low binding energy will cause polymer as heat insulators.

In order to maximize the thermal conductivity, the phonon scattering processes must be minimized. Phono travels at a speed of sound. The scattering of the phonon in composite materials is mainly due to interfacial thermal barrier resulting from acoustic mismatch and flaws associated with filler and matrix interface. For a given filler composition there are methods of increasing the thermal conductivity of the composites [110].

i. Forming conducting networks, through the appropriate packing of the filler in the matrix.
ii. Decreasing the amount of thermally resistant junctions involving a polymer layer between the filler unit by using the large filler unit with little or no defects.
iii. Decreasing the thermal contact resistance at the filler matrix interface by minimizing the interfacial flaws.

5. Environmental Prospective

The use of electronic component at elevated temperature often causes problem and limit its application. The use of electronic control unit (ECU) is expected to increase exponentially in the near future. The use of ECU in automotive and other engines should be green and free from the pollutant that cause by using of conventional fuels. The introduction of European Union legislations (Euro 5 and 6 standards), put more restriction on emission of pollutants from vehicles further increased the demand of electronic control vehicles and ECU [111]. The legislation; Euro 6 standard (already enforced from September 2014) stringently restrict the emission of hydrocarbon, nitrogen oxides from the diesel engines. The new legislation of European Union is on alternative fuels to minimize dependence on oil and to mitigate the environmental impact of transport. It is evident the use of electronic vehicles is prominent [112] These levels of restriction from European Union on gas emission and use of alternative of fossil fuels; substantially replace the conventional carburetor with fuel injection [113].Fuel injection is based on electron method to meter fuel to engine more efficiently, that why named as Electronic Fuel Injection (EFI). The ECU controls the EFI, which monitors mixture formation, combustion and exhaust gas treatment. Besides that ECU also look after ignition system, air compression, air bag, engine brake and alternator. The use of environmental friendly power electronic is critical demand of the day. The high proportion of electronic devices in automobiles and others system requires properly heat management, otherwise life expectancy and reliability of these units will suffer. Typical heat dissipation in automotive electronic systems is shown in Figure 11.

The performance of the electronic device will alter as the temperature increases in the surrounding, that the removal of dissipated heat is important for electronic system and reliability. This can be possibly achieved through Thermal interface materials (TIMs). The thermal operating environment for various electronic systems is listed in Table 3.

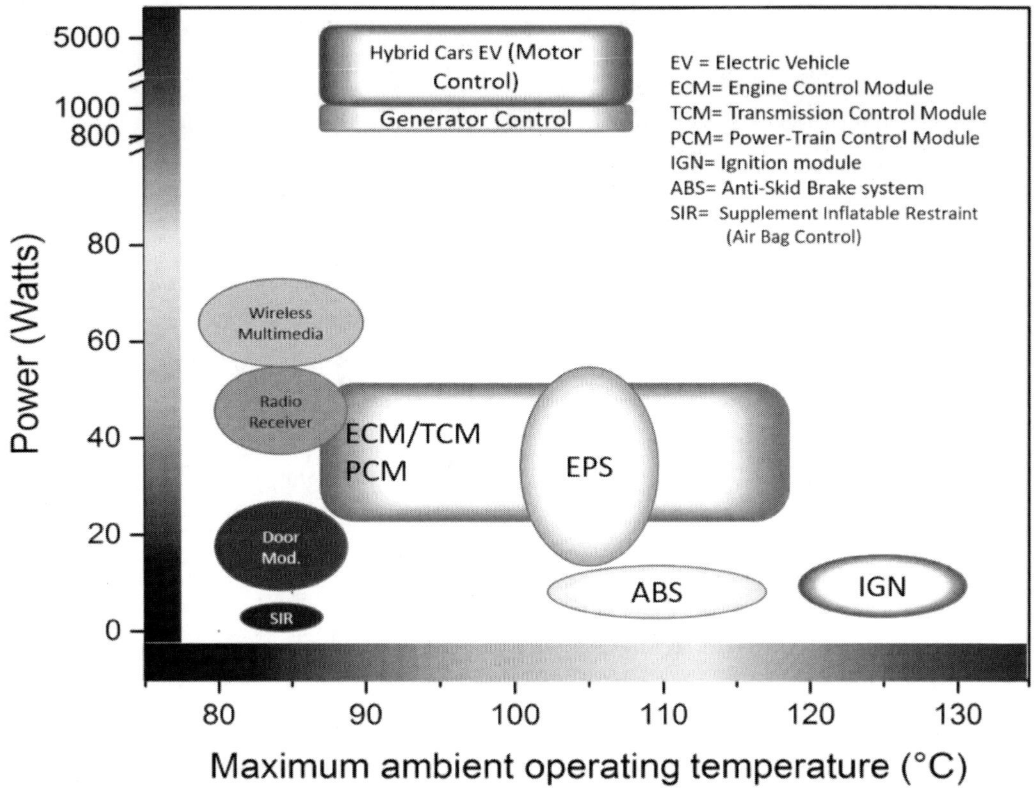

Figure 11. Thermal power dissipation in electronic system.

Table 3. Thermal operating environment for various electronic systems

Electronics	Operating Temperature (°C)
Consumers	0 ~ 70
Industry	-40 ~ +85
Automotive	-40 ~ +125
Military	-55 ~ +125

In addition; conventional thermal grease for TIMs has also bad impact on the environment due to pump-out. Thermal grease is messy and in a manufacturing environment can pollute thermal resistance assemblies, which at the elevated temperature they vaporized and contaminated the surrounding and indirectly affect the environment. Polymer filled high thermal conductive particles is a better choice, but the use of Cadmium, lead and mercury based alloys are generally avoided because of the environmental issues related to them. The use of organic and inorganic based filled and their hybridization for thermal effective filler is crucial and should be good hands in the development of green TIMs for environmental friendly electronic control units.

Conclusion

Thermal interface materials (TIMs) still requires a sufficient amount of research to meets the challenges of next-generation materials. As the demands of electronic devices are escalating, the problem associated with heat dissipation is proportionally increasing. This chapter reviews on the basic classification of polymer composites that could be possibly used for thermal interface materials (TIMs) and specifically thermal conductive fillers that plays a major role in thermal conductivity enhancement. To optimize the performance of TIMs, it is more than important to find out the right direction between the reliability and performance of these materials. This chapter emphasizes that even after extensive research has been made to improve the TIMs performance conventional TIMs are unable to meet the demand of modern electronic packaging. This chapter also presents a different classification of thermally conductive fillers that might be useful to improve thermal performance of the polymer composite. The enhancement in the thermal conductivity by incorporation of thermal conductive fillers is still far beyond from the theoretical predictions. It is concluded; better interface between the filler and the matrix and highly packed structure will improve the phonon conduction and escalate the thermal performance of TIMs. The use of hybrid filler shows the synergistic effect and could be possible future candidate as thermal conductive fillers. From the comprehensive, literature surveys following future recommendations are suggested.

- Homogenous dispersion of filler and highly compact structure is beneficial in thermal conductivity enhancement of TIMs, which can potentially be achieved through surface modification of fillers.
- Surface treatment improves interfacial interaction between filler and matrix, however attachment of functional moieties on the filler surface lead to decrease in their physical and chemical properties; during functionalization, one might consider minimal alteration in filler properties, which possibly attained with mild chemical treatments such as non-covalent attachments.
- Hybridization of the fillers improves thermal performance due to a synergistic effect. Further modification of these hybrid fillers via functionalization to developed strong interface between filler and matrix certainly enhance thermal properties of the polymer composite.
- The junctions between hybrid filler are a weak point, and a major source of phonon scattering, doping of highly conductive metal nano particles at these junctions acts as a bridging agent. These bridging agents serve as phonon transporter to adjacent filler, which surely minimizes the losses at the junctions result in better thermal conduction.

Taking this recommendation in consideration, the improvement in thermal conductivity in is accomplished and effective thermal performance of TIMs is further improved. There is still enough space to further improvement in thermal conductive fillers and their better interfacial connection between the matrix and the filler to develop TIMs with more reliability and better performance for their application as thermal interface materials in electronic packaging.

REFERENCES

[1] W. Yu, J. Zhao, M. Wang, Y. Hu, L. Chen, H. Xie, Thermal conductivity enhancement in thermal grease containing different CuO structures, *Nanoscale Res. Lett.* 10 (2015) 113.

[2] H. Chen, H. Wei, M. Chen, F. Meng, H. Li, Q. Li, Enhancing the effectiveness of silicone thermal grease by the addition of functionalized carbon nanotubes, *Appl. Surf. Sci.* 283 (2013) 525.

[3] C.P. Chiu, N. Sharaf, G. Solbrekken, C.D. Cooks, *Google Patents*. 2000.

[4] C.P. Chu, G.L. Solbrekken, V. LeBonheur, Y.E. Xu, Application of phase change materials in Pentium III and Pentium III Xeon processor cartridges, in: *International Symposium on Advanced Packaging Materials* (2000) 265.

[5] R.S. Prasher, J.C. Matayabas, Thermal contact resistance of cured gel polymeric thermal interface material, *IEEE. T. Comp. Pack. Tech.* 27 (2004) 702.

[6] R. Viswanath, V. Wakharkar, A. Watwe, V. Lebonheur, Thermal performance challenges from silicon to systems, *Intel Technology Journal Q3* (2000) 1.

[7] Y. Xu, X. Luo, D.D.L. Chung, Sodium Silicate Based Thermal Interface Material for High Thermal Contact Conductance, *J. Electron. Pack.* 122 (2000) 128.

[8] P. Khatri, J. Ziemski, "Dry-to-the-touch" thermal grease, *Proceedings International Symposium on Advanced Packaging Materials Processes, Properties and Interfaces* (IEEE Cat. No.01TH8562) (2001) 236.

[9] L. C. Sim, S. R. Ramanan, H. Ismail, K. N. Seetharamu, T. J. Goh, Thermal characterization of Al_2O_3 and ZnO reinforced silicone rubber as thermal pads for heat dissipation purposes, *Thermochim. Acta.* 430 (2005) 155.

[10] S. C. Lin, C. C. M. Ma, W. H. Liao, J. A. Wang, S. J. Zeng, S. Y. Hsu, Y. H. Chen, S. T. Hsiao, T. Y. Cheng, C. W. Lin, P. Y. Hsiao, Preparation of a graphene–silver nanowire hybrid/silicone rubber composite for thermal interface materials, *J. Tai. Ins. Chem. Eng.* 68 (2016) 396.

[11] J. P. Gwinn, R. L. Webb, Performance and testing of thermal interface materials, *Microelectron. J.* 34 (2003) 215.

[12] B. Rauch, Understanding the performance characteristics of phase-change thermal interface materials, *ITHERM 2000 - The Seventh Intersociety Conference on*

Thermal and Thermomechanical Phenomena in Electronic Systems (Cat. No.00CH37069), IEEE (2000) 1.

[13] N. H. Mohamed, F. S. Soliman, H. El Maghraby, Y. M. Moustfa, Thermal conductivity enhancement of treated petroleum waxes, as phase change material, by α nano alumina: Energy storage, *Renewable Sustainable Energy Rev.* 70 (2017) 1052.

[14] C. Balian, S. E. Bergerson, G .C. Currier, *Google Patents*, 2006.

[15] B. Poudel, Q. Hao, Y. Ma, Y. Lan, A. Minnich, B. Yu, X. Yan, D. Wang,A. Muto, D. Vashaee, X. Chen, J. Liu, M. S. Dresselhaus, G. Chen, Z. Ren, High-Thermoelectric Performance of Nanostructured Bismuth Antimony Telluride Bulk Alloys, *Science*, 320 (2008) 634.

[16] R. L. Webb, J. P. Gwinn, Low melting point thermal interface material, ITherm 2002. *C. Th. Thermomechanical. Elect. Sys.* (2002) 671.

[17] R.S. Cook, K. H. Token, R.L. Calkins. A novel concept for reducing thermal contact resistance, *J. Sp. Rockets*. 21 (1984) 122.

[18] D. R. Frear, P. T. Vianco, Intermetallic growth rate and mechanical behaviour of low and high melting temperature solder alloys, *Metallurgical. Mater. Transactions A* 25 (1994) 1509.

[19] C. G. Macris, T .R. Sanderson, R .G. Ebel, C. B. Leyerle, Performance, reliability, approaches using a low melt allow as a thermal interface material, *Proceedings of IMAPS, 37th international symposium on microelectronics*, 2004.

[20] X. C. Tong, Thermal management Fundamentals and design guides in electronic packaging, *Adv. Mater. Th. Management Electron.* Packaging, Springer Series in Advanced Microelectronics 30, 2011.

[21] C.K. Roy, S.B. Havnani, M.C. Hamilton, R.W. Johnson, R.W. Knight, D.K. Harris, Thermal performance of low melting temperature alloys at the interface between dissimilar materials, *Appl. Therm. Eng.* 99 (2016) 72.

[22] G. Chen, *Google Patents*, 2005.

[23] E.C. Samson, S.V. Machiroutu, C. J. Young, I. Santos, J. Hermerding, A. Dani, R. Prasher, D.W. Song, "Interface Material Selection and a Thermal Management Technique in Second-Generation Platforms Built on Intel Centrino Mobile Technology, *J. Intel. Technology* 9 (2005) 75.

[24] Y. Li, C.P. Wong, Recent advances of conductive adhesives as a lead-free alternative in electronic packaging: Materials, processing, reliability and applications, *Mater. Sci. Eng. R* 51 (2006) 1.

[25] M. H. Bunyan, *Google Patents*, 2002.

[26] M. J. Yim, K.W. Paik, Recent advances on anisotropic conductive adhesives (ACAs) for flat panel displays and semiconductor packaging applications, *J. Adh. Adhesives* 26 (2006) 304.

[27] K. Pashayi, H.R. Fard, F. Lai, S. Iruvanti, J. Plawsky, T. Borca-Tasciuc, High thermal conductivity epoxy-silver composites based on self-constructed nanostructured metallic networks, *J. Appl. Phys.* 111 (2012) 104310.

[28] X. Lu, G. Xu, Thermally conductive polymer composites for electronic packaging, *J. Appl. Polym. Sci.* 65 (1997) 2733.

[29] C.C. Teng, C. C. M. Ma, C. H. Lu, M. Y. Yang, S. H. Lee, M. C. Hsiao, M. Y. Yen, K. C. Chiou, T. M. Lee, Thermal conductivity and structure of noncovalent functionalized graphene/epoxy composites, *Carbon.* 49 (2011) 5107.

[30] Y.P. Mamunya, V.V. Davydenko, P. Pissis, E.V. Lebedev, Electrical and thermal conductivity of polymers filled with metal powders, *Eur. Polym. J.* 38 (2002) 1887.

[31] I. Tavman, Thermal and mechanical properties of aluminum powder-filled high-density polyethylene composites, *J. Appl. Polym. Sci.* 62 (1996) 2161.

[32] M. Rusu, N. Sofian, D. Rusu, Mechanical and thermal properties of zinc powder filled high density polyethylene composites, *Polym. Test.* 20 (2001) 409.

[33] D. M. Bigg, Mechanical, thermal, and electrical properties of metal fiber-filled polymer composites, *Polym. Eng. Sci.* 19 (1979) 1188.

[34] J. W. Bae, W. Kim, S. H . Cho, S. H. Lee, The properties of AlN-filled epoxy molding compounds by the effects of filler size distribution, *J. Mater. Sci.* 35 (2000) 5907.

[35] C. Wong, R. S. Bollampally, Comparative study of thermally conductive fillers for use in liquid encapsulants for electronic packaging, *IEEE .Trans. Adv. Pack.* 22 (1999) 54.

[36] P. Gonon, A. Sylvestre, J. Teysseyre, C. Prior, Dielectric properties of epoxy/silica composites used for microlectronic packaging, and their dependence on post-curing, *J. Mater. Sc. Mater. Elect.* 12(2001)81.

[37] C .Wong, R S. Bollampally, Thermal conductivity, elastic modulus, and coefficient of thermal expansion of polymer composites filled with ceramic particles for electronic packaging, *J. App. Pol. Sci.*74(1999)3396.

[38] E. S. Lee, S. M. Lee, D.J. Shanefield, W.R. Cannon, Enhanced Thermal Conductivity of Polymer Matrix Composite via High Solids Loading of Aluminum Nitride in Epoxy Resin, *J. Am. Ceram. Soc.* 91 (2008) 1169.

[39] W. Zhou, C. Wang, T. Ai, K. Wu, F. Zhao, H. Gu, A novel fiber-reinforced polyethylene composite with added silicon nitride particles for enhanced thermal conductivity, *Com. A: App. Sci. Man.* 40 (2009) 830.

[40] B. Weidenfeller, M. Höfer, F. R. Schilling, Thermal conductivity, thermal diffusivity, and specific heat capacity of particle filled polypropylene, *Com. A: App. Sci. Man.* 35 (2004) 423.

[41] B. Weidenfeller, M. Höfer, F. T. Schilling, Thermal and electrical properties of magnetite filled polymers, *Com. A: App. Sci. Man.* 33 (2002) 1041.

[42] S. Kemaloglu, G. Ozkoc, A. Aytac, Properties of thermally conductive micro and nano size boron nitride reinforced silicon rubber composites, *Thermochim. Acta.* 499 (2010) 40.

[43] W. Zhou, S. Qi, Q. An, H. Zhao, N. Liu, Thermal conductivity of boron nitride reinforced polyethylene composites, *Mater. Res. Bull.* 42 (2007) 1863.

[44] H. Ishida, S. Rimdusit, Very high thermal conductivity obtained by boron nitride-filled polybenzoxazine, *Thermochim. Acta.* 320 (1998) 177.

[45] H.Y. Ng, X. Lu, S.K. Lau, Thermal conductivity of boron nitride-filled thermoplastics: Effect of filler characteristics and composite processing conditions, *Poly. Com.* 26 (2005) 778.

[46] C. Zhi, Y. Bando, T. Terao, C. Tang, H. Kuwahara, D. Golberg, Towards Thermoconductive, Electrically Insulating Polymeric Composites with Boron Nitride Nanotubes as Fillers, *Adv. Funct. Mater.* 19 (2009) 1857

[47] Y. Li, S. Wang, Y. Zhang, Y. C. Zhang, Carbon black-filled immiscible polypropylene/epoxy blends, *J. App. Poly. Sci.* 99 (2006) 461.

[48] N. Abdel-Aal, F. El-Tantawy, A. Al-Hajry, M. Bououdina, Epoxy resin/plasticized carbon black composites. Part I. Electrical and thermal properties and their applications, *Poly. Com.* 29 (2008) 511.

[49] V. Causin, C. Marega, A. Marigo, G. Ferrara, A. Ferraro, Morphological and structural characterization of polypropylene/conductive graphite nanocomposites, *Euro. Poly. J.* 42 (2006) 3153.

[50] H. Tu, L. Ye, Thermal conductive PS/graphite composites, *Poly. Adv. Tech.* 20 (2009) 21.

[51] S. Kim, L .T. Drzal, High latent heat storage and high thermal conductive phase change materials using exfoliated graphite nanoplatelets, *Sol. Energy Mater. Sol. Cells* 93 (2009) 136.

[52] A. Sarı, A. Karaipekli, Thermal conductivity and latent heat thermal energy storage characteristics of paraffin/expanded graphite composite as phase change material, *App. Th. Eng.* 27 (2007) 1271.

[53] B. Debelak, K. Lafdi, Use of exfoliated graphite filler to enhance polymer physical properties, *Carbon.* 45 (2007) 1727.

[54] A. Yu, P. Ramesh, M. E. Itkis, E. Bekyarova, R. C. Haddon, Graphite Nanoplatelet−Epoxy Composite Thermal Interface Materials, *J. Phys. Chem. C* 111 (2007) 7565.

[55] M. A. Raza, A. Westwood, A. Brown, N. Hondow, C. Stirling, Characterisation of graphite nanoplatelets and the physical properties of graphite nanoplatelet/silicone composites for thermal interface applications, *Carbon* 49 (2011) 4269.

[56] H. B. Cho, A. Konno, T. Fujihara, T. Suzuki, S. Tanaka, W. Jiang, H. Suematsu, K. Niihara, T. Nakayama, Self-assemblies of linearly aligned diamond fillers in

polysiloxane/diamond composite films with enhanced thermal conductivity, *Com. Sci. Tech.* 72 (2011) 112.

[57] W. S. Lee, J. Yu, Comparative study of thermally conductive fillers in underfill for the electronic components, *Diamond Relat. Mater.* 14 (2005) 1647.

[58] J. Fukai, M. Kanou, Y. Kodama, O. Miyatake, Thermal conductivity enhancement of energy storage media using carbon fibers, *Energy Conv. Manag.* 41 (2000) 1543.

[59] Y. Agari, A. Ueda, S. Nagai, Thermal conductivity of a polyethylene filled with disoriented short-cut carbon fibers, *J. Appl. Polym. Sci.* 43 (1991) 1117.

[60] A. S. Fleischer, R.D. Weinstein, T. Kopec, Analysis of transient thermal management characteristics of pcm with an embedded carbon fiber heat sink, *ITHERM 2006 - Thermal and Thermomechanical Proceedings of 10th Intersociety Conference on Phenomena in Electronics Systems*, 2006, pp. 4 - 1268.

[61] M. A. Raza, A. Westwood, C. Stirling, Effect of processing technique on the transport and mechanical properties of vapour grown carbon nanofibre/rubbery epoxy composites for electronic packaging applications, *Carbon* 50 (2012) 84.

[62] S. Berber, Y. K. Kwon, D. Tománek, Unusually High Thermal Conductivity of Carbon Nanotubes, *Phy. Rev. Lett.* 84 (2000) 4613.

[63] L. Qingwei, L. Changhong, W. Xueshen, F. Shoushan, *Nanotechnology* 20 (2009) 145702.

[64] A. Moisala, Q. Li, I. A. Kinloch, A .H. Windle, Thermal and electrical conductivity of single- and multi-walled carbon nanotube-epoxy composites, *Com. Sci. Tech.* 66 (2006) 1285.

[65] F.H. Gojny, M.H.G. Wichmann, B. Fiedler, I.A. Kinloch, W. Bauhofer, A.H. Windle, K. Schulte, Evaluation and identification of electrical and thermal conduction mechanisms in carbon nanotube/epoxy composites, *Polymer* 47 (2006) 2036.

[66] P. Chantrenne, J. L. Barrat, Analytical model for the thermal conductivity of nanostructures, *Sup. Micro.* 35 (2004) 173.

[67] H. Huang, C. H. Liu, Y. Wu, S. Fan, Aligned Carbon Nanotube Composite Films for Thermal Management, *Adv. Mater.*17 (2005) 1652.

[68] P. Bonnet, D. Sireude, B. Garnier, O. Chauvet, Thermal properties and percolation in carbon nanotube-polymer composites, *App. Phys. Lett.* 91 (2007) 201910.

[69] S.T. Huxtable, D.G. Cahill, S. Shenogin, L. Xue, R. Ozisik, P. Barone, M. Usrey, M.S. Strano, G. Siddons, M. Shim, P. Keblinski, Interfacial heat flow in carbon nanotube suspensions, *Nat. Mater.* 2 (2003) 731.

[70] F. Yavari, H .R. Fard, K. Pashayi, M. A. Rafiee, A. Zamiri, Z. Yu, R. Ozisik, T. Borca-Tasciuc, N. Koratkar, Enhanced Thermal Conductivity in a Nanostructured Phase Change Composite due to Low Concentration Graphene Additives, *J. Phy. Chem. C.* 115 (2011) 8753.

[71] Q. Li, Y. Guo, W. Li, S .Qiu, C. Zhu, X Wei, M. Chen, C . Liu, S. Liao, Y. Gong, A .K. Mishra, L. Liu, Ultrahigh Thermal Conductivity of Assembled Aligned Multilayer Graphene/Epoxy Composite, *Chem. Mater.* 26 (2014) 4459.

[72] S.Y. Fu, G. Xu, Y.W. Mai, On the elastic modulus of hybrid particle/short fiber/polymer composites, *Com B. Engineering* 33 (2002) 291.

[73] A. Yu, P. Ramesh, X. Sun, E. Bekyarova, M.E. Itkis, R.C. Haddon, Enhanced Thermal Conductivity in a Hybrid Graphite Nanoplatelet – Carbon Nanotube Filler for Epoxy Composites, *Adv. Mater.* 20 (2008) 4740.

[74] T.L. Li, S.L.C. Hsu, Enhanced Thermal Conductivity of Polyimide Films via a Hybrid of Micro- and Nano-Sized Boron Nitride, *J. Phy. Chem. B* 114 (2010) 6825.

[75] S. Choi, J. Kim, Thermal conductivity of epoxy composites with a binary particle system of aluminum oxide and aluminum nitride fillers, *Com. B: Engineering* 51 (2013) 140.

[76] J. P. Hong, S. W. Yoon, T. Hwang, J. S. Oh, S. C. Hong, Y. Lee, J. D. Nam, High thermal conductivity epoxy composites with bimodal distribution of aluminum nitride and boron nitride fillers, *Thermochim. Acta* 537 (2012) 70.

[77] G. Kalaprasad, P. Pradeep, G. Mathew, C. Pavithran, S. Thomas, Thermal conductivity and thermal diffusivity analyses of low-density polyethylene composites reinforced with sisal, glass and intimately mixed sisal/glass fibres, *Com. Sci. Tech.* 60 (2000) 2967.

[78] Z. Han, A. Fina, Thermal conductivity of carbon nanotubes and their polymer nanocomposites: A review, *Pro. Poly. Sci.* 36 (2011) 914.

[79] K.M.F. Shahil, A.A. Balandin, Graphene–Multilayer Graphene Nanocomposites as Highly Efficient Thermal Interface Materials, *Nano Letters* 12 (2012) 861.

[80] II. Im, J. Kim, Thermal conductivity of a graphene oxide–carbon nanotube hybrid/epoxy composite, *Carbon.* 50 (2012) 5429.

[81] Y. Lan, P. J. Sun, L .Y. Soo, L. C. Seung, S. Kwonwoo, M. Ho Jun, Y. C. Min, L. Y. Sil, H. Jong Hun, Carbon hybrid fillers composed of carbon nanotubes directly grown on graphene nanoplatelets for effective thermal conductivity in epoxy composites, *Nanotechnology* 2 (2013) 155604.

[82] S. Shenogin, L. Xue, R. Ozisik, P. Keblinski, D.G. Cahill, Role of thermal boundary resistance on the heat flow in carbon-nanotube composites, *J. App. Phys.* 95 (2004) 8136.

[83] D. H. Park, T. G. Kan, Y. K .Lee, W. N. Kim, Effect of multi-walled carbon nanotube dispersion on the electrical and rheological properties of poly(propylene carbonate)/poly(lactic acid)/multi-walled carbon nanotube composites, *J. Mater. Sci.* 48 (2013) 481.

[84] H. P. Xu, Z.M. Dang, Electrical property and microstructure analysis of poly (vinylidene fluoride)-based composites with different conducting fillers, *Chem. Phys. Lett.* 438 (2007) 196.

[85] S. Choi, H. Im, J. Kim, Flexible and high thermal conductivity thin films based on polymer: Aminated multi-walled carbon nanotubes/micro-aluminum nitride hybrid composites, *Com. A: App. Sci. Manufacturing* 43 (2012) 1860.

[86] Y.X. Fu, Z.X. He, D.C. Mo, S.S. Lu, Thermal conductivity enhancement with different fillers for epoxy resin adhesives, *App. Th. Eng.* 66 (2014) 493.

[87] T. Zhou, X. Wang, X. Liu, D. Xiong, Improved thermal conductivity of epoxy composites using a hybrid multi-walled carbon nanotube/micro-SiC filler, *Carbon* 48 (2010) 1171.

[88] C.C. Teng, C.C.M. Ma, K.C. Chiou, T.M. Lee, Y.F. Shih, Synergetic effect of hybrid boron nitride and multi-walled carbon nanotubes on the thermal conductivity of epoxy composites, *Mater. Chem. Phys.* 126 (2011) 722.

[89] K. Sanada, Y. Tada, Y. Shindo, Thermal conductivity of polymer composites with close-packed structure of nano and micro fillers, *Composites Part A: App. Sci. Manufacturing* 40 (2009) 724.

[90] H.Y. Ng, X. Lu, S.K. Lau, Thermal conductivity of boron nitride-filled thermoplastics: Effect of filler characteristics and composite processing conditions, *Poly. Com.* 26 (2005) 66.

[91] V.A. Harmandaris, K.C. Daoulas, V.G. Mavrantzas, Molecular Dynamics Simulation of a Polymer Melt/Solid Interface: Local Dynamics and Chain Mobility in a Thin Film of Polyethylene Melt Adsorbed on Graphite, *Macromolecules* 38 (2005) 5796.

[92] F.W. Starr, T.B. Schrøder, S.C. Glotzer, Molecular Dynamics Simulation of a Polymer Melt with a Nanoscopic Particle, *Macromolecules* 35 (2002) 4481.

[93] J. Kim, B. S. Yim, J. M. Kim, J. Kim, The effects of functionalized graphene nanosheets on the thermal and mechanical properties of epoxy composites for anisotropic conductive adhesives (ACAs), *Microelectronics Reliability* 52 (2012) 595.

[94] S. Ganguli, A.K. Roy, D.P. Anderson, Improved thermal conductivity for chemically functionalized exfoliated graphite/epoxy composites, *Carbon* 46 (2008) 806.

[95] S.M .Yuen, C.C.M. Ma, C.L. Chiang, J.A. Chang, S.W. Huang, S.C .Chen, C. Y. Chuang, C.C. Yang, M.H. Wei, Silane-modified MWCNT/PMMA composites – Preparation, electrical resistivity, thermal conductivity and thermal stability, *Com. A: App. Sci. Manufacturing* 38 (2007) 2527.

[96] K. Yang, M. Gu, Y. Guo, X. Pan, G. Mu, Effects of carbon nanotube functionalization on the mechanical and thermal properties of epoxy composites, *Carbon.* 47 (2009) 1723.

[97] M.W. Akhtar, Y.S. Lee, C.M. Yang, J.S. Kim, Functionalization of mild oxidized graphene with O-phenylenediamine for highly thermally conductive and thermally stable epoxy composites, *RSC. Adv.* 6 (2016) 100448.

[98] M. W. Akhtar, C. W. Park, Y. S. Kim, J. S. Kim, Facile Large Scale Production of Few-Layer Graphene Sheets by Shear Exfoliation in Volatile Solvent, *J. Nanosci. Nanotechnol.* 15 (2015) 9624.

[99] S. Y. Yang, C.C.M. Ma, C,C. Teng, Y.W. Huang, S. H. Liao, Y. L. Huang, H.W. Tien, T. M. Lee, K. C. Chiou, Effect of functionalized carbon nanotubes on the thermal conductivity of epoxy composites, *Carbon* 48 (2010) 592.

[100] J. Hou, G. Li, N. Yang, L. Qin, M. E. Grami, Q. Zhang, N. Wang, X. Qu, Preparation and characterization of surface modified boron nitride epoxy composites with enhanced thermal conductivity, *RSC. Adv.* 4 (2014) 44282.

[101] H. Dong, L. Fan, C. Wong, *Proceedings of the* 55th *Electronic Components Technology Conference* (2005) pp. 1451.

[102] R. Qian, J. Yu, C. Wu, X. Zhai, P. Jiang, Alumina-coated graphene sheet hybrids for electrically insulating polymer composites with high thermal conductivity, *RSC. Adv.* 3 (2013) 17373.

[103] X. Lou, R. Daussin, S. Cuenot, A.S. Duwez, C. Pagnoulle, C. Detrembleur, C. Bailly, R. Jérôme, Synthesis of Pyrene-Containing Polymers and Noncovalent Sidewall Functionalization of Multiwalled Carbon Nanotubes, *Chem. Mater.* 16 (2004) 4005.

[104] Q. Yang, X. Pan, F. Huang, K. Li, Fabrication of High-Concentration and Stable Aqueous Suspensions of Graphene Nanosheets by Noncovalent Functionalization with Lignin and Cellulose Derivatives, *Phy. Chem. C* 114 (2010) 3811.

[105] I. H. Choi, M. Park, S.S. Lee, S.C. Hong. Pyrene-containing polystyrene segmented copolymer from nitroxide mediated polymerization and its application for the noncovalent functionalization of as-prepared multiwalled carbon nanotubes, *Euro. Poly. J.* 44 (2008) 3087.

[106] L. Cao, X. Liu, H. Na, Y. Wu, W. Zheng, J. Zhu, How a bio-based epoxy monomer enhanced the properties of diglycidyl ether of bisphenol A (DGEBA)/graphene composites, *J. Mater. Chem. A* 1 (2013) 5081.

[107] S. H. Song, K.H. Park, B. H. Kim, Y .W. Choi, G. H. Jun, D. J. Lee, B. S. Kong, K. W. Paik, S. Jeon, Enhanced thermal conductivity of epoxy-graphene composites by using non-oxidized graphene flakes with non-covalent functionalization, *Adv. Mater.* 25 (2013) 732.

[108] Y. Wang, C. Yang, Y. W. Mai, Y. Zhang, Effect of non-covalent functionalization on thermal and mechanical properties of graphene-polymer nanocomposites, *Carbon* 102 (2016) 311.

[109] J. H. Lee, S.H. Lee, C. Choi, S. Jang, S. Choi, A Review of Thermal Conductivity Data, Mechanisms and Models for Nanofluids, *Inter. J. Micro-Nano Scale Trans.* 1 (2010) 269.

[110] Y. Xu, D. D. L. Chung, C. Mroz, Thermally conducting aluminum nitride polymer-matrix composites, *Com. Part A: App. Sci. Manufacturing* 32 (2001) 1749.

[111] *Reducing pollution from light motor vehicles* (Euro 5 and Euro 6) http://eur-lex.europa.eu/legal-content/EN/TXT/?uri=URISERV:l28186 (2010).

[112] *European Parliament CotEU On the deployment of alternative fuels infrastructure.* http:// eurlex.europa.eu/ legalcontent/ EN/ TXT/ ?qid= 1489978277787 &uri= CELEX:32014L0094 (2010).

[113] E. Gutiérrez González, J. Alvarez Flórez, S. Arab, Development of the management strategies of the ECU for an internal combustion engine, *Mech. Sys. Sig. Process.* 22 (2008) 1356.

[114] G.W. Lee, M. Park, J. Kim, J. I. Lee, H.G. Yoon, Enhanced thermal conductivity of polymer composites filled with hybrid filler, *Com. A: App. Sci. Manuf.* 37 (2006) 727.

[115] Hu, E.J., D.S. Zhu, X.Y. Sang, L. Wang, Y.K. Tan, Enhancement of Thermal Conductivity by Using Polymer Zeolite in Solid Adsorption Heat Pumps, *J. Heat Transfer - Transactions of The ASME*, 119, (1997) 627.

In: Handbook of Remediation ...
Editors: Sadia Ameen, M. Shaheer Akhtar et al.
ISBN: 978-1-53615-118-3

Chapter 7

NANOPARTICLES: THE BROADER PERSPECTIVES

Saima Amin[*]
Department of Pharmaceutics, School of Pharmaceutical Education and Research,
Jamia Hamdard, New Delhi, India

ABSTRACT

Nanotechnology, a science and engineering related to generation of nano-sized particles, has been increasingly employed to explore avenues in medical sciences. It has been considered as a systematic and standardized technology to study the changes in the particle properties after nanonization. The crucial properties of material such as size, shape, purity and surface charges are changed with its development as nanoparticles. In the current chapter an approach is made to understand the application of nanotechnology, particularly in combating diseases, bacterial growth, cellular imaging and diagnosis. The authors have also summarized the other facets of nanotechnology particularly in water purification, waste treatment, food packaging, electronics and biosensors.

Keywords: nanoparticles, medical science, diagnosis, microorganism growth

INTRODUCTION

Nanotechnology refers to amalgamation of science and engineering where interaction of cellular, molecular and engineered materials such as clusters of atoms, molecules, and molecular fragments result in small particles ranging from 1 to 100 nm. The word nanotechnology has originated from Greek which means dwarf. Generation of nanoparticles through this technology has imparted new physical, chemical, optical, electronic and structural properties that are not available either in parent molecules or bulk

[*] Corresponding Author's Email: samin@jamiahamdard.ac.in.

solids. This newer concept of nanotechnology has led to the development of pharmaceutical nanoparticles engineered for the targeted delivery of anticancer drugs or metal nanoparticles gaining recognition in various fields, such as theranostics, photo catalysis, lithography, electronics and optical biosensors. These nano-constructs are also customized for targeted drug delivery to ferry large doses of drugs or therapeutic genes to distant cells [1].Some of the most common applications of nanoparticles are in medicine (Au, CeO_2), food (Ag, TiO_2, SiO_2, Fe), cosmetics (TiO_2, ZnO), sports clothes (Ag), electronics, energy and fuel additives (CeO_2), nutraceuticals and pharmaceuticals (silicate and alumina silicate) and waste water treatment (Fe,TiO_2) [2]. Figure 1 provides a gist of the production of nanomaterials over the last 70 years through both the technologies to generate a stable and robust nanosystem.

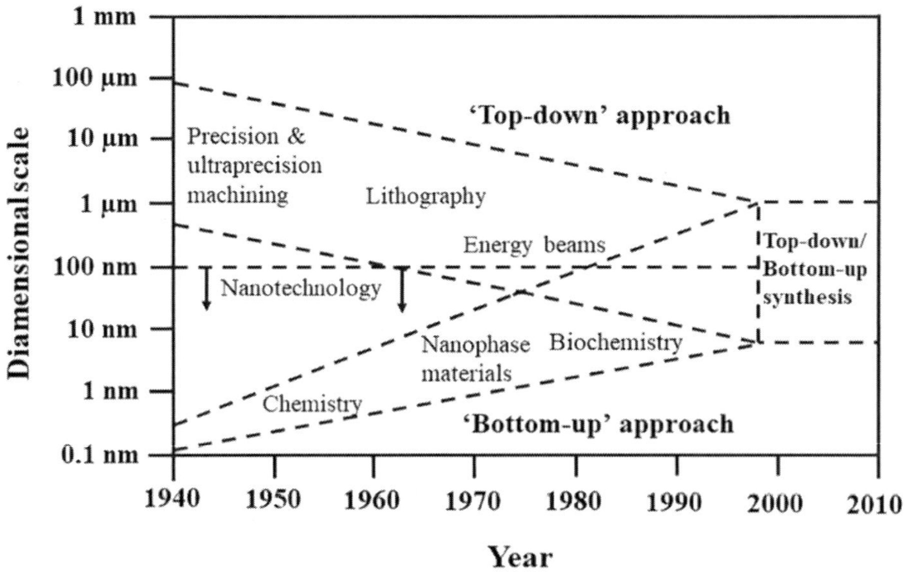

Figure 1. Top down and bottom up techniques for nanoparticles.

Production and Process Optimization

Nanoparticles can be synthesized via different routes such as chemical reduction, oxidation, gas phase condensation, plasma techniques and laser technology but nano-pharmaceuticals are generated either through top-down approach or bottom up approach. In top down approach, nanoscale material is literally created by physical or chemical breaking down of larger materials (Figure 2). Some of the techniques used for top-down approach are non-imprint lithography, step and flash lithography (S-FIL) and particle replication in non-wetting templates (PRINT) [3]. In nano-pharmaceuticals, the top down technology involves high pressure homogenization and milling.

Figure 2. Top down approach for generation of nanoparticles. Reprinted with the permission from R. Misra, M. Upadhyay, S. Mohanty (2014). Design considerations for chemotherapeutic drug nanocarriers. Reprinted with the permission from Pharm Anal Acta 5: 279. doi: 10.4172/2153-2435.1000279.

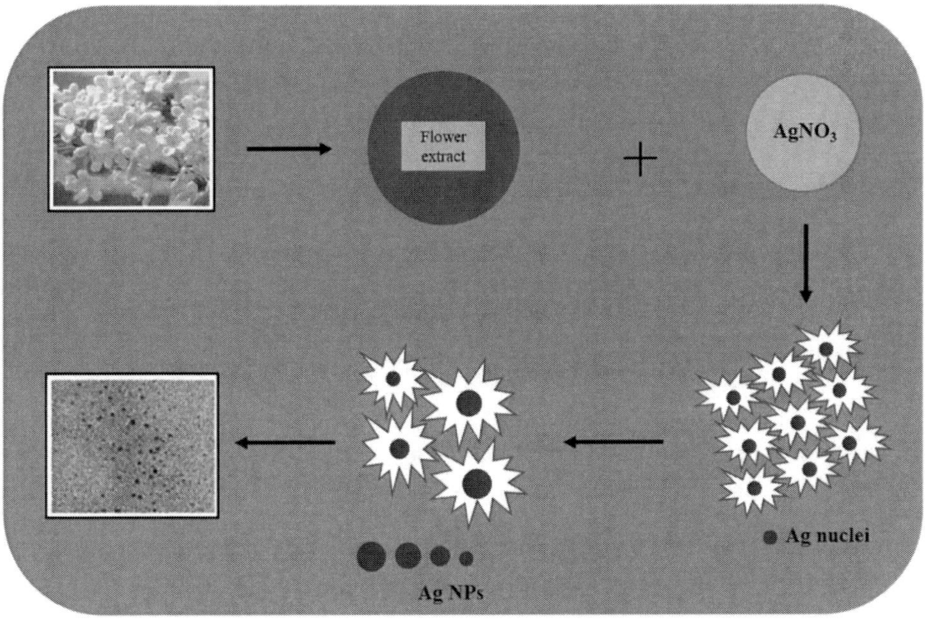

Figure 3. Synthesis of silver nanoparticles through green route (Bottom up approach).

In top down approach, nanoscale material is literally created by physical or chemical breaking down of larger materials. Some of the techniques used for top-down approach are non-imprint lithography, step and flash lithography (S-FIL) and particle replication in non-wetting templates (PRINT) (Figure 2) [3]. In nano-pharmaceuticals, the top down technology involves high pressure homogenization and milling.

In bottom –up approach, assembly of nano-materials atom by atom is taken place. This approach has been adopted for polymeric nanoparticles, micelles, liposomes, nanoemulsions and other scalable nanocarrier system [3] (Figure 3). Precipitation has been applied for many years for preparation of fine particles through bottom up approach, particularly in development of photographic film, and lately for preparation of Sandoz (Novartis product) and Nanomorph (Abbott product). The drug is dissolved in a suitable solvent and the solution is subsequently added to a non-solvent. This results in high super saturation, rapid nucleation and the formation of many small nuclei. Upon solvent removal, the suspension is filtered and lyophilized [4]. The mixing process in such a system may vary considerably but through careful control of this addition process, it is possible to obtain particles with a narrow size distribution. In case of Nanomorph, amorphous drug nanocrystals are produced to further enhance its dissolution and solubility. The drawback of this technique is formation of a metastable product and a wider size distribution. Since the process of manufacturing requires several variables to be controlled effectively, optimization of process variables is equally required for a robust formulation.

Optimization of a pharmaceutical formulation or process using traditional approach involves changing one variable at a time. Thus using this approach, solution for a specific problematic formulation or characteristics can be achieved. Generally, optimization approaches employ systematic design of experiments (DOE). DOE represents effective and cost effective yield in terms of best solution for a particular problem. This DOE optimization methodology encompasses planning the study object, screening of influential variables, experimental designs, postulation of mathematical models for various chosen response characteristics, fitting experimental data into these models(s), mapping and generating graphic outcomes and design validation using model-based response. Ahmad et al., have optimized paclitaxel loaded self emulsifying system by Box–Behnken statistical design to optimize the concentration of oil, surfactant and co-surfactant to develop a nano-sized formulation for targeting to breast cancer [5]. Also, the same approach has been used for optimizing nanostructured lipid carriers through optimization of process variables such as lipid concentration, sonication time, and surfactant concentration [6].

NANOMEDICINES

Nobel Laureate, Paul Elrich in 1908 introduced the concept of 'magic bullet' for targeting the diseased tissue or even cells leaving the healthy cells unharmed. But later the concept was realised for other nanocarriers such as liposomes, solid lipid nanoparticles, nanoemulsions etc. These pharmaceutical nanocarriers are designed to resolve the fundamental problems of conventional drugs such as poor solubility, low bioavailability, enhanced toxicity, high doses and poor specificity for the tissues, for example, drugs such as paclitaxel has been designed as Abraxane; a protein bound nanoparticle of paclitaxel to improve bioavailability of the free drug in breast cancer while brentuximab viditonin is delivered as antibody drug conjugate for Hodgkin Lymphoma because of higher toxicities of the plain drug. DaunoXome is yet another anticancer formulation of docetaxel developed to overcome the high dose related side effects of the free drug. Therefore, nanocarrier based drug delivery strategies leverage multiple aspects of nanoparticle [3].

Nanomedicines have been designed using carriers such as liposomes, niosomes, polymer conjugates, polymeric micelles, nano-emulsions, nanocrystals and as solid lipid nanoparticles. The drug nanocarriers are designed with the intention to improve the biopharmaceutics, that is, absorption, distribution, metabolism and excretion of the drug. The vital consideration for these engineered nanoparticles involve prolonged circulation in blood, high absorption through lumen for orally administered therapeutics and reduce drug resistance. The successful design of nanoparticles requires thorough understanding of certain physico-chemical characteristics such as particle size, shape, solubility, pH etc.

Naturally, the prerequisite for nano-particulate drug delivery systems to be effective in treating diseases like solid tumors is drug accumulation in the affected area and the effective release at the target site. A tremendous amount of nanoparticle based drug release strategies have been conceptualized and are listed in Table 1 and are illustrated through (Figure 4).

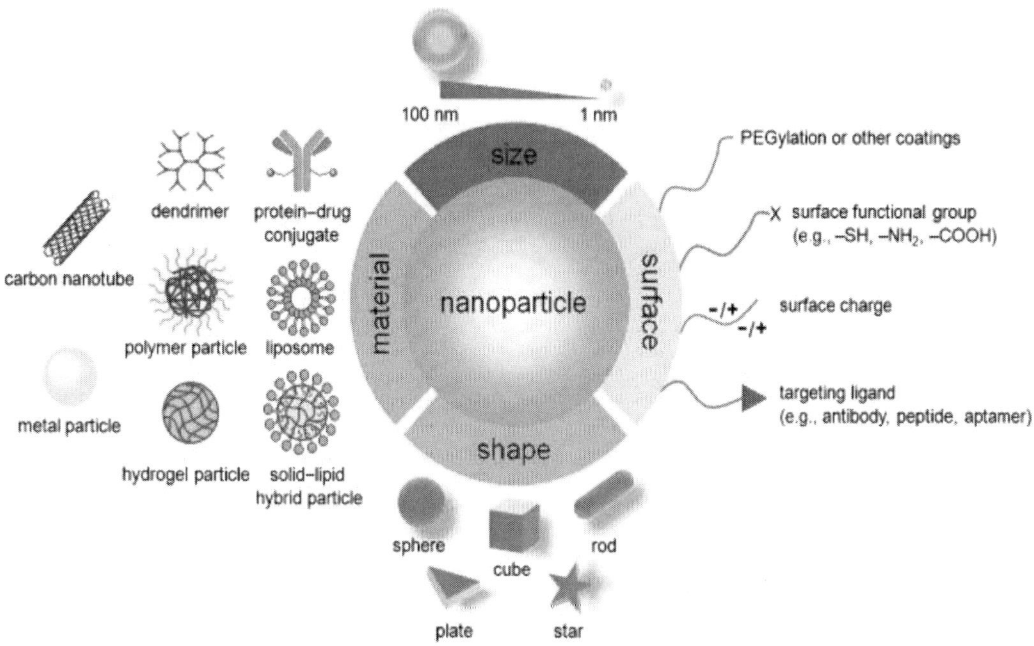

Figure 4. Various types of nanoparticles generated through nanotechnology. https://mail.google.com/mail/u/0/?ui=2&ik=8063618991&view=att&th=15a401a8f160e61c&attid=0.1&disp=safe&realattid=f_iz6h7stq0&zw

Table 1. Different types of market products based on nanotechnology

Product name/ Nanomedicine	API	Description	Company	Indication and safety
Polymer based				
Eligard®	Leuprolide acetate	Polymeric matrix formulation of Leuprolide acetate, a synthetic nonpeptide analogue of GnRH loaded in nanoparticles comprised of biodegradable copolymer poly (DL-lactide-co-glycolide) PLG.	Tolmar Pharmaceutical	In the treatment of advance prostate cancer, suppress steroidogenesis in testis and ovary
Renagel®	Sevelamer	Poly (allylamine-co-N,N'-diallyl-1,3-diamino2-hydroxypropane) hydrochloride with hypromellose, diacetylated monoglyceride.	Sanofi	Indicated for maintenance of serum phosphate in patient with kidney ailments. Contraindicated in hypersensitive patients
Opaxio®	Paclitaxel	Polyglutamate nanoparticle	CTI BioPharma	In the treatment of glioblastoma
Zinostatin stimalamer	Zinostatin	Conjugation product of poly (styrene-co-maleic acid) and antitumor protein neocarzinostatin.	Dainippon Sumitomo Pharma	In the treatment of hepatocellular carcinoma
Copaxone®	Glatiramer acetate	Polypeptides of four amino acid (L-glutamic acid, L-alanine, L-lysine, and L-tyrosine	Teva Pharmaceutical Industries	Relapsing multiple sclerosis, local injection causes erythemia and induration.
Nanocrystals based				
Rapamune	Sirolimus	A nanocrystal colloidal dispersion with α-tocopherol (Vitamin E USP) tablet formulation	Pfizer Inc	Immunosuppressant, preventing organ transplant rejection, safety profile of rapmune has not been established in liver or lung transplant
Invega Sustenna®	Paliperidone palmitate	Extended release injectable nanocrystal	Johnson & Johnson	Indicated in acute and maintenance therapy of Schizophrenia, not approved for dementia related psychosis in elder patients
Megace ES®	Megestrol acetate	Liquid dispersion nanocrystal	Par Pharma	Indicated in anorexia and cachexia, or unknown weight loss in AIDS, contraindicated in hypersensitive patients
Semapimod®	Guanylhydrazone	Nanocrystal formulation	Cytokine PharmaSciences	For chron's disease and inflammatory disorder under clinical trial
Theodur®	Theophylline	-	Mitsubishi Tanabe Pharma	Bronchial dilation

Product name/ Nanomedicine	API	Description	Company	Indication and safety
Nanoparticle based				
Abraxane®	Paclitaxel	Powder suspension for intravenous infusion	Celgene Europe Ltd.	Breast neoplasma
Taxol®	Paclitaxel	Paclitaxel was solubilized in a Cremophor®EL and ethanol mixture	Corden Pharma	As adjuvant therapy in breast cancer, therapy closely monitored for blood cell count, hepatic and cardiac function
Focalin	Dexmethyl Phenidate	-	Novartis/Elan Nanosystems	Attention Deficit Hyperactivity Disorder (ADHD).
Ritalin	Methyl Phenidate	Available as tablet dosage form	Novartis/Elan Nanosystems	CNS Stimulant
Emend®	Dexamthasone	5-HT_3 antagonist	Merck Sharp & Dohme	Prevention of nausea and vomiting in emetogenic cancer chemotherapy
Avinza®	Morphine sulphate	extended-release capsules active against mu-opioid receptor	Pfizer	Indicated in moderate to severe pain
Zanaflex	Tizanidine	Central alpha 2- adrenergic agonist	Acorda/Elan Nanosystems	Muscle Relaxant
Y-ibritumomab tiuxetan	ibritumomab tiuxetan, monoclonal antibody	Intravenous solution	Bayer Pharma AG	Follicular lymphoma
Metallic Nanoparticle				
Nanotherm®	Iron oxide	superparamagnetic iron oxide nanoparticle with aminosilane coating	MagForce AG	Nanotherm/thermal ablation therapy in cancer due to depot formation
Feraheme™	Ferumoxytol	Dextran coated superparamagnetic iron oxide nanoparticle	AMAG Pharmaceuticals	Given in patient with iron deficiency or anaemia.
Feridex®	Iron oxide	superparamagnetic iron oxide nanoparticle with dextran coating	Bayer Health Care Pharmaceuticals	Manufacturing forbidden in 2008 due to spleen or liver lesion observed in MRI
Tricor® and Triglide®	Fenofibrate	Fenofibrate tablet comprised of phospholipid surface-modifying agents	Abbott Laboratories	Hypercholesterolemia and hypertriglyceridemia
Virus-like particles (VLP)	Nanovaccine	Self-assembling nanoparticles formed by self-assembled biocompatible capsid protein	-	Induce potent immune responses, First virus like vaccine for hepatitis B virus
Biosensor/Dendrimer based				
PAMAM	antibody	Binding, anthrax detection	Alert Ticket™ (by US Army)	Antibody conjugate

Table 1. (Continued)

Product name/ Nanomedicine	API	Description	Company	Indication and safety
PAMAM	Monoclonal sheep antibody	-	Siemens Healthcare)	Antibody conjugate
PAMAM	DNA	-	Starpharma) PrioFect™	Intracellular gene delivery and transfecting agent
DEP™-Docetaxel DTX-SPL8783	Docetaxel	Dendrimer-Based Conjugate	Starpharma	Treatment of advanced or metastatic cancer (under phase I clinical trial (ACTRN12614000171617)
VivaGel® SPL7013	Anti-HIV agent	mucoadhesive gel formulation, Dendrimer-Based Topical preparation	Starpharma Pty Ltd.	In the treatment of bacterial vaginosis and prevention of sexually transmitted disease (Clinical trial Phase II)
Stratus® CS Analyzer	Dendrimer conjugated with monoclonal antibody	Hybrane® hyper branched polyester amides	US hospitals	In acute diagnostic care for heart patient especially detection of myocardial ischemia
Gadomer-17	Gadolinium-based extracellular fluid contrast agents	DOTA-functionalized poly-L-lysin arms has central core of 1, 3, 5–benzenetricarboxylate	Bayer Schering Pharma,	Used as image contrasting agents in lymph node carcinoma, pulmonary embolism and blockage in microvasculature
Superfect™	activated-dendrimer molecules	protein overexpression, plasmid transfection based on dendrimer technology	***Quiagen***	Gene transfecting agent in different cell lines, gene function and level of expression, drug discovery and developmental process
Micelles based				
Fungizone®	Amphotericin B	Lyophilized Amphotericin B powder with sodium deoxycholate which forms colloidal (micellar) dispersion when subjected to reconstitution.	Abbott Healthcare	Indicated for systemic fungal infections
Estrasorb™	Estradiol	Drug emulsion in soybean oil, polysorbate 80 or estradiol micellarnanoparticle	Novavax	Transdermaly for Hormone replacement therapy in case of menopause
Diprivane®	Propofol	O/W emulsion in soybean oil/glycerol/egg lecithin	AstraZeneca	Induce and maintain sedation/hypnosis in surgical anesthesia (IV)
Genexol-PM	Paclitaxel	poly(lactic acid)-block-poly(ethylene glycol) micelle formulation,	Samyang	treatment of breast and lung cancer
Paclical	Paclitaxel	Paclitaxel loaded micelles	Oasmia Pharmaceutical AB	Ovarian cancer therapy (Under phase III clinical trial)

Product name/ Nanomedicine	API	Description	Company	Indication and safety
Drug conjugate based				
Abraxane®	Paclitaxel	Nanoparticles (130 nm) formed by albumin with conjugated paclitaxel 44,45	American Bioscience	Metastatic breast cancer, non-small-cell lung cancer (IV)
Ontak®	Antibody fragment	Recombinant fusion protein of fragment A of diphtheria toxin and subunit binding to interleukin-2 receptor	Eisai Medical Research	Persistent or recurrent CD-25 positive cutaneous T-cell lymphoma.
Kadcyla®	Ado-rastuzumab emtansine	Immunoconjugate. Monoclonal antibody against EGFR-2)–drug conjugate, linked via thioether. Antibody and chemotherapy taken together	Genentech	Indicated in HER2- targeted Metastatic breast cancer; liver complication may be life threatening
Oncaspar®	Pegaspargase	PEG-asparaginase	Enzon Pharmaceutical	Acute lymphoblastic leukaemia; Hyperlipidemia, anaphylaxis, hypersentivity, allergic reaction should be monitored
Peg-intron®	Alfa interferon	Conjugation product of recombinant alfa interferon with PEG of Mw 12000 D developed by recombinant DNA technology	Schering-Plough	Indicated in hepatitis C; It may worsen the patient condition and stop it life threatening symptom come out
Neulasta™	***Pegfilgrastim***	Recombinant human PEG granulocyte colony-stimulating factor (G-CSF) analogous to ***filgrastim***	Amgen	Chemotherapy-induced neutropenia
Adagen®	Pegademase bovine	PEG-adenosine deaminase	Enzon	Severe combined immune deficiency syndrome
Somavert®	Pegvisomant	PEG-HGH antagonist	Pfizer	Acromegalia
Mircera®		PEG-EPO (polyethylene glycol-epoetin beta)	Roche	Treatment of anemia associated with chronic kidney disease
KrystexxaTM	Pegloticase	PEGylated recombinant uricase	Crealta Pharmaceuticals	Chronic gout
Cimzia®	Certolizumab pegol	Antibody Fab fragment specific for human tumor necrosis factor alfa (TNFα)	UCB care	Indicated for crohn's disease, psoriatic and rheumatoid arthritis. Patient prone to develop microbial infection due to immunosuppression

Characterization of Nanoparticles

Particles transformed to nanonized structures involve change in the fundamental properties such as particle size, surface morphology, optical property etc. therefore their thorough characterization is required for novel drug delivery.

Particle Size

The functionality of nanoparticles is affected by particle size. The important pharmaceutical functions such as adherence to the surface, degradation and drug residence are altered due to change in particle size.

The tissue or cell internalization is dependent on particle size. Smaller particles tend to show *in-vivo* circulation and accumulation [3]. Greater circulation time is seen with particles of size less than 70nm however, particles of size more than 200 nm, show rapid clearance [7].

Surface Charge and Shape

Nanoparticles tend to show their effect through cell adhesion or interaction with the biological membrane. It is observed that positive charged nanoparticles such as super paramagnetic iron oxide particles and silver nanoparticles are taken up by the cells at a higher level than anionic nanoparticles [8]. The cellular uptake of chitosan nanoparticles indicated a positive relation with respect to surface charge [9]. Carboxyl quantum dots conjugated with amino-polyethylene glycol quantum dots forming amine terminated quantum dots of 100 nm size showed higher internalization in various cell lines with high negative zeta potential [10]. The particle shape and geometry also affect the cellular adhesion and uptake. It is observed that non-spherical particles tend to show slower rate of degradation due to areas of varying thickness. A positive correlation is observed between blood circulation of nanoparticles and their length to width ratio [11]. The non-spherical particles tend to tumble when flowing to RES rich organs such as liver, spleen or bone marrow and also when encountered with other obstacles in the blood vessels [12]. Positively charged nanoparticles show higher phagocytic uptake than negatively charged particles [13]. Less negatively charged particles and more positively charged particles show high cellular uptake. Negatively charged nanoparticles show high distribution due to longer blood circulation while positively charged particles through leaky microvasculature show high internalization into tumor cells [14].

Polydispersity Index

Polydispersity index (PDI) is a parameter to define the particle size distribution of nanoparticles obtained from photon correlation spectroscopic analysis. It is a dimensionless number extrapolated from the autocorrelation function and ranges from a value of 0.01 for mono dispersed particles and up to value of 0.5 – 0.7 for broad particles. Samples with very broad size distribution have polydispersity index value greater than 0.7. In pharmaceutical research, PDI value less than 0.5 favors narrow distribution of particles.

Zeta Potential

Zeta potential is an abbreviation for electrokinetic potential in colloidal systems. Zeta potential is electric potential in the interfacial double layer at the location of the slipping plane versus a point in the bulk fluid away from the interface. Zeta potential is not directly measurable but it can be calculated using theoretical models and an experimentally determined electrophoretic mobility or dynamic electrophoretic mobility. Mobility is defined as the velocity of a particle per electric field unit and is measured by applying an electric field to the dispersion of particles and measuring their average velocity. This velocity can be determined by measuring the dropper shift of laser light scattered off the moving particles. Values of zeta potential are dependent on the net charge carried by the polymer used, for example, with chitosan, zeta potential is high due to cationic nature of chitosan [15] while poly hydroxybutyric acid (PHB), polycaprolactone (PCL) and polylacticacid (PLA) are acidic in character due to surface oxidation resulting in the carboxylic groups formation, and/or hydroxyl ion adsorption, as suggested by Beattie [16]. Negative zeta potential observed in case of lipids used in the liposomes is due to the influence of composition and ionic strength of the buffer solution used to hydrate and suspend the liposomes on the surface charge density of the particles [17]. However, Voget et al., have reported that both positive and negative zeta potential for polybutylcyanoacrylate nanoparticles showed passage to blood brain barrier [18]. Generally, all living cells are negatively charged at neutral pH [19], therefore nanoparticles with cationic character shows better cell adhesion while particles with Zeta potential below 15mV escapes phagocytic uptake.

Shape and Surface Morphology

Shape and surface morphology of nanoparticles is done by Scanning Electron Microscopy (SEM). It has been used to determine surface topography, texture and to examine the morphology of fractured surface. Small volume of nanoparticle suspension is placed on an electron microscope brass stub. The stubs are placed briefly in a drier and then coated with gold in an ion sputter. Pictures of nanoparticles are taken by random scanning of the stub, then the shape and surface morphology of the nanoparticles is determined from the photomicrographs. SEM analysis is a vital tool to analyze the surface morphology of nanoparticles because the drug molecule is usually nonspherical but is turned to spherical morphology through nanonization [20]. If nanoparticles make cluster in micro size by agglomeration, then existed clusters can be identified by SEM. Transmission electron microscopy is also a vital technique to analyze shape of nanoparticles but after irradiation with light. TEM samples are prepared using a drop cast technique on a 200 mesh copper grid coated with a thick carbon layer then a very dilute drop of nanoparticle sample is placed on the top of the copper grid and left to dry for 24 hours. The next day images are taken at 100 kv. It is observed that particle size measurement through SEM is less than TEM as the former analyses the images of nanoparticles in a dry state than TEM which uses dispersion of sample [21].

Surface Modification

Fundamental concepts with the promising characteristics of materials at their nano-scale are mainly attributed to increased surface area, as well as quantum effects which further exhibit relatively unique reactivity, strength and electrical characteristics to the materials. Since the chemical reaction occurs at the surface, higher surface area per unit of mass favors the kinetics of reaction more. In nano-particulate form of materials, a greater proportion of atoms are distributed at the surface interface from the inner side. For example, in materials with 30 nm of average particle distribution, nearly 5% of its atoms are distributed at its surface while 20% of atoms are arranged on the surface with 10 nm of particle size, however when size of particles is further reduced to 3 nm, approximately, 50% of its atoms are distributed at the surface. This fundamental principle is used in designing targeted drug carriers for site specific delivery. In nanomedicines, the surface engineered particles are designed to impart specific characteristics such as stealth or prolonged circulation. If the surface of nanoparticles is hydrophobic, more of blood proteins such as opsonin tend to adsorb making the particle vulnerable to phagocytosis. Also incorporation of hydrophilic polymer such as polyethylene glycol makes the particles hydrophilic which enhances the circulation time. This is called as PEGylation. PEGylation also stabilizes the nanoparticles [22]. Due to high surface to volume ratio, these engineered nanoparticles tend to offer variety of advantages such as low dose as distribution volume of drug is reduced, therefore improving the pharmacokinetics and the biodistribution of the drug to specific organs. They are specific thus lower the cytotoxicity to healthy tissues. Most of the nanocarriers are biodegradable and offer better efficacy. They act as a vehicle for protected transport of drugs to affected sites and their controlled release in the body.

But the disadvantages associated with them are: after systemic administration they get diluted with blood leading to sub therapeutic dose at receptor level and hydrophobic particles do not circulate for long than hydrophilic particles (because of opsonization).

Surface modification is a widely known technique for targeted drug delivery of nanoparticles especially in cancer treatment. Such nanoparticles have provided an effective platform for a better and more specific delivery of cancer therapeutics. Ahead of targeted nanoparticles, are multifunctional nanoparticles that have been designed not only to detect but cure the affected tissue without associated problems of conventional drug delivery [23]. Researchers have experimented aerosol delivery of drug nanoparticles by inhalation to target cancerous tissues of lung [24]. A combinatorial approach where targeting and tracking of nanocarrier has been adopted for targeted delivery to lung cancer cells, was accomplished through flexible designing (surface modification) of the carrier [25]

Solid lipid nanoparticles (SLNs) and nanostructured lipid carriers (NLC) are common forms of lipid nanoparticles which are widely used in cancer chemotherapeutic delivery. These are highly promising due to their composition and potential for scale up technology. SLNs have been used as effective carriers for drugs but their widespread use is limited due to difficulties in formulation, erratic gelation, phase transition of lipids, crystallization during storage, low encapsulation and eviction of payload. It has been reported that NLC has overcome these issues due to the presence of liquid lipids along with solid lipids in an optimum ratio [26]. Iron oxide nanoparticles have been widely investigated for spatial drug delivery. The magnetite shows its magnetic behavior under an applied magnetic field. Also these magnetic nanoparticles tend to generate heat from magnetic materials thus when applied to cancer tissues, the temperature of cancer tissue rises to 42°C to 46°C which eventually leads to apoptosis [27]. Monodisperse magnetite nanospheres with hollow interior and porous shell structure have been reported for hyperthermia based treatment to cancer cells. The nanospheres have been synthesized through one-pot solvo thermal process where chemical conversions of the Fe (III) compounds is carried out to generate Fe_3O_4 which is then subjected to Ostwald ripening leading to formation of magnetite porous/hollow nanostructure with excellent heat generation for several continuous cycles of applied field [28]. Super-paramagnetic iron oxide nanoparticles (SPIONs) -drug conjugate with a magnet allows site-specific therapeutic effects. The prepared particles were found to be biocompatible and biodegradable as iron can be recycled through normal biochemical pathway for iron metabolism. The nanoparticles showed good cellular uptake as were coated with fatty acids for magnetic drug targeting. Cytotoxic drugs such as mitoxantrone conjugated with SPIONs showed adsorption on the protein shell with an eventual slow release at zero order kinetics, therefore assured site specific effects [29].

Entrapment Efficiency and Loading Efficiency

In medical science, drug entrapment into the polymeric carrier or onto metallic surface is important as it offers the payload for the drug required for targeting to tissue or cell. Drug encapsulation or loading is determined by drug content analysis using high pressure liquid chromatography (HPLC) [30].The nanoparticles are dissolved in suitable aqueous or organic solvent and then filtered through membrane filters. The resultant sample is then analysed by HPLC. The total concentration of drug (initial drug loaded in mg), volume of supernatant (ml), concentration of drug in supernatant (mg/ml); and weight of nanoparticles (mg) are required for entrapment efficiency and drug loading in nanoparticles.

Turbidity, Transmittance and pH

Nanoparticles have optical properties that are sensitive to size, shape, concentration, agglomeration state, and refractive index near the nanoparticle surface, which makes UV/Vis/IR spectroscopy a valuable tool for identifying, characterizing, and studying these materials. Nanoparticles made from certain metals, such as gold and silver, strongly interact with specific wavelengths of light and the unique optical properties of these materials is the foundation for the field of plasmonics. The transmittance of a sample (T) is defined as the fraction of photons that pass through the sample over the incident number of photons. UV/Vis spectroscopy measures those photons that are not absorbed or scattered by the sample. It is common to report the absorbance (A) of the sample, which is related to the transmittance (T) by expression $A = -\log 10\ (T)$. The relative percentage of scatter or absorption from the measured extinction spectrum depends on the size, shape, and composition and aggregation state of sample. Nanoparticle may absorb light, scatter light, or both. As a general rule, smaller particles have a higher percentage of their extinction due to absorption. For example, light scattering by small (2 nm diameter) gold particles is negligible; the optical spectrum of such nanoparticles is almost entirely due to photon absorption by the metal with negligible scattering. The diluted solution of gold nanoparticle appears brownish-red since light is absorbed across the entire visible spectrum. In case of amorphous silica, the nanoparticles do not absorb photons in the visible portion of the spectrum and large silica colloids scatter light (known as Rayleigh scattering). The curves related to optical density of both the metallic nanoparticles are same but the mechanism for extinction is different which is reflected in the strong differences in the visible appearance of these two materials. In nano-pharmaceuticals, especially in self-emulsifying lipid system, determination of percentage transmittance is carried out to understand the optical properties which may get altered during processing, thereby affecting the stability [31]. Light and ultrasound sensitive systems, and in particular, pH sensitive systems, the release of drug from nanocarriers is dependent on such stimuli. Chitin based nanogels have shown pH responsive swelling and drug release at acidic pH [32, 33]. This property is used to treat various types of cancers where due to tumour micro environment, complete internalization of drug is not possible. The rate and amount of doxorubicin released from gold nanoparticles was observed to be strongly dependent on the pH of the medium. The pH dependent release may help to improve the efficiency of the system without affecting the normal cells [34].

NANOSENSORS

A sensor, whether a chemisensor or a biosensor, is an integrated device capable of providing analytical information (such as concentration of an analyte) using a chemical or a biological recognition element [35]. Metal nanoparticles have drawn a great deal of attention due to their remarkable optical, physical and chemical properties, which largely differ from their bulk ones. Metal nanoparticles are more chemically reactive than their bulk solid counterparts and they are widely used as sensor materials in electronic devices. As a consequence, fabrication and preparation of metal nanoparticles have become increasingly popular in recent years. Among the metal nanoparticles, silver nanoparticles have got much recognition, because they have a number of superior properties and are widely applied in various fields, such as medicine, photo catalysis, lithography, electronics, surface-enhanced resonance Raman scattering and optical biosensors, etc. But the fundamental properties of these nanosensors depend on the size, size distribution, shapes and stability in aqueous solutions. Therefore their synthesis should investigate not only unusual properties of nanoparticles but also to adjust their physical properties and chemical properties as desired. Nanoparticles used as sensors show change in their conducting properties due to change in charge, for example, nitrogen dioxide if present, takes an electron from carbon nanotube making it less conductive, but if ammonia is present it reacts with water vapor and donates an electron to the carbon nanotube, making it more conductive. These engineered nanomaterials are classified in terms of their main component with a base of carbon as in nanotubes, fullerenes, nanoglobules, nanospheres, nanoparticles of latex, graphene and with a base of metal oxides such as titanium oxide, zinc oxide, iron oxide, silicon oxide or base of metals such as gold silver or iron. These nanosensors are used in food product packaging due to their advanced functional properties. Nanoparticles of silver or titanium dioxide are used in packaging to prevent spoilage of foods. These act as antimicrobials. Such a nanotechnology-enabled food packaging is generally divided into three main categories [36, 37, 38].

Improved Packaging

The nanomaterials are mixed into the polymer matrix to improve the gas barrier properties, as well as temperature and humidity resistance of the packaging.

Active Packaging

The nanomaterial interacts directly with the food or the environment to allow better protection of the product such as silver nanoparticles used as antimicrobials.

Intelligent Packaging

The nanomaterials are designed for sensing biochemical or microbial changes in the food, for example detecting specific pathogens developing in the food, or specific gases evolving from food spoiling.

These nanosensors are able to respond to environmental changes inside the package such as temperature, humidity and level of oxygen exposure. Therefore condition of food can be easily determined. They are also able to communicate the degradation of product or microbial contamination [39] as well as provide history of storage and duration of storage of food product. Nano plastic wrap (from Song Sing Nano Technology Co., Ltd in Taiwan) contains nanozinc oxide as a light catalyst to sterilize. The group has reported that nano zinc oxide coated films on a packaging material exhibited antimicrobial effects against L.monocytogenes and S. enteritidis in liquid egg white and in culture media packaging. Some of the nanomaterials have been added into food bottles and packages to make them light and fire resistant and impart greater mechanical strength [40, 41].

Biosensors

Biosensor is an analytical device which converts a biological response into an electrical signal. It is used to detect any biomolecule such as glucose, gonadotropic hormone, insulin, antigen or antibody etc. Such biochemical interactions of molecule with the metallic film of gold or silver result in a signal that is translated by a physical transducer. This happens due to optical property of these noble metals [42]. Although a lot of research is already devoted to develop biosensors based on this signal transduction methodology, for diagnosis and detection of several diseases, but the simplest kind of biosensors are the diagnostic kits for estimation of glucose or the pregnancy test kit where the gonadotropin hormone (HcG) protein is detected in the urine of pregnant women. The principle is based on a very attractive optical property of metals or nanostructures metals i.e., localized surface plasmon resonance (LSPR), which is generated by collective oscillations of the electron gas in metal nanostructures surrounded by a dielectric. When these nanostructures interact with beam of light, part of incident light as photons is absorbed, and a part is scattered in different directions. Both absorption of photon and scattering is enhanced when LSPR is excited [43]. Liedberg et al., [44] recognized that immunoglobulin antibodies when bind selectively on a gold film, there were shift in the critical angle values (the red shift of LSPR), changing the refractive index of the medium. This led to further development of several other biosensors. One such type is the ultrasensitive first clinical lab diagnostic, based on LSPR nanosensor, to detect ADDLs/anti-ADDLs antibody to diagnose Alzheimer's disease.

WASTE TREATMENTS

Pharmaceuticals have now been identified as environmental pollutants as they are partly excreted unchanged in waste water. Older techniques involved degradation of model pollutants but a thorough investigation is required for their complete degradation. Rice husk, groundnut shell, bamboo leaves and sugarcane bagasse are major waste materials which are burnt to get nutrients for the next crop. Burning turns the carbon, oxygen and hydrogen elements of the waste material into flammable gases such as carbon monoxide, hydrogen, methane and ash. This ash is rich in silica and carbon and it has no exploitation yet. This silica is used as silica nanoparticles in many fields such as ceramics, chromatography, anti-corrosion agent and in catalysis.

Disposal of rubber tires is an environmental challenge in many countries. Recently a new technology to use rubber as a source of zinc (Zn) for crops has been used. The rubber particles size is a key factor determining the rate of nutrient release from the rubber wastes. The efficiency of rubber nano-particles as Zn source for cucumber in nutrient solution culture was compared with a commercial Zn-sulfate fertilizer. In hydroponic experiment, ground tire rubber and nano-particles of tire rubber and rubber ash were used. The cucumber seedlings grown in the nutrient solution containing nano-particles of tire rubber and rubber ash had greater shoot and fruit yield and also accumulated higher amount of Zn in their tissues as compared with those grown in Zn sulfate-containing nutrient solution. Thus, nano-particles synthesized from waste tire rubber and rubber ash offer a strong value as Zn fertilizer for cucumber grown in the nutrient solution culture [45].

The solid waste such as extract of black tea solid has been treated with Zn acetate dehydrate to form zinc oxide nanoparticles which have been used for removal of pharmaceutical compounds such as ibuprofen, propranolol and ephedrine from human urine [46]. A unique green process is adopted for the synthesis of silver nanoparticles where silver nitrate is treated with aqueous extract of Ammonia baccifera. The phyto-synthesized silver nanoparticles have been tested for minimum inhibitory concentration and minimum bactericidal concentration against microorganism involved in infections in burns [47, 48].

China's largest coal company (Shenhua Group) in technical collaboration of licensed technology of Hydrocarbon Technologies started to liquify coal and turned it into fuel gas, reducing the pollution, acid rain and sulfur related pollution, using a gel-based nanoscale catalyst in an efficient and cost effective manner.

THERANOSTICS

Theranostic is a novel term coined for medicines that are used for simultaneous diagnosis and treatment. The primary goal of theranostics is to selectively target the disease confined to tissue or cell in order to increase diagnostic and therapeutic selectivity that makes the treatment shorter, safer, effective and inexpensive. Theranostic nanomedicines can be developed through various techniques. (i) therapeutic agents (e.g., anticancer drugs and photosensitizers) conjugated or loaded to imaging nanoparticles such as quantum dots, magnetic nanoparticles, and gold nanocages. (ii) tagging of imaging contrast agents, such as fluorescent dyes, optical or magnetic nanoparticles, and many other radioisotopes, to therapeutic nanoparticles. In addition, encapsulating both imaging and therapeutic agents together in biocompatible nanoplatforms such as lipid nanoparticles (solid lipid nanoparticles, nanostructured lipid carriers, liposomes), polymeric nanoparticles, ferritin nanocages, and porous silica nanoparticles [6]. Nanoparticles have also been used in radiotherapy as radiosensitizers and in imaging as contrast agents. Many studies have performed *in vitro* and/or *in vivo* experiments on these particles in human and animal cell lines. Complete imaging of hepatic cells is possible due to ready uptake of SPIONs by normal liver cells turning the image dark while the cancer cells are no able to take iron nanoparticles thus appear bright [49].

MISCELLANEOUS

Nanomaterials have been widely used in cosmetics, for example, nanoemulsions are used to increase the content of nutritious oils which are used to protect fragile active ingredients, like vitamins, which are released upon contact with the skin. Nanoemulsions do not cross the skin barrier and are recognized as safe but still for cosmetic products, the regulation of European Commission (EC) No 1223/2009 has established that a report concerning the nanomaterial-containing products should be submitted to the European Commission six months before their release into the market and the label should include the word 'nano' in brackets [50]. Nanopigments in cosmetic products are essentially used as UV-filters as they are able to reflect and scatter UV light. Nanosilver is used in soaps, toothpastes, wet wipes, deodorants, lip products, as well as face and body foams as has bactericidal and fungicidal properties [51, 27]. With the rapid development of nanotechnology in the past three decades, titanium dioxide (TiO_2) nanoparticles are frequently used in consumer products, food additives, cosmetics as skin whitener but once human beings are unintentionally exposed to nanoparticles they could be absorbed, and then accumulated in brain regions by passing through the blood–brain barrier or through the nose-to-brain pathway, potentially leading to dysfunctions of brain [52]. For the moment, only calcium carbonate (E170) and vegetable carbon (E153) are allowed in nanoform in foods, but their particle size should be included in the specifications of the product.

The application of a weak electric field than electroporation has been attempted for targeted delivery of vaccine loaded liposomes as nanoparticles. Oval albumin-loaded liposomes along with silver nanoparticles have been developed for transcutaneous immunization. The antigen showed deeper penetration into the viable epidermis thus offers a better alternative for needle-free vaccination [53]. Functionalization of textile materials using fabric dyes modified with silver or zinc oxide nanoparticles have been reported to impart antimicrobial properties to the textile [54].

Eddie Bauer Nano-CareTM developed khakis having properties of stain-repellent due to teflon coating of each fiber surface (10 to 100 nanometers) using a process called nano-whiskers. Similarly, Dockers also launched the khakis shirt and tie which were treated at nanoscale to make it "Stain Defender".

Wilson's Double Core tennis balls and official Davis cup ball, have developed a strategy to coat the ball with a unique nanocomposite made up of mix of butyl rubber, intermingled with nanoclay particles, which lock in air to enable more bouncing, lighter and flexible with long shelf life.

OLED (organic light emitting diodes) color screens developed by Kodak are based on nanostructured polymer films due to its thinner size, lighter weight, more flexibility, less power consuming, being preferred in various displays, and consumer devices such as cameras, laptops and televisions [55]. Similarly, the packaging industry is now shifting its trend from glass and aluminum to nanoclay treated plastic as packaging material because of low weight of packaging material, better protection against the external and higher shelf-life of stored product [56]. GM Safari and Astro Vans are using a light-weight but highly durable plastic nanocomposite which is known to enable it a scratch-resistant and rust-proof material which saves the fuel and increase longevity [57]. Toyota has also started using nanocomposites material in its bumper to reduce the weight by 60% while double the protection against denting and scratching [58].

Conclusion

Nanotechnology has emerged as a potential technique to both medical and non-medical research. With the advent of nanoparticles in biomedical research, hope for future in substantial drug delivery is seen. Being surface engineered, stable and compatible nanoparticles, they assure treatment to life threatening diseases such as cancer. Also the concept of nanotechnology allows the feasible detection of metal ions in water contamination, environmental toxicity, waste management and better human health through sound diagnosis. Nanosensors, nanochips and nano-spray paints are some of the other potential applications of nanotechnology where particle size, morphology and optical properties have given an edge to such nanomaterials. Therefore, the conclusion is, nanotechnology is a field where a physicist, a biologist and a chemist play an interactive role for a common goal-people comfort at an effective cost.

References

[1] R. Sinha, G. J. Kim, S. Nie, D. M. Shin, Nanotechnology in cancer therapeutics: bioconjugated nanoparticles for drug delivery. *Mol. Cancer. Ther.*, 8 (2006) 1909.

[2] G. Seeta, R. Raju, L. Benton, E. Pavitra, J. S. Yu, Multifunctional nanoparticles: recent progress in cancer therapeutics. *Chem. Commun.*, 51 (2015) 13248.

[3] R. Misra, M. Upadhyay, S. Mohanty, Design Considerations for Chemotherapeutic Drug Nanocarriers. *Pharm. Anal. Acta.*, 5 (2014)

[4] D. Nguyen, K. Kim, Controlled synthesis of monodisperse magnetite nanoparticles for hyperthermia-based treatments, *Powder Technol.*, 301 (2016) 1112.

[5] K. T. Nguyen, Targeted Nanoparticles for Cancer Therapy: Promises and Challenges, Nanomedic. *Nanotechnol.*, 2 (2011) 1.

[6] J. Ahmad, S. Akhter, M. Rizwanullah, S. Amin, M. Rahman, M. Z. Ahmad, M. A. Rizvi, M. A Kamal, F. J. Ahmad, Nanotechnology-based inhalation treatments for lung cancer: state of the art, *Nanotechnol. Sci. Appl.*, 8 (2015) 55.

[7] M. Rizwanullah, J. Ahmad, S. Amin, Nanostructured Lipid Carriers: A Novel Platform for Chemotherapeutics, *Curr. Drug Deliv.*, 13 (2016) 4.

[8] L. Q. Shao, B. Song, Toxicity of Titanium Dioxide Nanoparticles on Brain. In: X. Jiang and H. Gao (Eds). *Neurotoxicity of Nanomaterials and Nanomedicine*, (2017) 99.

[9] D. S. Bernardi, C. Bitencourt, D. S. D. Silvera, E. L. D. Cruz, M. A. Pereira-da-Silva, L. H. Facciolia, R. F.V. Lopez, Effective transcutaneous immunization using a combination of iontophoresis and nanoparticles. *Nanomed. Nanotechnol.*, 12 (2016) 2439.

[10] J. Zaloga, M. Pottler, G. Leitinger, R. P. Friedrich, G. Almer, S. Lyer, E. Baume, Tietze, R. Heimke-Brinck, H. Mangge, F. Dorje, G. Lee, C. Alexiou, Pharmaceutical formulation of HSA hybrid coated iron oxide nanoparticles for magnetic drug targeting. *Eur. J. Pharm. Biopharm.*, 101 (2016) 152.

[11] J. Pulit-Prociak, J. Chwastowski, A. Kucharski, M. Banach, Functionalization of textiles with silver and zinc oxide nanoparticles. *Appl. Surf. Sci.*, 385 (2016) 543.

[12] D. T. Nguyen, K. Kim, Controlled synthesis of monodisperse magnetite nanoparticles for hyperthermia-based treatmenrs. *Powder Technol.*, 301 (2016) 1112-1118.

[13] S. S. M. Hassan, H. I. Abdel-Shafy, M. S.M. Mansour, Removal of pharmaceutical compounds from urine via chemical coagulation by green synthesized ZnO-nanoparticles followed by microfiltration for safe reuse. *Arab. J. Chem.*, (2016) http://dx.doi.org/10.1016/j.arabjc.2016.04.009.

[14] K. Jadhav, D. Dhamecha, D. Bhattacharya, M. Patil, Green and ecofriendly synthesis of silver nanoparticles: Characterization, biocompatibility studies and gel formulation for treatment of infections in burns, *J. Photoch. Photobio.*, 155 (2016) 109

[15] V. Vaibhav, U. Vijayalakshmi, S. M. Roopan, Agricultural waste as a source for the production of silica nanoparticles Spectrochim. *Acta. A.*, 139 (2015) 515

[16] S. Moghaddasi, A. H. Khoshgoftarmanesh, F. Karimzadeh, R. L. Chaney, Preparation of nano-particles from waste tire rubber and evaluation of their effectiveness as zinc source for cucumber in nutrient solution culture, *Sci. Hortic.*, 160 (2013) 398.

[17] N. Chaniotakis, R. Buiculescu, Semiconductor quantum dots in chemical sensors and biosensors. In: *Nanosensors for Chemical and Biological Applications. Nanosensors for Chemical and Biological Applications Sensing with Nanotubes, Nanowires and Nanoparticles* ed. K. C. Honeychurch. (2014) 267.

[18] P. Pandey, M. Datta, B. D. Malhotra, Prospects of nanomaterials in biosensors *Anal. Lett.*, 41 (2008) 159.

[19] M. Cushen, J. Kerry, J. Morris, M. Cruz-Romero, E. Cummins, Evaluation and simulation of silver and copper nanoparticle migration from polyethylene nanocomposites to food and an associated exposure assessment. *J. Agric. Food. Chem.*, 62 (2014) 1403.

[20] C. Silvestre, D. Duraccio, C. Sossio, Food packaging based on polymer nanomaterials. *Prog. Polym. Sci.*, 36 (2011) 1766.

[21] T. V. Duncan, Applications of nanotechnology in food packaging and food safety: Barrier materials, antimicrobials and sensors. *J. Colloid Interface Sci.*, 363 (2011)1.

[22] D. L. Thorek, A. Tsourkas, Size, charge and concentration dependent uptake of iron oxide particles by non-phagocytic cells. *Biomaterials.*, 29 (2008) 3583.

[23] Z. G. Yue, W. Wei, P. P. Lv, H. Yue, L. Y. Wang, Z. G. Su, G. H. Ma, Surface Charge Affects Cellular Uptake and Intracellular Trafficking of Chitosan-Based Nanoparticles. *Biomacromolecules.*, 12 (2011) 2440.

[24] T. A. Kelf, V. K. A. Sreenivasan, J. Sun, E. J. Kim, E. M. Goldys, A. V. Zvyagin, Non-specific cellular uptake of surface-functionalized quantum dots. *Nanotechnology*, 21 (2010) 1.

[25] S. M. Moghimi, A. C. Hunter, J. C. Murray, Long-circulating and target-specific nanoparticles: theory to practice. *Pharmacol. Rev.*, 53 (2001) 283-318.

[26] Y. Geng , P. Dalhaimer, S. Cai, R. Tsai, M. Tewari, Shape effects of filaments versus spherical particles in flow and drug delivery. *Nat. Nanotechnol.*, 2 (2007) 249.

[27] A. R. Nicholas, M. J. Scott, N. I. Kennedy, M. N. Jones, Effect of grafted polyethylene glycol (PEG) on the size, encapsulation efficiency and permeability of vesicles. *Biochim. Biophys. Acta.*, 1463 (2000) 167.

[28] I. D. L. Calle, M. Menta, F. Seby, Current trends and challenges in sample preparation for metallic nanoparticles analysis in daily products and environmental samples: A review. *Spectrochim. Acta. B.*, 125 (2016) 66.

[29] European Commission Regulation no 1169/2011 of the European parliament and of the council of 25 October 2011 on the provision of food information to consumers, amending Regulations (EC) No 1924/2006 and (EC) No 1925/2006 of the European Parliament and of the Council, and repealing Commission Directive 87/250/EEC, Council Directive 90/496/EEC, Commission Directive 1999/10/EC, Directive 2000/13/EC of the European Parliament and of the Council, Commission Directives 2002/67/EC and 2008/5/EC and Commission Regulation (EC) No 608/2004. *Off. J. Eur. Union* (2011) (304/18-304-63).

[30] S. Li, A. Wang, W. Jiang, Z. Guan, Pharmacokinetic characteristics and anticancer effects of 5-Fluorouracil loaded nanoparticles. *BMC Cancer.*, 8 (2008) 1.

[31] R. Al-Kassas, J. Wen, A. E. M. Cheng, A. M. J. Kim, S. Z. M. Liu, J. Yu, Transdermal delivery of propranolol hydrochloride through chitosan nanoparticles dispersed in mucoadhesive gel. *Carbohyd. Polym.*, 153 (2016) 176.

[32] D. S. Bernardi , C. Bitencourt, D.S.C. da Silveira, E. L.C.M. da Cruz, M. A. Pereira-da-Silva, L. H. Faccioli, R. F.V. Lopez Effective transcutaneous immunization using a combination of iontophoresis and nanoparticles. *Nanomed. Nanotechnol Nanomedicine: NBM,*12 (2016) 2439.

[33] N. Voigt, P. Henrich-Noack, S. Kockentiedt, W. Hintz , J. Tomas, B. A. Sabel, Surfactants, not size or zeta-potential influence blood-brain barrier passage of polymeric nanoparticles. *Eur. J. Pharm. Biopharm.*, 87 (2014) 19.

[34] J. M. Kleijn, H.P. van Leeuwen, in: A. Bazkin, W. Norde (Eds.), *Physical Chemistry of Biological Interfaces*, Marcel Dekker, New York, 2000 (Chapter 2).

[35] J. K. Beattie, The intrinsic charge on hydrophobic microfluidic substrates, *Lab Chip* 6 (2006) 1409.

[36] B. Kanga, J. Choia, S. Leea, J. Leea, T. Kima, W. Janga, A. Tunsirikongkonb, J. Kim, J. Park, Enhancing the *in vitro* anticancer activity of albendazole incorporated into chitosan-coated PLGA nanoparticles. *Carbohydr. Polym.*, 159 (2017) 39.

[37] E. M. Huseynov, Investigation of the agglomeration and amorphous transformation effects of neutron irradiation on the nanocrystalline silicon carbide (3C-SiC) using TEM and SEM methods. *Physica. B*, 510 (2017) 99.

[38] A. Pardakhty, E. Moazeni, J. Varshosaz, V. Hajhashemi, A. R. Najafabadi. Pharmacokinetic study of niosome-loaded insulin in diabetic rats. *Daru.*, 19 (2011) 404.

[39] N. S. Rejinold, A. Nair, M. Sabitha, K.P. Chennazhi, H. Tamura, S.V. Nair, R. Jayakumar, Synthesis, characterization and *in vitro* cytocompatibility studies of chitin nanogels for biomedical applications. *Carbohydr Polym.*, 87 (2012) 1, 943.

[40] T. Zhou , C. Xiao , J. Fan , S. Chen , J. Shen , W. Wu , S. Zhou A nanogel of on-site tunable pH-response for efficient anticancer drug delivery. *Acta Biomater.*, 9(1) (2013) 4546

[41] D. E. Owens, N. A. Peppas, Opsonization, biodistribution, and pharmacokinetics of polymeric nanoparticles. *Int. J. Pharm.*, 307 (2006) 93

[42] J. Ahmad, K. Kohli, S. R. Mir, S. Amin, Effect of oil and co-surfactant on the formation of Solutol HS 15 based colloidal drug carrier by Box–Behnken statistical design. *Colloids Surf.*, A 453 (2014) 68.

[43] M. Rizwanullah, S. Amin, J. Ahmad, Improved pharmacokinetics and antihyperlipidemic efficacy of rosuvastatin-loaded nanostructured lipid carriers. *J. Drug Target*, 25 (2017) 58.

[44] J.E. Kipp, J.C.T Wong, M.J. Doty, C.L Rebbeck, *United States Patent 6,607,784*, Baxter International Inc. (Deerfield, IL), (2003). USA.

[45] A. Madhusudhan, G. B. Reddy, M. Venkatesham, G. Veerabhadram, D. A. Kumar, S. Natarajan, M. Yang, A. Hu, S. S. Singh. Efficient pH Dependent Drug Delivery to Target Cancer Cells by Gold Nanoparticles Capped with Carboxymethyl Chitosan. *Int. J. Mol. Sci.*, 15(2014) 8216.

[46] B. Liedberg, C. Nylander, I. Lundstrom, Biosensing with surface plasmon resonance--how it all started. *Biosens. Bioelectron.* 10 (1995)

[47] B. Sepulveda, P.C. Angelome, L.M., Lechuga, L.M. Liz-Marzan. LSPR-based nanobiosensors *Nano Today.*, 4 (2009) 244.

[48] V. Ayala, A. P. Herrera, M. Latorre-Esteves, M. Torres-Lugo, C. Rinaldi, Effect of surface charge on the colloidal stability and in vitro uptake of carboxymethyl dextran-coated iron oxide nanoparticles. *J. Nanopart. Res.*, 15(2013) 1874.

[49] E. Frohlich, The role of surface charge in cellular uptake and cytotoxicity of medical nanoparticles. *Int. J. Nanomedicine.* 7 (2012) 5577.

[50] S. Gajbhiye, S. Sakharwade, Silver Nanoparticles in Cosmetics, *JDSCA*, 6 (2016) 48-53.

[51] J. Huang, Y. Li, A. Orza, Q. Lu, P. Guo, L. Wang, L. Yang, H. Mao, Magnetic Nanoparticle Facilitated Drug Delivery for Cancer Therapy with Targeted and Image-Guided Approaches. *Adv. Funct. Mater.* 26 (2016) 3818.

[52] R. V. Kurahatti, A. O. Surendranathan, S. A Kori, N. Singh, A. R. Kumar, S. Srivastava, Defence Applications of Polymer Nanocomposites. *Def. Sci. J.*, 60 (2010) 551.

[53] Lloyd, M. Shannon, L. B. Lave, Life Cycle Economic and Environmental Implications of Using Nanocomposites in Automobiles. *Environ. Sci. T.,* 37 (2003) 3458.

[54] A. Wong, S. F. Wijnands, T. Kuboki, C. B. Park, Mechanisms of nanoclay-enhanced plastic foaming processes: effects of nanoclay intercalation and exfoliation. *J. Nanopart. Res.* 15 (2013) 1815.

[55] M. S. AlSalhi, J. Alam, L. A Dass, M. Raja, Recent advances in conjugated polymers for light emitting devices. *Int. J. Mol. Sci.* 12 (2011) 2036.

[56] A. Babu, A. K. Templeton, A. Munshi, R. Ramesh. Nanoparticle-Based Drug Delivery for Therapy of Lung Cancer: Progress and Challenges. *J. Nanomater.*, (2013)1.

[57] H. Bouwmeester, S. Dekkers, M. Y. Noordam, W. I. Hagens, A. S. Bulder, C. De Heer, A. J. Sips, Review of health safety aspects of nanotechnologies in food production. *Regul. Toxicol. Pharmacol.* 53 (2009) 52.

[58] S. Bajpai, N. Chand, V. Chaurasia, Nano Zinc Oxide-Loaded Calcium Alginate Films with Potential Antibacterial Properties, *Food and Bioprocess Tech.* 5 (2012)1871.

[59] T. Jin, D. Sun, J. Y. Su, H. Zhang, H. J. Sue, Antimicrobial efficacy of zinc oxide quantum dots against Listeria monocytogenes, Salmonella Enteritidis, and Escherichia coli O157:H7. *J. Food Sci.* 74 (2009) 46.

In: Handbook of Remediation ...
Editors: Sadia Ameen, M. Shaheer Akhtar et al.
ISBN: 978-1-53615-118-3

Chapter 8

TOXICOLOGICAL AND ANTIMICROBIAL PROPERTIES OF NANOSTRUCTURED METAL OXIDES

Z. A. Ansari[1], S. G. Ansari[1,*], Rajesh Kumar[2], Girish Kumar[2], Ahmad Umar[3,4], Sang-Hoon Kim[3,4,5] and Hamed Algarni[6]

[1]Centre for Interdisciplinary Research in Basic Sciences, Jamia Millia Islamia, New Delhi, India
[2]PG Department of Chemistry, JCDAV College, Dasuya, Punjab, India
[3]Promising Centre for Sensors and Electronic Devices, Najran University, Najran, Kingdom of Saudi Arabia
[4]Department of Chemistry, Faculty of Sciences and Arts, Najran University, Najran, Kingdom of Saudi Arabia
[5]BND Networks Co., LTD., Business Incubator Center 108-Ho, Iksan, Jeonbuk, Republic of Korea
[6]Department of Physics, Faculty of Science, King Khalid University, Abha, Kingdom of Saudi Arabia

ABSTRACT

Safety from nanomaterials including metals and metal oxides is of great importance in applied nanotechnology especially when these are used in biomedical applications. Scientists and engineers have expressed considerable concern for cytotoxicity and biosafety of utilizing these materials with bio-molecules in biomedical applications. Therefore, evaluation of safety has to be considered from the interface point of view, including their interactions with proteins, DNA, lipids, membranes, organelles, cells, tissues, and biological fluids, as against the traditional toxicological tests. In this review, we report the studies on cytotoxic and antibacterial effects of zinc oxide (ZnO), magnesium

* Corresponding Author's Email: piezo2k@gmail.com.

oxide (MgO) and surface functionalized ZnO to cancerous cells, human cells and gram positive and negative bacterial cells. Nanoparticles were synthesized by sol-gel method. Structural and morphological characterization of nanoparticles were carried out using X-ray diffraction technique (XRD), Fourier transform infrared spectroscopy (FTIR), high resolution and Bio-transmission electron microscopy (HRTEM) techniques. The cytotoxic assays of metal oxide nanoparticles were examined using malignant human T98G gliomas, KB epithermoids and HEK normal non-malignant kidney cells, human intestinal cell lines (INT 407) and human cervical cancer cell lines ***(SiHa)*** as a function of nanoparticle concentration. The antibacterial properties of the synthesized nanoparticles were investigated using Vibrio Cholerae, *E. coli, and Bacillus Subtilis* bacterial system where significant inhibition of bacterial growth is noticed. The results demonstrated that treatments with ZnO nanoparticles sensitize cancerous cells by increasing both the mitotic (linked to cytogenetic damage) and interphase (apoptosis) death. It is expected that the nanoparticles can behave as genotoxic drugs, since they induce a micronucleus formation in the cells and could be helpful in designing more potent anticancer agents for therapeutic use.

Keywords: nanomaterials, metal oxides, biomedical applications, antibacterial activity, anticancer activity

1. INTRODUCTION

Owing to the wide range of potential applications, e.g., optoelectronics, medicine, catalysts, information technology, energy sector, etc. nanotechnology has emerged as one of the most important branches of science and technology [1, 40]. One of the interesting features of nanotechnology is its multidisciplinary nature that fosters on the developments in many fields, including applied physics, materials science, colloid and interface science, chemical science and all branches of engineering, etc. In today's world, nanomaterials have been applied to nano-photonics/laser, nanoelectronics, solar cells, resonators and high sensitivity sensors, high efficiency photovoltaics, and catalysts, functional coatings, energy storage, drug delivery and biomedicines [1, 16], etc. This becomes more interesting, when the frontier between the nano science and life science starts exchanging knowledge and innovation. Nanotechnology has set high expectation in biological and medical sciences to solve key questions concerning biological system that transpire at the nanoscale. Nanomedicine, the approach of science and engineering at the nanometer scale towards biomedical applications, has been drawing considerable attention [17]. Living organisms are built of cells that are typically of the size ~ 10 μm. However, the cell parts are much smaller and are in the sub-micron size domain. Proteins are even smaller having typical size of 5 nm, which is comparable with the dimensions of smallest manmade nanoparticles. This size comparison gives an idea of using nanoparticles as very small probes that would allow to spy at the cellular machinery without introducing too much interference [18]. It is well known that nanomaterials have novel electronic, optical, magnetic and structural properties different from individual molecules and bulk materials, which can be tuned precisely to explore biological phenomena through numerous innovative techniques. For such bio-applications, the nanomaterials: (i) must be designed to interact with proteins and cells without interfering with their biological activities; (ii) must be nontoxic and (iii) must maintain their properties after surface modifications. The major areas in which nanoparticles are applicable are: detection of bio-molecules, bio-imaging, cancer diagnostics and therapeutics, drug/gene delivery, bio-sensing and bactericidal

The key problem in the exploitation of nanoparticles in biology is to stabilize them in high ionic strength aqueous buffers and to couple targeting moieties or bio-molecules to their surfaces [19, 20]. This involves an initial ligand exchange step or may require different procedures depending on the nature of nanoparticle. The conjugation of nanoparticles with bio-molecules such as proteins, DNA can be done either by direct covalent linkage or by non-covalent interactions [21, 24]. The covalent attachment can be achieved either through chemisorption of the bio-molecule to the particle surface or through the use of hetero-bifunctional linkers [25]. For example, chemisorption of proteins on nanoparticles (Au, ZnS, CdS, and CdSe/ZnS) surface can be done through cysteine residues present at protein surface (e.g., oligopeptide, serum albumin) [26], or by using chemical reagent 2-iminothiolane [27]. Bio-molecules are often covalently linked to ligands on the nanoparticle surface via traditional coupling strategies such as carbodiimide-mediated amidation and esterification [28].

Nanostructured semiconducting oxides are the promising agents with applications in majority of the fields of science and technology including biomedical sciences leading to novel developments. New innovations are being made in various disciplines of medical science and biology by utilizing/exploring combination of nanomaterials and biomaterials. Nanoparticles of metal oxides are being investigated for diverse areas of medical and biological sciences for example cosmetics and skin care products, drug delivery, nano-medicines, molecular biology, bio-markers, fillers, catalysts, tissue engineering, non-viral gene carriers, clinical bio-analytical diagnostics, detection of DNA, intracellular labelling, imaging, therapeutics, clinical bio-analytical diagnostics and are largely used to diagnose and treat different types of cancers due to their small size and functionality which are different from their bulk counterparts [29, 38]. Among various areas, the utilization of nanomaterials for public health and safety is one of the most important and serious concern of the scientific community. Nanomaterials are being used for various purposes such as thc early detection of diseases in human beings, sensing and detection of biological molecules and systems along with prohibition, inhibition and cure of various diseases [29, 37].

Other than metals and carbon nanotubes, researchers have focussed on to explore the nanostructured metal oxides (NMOx) such as copper oxide (CuO), iron oxide (Fe_2O_3), zinc oxide (ZnO), aluminium oxide (Al_2O_3), cerium oxide (CeO_2), titanium oxide (TiO_2), silicon oxide (SiO_2) and zirconium oxide (ZrO_2), etc. for tissue engineering and biomedical diagnostics due to their tunable size and large surface-to-volume ratio [39, 42]. ZnO has a rich family of the nanostructures with widespread applications in sensors, solar cells, lasers, optical devices, etc. and is one of the important biocompatible oxides [43, 47]. Synthesis of zinc oxide nanostructures require a controlled conditions as they are prone to the defects. ZnO can be used as a suitable matrix to conjugate with bio-molecules due to IEP and fast electron transfer properties [43, 47]. ZnO nanoparticles can be synthesized in such a manner that their charge on the surface can be controlled by using the surfactants which prevents agglomeration, stabilizing them in sol either by steric repulsion or by electrostatic repulsion [48]. For bio-applications, different amines like iso-propylamine (IPA), diethylamine (DEA) and triethylamine (TEA) can be used as surfactant during the synthesis to get amine functionalized ZnO NPs/QDs where the amine groups can help to conjugate with bio-molecules. Such amine functionalized ZnO NPs can be used to study/manage the oxidative stress in the biological systems, where probably double effect can be observed.

Magnesium oxide (MgO) is known for its applications in catalysis, antibacterial activity, toxic-water remediation, high adsorption capability, good chemical stability and high isoelectric point ~ 12 [49, 52]. However, both the beneficial and toxic effects of MgO nanoparticles in biological systems are notable with potential applications [53]. MgO nanoparticles are well known for their biological applications in relief of heartburn and bone regeneration and cancer therapy [54].

Recently, some studies about the cytotoxicity effect of various inorganic NMOx such as CuO, Fe_2O_3, ZnO, Al_2O_3, CeO_2, TiO_2 and SiO_2 have been carried out using airway epithelial cells, human lung epithelial cells, human lung cancer cells, human neural cells and human fetal lung fibroblasts [55, 63]. Ge et al. have reported cytotoxicity of MgO nanoparticles on human umbilical vein endothelial cells cultured *in vitro* by MTT assay and found that MgO is non-cytotoxic at lower concentrations of 200 μg/ml whereas at higher concentration of 500 μg/ml, it yields relatively lower growth rate than control [53]. Sun et al. examined the cytotoxicity of several metal oxide nanoparticles including MgO on human cardiac microvascular endothelial cells [64]. Fischer et al. have reported the cytotoxicity of magnesium (Mg) material onto human osteosarcoma cell line (MG63) and found less cytotoxicity (26 - 40%) as compared to control [65]. The cytotoxic effects of MgO nanoparticles against normal lung fibroblast cells and different types of cancerous cells such as HeLa, AGS and SNU-16 cells have also been reported [65]. Thus, there is perhaps a considerable scope to study the toxicity of MgO nanoparticles for *in vitro* and *in vivo* applications.

Cancer is a complex disease which is caused due to the cellular malfunction. The cellular malfunction occurs due to two reasons; the first is due to a carcinogen-induced or natural mutation, gene amplification or chromosomal translocation while the second is *apoptosis complex* which is caused due to genetic mutations [66, 70]. The cancer cells develop due to the above stated factors in the same cell at the same time. Therefore, the transformation of normal to cancerous cell involves three distinct phases i.e., initiation, promotion and progression [71, 74]. Cancer is a serious disease and according to a recent report of National Institutes of Health, USA, published in 2010 that more than 1.5 million new cases have been diagnosed in USA with overall cost of US $ 263.8 billion [67]. Due to the above given facts, the serious and complex disease 'cancer' is one of the most widely investigated disease by medical practitioners and scientists.

Cervical cancer in human beings is one of the leading diseases worldwide, in which a group of cells shows uncontrolled growth [75]. Particular human papillomavirus (HPV) types are carcinogenic in humans and more than 99% of cervical cancers are initiated by HPV infection [76]. India with a population of 366.58 million women having age of 15 years and older are at risk of developing cervical cancer. Current estimates indicate that every year 1,34,420 women are diagnosed with cervical cancer and 72,825 die from the disease [77]. The present treatments in cancer therapy, including surgery, radiation, photodynamic therapy and conventional chemotherapy, have many limitations for example they can affect all the cells in the body [78, 79]. In this context, nanomaterial based therapy would be very effective to prevent uncontrolled growth of the cells due to its nanoscale dimensions which are comparable to that of biological cells. The nanoparticles are being utilized in the area of nanomedicine as new therapeutic tools for the treatment of human diseases [80]. The nanoparticles interact to the cellular components such as the plasma membrane, organelles, or macromolecules and create genotoxic effect which leads to the DNA damage [81].

With the fast growth of nanotechnology in multidisciplinary fields, scientists are working extensively to apply this extremely wide and versatile field of nano-science for the cancer treatment and it is expected that nanotechnology will bring new methods and approaches for cancer diagnosis, treatment, and prevention [82, 87]. Working in this direction, scientists revealed that the nanoscale materials which exhibit unique physicochemical characteristics due to their size, surface area and enhanced reactivity can easily enter in cells, protein, etc. and hence can effectively participate in various biological systems [82, 88]. Among various nanomaterials, ZnO nanostructures possess excellent properties and wide applications. Due to excellent properties of ZnO, such as wide band gap, high exciton energy, low cost synthesis, high specific surface area, high conductivity, bio-compatibility and so on, it is widely used for variety of applications for example photocatalysis, sunscreen, cosmetic products, optoelectronic devices, solar cells, chemical, bio, and gas sensors, piezoelectric devices, light emitting diodes, several biological applications such as antibacterial, antifungal agents, etc. [89, 105]. Over other applications, it is also used as photosensitizer (PS) in photodynamic therapy (PDT) as their light energy usually matches with the maximum absorption of wave length to the targeting drugs. The molecules absorb this energy and transform to an excited singlet state and again return to their native or original state. This is followed by the transfer of energy to the oxygen molecule or to other substrate molecule in the surroundings which then react with the oxygen [106, 108]. Recently, few reports have appeared in the literature related to the utilization of ZnO nanomaterials for cancer detection and therapy. Wahab et al. have reported the effect of ZnO nanoparticles on CloudmanS91 melanoma cancer cells [108]. Park et al. have demonstrated the toxic effect of ZnO nanomaterials by physicochemical characterization and reactive oxygen species (ROS) [109]. Ostrovsky et al. have demonstrated the cytotoxic effects of 60 nm sized ZnO nanoparticles, prepared by ultrasonic process, towards human glioma cell lines U87, A172, U251, LNZ 308, LN18, LN 229, human normal astrocytes as well as on breast and prostate cancer cells [110]. After *in-vitro* studies on various animal and human cells, the effect of ZnO nanoparticles on genetic material or chromosome damage was important to study. Therefore, this study deals with clono-genicity and genotoxicity analyses of ZnO nanoparticles in cancer cells.

Recent studies have shown that ZnO nanoparticles can be toxic to a wide variety of bacteria [111]. It was found to be more toxic to *Escherichia coli* (*E. coli*) compared to other metal oxide nanoparticles such as Fe_2O_3, Y_2O_3, TiO_2, and CuO [112]. One possible reason for the observed toxicity is that nanosized ZnO is an excellent photocatalyst with band gap of 3.3 eV and high excitonic binding energy of 60 meV in which free electrons and holes could be generated by light equal to or higher than its band gap energy. These electron-hole pairs diffuse out to the surface and transform the surrounding oxygen or water molecules into hydroxyl radicals via strong oxidization, which kills bacteria [113]. Among the metal oxides nanoparticles, TiO_2, MgO, and ZnO are reported to be superior in terms of safety, durability and heat resistance, when compared with conventional organic antibacterial agents [114, 115]. Further it has been established that identical ZnO nanoparticles had minimal effects on primary human T cell viability at concentrations toxic to both gram negative and gram positive bacteria exhibiting the selective toxicity of ZnO nanoparticles towards prokaryotic as compared to eukaryotic system [116]. In recent studies, ZnO QDs have been reported to show antibacterial activities against major food borne pathogens like Escherichia coli (E. coli), Salmonella, Listeria monocytogenes, and Staphylococcus aureus [117, 118]. Also, ZnO QDs increase the sensitivity of laboratory *E. coli* to ciprofloxacin in disc diffusion assays [119]. Therefore, ZnO exhibiting high antibacterial activity is advantageous even for *in-vivo* administration since it contains mineral elements essential for human being. Owing to these properties, extensive efforts have been made towards the study of antibacterial activity of ZnO [117, 120, 128].

Metal oxide nanomaterials constitute one of the most important classes of materials which are efficiently used in a variety of applications due to exotic properties such as non-toxicity, high chemical and physical stability, efficient biological properties and so on [129]. Among several metal oxide nanomaterials, ZnO possesses special place due to its own properties and wide application [130]. The properties of ZnO include bio-compatibility, non-toxicity, chemical and photochemical stability, high specific surface area, electrochemical activities, stability under harsh processing conditions, etc. [11]. In addition to this, ZnO is also enlisted as a GRAS (generally recognized as safe) material by the US Food and Drug Administration (21CFR182.8991) [43, 131]. As a GRAS material, ZnO exhibits significant growth inhibition properties over a broad range of both Gram-positive and Gram-negative bacteria, to name a few, *Escherichia Coli, L. Monocytogenes, S. Enteritidis, B. Subtillis, S. Aureus, P. Aeruginosa, E. Faecalis*, and so on [44, 46]. It is found that the interaction of ZnO nanomaterials with the bacteria resulted in the disruption of cell membrane integrity. The exact mechanism for the antibacterial properties of ZnO is not very clear but the suggested mechanism is based on the catalysis of reactive oxygen species (ROS) from water and oxygen which leads to the disruption of cell membrane integrity [132]. Moreover, it was observed that the morphologies and surface area of ZnO nanomaterials have strong impact on the antibacterial properties as the catalysis of radical formation occurs on the surface of nanomaterials [45, 132]. Therefore, the surface area of the ZnO nanomaterial is related to the antibacterial activities, i.e., increasing the surface area enhances its antibacterial properties while decreasing the surface area leads to the deterioration of the antibacterial properties [46].

It is very well known that different amines in the biological system play a vital role in the functioning of the system which are nitrogen based and are used in intercellular communication. Additionally, amines have been found to have antioxidant effect and hence are used as commercial and industrial antioxidants [129]. A proton from amines can be easily donated to the chain propagating peroxy, alkoxy, and hydroxy radicals converting it to water or alcohol thereby stabilizing the system. The antioxidant functions as stabilizer by counteracting the oxidizing species generated during the processes mediated by free radicals [130]. In recent studies reported by Omer Kesmez et al. four different amines were tested as base catalyst in the synthesis of SiO_2 nanoparticles [131]. Oxidative stress is a biological phenomenon of various cellular systems and is important in cellular and physiological signaling but in excess can cause diseases and aging for both unicellular and multicellular organisms/biosystems. Oxidative stress in biological systems arises due to the imbalance between the generation of ROS and the natural ability of the systems to detoxify them. The ROS includes chemically reactive ions, radicals and molecules of oxygen such as $^{\bullet}O_2$, $^{\bullet}OH$, H_2O_2 which damage the cellular components like nucleic acid, proteins, enzymes, membranes, etc. causing oxidative stress [133, 134].

In general ROS are generated in electron transport chain of aerobic pathways and redox reactions involving different metabolites. Scavenging ROS has been a difficult task causing problems to aerobic cellular machinery survival due to ROS induced non-pathogenic disorders like cancer, arthritis, inflammation, myocardial disorders, and neural degenerative disorders in humans and other organisms. Biological systems have various mechanisms to overcome the effect of ROS by producing number of reducing equivalents [135, 140]. On the other hand, ROS also play an important role in sustaining life through cellular signaling and coordinating with various components of cells in both unicellular and multicellular biological systems. Therefore, it becomes an important issue to maintain the balance between the ROS generation and antioxidation of excess ROS in the cellular system. Various antioxidants such as bio-molecules, enzymes and vitamins are provided naturally in the biological systems to counter the negative effects of reactive radicals and agents. Some of the known antioxidants are namely glutathione, carotenoids, flavnoids, curcumin, vitamin C, vitamin E, vitamin A, peroxidases, catalase, superoxide dismutase, etc. [132].

In this report, we present the review of the work carried out on cytotoxicity/genotoxicity/antibacterial studies against Human T98G Gliomas (Brain cancer), KB epidermoid (Skin Cancer) and HEK normal kidney cells, human cervical cancer (SiHa) cell lines with normal cell lines (INT 407), *Vibrio Cholerae* (gram negative), *E. coli* (gram negative) and *S. aureus* (gram positive) cells using ZnO, MgO, isopropylamine, di- and tri ethylamines funtionalized ZnO nanoparticles. These nanoparticles were used for understanding cytotoxicity. Cell viability, morphologies, clonogenicity and micronucleus formation in normal/malignant cells at different incubation time were also studied. Detailed experiments were performed to determine the minimum inhibitory concentration (MIC) of MgO and ZnO. Additionally the antioxidant properties of sol-gel synthesized amine functionalized ZnO-NPs were carried out for *E. coli* and *S. aureus* cells. Three different amines viz. isopropylamine, di- and tri ethylamines were used as surfactants for ZnO synthesis. The nanoparticles were synthesized using sol-gel method and characterized for morphological, structural, compositional and optical properties using various techniques. A mechanism for activity of the nanoparticles on the investigated cells and bacteria has been proposed.

2. SYNTHESIS AND CHARACTERIZATIONS OF NANOPARTICLES

For a set of experiments on induction of cell death in malignant human T98G gliomas, KB epithermoids and HEK normal non-malignant kidney cells, ZnO was synthesized using zinc acetate dihydrate ($Zn(CH_3COO)_2 \cdot 2H_2O$; $Zn(OAc)_2$) and sodium hydroxide (NaOH). In a typical reaction, 0.01M $Zn(OAc)_2$ was dissolved in 100 ml of distilled water (DW) in a round bottom flask. An appropriate amount of NaOH was dissolved in 50 ml DW and then it was added drop-wise to $Zn(OAc)_2$ solution under continuous stirring for 10 min. After stirring, the resultant solution was refluxed at 90°C for desired time. White coloured precipitate was obtained which was washed several times with water, ethyl alcohol and acetone, sequentially and dried for few hours at 50°C. The synthesized powder was characterized for its morphological, structural and compositional properties. This was named as ZNO [142].

For another set of experiments, MgO was synthesized using Magnesium nitrate and oxalic acid (4:1 M) solution via sol-gel (co-precipitation) method. Solution pH was neutralized by washing several times with distilled water. The solution was then evaporated at 80ºC in a thermostatic water bath followed by drying at 120ºC in a vacuum furnace. The dried gel was calcined at 400ºC for about 3 h to obtain MgO powder [143].

For antibacterial studies of ZnO nanoparticles, it was synthesized by non-protonated process using p-xylene ($C_6H_4(CH_3)_2$) and zinc acetate di-hydrate ($Zn(CH_3COO)_2{\cdot}2H_2O$) both AR grade (Aldrich Chemical Co. Ltd.). In a typical reaction, 0.3M zinc acetate di-hydrate and p-xylene were mixed well in methanol having ratio 1:9 respectively, under continuous stirring for 30 min. Colorless solution was obtained with pH of 6.5 which was then refluxed at ~ 65°C for 6 h. After completing the reaction, the refluxing pot was cooled to room-temperature. White colored precipitate was obtained which was washed by methanol and acetone, sequentially and dried at room-temperature. This was named as ZNOX. Finally, the obtained powder was characterized for its morphological, compositional and structural properties [144].

For another experiment, surface functionalized ZnO powder was synthesized via non-aqueous sol-gel method using zinc acetate dihydrate as a source material. Different amines such as IPA, DEA and TEA were used as both catalyst and surfactant. In a typical reaction, 10 mM zinc acetate dihydrate was dissolved in methanol by continuous stirring using magnetic stirrer for 30 minutes to get homogenous transparent solution. 1 ml of amine (e.g., IPA, DEA, TEA) was added to zinc acetate solution while stirring. The pH of the solution was monitored using digital pH-meter (Hana Instruments). The final solution was post stirred for 20 min at room temperature followed by refluxing in a three necked flask at ~ 70°C for 6 hrs. The resultant precipitate obtained was washed thoroughly with methanol until the pH was neutralized. The precipitate was then decanted and dried at 40°C before using for various bacterial culture experiments. These were named as IPAZNO, DEAZNO and TEAZNO [145, 146].

The morphology of the prepared powder was examined by field emission scanning electron microscope (FESEM), transmission electron microscope (TEM) and high-resolution transmission electron microscope (HRTEM), Bio-TEM. The structural properties were characterized by X-ray diffraction technique (XRD) with Cu-Kα radiation (λ = 1.54178 Å) in the range of 20° to 80°. The chemical composition of as-synthesized material was studied by Fourier transforms infrared (FTIR) spectroscopy in the range of 4000 - 450 cm^{-1}. Optical properties were investigated using UV-Visible spectrophotometer.

2.1. Effect of Nanoparticles on Cancer Cells

2.1.1. Interaction of ZnO Nanoparticles with DNA and Cancer Cells

To understand the induction of synthesized ZnO nanoparticles, the fluorescence quenching titration with increasing ZnO concentration with ctDNA (complementary target DA) at molar ratios was performed. For example, at fixed concentration of ZnO-NPs (50 μM) various concentrations of ctDNA (5-50 μM) were added in 10 mM Tris-HCl buffer at ambient temperature. The fluorescence spectra were recorded under subdued light.

Fluorescence measurements were carried out using Shimadzu spectrofluorometer (model RF5301PC) by a quartz cell of 1 cm path length. The excitation and emission slits were set at 3 nm and 10 nm, respectively. The excitation and emission wavelengths for recording fluorescence of ZnO-NPs were 350 nm and 430 nm, respectively.

2.1.2. Human Cell Culture

T98G, KB and HEK cells used in the present studies were purchased from Korean cell line bank (SNU, Seoul). These cell lines were cultured in 75 cm^2 culture flasks (Corning, USA) using Dulbecco's modified Eagle's medium (DMEM) supplemented with 10% fetal bovine serum, 1% nonessential amino acids, 1% glutamine, penicillin (100 IU/ml) and streptomycin (100 mg/ml) (all from Euroclone, UK). All cultures were maintained at 37°C, 95% relative humidity and 5% CO_2. The growth medium was changed every other day until the time of use. T98G, KB and HEK cells passage number 25, 12, and 19, respectively were used in the cytotoxicity tests. Prior to each cytotoxicity test, the cells were harvested using trypsin–ethylenediaminetetraacetic acid (EDTA)–PBS solution (with 0.25% trypsin - 0.05 mM according to the distributor's instructions) and diluted at a density of 5×10^5 cells/ml in MTT assays. Stock cultures were passaged every third day after harvesting the cells with 0.05% trypsin and seeding 8×10^3 cells/cm^2 in tissue culture flasks to maintain the cells in the exponential phase. All experiments were carried out in exponentially growing cells. The cell suspension was seeded into 96-well plates (Corning, USA) at 100μl/well, and incubated for approximately 20 - 24 h before tests in order to reach confluences. Before the cells were seeded into 96-well plates, the plates were treated with 0.01% poly-D-lysine solution (Sigma-Aldrich, Germany).

2.1.3. In Vitro Cell Viability Assay

Cells were seeded in 96-well plates at a concentration of 2×10^3 to 4×10^3 cells/well in 200μL of complete media and incubated for 24 h at 37°C in 5% CO_2 atmosphere to allow for cell adhesion. Stock solutions (4 mg/ml) of the nanoparticles made in PBS, and then diluted upto 2 mg/ml in incomplete media. This solution was further diluted upto 0.45 μg/ml, in incomplete media for treatment against cell lines, where 100μL of nanoparticle solutions added to 100 μL respectively, of fresh medium in wells to give final concentrations of 2000–0.45 μg/ml. All assays were performed in two independent sets of quadruplicate tests. Control group containing no structures was run in each assay. After 48 h of exposure of cells to drug, each well was carefully rinsed with 200 μL PBS buffer. Cytotoxicity was assessed using MTT (3-[4, 5-dimethyl thiazol-2yl]-2, 5-diphenyltetrazolium bromide). MTT solutions 20μL (5 mg ml^{-1} dd H_2O) along with 200 μL of fresh complete media were added to each well and plates were incubated for 4 h. After incubation, the medium was removed and the purple formazan precipitate in each well was sterilized in 200 μL DMSO [24, 25]. Absorbance was measured using Techano microplate reader (molecular device) at 570 nm and results are expressed as IC_{50} evaluated form % viability which is directly proportional to metabolic active cell number. Percentage (%) viability was calculated from the equation mentioned below [142]:

$$\% \text{ Viability} = \text{OD in sample well/OD in control well} \times 100 \quad (1)$$

2.1.4. Growth Kinetics Assay

T98G, KB and HEK cells were seeded at 7000-8000 cells/cm^2 and their proliferation kinetics was studied at 24, 48 and 72 h intervals following trypsinization and counted for total cells per plate using a haemocytometer [141].

2.1.5. Clonogenic Assay

Clonogenic assay is the method of choice to determine cell reproductive death after treatment with ionizing radiation, but can also be used to determine the effectiveness of other cytotoxic agents [141]. Only a fraction of seeded cells retains the capacity to produce colonies. Before or after treatment, cells are seeded out in appropriate dilutions to form colonies in 1 – 3 weeks. After harvesting with 0.05% trypsin, 150 – 400 (depending on the treatment) cells were plated 10 – 14 h before treating with varying concentrations of drugs in DMEM at 37°C. Cultured cells were treated with doses (7.8 to 125 μg/ml) of nanoparticles. After the treatment, cells were incubated in dark and humid atmosphere with 5% CO_2 at 37°C for 8 – 10 days to allow colony formation. Colonies were fixed with methanol and stained with 1% crystal violet. Colonies of more than 50 cells were counted and the surviving fraction (SF) was calculated. Clonogenic survival curves were constructed from three independent experiments by least-squares regression fitting average of survival levels.

2.1.6. Micronuclei Formation (MNi)

Air-dried slides containing acetic acid–methanol (1:3 V/V) fixed cells were stained with Hoechst-33342 (10 μg/ml in PBS (0.1M), disodium phosphate (0.45 M) buffer containing 0.05% Tween-20 detergent. Slides were examined under the fluorescence microscope using an UV excitation filter. Fluorescent nuclei were visualized using a blue emission filter. Cells containing micronuclei were counted from > 1000 cells by employing the criteria of Countrymen and Heddle. The fraction of cells containing micronuclei, called the M-fraction (%) or MN frequency was calculated as follows:

$$\text{M-fraction } (\%) = Nm/Nt \times 100 \quad (2)$$

Where *Nm* is the number of cells with micronuclei and *Nt* is the total number of cells analyzed. Since, micronuclei formation is linked to cell proliferation, the micronuclei frequencies were normalized with respect to the cell numbers.

2.1.7. Detection of Apoptotic Cells

The percentage of cells undergoing apoptosis was determined microscopically using Hoescht 333242 labelled cells. At least 1000 cells were counted and the percentage of apoptotic cells were determined from the slides prepared for micronuclei formation [142].

3. Cytotoxicity Studies of MGO

Human intestinal cell lines (INT 407) and cervical cell lines (SiHa) were procured from the repository of National Centre for Cell Sciences (NCCS), Pune, India. Cell line were grown as monolayer cultures in DMEM containing 10% fetal bovine serum and antibiotic (100 units/ml penicillin and 100 μg/ml streptomycin) in a humidified atmosphere of 5% CO_2 at 37°C and is sub-cultured twice a week. Cytotoxicity of MgO nanoparticles was evaluated via assessment of its *in vitro* impact on both cell lines. 2×10^4 cells/well were seeded in 96 well plates, each having 200 μg/well. After 24 h, cells were treated with increasing concentration of MgO nanoparticles from 0 - 350 μg/ml for 48 h. The medium was then removed and cells were incubated with 20 μl of MTT (5 mg/ml in PBS) in a fresh medium at 37°C for 4 h. Farmazon crystals, formed by mitochondrial reduction of MTT, were stabilized in DMSO (150 μl/well) and the optical absorbance was measured at 570 nm after 10 minutes of incubation on the *i*Mark Micro plate reader (Bio-Rad, USA). The cytotoxicity was calculated as a fraction of control without MgO nanoparticles (untreated). The inspection of cell morphology was carried out using an inverted microscope Motic, AE31 [143].

3.1. Effect of Nanoparticles on Bacterial Cells

The antibacterial activities of MgO nanoparticles were studied directly using Gram-negative bacteria *V. Cholerae.* A bacterial culture was grown overnight on a peptone broth medium at 28 ± 2°C in a rotary shaker at 150 rpm. The culture (100 μl) was spread on a Müller-Hinton agar plate, and wells (2 mm diameter) were made with the sterile borer and antimicrobial activity of nano-sized MgO was measured [143]. In another set of experiments, ZnO (ZNOX) was used to understand the microbial activity for Gram negative *E. coli* AB 1157 (w.t.) and Gram positive *Bacillus subtilis strains.* These were harvested from exponential grown cultures, and the pellets were suspended in NaCl. The cells were treated with increasing concentrations (0 to 100, μg/ml) of ZNOX nanoparticles for 6 h. The cells were again centrifuged and pellet was suspended in NaCl. The treated cells along with untreated control were serially diluted and plated on nutrient agar plates to assay the colony forming ability. Plates were incubated overnight at 37°C and the colonies were counted. The percentage survival was determined and plotted as a function of nanoparticles concentrations for kinetic studies [144].

Antibacterial properties of amine functionalized ZnO (IPAZNO, DEAZNO and TEAZNO) were studied by preparing 10 ml of stock culture of both bacterial strains i.e., *E. coli* and *S. aureus* which were cultured overnight at 37°C in Luria–Bertani (LB) media. All media for inoculation were prepared and autoclaved for 15 min at 121°C and 15 psi. The stock cultures were maintained frozen at - 20°C in cryovials containing anti-freezing agent, glycerol, to preserve the viability of the cells during storage and were used for further experiments to analyse ROS activity of amine functionalized ZnO-NPs. Bacterial culture of *E. coli* and *S. aureus* were streaked from glycerol stocks onto agar medium plates individually and grown overnight at 37°C to get the secondary culture. Then 2 μl of each secondary culture was used to inoculate 100 ml of LB media with different amine functionalized ZnO-NPs. The culture experiments were repeated for various concentrations (0.5 mM until MIC value) of each amine functionalized ZnO-NPs. For each experiment, the cultures were incubated at 160 RPM at 37°C during constant shaking. Optical Density (OD) was measured at 600 nm after every two hours using UV-visible spectrophotometer (Labtronics, LT 2800, India) [145, 146].

4. Morphological, Structural and Compositional Properties of Nanoparticles

Observations by electron microscope confirmed that the synthesized ZnO particles exhibit mostly spherical shapes with the average diameters in the range of 20 ± 5 nm. The crystallinity was confirmed with X-ray diffraction measurements and the pattern match well with the wurtzite hexagonal phase of pure ZnO (JCPDS 36-1451) [92].

Functional groups of the as-synthesized ZnO nanoparticles were examined by FTIR technique where several well-defined bands were observed related to metal-oxygen (Zn-O) bond at 438 cm^{-1} indicating the formation of ZnO. The bands related to C-O, C=O, O-H are observed at 1430 cm^{-1}, ~ 1579 cm^{-1} and 3300 cm^{-1} respectively [147, 152].

While for MgO, the diffraction peaks matched well with the typical face centred cubic phase of MgO (JCPDS Card No. 89-7746) with intense peak indicating the crystalline nature of the synthesized MgO nanoparticles. The average crystallite size of MgO is estimated to be ~ 16 nm using Scherrer's formula. From FTIR studies, the band observed at ~ 493 cm^{-1} corresponds to vibrational band of Mg-O in the finger print region. The bands observed at 1326 cm^{-1} and 810 cm^{-1} are ascribed to the presence of NO_3^- group since MgO was synthesized using magnesium nitrate. The broad band around 3494 cm^{-1} indicates the presence of functionalized OH groups on the MgO surface [143].

In case of ZnO functionalized with IPA, DEA and TEA, spherically shaped particles are observed with uniform size distribution ~ 10 nm with all the three surfactants. It is expected that smaller particle size can result in better management of the oxidative stress in the biological systems due to size compatibility. FTIR observations of different amine functionalized ZnO NP, showed a broad band around 450 cm^{-1} which is in good agreement with the reported finger print region of metal oxides [44, 152]. Two standard amide bands are observed at 1569 and 1419 cm^{-1}. Two additional bands observed at 1348 cm^{-1} and 1027 cm^{-1} are related to C-N and C-C vibrations, respectively. The absorption band observed between 3000 to 3600 cm^{-1} corresponds to N–H stretching vibration which overlaps with the band of hydroxyl group i.e., O–H mode of vibration. The band observed at 1600 cm^{-1} corresponds to the stretching mode of C-O and C=O. When TEA is coated on ZnO, a broad band is observed from 1450 to 1650 cm^{-1} with a shoulder at 1600 cm^{-1} indicating that the C=O, C-H and C-O bands are merged. The presence of the amine I and amine II bands in the FTIR spectra of all three samples confirm the amine adsorption or functionalization on ZnO-NPs [48, 152].

The UV-Visible spectroscopic measurement of the amine functionalized ZnO-NPs showed a blue shift in the absorption peak with increasing amounts of amines i.e., from IPA to TEA, which could be due to the changes in functional properties of ZnO [146].

4.1. Effect of Nanoparticles on Cancer Cells

4.1.1. Interaction of ZnO Nanoparticles with DNA and Cancer Cells

The interaction of ZnO nanoparticles with DNA was examined by monitoring the changes in the intrinsic fluorescence of the ZnO products at varying concentrations of the DNA. Figure 1 (a) shows the fluorescence emission spectra of ZnO nanoparticles (50 μM) with varying DNA concentrations, i.e., 5 to 50 μM with a step concentration of 5 μM. An emission maximum is noticed at 430 nm, which gradually decreases with increasing concentrations of DNA indicating ZnO fluorescence quenching upon binding with DNA. The fluorescence quenching constant (Ksv) can be evaluated using the Stern–Volmer equation [153]

$$F_0/F = 1 + Ksv\ [DNA] \qquad (3)$$

where F_0 and F are the fluorescence intensities in the absence and presence of DNA respectively, K_{SV} is the Stern–Volmer quenching constant, which is a measure of the efficiency of quenching by DNA. A plot of F_0/F versus [DNA] concentration is shown in Figure 1 (b).

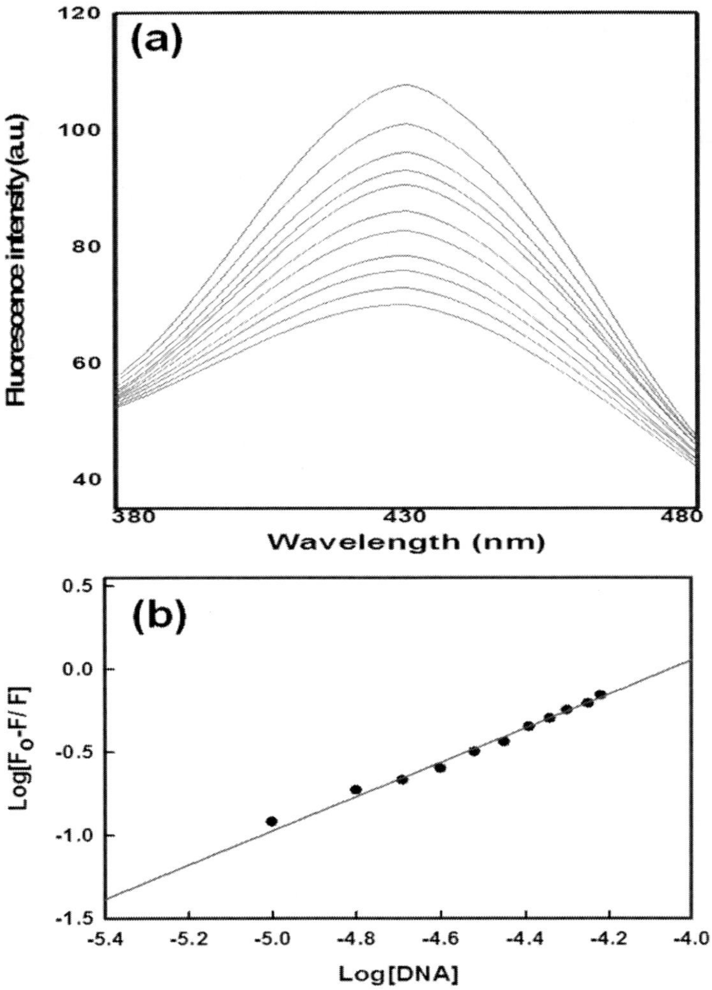

[Reprinted with permission from ref. (142), R. Wahab, N. K. Kaushik, N. Kaushik, E. H. Choi, Ahmad Umar, S. Dwivedi, J. Musarrat, A.A. Al-Khedhairy, ZnO nanoparticles induces cell death in malignant human T98G gliomas, KB and non-malignant HEK cells, J Biomed Nanotech. 9, 1181-9, (2013)., Copyright @ American Scientific Publishers, USA].

Figure 1. Representative fluorescence spectra of ZnO nanoparticles in the absence and presence of various concentrations of DNA; (a) shows the emission spectra of ZnO nanoparticles (50 μM) with varying DNA concentrations, i.e., 5, 10, 15, 20, 25, 30, 35, 40, 45 and 50 μM and (b) exhibits the binding isotherm of log [(F_0-F)/F] versus log [DNA].

The slope of this linear plot (Figure 1 (b)) is estimated to be 4.5×10^4 $Lmol^{-1}$ which is nothing but Ksv. The linearity indicates that only one type of quenching process occurs, i.e., either static or dynamic quenching [154]. Moreover, when the molecules bind independently to a set of equivalent sites on a macromolecule, the equilibrium between free and bound molecules is given by the equation mentioned below [153]

$$\log [(F_0 - F)/F] = \log K + n \log[Q] \quad (4)$$

where *K* and *n* are the binding constant and the number of binding sites, respectively. Thus, a plot of log (F_0 -*F*)/*F* versus log [DNA] (Figure 1 (b)) can be used to determine *K* as well as *n*.

4.1.2. Inhibitory Effect of Nanoparticles on Cancer and Normal HEK Cells

A dose (upto 1000 μg/ml) and time dependent study was carried out and data was obtained by MTT assay which showed that ZnO nanoparticles had inhibitory effects on the growth of T98G gliomas, KB epidermoid cancer and HEK normal cells. 24 hours of treatment resulted in the inhibition of T98G cell growth with an IC_{50} (inhibitory concentration) value of 16.8 (± 2.2) μg/ml. Similar study was carried out with KB and HEK cells growth where IC_{50} values were 38.4 μg/ml and 52.9 μg/ml, indicating the cytotoxic effects of ZnO on these cells. The ZnO nanostructures are found less cytotoxic on HEK (normal) cells in comparison with T98G and KB cells. Effect of ZnO nanoparticles on cell viability of T98G, KB and HEK cells via MTT assay can be seen in Figure 2 (a) as a function of ZnO concentration per ml. The graphs clearly show that ZnO is toxic against T98G and less toxic to normal HEK cells [142].

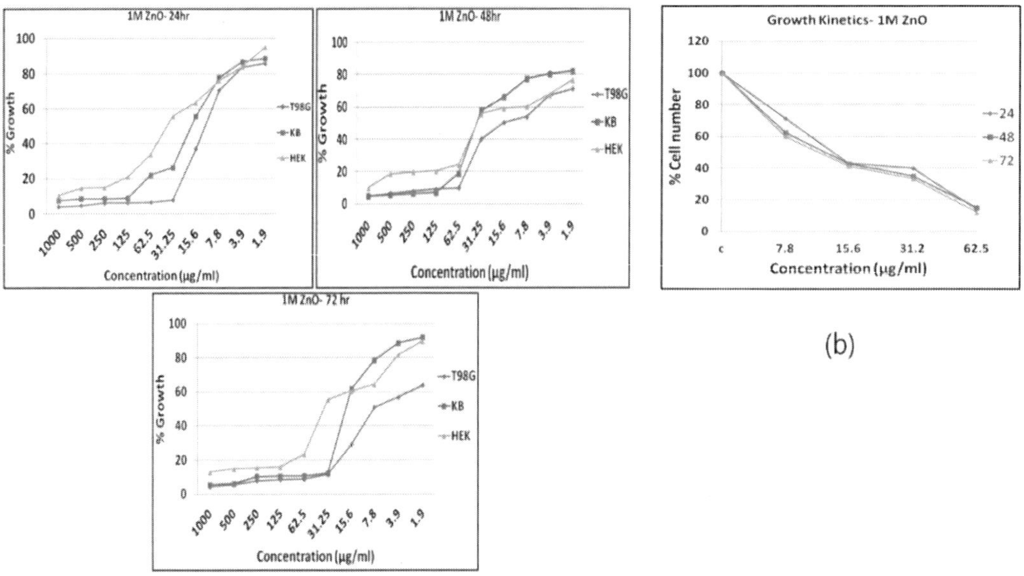

(a)

[Reprinted with permission from ref. (142), R. Wahab, N. K. Kaushik, N. Kaushik, E. H. Choi, Ahmad Umar, S. Dwivedi, J. Musarrat, A. A. Al-Khedhairy, ZnO nanoparticles induces cell death in malignant human T98G gliomas, KB and non-malignant HEK cells, J Biomed Nanotech. 9, 1181-9, (2013)., Copyright @ American Scientific Publishers, USA].

Figure 2. (a) Effect of nanoparticles on cell viability of T98G, KB and HEK cells. Results are given as percentage mean growth, where n = 3. (b) Effect of zinc oxide nanoparticles on the growth kinetics of T98G cells as a function of time and concentration.

The effects of these zinc oxide nanoparticles on the growth kinetics of T98G cells were also investigated. The cells were incubated with ZnO nanoparticles (1 M) for 24, 48 and 72 h and counted by heamocytometer. A gradual reduction in cell number was noticed with increasing incubation time and concentration which can be observed in Figure 2 (b).

The corresponding changes in the cells were observed using a microscope and shown in Figure 3 at different concentrations and time intervals. The images clearly show that the number of T98G cells have greatly reduced with increasing concentration indicating the toxic effects of ZnO nanoparticles [142].

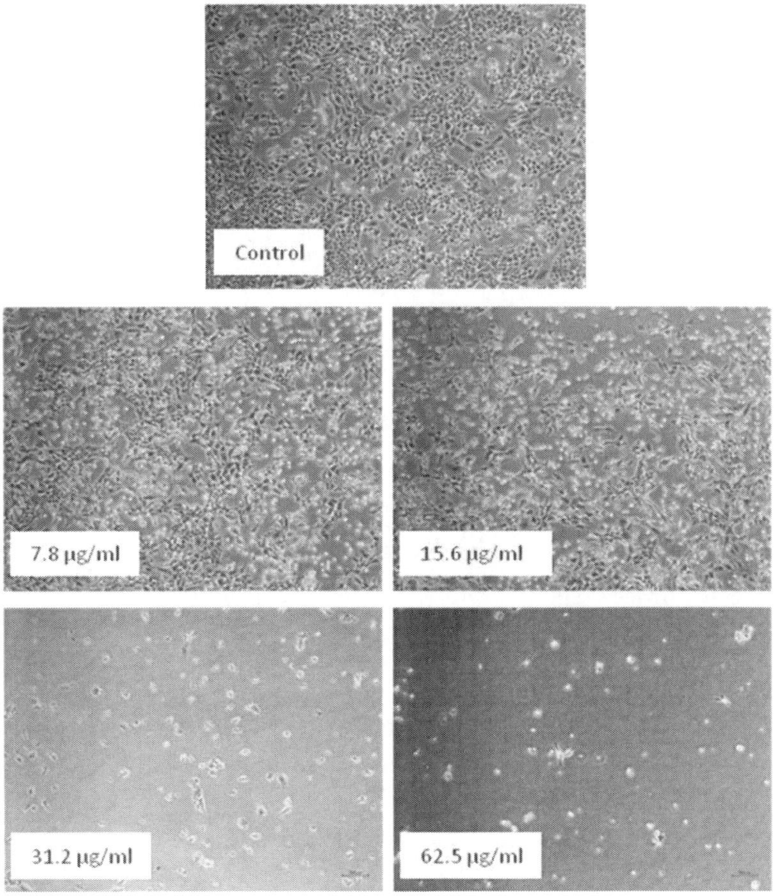

[Reprinted with permission from ref. (142), R. Wahab, N. K. Kaushik, N. Kaushik, E. H. Choi, Ahmad Umar, S. Dwivedi, J. Musarrat, A. A. Al-Khedhairy, ZnO nanoparticles induces cell death in malignant human T98G gliomas, KB and non-malignant HEK cells, J Biomed Nanotech. 9, 1181-9, (2013)., Copyright @ American Scientific Publishers, USA].

Figure 3. Micrographic images for the effect of nanoparticles on the growth kinetics of T98G Cells. Cells were grown in the presence of nanoparticles and number of proliferating cells were counted using homocytometer at different concentrations and time intervals (24, 48, and 72 hrs). Results are shown in % cell number.

A macro colony assay (clonogenecity) technique was used to understand the capacity of colony formation and survival of T98G cells with ZnO NP. The measurements were carried out using three independent observations. It was observed that survival was very less at higher concentrations, analogous to MTT and growth kinetic results. It was noticed that upto 15.6 μg/ml, the cells were able to survive sufficiently but treatment above this concentration resulted in increased toxicity thereby declining the cell survival almost to zero. The results clearly displayed the significant effect of ZnO NP on inhibiting the colony forming capacity of single cell in the T98G cancer cells [142].

To investigate the cytogenetic damage by the ZnO nanoparticles, the treatment-induced micronuclei (MNi) formation in T98G cells was carried out up to 24 h of post treatment as the cell proliferation influences treatment induced micronuclei expression. Untreated T98G cells exposed to vehicle (PBS) had normal micronuclei frequency which is 4.6% and apoptotic frequency of 2.3% cells. With increasing concentrations of ZnO nanoparticles (7.8 to 62.5 μg/ml), the frequency of cells with micronuclei increases from 8.4%, to 21.64%, which is 2 - 4 folds greater than the control group (untreated), as shown in Figure 4. The apoptotic mode of the cell death was further confirmed by flow cytometry and the corresponding images are shown in Figure 5.

Untreated T98G cells (negative control) exposed by vehicle (PBS) have 2.3% apoptotic frequency, while for treated cells with increasing concentrations of ZnO nanoparticles, the apoptotic frequency increased from 5.4%, to 15.6% (2 – 3 fold increase) at 24 h [142].

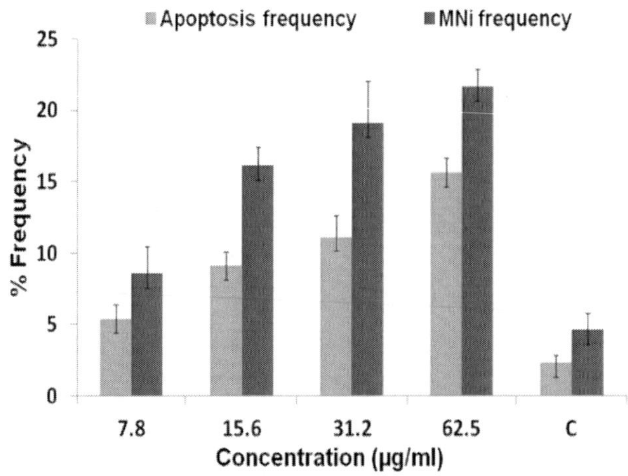

[Reprinted with permission from ref. (142), R. Wahab, N. K. Kaushik, N. Kaushik, E. H. Choi, Ahmad Umar, S. Dwivedi, J. Musarrat, A. A. Al-Khedhairy, ZnO nanoparticles induces cell death in malignant human T98G gliomas, KB and non-malignant HEK cells, J Biomed Nanotech. 9, 1181-9, (2013)., Copyright @ American Scientific Publishers, USA].

Figure 4. Bar diagram for comparative study between micronucleus and apoptosis frequencies in human T98G gliomas treated with various concentrations (7.8, 15.6, 31.2 and 62.5 μg/ml) of ZnO nanoparticles.

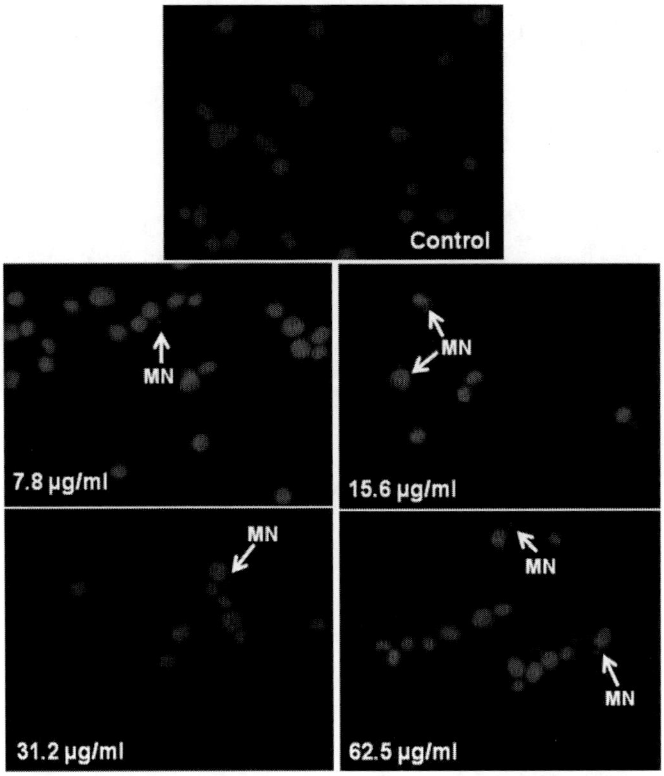

[Reprinted with permission from ref. (142), R. Wahab, N. K. Kaushik, N. Kaushik, E. H. Choi, Ahmad Umar, S. Dwivedi, J. Musarrat, A. A. Al-Khedhairy, ZnO nanoparticles induces cell death in malignant human T98G gliomas, KB and non-malignant HEK cells, J Biomed Nanotech. 9, 1181-9, (2013)., Copyright @ American Scientific Publishers, USA].

Figure 5. Comparison between micronucleus and apoptosis frequencies in human T98G gliomas treated with nanoparticles at concentrations of 7.8, 15.6, 31.2 and 62.5 μg/ml. The 'Control' is untreated control. Results are measured in-vitro during G1/S/G2 phase, both in mono-nucleated cells and bi-nucleated cells in cultures. The cells shown with one or many micronucleus were scored for MN and apoptotic frequency. The data represents mean ± SEM, n = 3. Morphologically, marked condensation and margination of chromatin, fragmentation of nuclei and cell shrinkage characterize apoptotic cells. A good correlation between these morphological changes and DNA fragmentation (ladder) as hallmarks has been demonstrated.

4.1.3. Cytotoxicity Studies of MgO

As per the protocol mentioned in section 2.2.2, INT 407 cells were treated with difference concentration (0 - 350 μg/ml) of MgO nanoparticles for 48 h and the images were taken using an inverted microscope at different time intervals. The INT 407 cells morphology is shown in Figure 6 (a), with control and cells treated with MgO nanoparticles ranging from 150 to 350 μg/ml concentrations. MTT assay results indicated that MgO nanoparticles do not have significant toxic effect on INT 407 cell line within the studied concentration range as the cell viability reduces by about 8% of the control data as shown in Figure 6 (b) [143].

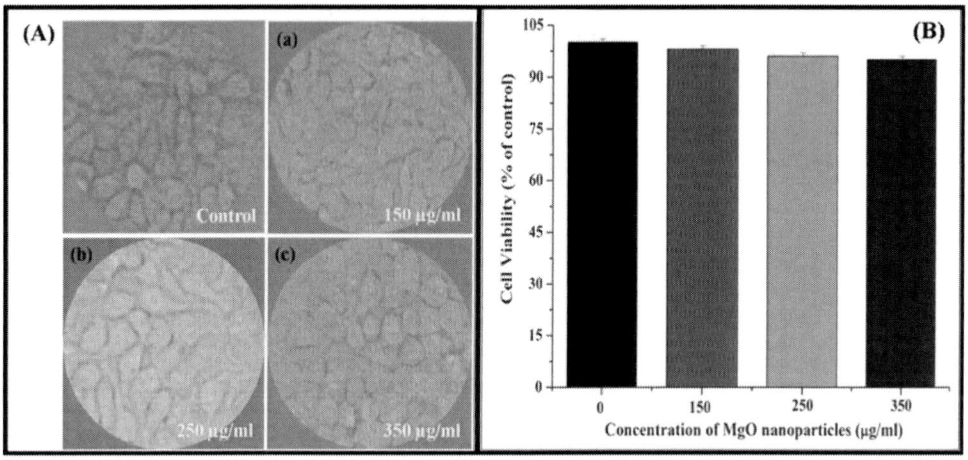

[Reprinted with permission from ref. (143), M. K. Patel, Md. Zafaryab, M. M. A. Rizvi, V. V. Agrawal, Z. A. Ansari, B. D. Malhotra, S. G. Ansari, Antibacterial and Cytotoxic effect of Magnesium Oxide nanoparticles on bacterial and human cells, J. Nanoeng. Nanomanuf, 3, 162-166, (2013), Copyright @ American Scientific Publishers, USA].

Figure 6. (a) Morphology of human intestinal cell line (*INT 407*) with diffrent concentration of MgO nanoparticles and (b) cell viability as function of MgO concentrations with respect to Control.

Similar studies were carried out on human SiHa cells line and concentration of MgO nanoparticles was varied from 150 to 350 µg/ml. There are no remarkable changes in cell morphology upto 200 µg/ml concentration of MgO. However, at higher concentrations from 250 to 350 µg/ml, the changes in cell morphology were significant with time as shown in Figure 7 (a). This result clearly indicates that MgO above certain concentrations are toxic on human SiHa cells.

S. Ge et al., also reported that MgO nanoparticles below 200 µg/ml do not show any cytotoxic effect of non-cancerous cell line [53]. The results indicates that MgO nanoparticles can be used to kill cancer cells upto a concentration level that is safe for normal cells [143]. It is widely reported that free radicals and reactive oxygen species (ROS) generated in aerobic organisms is a common feature of abnormalities in them. In humans these ROS can lead to disorders and patho-physiological events like cancer, arthritis, inflammation, myocardial disorders, and neural degenerative disorders other than their role in inflammatory and nutritional diseases.

Few metal oxide (MOx) show their antimicrobial activity by the generation of hydrogen peroxide and ROS from their surface as well as by adhering to the cellular membrane and internalization into the cells [155]. Therefore, the cytotoxic activity of ZnO/MgO nanoparticles is perhaps due to the electrostatic condensation of these nanoparticles that adhere to the cell walls resulting in negatively charged cell surface at biological pH. The oppositely charged electrostatic condensation is perhaps the reason for the bactericidal or cytotoxic effect of MOx nanoparticles [143].

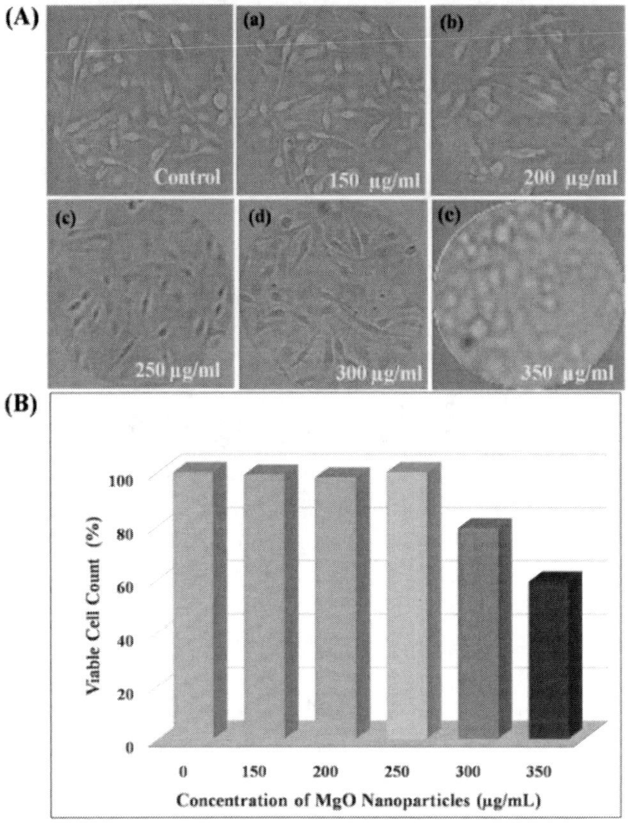

[Reprinted with permission from ref. (143), M. K. Patel, Md. Zafaryab, M. M. A. Rizvi, V. V. Agrawal, Z. A. Ansari, B. D. Malhotra, S. G. Ansari, Antibacterial and Cytotoxic effect of Magnesium Oxide nanoparticles on bacterial and human cells, J. Nanoeng. Nanomanuf, 3, 162-166, (2013), Copyright @ American Scientific Publishers, USA].

Figure 7. (a) Cytotoxic effect of MgO nanoparticles on Human cervical cancer (SiHa) cell line. Cell was cultured in 10% FBS medium, treated with 15, 20, 25, 30, 35 µg/ml MgO nanoparticles for 4 h, and monitored by MTT assay and (b) percentages of viable cells in comparison to untreated cells (control).

4.2. Effect of Nanoparticles on Bacterial Cells

The antibacterial activities of MgO nanoparticles were studied directly using Gram-negative bacteria *V. Cholerae*, as a model bacterial system and antimicrobial activity of MgO nanoparticles was measured following the protocol mentioned in section 2.3. 10 µl of separate aqueous solutions comprising of different concentrations (0.001, 0.01, 0.1 to 1 mg/ml) of MgO nanoparticles was added to the wells and the plate was incubated at 28 ± 2°C. After ~ 48 h of incubation, the zone of bacterial inhibition was found to be ~ 6 ± 1 mm which is around 60% more than the diameter of the wells. Figure 8 indicates the antimicrobial activity of MgO nanoparticles [143].

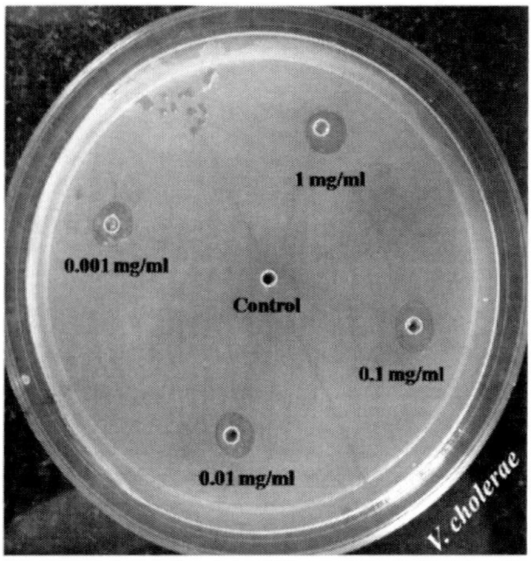

[Reprinted with permission from ref. (143), M. K. Patel, Md. Zafaryab, M. M. A. Rizvi, V. V. Agrawal, Z. A. Ansari, B. D. Malhotra, S. G. Ansari, Antibacterial and Cytotoxic effect of Magnesium Oxide nanoparticles on bacterial and human cells, J. Nanoeng. Nanomanuf, 3, 162-166, (2013), Copyright @ American Scientific Publishers, USA].

Figure 8. Microbial assay of MgO nanoparticles on *Vibrio cholerae* culture.

In order to understand the antibacterial properties of ZnO (ZNOX) against *E.coli* AB 1157 (w.t.) and *Bacillus subtilis* strains, a protocol mentioned in section 2.3 was followed and the cells were treated with increasing concentrations (0 to 100, μg/ml) of ZnO nanoparticles for 6 h.

It is found that ZnO nanoparticles show significant antibacterial activities towards both the strains. The antibacterial activity for *E.coli* AB 1157 was higher than *Bacillus subtilis* strain, as can be seen from the bacterial survival data presented in Figure 9 as a function of ZnO NP concentration [144].

To investigate the antibacterial activity by agar plate assay for colony forming ability, the treated cells along with untreated control were serially diluted and plated on nutrient agar plate assay to test the colony forming ability. After 6 h of incubation, the diameter of bacterial inhibition zone was found to be ~ 62% and 75%, for *E. Coli* AB 1157 and *Bacillus Subtilis* respectively, lesser than the control. Figure 10 indicates the antimicrobial activity [144].

4.3. Effect of Amine Functionalized ZnO Nanoparticles on Bacterial Cells

Antibacterial activities of amine functionalized ZnO-NPs were evaluated for antioxidative studies on *E. coli* and *S. aureus*. For this study, these nanoparticles were dispersed thoroughly in Luria–Bertani culture medium in a dose-dependent manner and the dose–response relationship was determined. All media were prepared fresh and autoclaved for 15 min at 121°C under 15 psi for inoculation. Refer Figure 10.

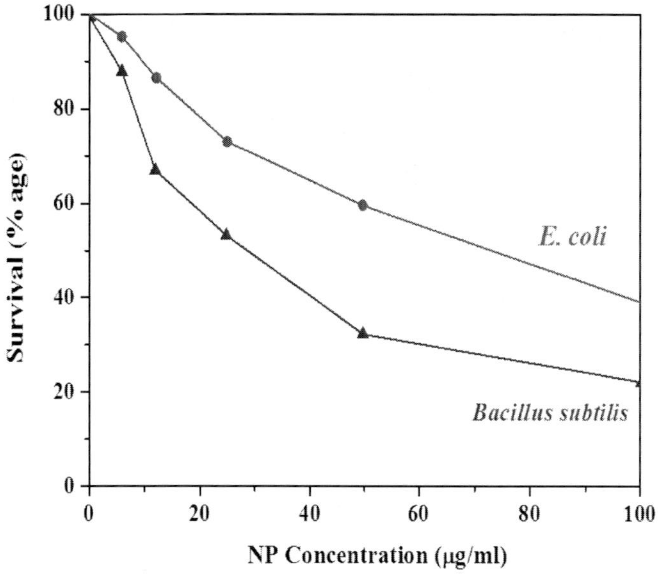

[Reprinted with permission from ref. (144), R. Wahab, Ahmad Umar, S. Dwivedi, K. J. Tomar, H. S. Shin, and I. H. Hwang, ZnO Nanoparticles: Cytological Effect on Chick Fibroblast Cells and Antimicrobial Activities towards Escherichia Coli and Bacillus Subtilis, Sci. Adv. Mater, 5, 1571-1580, (2013), Copyright @ American Scientific Publishers, USA].

Figure 9. Bacterial growth curves of *E.coli* and *B. Subtilis* at various concentrations of ZnO nanoparticles.

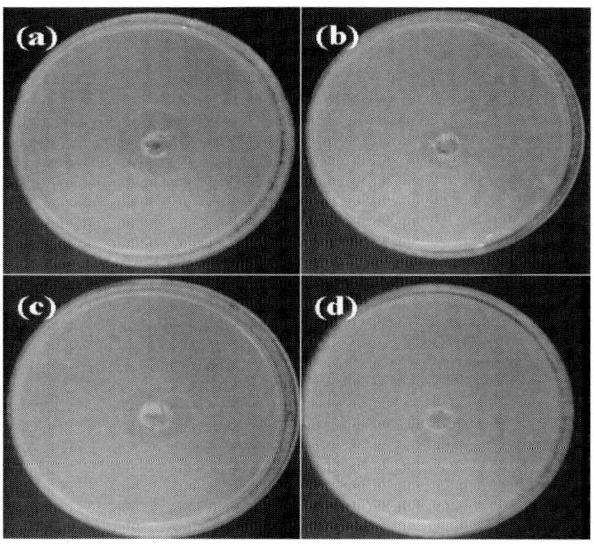

[Reprinted with permission from ref. (144), R. Wahab, Ahmad Umar, S. Dwivedi, K. J. Tomar, H. S. Shin, and I. H. Hwang, ZnO Nanoparticles: Cytological Effect on Chick Fibroblast Cells and Antimicrobial Activities Towards Escherichia Coli and Bacillus Subtilis, Sci. Adv. Mater, 5, 1571-1580, (2013), Copyright @ American Scientific Publishers, USA].

Figure 10. Bacterial growth plate counts: (a) Only *E.coli* and (b) *E.coli* with ZnO nanoparticles (100 μg/ml); (c) only *B. Subtilis*, and (d) *B. Subtilis* with ZnO nanoparticles (100 μg/ml).

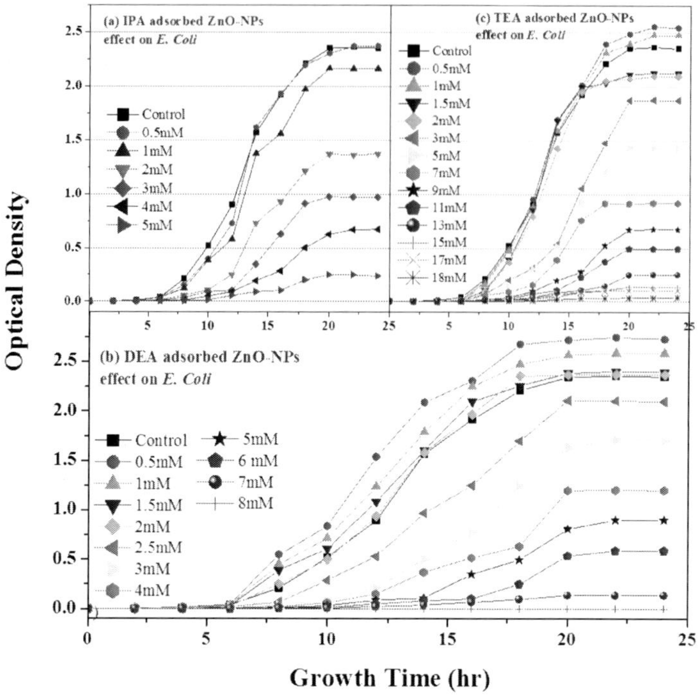

[Reprinted with permission from ref. (146), A. Kumar, Z. A. Ansari, H. Fouad, Ahmad Umar, and S. G. Ansari, Oxidative stress control in E. Coli and S. Aureus cells using amines adsorbed ZnO, Sci. Adv. Mater, 6, 1236-1243, **(2014)**, Copyright @ American Scientific Publishers, USA].

Figure 11. Growth curve for *E. coli* at various concentrations of amine functionalized ZnO.

Amine functionalized ZnO-NPs concentration in the nutrient culture media was varied from 0.5 mM to the minimum inhibitory concentration (MIC) of each bacterial strain. The growth kinetics of bacteria in the media was measured by monitoring the change in Optical Density (OD) at 600 nm after every two hours for which bacterial culture has maximum absorbance [145, 146]. Figure 11 and Figure 12 show the growth curves for the two bacterial strains with amine functionalized ZnO-NPs. The growth curve of bacteria is distinctly divided into its three phases namely; lag phase, logarithmic (log) phase and stationary phase. From these curves, growth kinetics and minimum inhibitory concentration (MIC) were determined.

In case of *E. coli* (Figure 11 (a) and (c)), which have thin peptidoglycan layer outside its cellular membrane, IPA functionalized ZnO-NPs show lower bacterial growth as well as lower reduction in oxidative stress with only 0.5 mM of it being capable of having higher growth than the control while DEA functionalized ZnO-NPs show higher bacterial growth than the control at lower concentrations viz. 0.5 mM, 1 mM and 1.5 mM (Figure 11 (b)). Similar results are observed for TEA functionalized ZnO-NPs (0.5 mM and 1 mM). However, higher than the prescribed concentrations, the amine functionalized nanoparticles served as anti-bacterial agent for *E. coli* reaching MIC value at 6 mM, 9 mM and 12 mM for IPA, DEA and TEA functionalized ZnO-NPs, respectively.

A prolonged exposure of the *E. coli* bacterial cultures to higher concentration upto MIC of the ZnO-NPs resulted in the total inhibition of *E. coli* cell growth as shown in Figure 11 (a-c). The lower concentrations served to be better in managing the bacterial growth and prevented metabolic free radicals from affecting the viability of cells. At higher concentrations of the amine functionalized ZnO-NPs in bacterial culture, the cytotoxic effects of the nanoparticles were highly synergistic and mainly mediated through cell wall and cell membrane rupture increasing the permeability of the membrane leading to death of the bacterial cell [156].

Similar results were obtained for Gram positive bacteria *S. aureus* which has a thicker peptidoglycan layer over the cell membrane. At lower concentrations of all the amine functionalized ZnO-NPs viz. 0.5 mM of IPA, 2 mM of DEA and 1 mM of TEA, higher bacterial growth was observed than the control as shown in Figure 12 (a-c). However, at relatively higher concentrations, anti-bacterial effect leading to MIC value at 14 mM for IPA, 15 mM for DEA and 18 mM for TEA functionalized ZnO-NPs were observed [146].

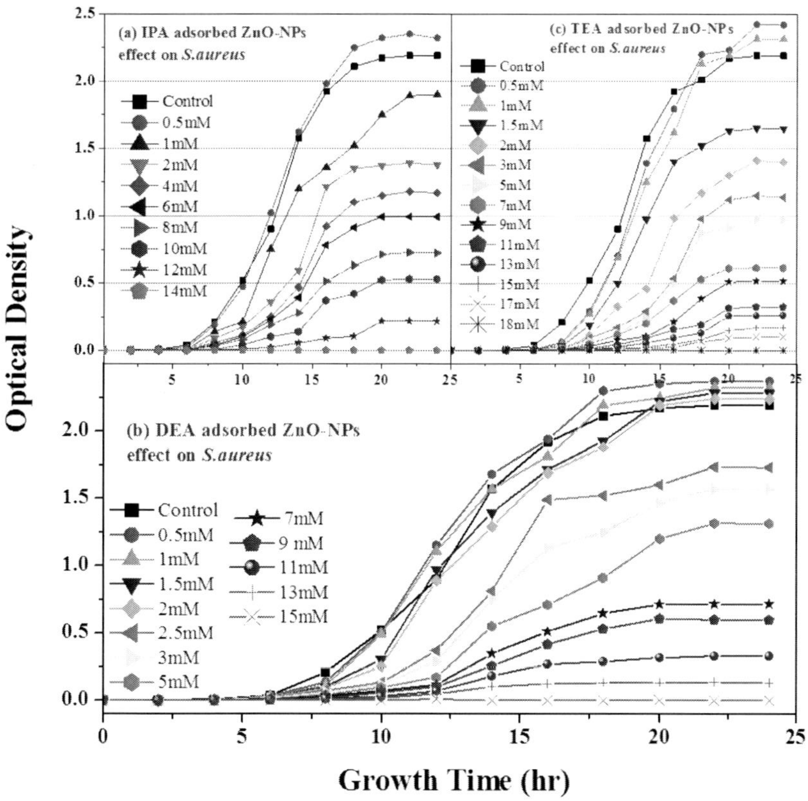

[Reprinted with permission from ref. (146), A. Kumar, Z. A. Ansari, H. Fouad, Ahmad Umar, and S. G. Ansari, Oxidative stress control in E. Coli and S. Aureus cells using amines adsorbed ZnO, Sci. Adv. Mater, 6, 1236-1243, (2014), Copyright @ American Scientific Publishers, USA].

Figure 12. Growth curve for *S. aureus* at various concentrations of amine functionalized ZnO.

Figure 13 and 14 show the specific effect of 0.5 mM and 1 mM concentration of amine functionalized ZnO-NPs on both *E. coli* and *S. aureus*. It is observed that 1 mM concentration of ZnO-NPs managed the stress better than 0.5 mM. Diethylamine functionalized ZnO-NPs have the best effect on both *E. coli* and *S. aureus* followed by TEA and IPA functionalized ZnO-NPs.

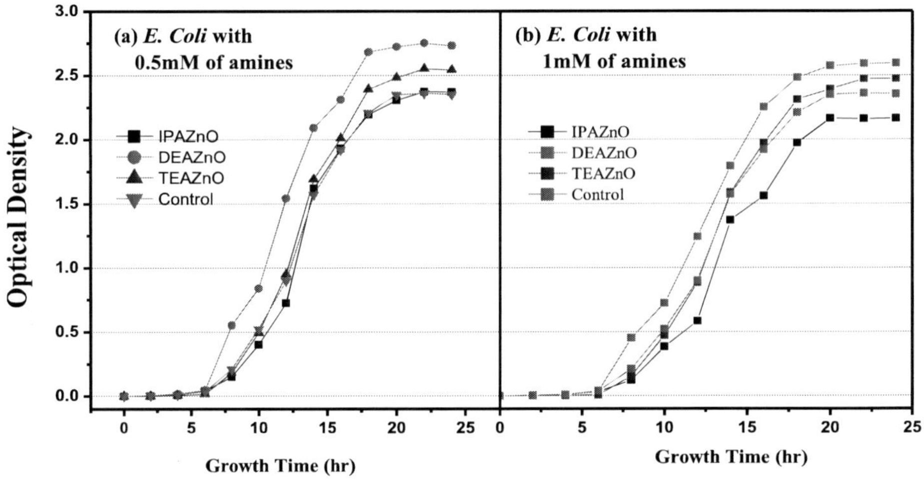

[Reprinted with permission from ref. (146), A. Kumar, Z. A. Ansari, H. Fouad, Ahmad Umar, and S. G. Ansari, Oxidative stress control in E. Coli and S. Aureus cells using amines adsorbed ZnO, Sci. Adv. Mater, 6, 1236-1243, (2014), Copyright @ American Scientific Publishers, USA].

Figure 13. Growth curve for *E. coli* indicating greater growth than the control at (a) 0.5 mM and (b) 1 mM concentration of amines.

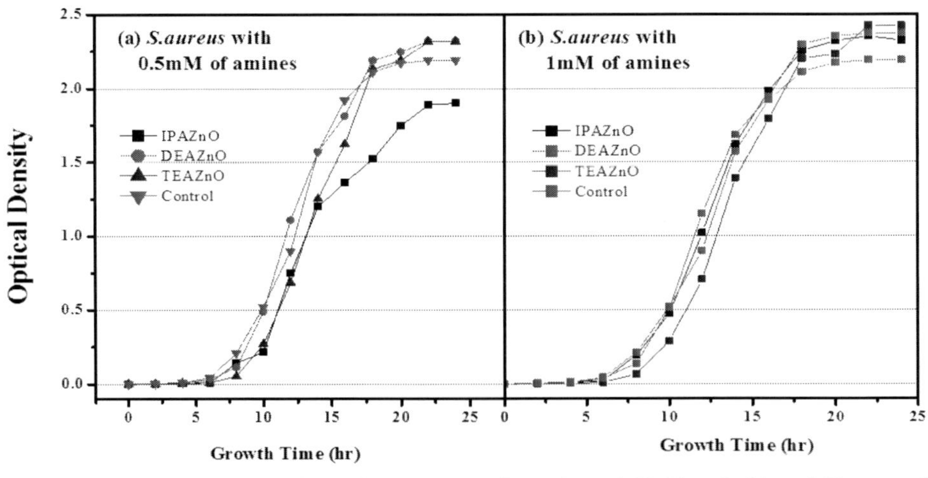

[Reprinted with permission from ref. (146), A. Kumar, Z. A. Ansari, H. Fouad, Ahmad Umar, and S. G. Ansari, Oxidative stress control in E. Coli and S. Aureus cells using amines adsorbed ZnO, Sci. Adv. Mater, 6, 1236-1243, (2014), Copyright @ American Scientific Publishers, USA].

Figure 14. Growth curve for *S. aureus* indicating greater growth than the control at (a) 0.5 mM and (b) 1 mM concentration of amines.

During the initial growth kinetics i.e., the lag phase, the cell division is lower than the control due to the time taken in adapting to new nutrient environment which has become rich by mixing of nanoparticles in media. However, during the log phase, the bacterial growth is exponential until it reaches its stationary phase. In both *E. coli* and *S. aureus* bacterial culture, the lag phase of bacteria with ZnO-NPs was prolonged due to initial stress surmounting lower concentrations. The growth rapidly increased thereafter which was the result of the nutritive environment and reduced toxicity from the metabolic waste [146].

4.4. Expected Mechanism of action of ZnO on Bacteria

It is assumed that ZnO nanoparticles, can easily attach and enter the inner surface of the microorganism as the size of each microorganism is about 200 - 250 times larger than the synthesized nanoparticles. Therefore, it is easier for the nanoparticles to form a film over microorganism and easily enter into their cell wall. The ZnO nanoparticles react with the microorganism due to high surface area, causing its destruction and lead to the death of the cell [157, 159].

The justification of the interaction between ZnO nanoparticles and pathogens *(E. Coli and B. Subtilis)* was further clarified by imaging with bio-transmission electron microscope (Bio-TEM). Figure 15 shows the typical bio-TEM images of the bacteria before and after treatment with ZnO nanoparticles. It shows the bio-TEM images of *E. Coli* (Figure 15 (a)) and *E. Coli* with ZnO nanoparticles at 100 μg/ml concentration (Figure 15 (b)) after incubation. From these images, agglomerated ZnO NPs are seen attached on the surface of the *E. Coli* that moved into the cytoplasm of the *E. Coli* through the membrane and causes damage to the cells and their cell walls which can be easily seen in the image. It is also seen from the micrograph that the shape of the *E. Coli* has changed significantly Figure 15 (b) [160].

Similar observations were observed in case of *B. Subtilis* bacterial strain, however penetration and shape change is much lesser E. coli as shown in Figure 15 (c) and (d). This could be due to the fact that the Gram +ve bacteria *(B. Subtilis)* has one cytoplasmic membrane and thick wall composed of multilayer of peptideglycan. Hence it indicates that ZnO nanoparticles damage cell membranes, leading to cell death (Figure 15).

An earlier study suggests that the ions are responsible for the change in membranes [161, 162]. It is well known that the zinc oxide exhibits stable oxidation state; however when it is being dissolved in a concentrated medium or alkaline medium, it can easily form Zn^{2+} ions. Some reports describe that the photo-catalytic reactions are responsible for membrane damage which is caused by the release of H_2O_2 [163, 165].

The ROS is another factor, which is produced in the bacterial solution with nanoparticles and are responsible to form the free radicals in the solution and these free radicals penetrate the outer wall of the cells and enter the inner wall of the membrane. The free radicals when react with the cytoplasm of the bacteria, enzymatic changes take place and it leads to the disorganization of the cell which further cause leakage of the cell contents. Although intensive biochemical studies are needed to establish the exact relation between growth inhibition activities of bacteria with nanoparticles, the present studies can contribute sufficiently to understand the process to inhibit the growth of bacteria [144].

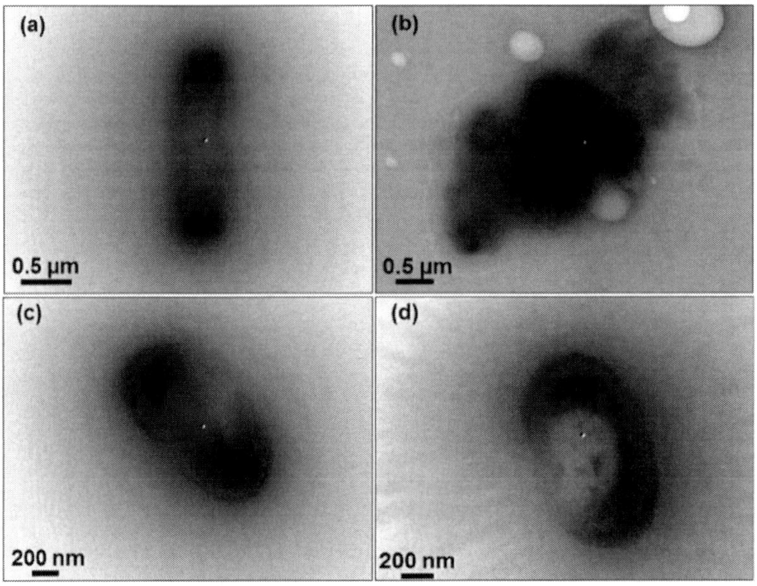

[Reprinted with permission from ref. (144), R. Wahab, Ahmad Umar, S. Dwivedi, K. J. Tomar, H. S. Shin, and I. H. Hwang, ZnO Nanoparticles: Cytological Effect on Chick Fibroblast Cells and Antimicrobial Activities Towards Escherichia Coli and Bacillus Subtilis, Sci. Adv. Mater, 5, 1571-1580, (2013), Copyright @ American Scientific Publishers, USA].

Figure 15. Typical Bio-TEM images of (a) only *E. coli*, (b) ZnO nanoparticles with *E. coli;* (c) only *B. Subtilis*, and (d) *B. Subtilis* with ZnO nanoparticles.

4.5. Amine Functionalized ZnO

Literature review suggests that the antibacterial activity of ZnO-NPs in the liquid culture media are due to serious cell membrane damage by lipid peroxidation and internalization leading to bacteriostatic or bacteriocidal effects in both *E. coli* and *S. aureus* [22, 24]. The lower MIC for Gram-negative *E. coli* could be ascertained to relatively thinner peptidoglycan cell wall than the Gram-positive *S. aureus* which leads to more damage to the cell membrane at relatively lower concentrations [146].

It is also observed that these secondary (DEA) and tertiary amine (TEA) functionalized ZnO-NPs have enhanced biocompatibility and reduced the cytotoxicity that helped in reducing the oxidative damage by ROS in bacterial culture. The amine functional groups on the ZnO-NPs surface play an important role at specific concentrations in neutralizing the electron rich ROS which could alter the release of Zn^{2+} ions and the production of reactive oxygen species (ROS), as discussed below.

In case of amine functionalized ZnO-NPs, ZnO are bound by H-bond leaving small alkyl chains which can bind well with the organic cell membrane and penetrate inside. Growth of bacterial culture is affected by free radicals generated from mechanical stress, metabolic stress, temperature and light leading to oxidation or degradation of biological components in various ways. Unchecked free radical action would lead to inhibition of growth by damaging membrane, enzyme systems, proteins, nucleic acid, etc. of the cellular system [166, 167].

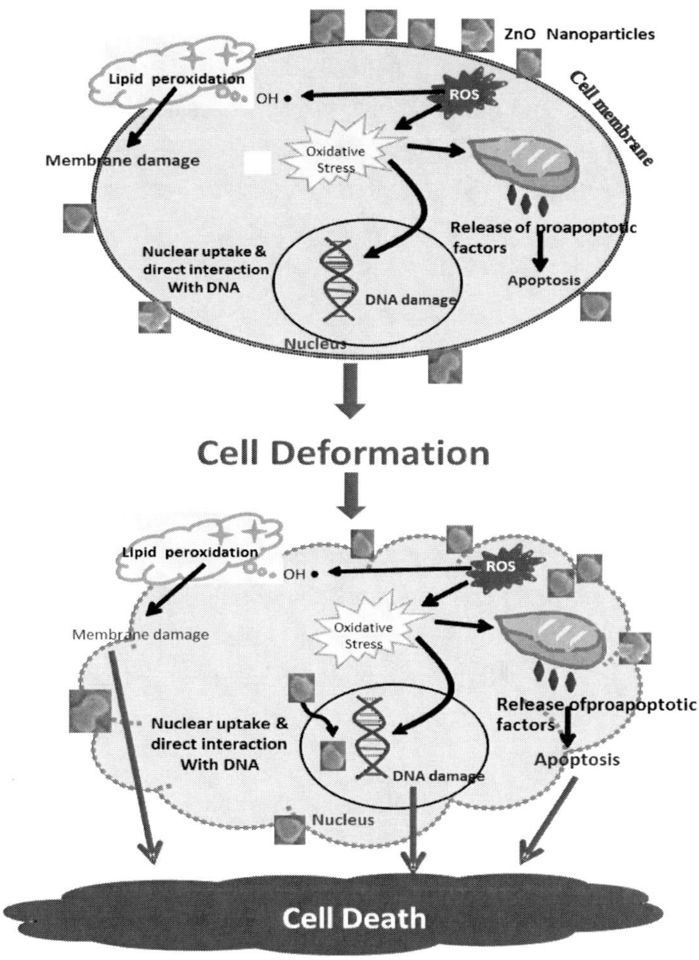

Figure 16. Mechanism of ZnO generating ROS induced toxicity in the bacterial cell.

Amine functionalized ZnO-NPs mediated antioxidant traps the oxygen free radicals and transform them to water or organic hydroxylamine which itself act as enzyme reactivating agent as well as an antioxidant. Thus, metabolic oxidative stress is sufficiently reduced by scavenging free radicals and enzyme systems are rejuvenated at controlled concentration of amine functionalized ZnO-NPs [168]. The reduction in oxidative stress due to availability of amine groups functionalized on ZnO-NPs is indicative of amine group accepting oxygen ($^{\bullet}O_2$) radicals or hydroxide ($^{\bullet}OH$) radicals which are being reduced in the redox reaction taking place. The antioxidant molecules work by getting themselves oxidized and acting as reducing agent. A detailed study in any case will be required to go through the phenomenon of reduction in oxidative stress via these amine functionalized ZnO-NPs [146].

4.6. ZnO Induced Toxicity: Generation of ROS

A recent study on human lung epithelial cells demonstrated the generation of reactive oxygen species (ROS) by ZnO nanoparticles [169]. These ROS exerts oxidative stress, which leads to internalization of lipid peroxidation and subsequent DNA damage [169]. The ability of ZnO to generate ROS is due to its wide and direct band gap of ~ 3.3 eV.

The radiation having equal or higher energy than the above value can promote the electrons from the valence band (VB) to the conduction band (CB) of ZnO, generating electrons (e^-) hole pairs as carriers. Electrons and holes often recombine quickly, but they can also migrate to the particle surface, where they react with functionalized species. Hence, two processes occur (i) electrons react with oxygen and (ii) holes react with hydroxyl ions or water to form superoxide and hydroxyl radicals. The highly active surface of ZnO nanoparticles is known to be susceptible to various surface defects and this tendency increases with decreasing the particle size [170]. Greater the defects higher the production of ROS and H_2O_2 from the nanoparticles surface, leading to lipid peroxidation of the cell wall. Lipid peroxidation increases membrane's permeability and hence the cellular internalization of ZnO kills the bacteria [170, 171]. The results show marked difference in the antibacterial activity of ZnO with different surfaces. The anti-bacterial activity can be attributed to the high surface area and defect density with decreasing particle size or surface functionalization.

ROS may result in oxidative stress, inflammation and consequent damage to proteins, membranes and DNA [172, 173]. ROS particularly the hydroxyl radicals, can react with all biological macromolecules (lipids, proteins, nucleic acids and carbohydrates) [172, 173]. Metal-catalyzed protein oxidation results in the addition of carbonyl groups, cross-linking or fragmentation of proteins [172]. Lipid (peroxidation) aldehydes can react with sulfhydryl (cysteine) or basic amino acids (histidine, lysine).

Similarly, modification of individual nucleotide bases, single-strand breaks and cross-linking are the typical effects of ROS on nucleic acids [172]. ROS are known to react with DNA molecule causing damage to both purine and pyrimidine bases as well as DNA backbone [174]. Another important outcome of ROS production is lipid peroxidation, which generates variety of products that are reactive towards the cellular macromolecules including DNA [172, 174]. Therefore, ZnO QDs induced genotoxicity, which is mediated by oxidative stress, can further be linked to different pathways [175]. The mechanisms of ZnO induced cellular toxicity is schematically shown in Figure 16 explaining the penetration of nanoparticles, their effect on cell morphology by generating ROS and causing cell death by spoiling various cell cycles.

CONCLUSION

The physico-chemical and structural properties of sol-gel synthesized nanostructured metal oxides greatly affect their biological and toxicological properties. The surface functionalization helps in managing the oxidative stress generated due to the interaction and helps cell survival. Higher concentration of nanoparticles induced ROS that disturbs the cell cycle and leads to cell killing. The bacteriocidal effect at higher concentration is due to damage to cell wall and cell membrane via lipid peroxidation. Amine coating of ZnO-nanoparticles helps managing the oxidative stress/effect at lower concentration. The antibacterial activity in both gram positive and gram negative bacteria is affected by surface modifications and hence proper surface functionalization can help in better management of nanoparticles for such applications.

REFERENCES

[1] (a) Ed. H. S. Nalwa, *Encyclopedia of Nanoscience and Nanotechnology*, American Scientific Publishers, Los, Angeles, USA,11(2011), (b) H. Zhao, Y. Ning, Techniques used for the preparation and application of gold powder in ancient China, *Gold Bull*, 33 (2000) 103 (c) C. Lee, Y. Ko,Y. Kim, Device Performances of Organic Light-Emitting Diodes with Indium Tin Oxide, Gallium Zinc Oxide, and Indium Zinc Tin Oxide Anodes Deposited at Room Temperature. *J. Nanosci. Nanotechnol.* 13(2013) 8011, (d) S. Won, J. Heo, S. Joung, Design of magnetic resonance force microscopy probe based on a commercially available charge coupled device camera. *J. Nanosci. Nanotechnol.* 13(2013) 8086.

[2] (a) M. Faraday, *Philosophical Transactions of the Royal Society of London, X. The Bakerian Lecture.* Experimental relations of gold (and other metals) to light 147 (1857) 145; (b) H. W. Hsu, H. S. Huang, C. C. Lee, S. Y. Chen, H. H. Teng, M. R. Peng, M. C. Wang, C. H. Liu, Comparison of NMOSFET and PMOSFET devices that combine CESL stressor and SiGe channel. *J. Nanosci. Nanotechnol*, 13 (2013)8127;(c) Y. Liu and K. Xie,. Nano-Array Electrodes for Next-Generation Lithium-Ion Batteries. *Sci. Adv. Mater.* 6 (2014) 863.

[3] (a) R. P. Feynman, *Philosophical Transactions*, Reinhold, New York (1961). (b) S. Y. Woo, Y. J. Yoon, S. Cho, J.-H. Lee, I.M. Kang, Analysis on RF parameters of nanoscale tunneling field-effect transistor based on InAs/InGaAs/InP heterojunctions. *J. Nanosci. Nanotechnol*, 13(2013)8133.

[4] H.K. Henson, K.E. Drexler, 4 (1984) 480; (b) S. H. Kim, J. W. Lee, I. Choi, Y.-C. Kim, J. B. Lee, J. H. Sung, A microfluidic device with 3-d hydrogel villi scaffold to simulate intestinal absorption. *J. Nanosci. Nanotechnol*, 13(2013)7220.

[5] K.E. Drexler, Mightier Machines from Tiny Atoms May Someday Grow, *Smithsonian*, 13 (1982)145, (b) Sohee Lee, Chaejeong Heo, Minah Suh, Young Hee Lee, Nanoscale Live Cell Optical Imaging of the Dynamics of Intracellular Microvesicles in Neural Cells. *J. Nanosci. Nanotechnol*, 13 (2013)7229, (c) Jihea Moon, Giyoung Kim, Sangdae Lee, Development of nanogold-based lateral flow immunoassay for the detection of ochratoxin A in buffer systems. *J. Nanosci. Nanotechnol*. 13, (2013) 7245, (d) Duck-Gun Park, Hoon Song, M. B. Kishore, G. Vértesy, Duk-Hyun Lee, Detection of gamma-irradiation effect on DNA and protein using magnetic sensor and cyclic voltammetry. *J. Nanosci. Nanotechnol.* 13 (2013)7250.

[6] J. H. Heo, K.-I. Kim, M. H. Lee, J. H. Lee, Stability of a gold nanoparticle-DNA system in seawater. *J. Nanosci. Nanotechnol.* 13(2013)7254.

[7] M.A. Zimmler, J. Bao, F. Capasso, S. Mller, C. Ronning, Laser action in nanowires: Observation of the transition from amplified spontaneous emission to laser oscillation, *Appl. Phys. Lett*, 93 051101, (2008).

[8] (a) A. Javey, S. Nam, R.S. Friedman, H. Yan, C. M. Lieber., Layer-by-layer assembly of nanowires for three-dimensional, multifunctional electronics. *Nano Lett*,7 (2007)773, (b) M. Sivasubramanian, J. Y. Lee, K. J. Kim, G. Saravanakumar, Y. M. Kang, J. H. Park, Cyclodextrin-based nanocomplexes for sustained delivery of human growth hormone. *J. Nanosci. Nanotechnol*, 13, (2013)7306.

[9] L. Tsakalakos, J. Balch, J. Fronheiser, B.A. Korevaar, O. Sulima, J. Rand,silicon nanowire solar cell. *Appl. Phys. Lett*, 91 (2007)233117.

[10] (a) S.M. Tanner, J.M. Gray, C. T. Rogers, K.A. Bertness, N.A. Sanford, High-Q GaN nanowire resonators and oscillators, Appl. Phys. Lett,91,203117, (2007), (b) V. G. Deepagan, Thavasyappan Thambi, Hyewon Ko, Young Mo Kang, Jae Hyung Park, Amphiphilic Polysialic Acid Derivatives: Synthesis, Characterization, and In-Vitro Cytotoxicity. *J. Nanosci. Nanotechnol.* 13(2013) 7312.

[11] (a) Z.Y. Fan, J.G. Lu, Gate-refreshable nanowire chemical sensors., *Appl. Phys. Lett.*, 86 (2005)123510 (b) D. H. Nam, S. Il Cha, Y. J. Jeong, S. H. Hong. Enhanced Graphitization of Carbon Around Carbon Nanotubes During the Formation of Carbon Nanotube/Graphite Composites by Pyrolysis of Carbon Nanotube/ Polyaniline Composites. *J. Nanosci. Nanotechnol*, 13 (2013)7365.

[12] (a) M. Haruta, S. Tsubota, T. Kobayashi, H. Kageyama, M.J. Genet, B. Delmon Low-Temperature Oxidation of CO over Gold Supported on TiO_2, α-Fe_2O_3, and Co_3O_4., *J. Catal.*, 144(1993)175., (b) Y. J. Yun K. Bong. Preparation and Characterization of Graphene Oxide Encapsulated Gold Nanoparticles., *J. Nanosci. Nanotechnol.* 13 (2013)7376.

[13] (a) E.C. Garnett, W. Liang, P. Yang. Growth and Electrical Characteristics of Platinum. *Nanowires Adv. Mater*, 19(2007)2946., (b) A. S. Roy, S. Gupta, S. Seethamraju, P. C. Ramamurthy, G. M. Fabrication of Poly(Vinylidene Chloride-Co-Vinyl Chloride)/TiO_2 Nanocomposite Films and Their Dielectric Properties, *Sci. Adv. Mater.* 6(2014) 946, (c) Jae-Hyun Lee, Yamujin Jang, Keun Heo, Jeong-Mi Lee, Soon Hyung Choi, Won-Jae Joo, Sung Woo Hwang, Dongmok Whang, Large-scale fabrication of 2-D nanoporous graphene using a thin anodic aluminum oxide etching mask. *J. Nanosci. Nanotechnol* 13(2013)7401.

[14] (a) J. J. Wang, J. Polleux, J. Lim, B. Dunn. Pseudocapacitive Contributions to Electrochemical Energy Storage in TiO_2 (Anatase) Nanoparticles. *J. Phys. Chem. C*, 111(2007) 14925.; (b) Rodrigo R. Resende, Emerson A. Fonseca, Fernanda M. P. Tonelli, Bruna R. Sousa, Anderson K. Santos, Katia N. Gomes, Silvia Guatimosim, Alexandre H. Kihara, Luiz Orlando Ladeira, *J. Biomed. Nanotechnol.* 10(2014)1157, (c) Hyung Bum An, Ji Man Kim, Jongsoo Jurng, Gwi-Nam Bae, Jong-Ki Jeon, Sung Hoon Park, Young-Kwon Park, Catalytic Oxidation of Benzene Using Mesoporous -Mn2O3, *J. Nanosci. Nanotechnol*, 13 (2013)7472.

[15] (a) C. Mah, I. Zolotukhin, T.J. Fraites, J. Dobson, C. Batich, B.J. Byrne, Improved method of recombinant AAV2 delivery for systemic targeted gene therapy. *Mol Therapy*, 1(2000)239. (b) Beatriz Ferreira de Carvalho Patricio, Marta de Souza Albernaz, Michelle Alvares Sarcinelli, Samira Marquesde Carvalho, Ralph Santos-Oliveira, Gilberto Weissmüller, *J. Biomed. Nanotechnol*, 10 (2014) 1242; (c) Adeel Afzal, Implantable zirconia bioceramics for bone repair and replacement: A chronological review. *Mater. Express*, 4 (2014)1, (d) Dong Hwan Kim, Ji Hyun Kim, Jin Seok Chung, Yunsang Lee, Control of the Visible Emission in the SrZrO3 Nano-Crystals with the Rare Earth Ion Doping. *J. Nanosci. Nanotechnol.* 13(2013)7572.

[16] (a) M. Pereiro, D. Baldomir, J. Botana, J.E. Arias, K. Warda, L. Wojtczak, Biomedical applications of small silver clusters. *J. Appl. Phys*, 103, 07A315, (2008); (b) Jianye Li, Meng He, One-Step *In Situ* Direct Growth of GaN Nanowire Devices. *Sci. Adv. Mater.* 6 (2014) 699 (c) Sang Hoon Kim, Ahmad Umar, M. S. Akhtar, G. N. Dar, Sang Woon Hwang, ZnO Balls Made of Intermingled Nanocrystalline Nanosheets for Photovoltaic Device Application, *Sci. Adv. Mater.* 6 (2014)562; (d) S. Valanarasu, I. Kulandaisamy, A. Kathalingam, Jin-Koo Rhee, T. A. Vijayan, R. Chandramohan. High-Performance Memory Device Using Graphene Oxide Flakes Sandwiched Polymethylmethacrylate Layers. *J. Nanosci. Nanotechnol*, 13 (2013) 6755.

[17] (a) S.M. Moghimi, A.C. Hunter, J.C. Murray, Nanomedicine: current status and future prospects. *FASEB J*, 19 (2005) 311, (b) Michael Loong Peng Tan, Huei Chaeng Chin, Li Len Lim, Weng Soon Wong, Eileen Lee Ming Su, Che Fai Yeong, Nanoscale Device Modeling and Circuit-Level Performance Projection of Top-Gated Graphene Nanoribbon Field-Effect Transistor for Digital Logic Gates. *Sci. Adv. Mater*. 6 (2014) 569;(c) Kin Fong Lei, Shih-I Yang, Bundled Carbon Nanotube-Based Sensor on Paper-Based Microfluidic Device. *J. Nanosci. Nanotechnol*, 13 (2013) 6917

[18] (a) T.A. Taton, Nanostructures as tailored biological probes. *Trends Biotechnol.*, 20(2002)277 (b) Taek Lee, Ajay Yagati Kumar, Si-Youl Yoo, Mi Jung, Junhong Min, Jeong-Woo Choi, Fabrication of Biomolecules Self-Assembled on Au Nanodot Array for Bioelectronic Device. *J. Nanosci. Nanotechnol.* 13 (2013) 6020.

[19] (a) R. Sinha, G.J. Kim, S. Nie, D.M. Shin, Nanotechnology in cancer therapeutics: bioconjugated nanoparticles for drug delivery. *Mol Cancer Ther*, 5 (2006) 1909, (b) Ke Fan, Renjie Li, Junnian Chen, Wenye Shi, Tianyou Peng, *Sci. Adv. Mater.*, 5 (2013) 1596; (c) Sungsu Kim, Kyoungah Cho, Kiyeol Kwak, Sangsig Kim, Memory Characteristics of Doubly Stacked Nano-Floating Gate Memory Devices with Channels of Single ZnO Nanowires. *J. Nanosci. Nanotechnol.* 13 (2013) 6196.

[20] (a) R.A. Sperling, W.J. Parak, Surface modification, functionalization and bioconjugation of colloidal n inorganic nanoparticles. *Phil. Trans. R. Soc. A*, 368 (2010) 1333. (b) Hsin-Chia Yang, Jui-Ming Tsai, Tsin-Yuan Chang, Wen-Shiang Liao, Sung-Ching Chi,The Effects of Thickness of Source/Drain Fin on *P*-Channel FinFET Devices and the Corresponding Quantum Effects. *Nanosci. Nanotechnol. Lett.* 6(2014)165; (c) Sang Wan Kim, Woo Young Choi, Min-Chul Sun, Byung-Gook Park, Investigation on the corner effect of L-shaped tunneling field-effect transistors and their fabrication method. *J. Nanosci. Nanotechnol*. 13 (2013) 6376.

[21] (a) C.M. Niemeyer Priv.-Doz. Dr., Nanoparticles, Proteins, and Nucleic Acids: Biotechnology Meets Materials Science. *Angewandte Chemie International Edition*, 40, (2001) 4128, (b) Jingyue Fang, Shiqiao Qin, Xueao Zhang, Jiandong Sun, Jiadong Li, Hua Qin, Dongming Wu, Liang Fang, Shengli Chang, Fabrication and Characterization of Single Electron Tunneling Device on Au/SiO_2 Nanocomposite Films. *J. Nanosci. Nanotechnol*, 13 (2013) 878.

[22] (a) M. Sastry, M. Rao, K.N. Ganesh, Electrostatic assembly of nanoparticles and biomacromolecules. *Acc. Chem. Res.*, 35 (2002) 847, (b) Jin-Han Lin, Yu-Rong Lin, Teng-Kai Wen, Rupesh S. Devan, Yung Liou, Yuan-Ron Ma, Nanoscale Dynamic Behavior of Surface Magnetic Domains on a La0.7Sr0.3MnO3 Thin Film. *J. Nanosci. Nanotechnol.* (2013) 888.

[23] E. Katz, I. Willner, Integrated nanoparticle-biomolecule hybrid systems: synthesis, properties, and applications. *Angewandte Chemie International Edition*, 43 (2004) 6042

[24] (a) A. Verma, V.M. Rotello, Surface recognition of biomacromolecules using nanoparticle receptors. *Chem. Commun.*, (2005) 303. (b) Yifeng Gu, Sannian Song, Zhitang Song, Yan Cheng, Xuyan Liu, Xiaofeng Du, Bo Liu, Songlin Feng, Reactive Ion Etching of SixSb2Te in CF4/Ar Plasma for Nonvolatile Phase-Change Memory Device. *J. Nanosci. Nanotechnol*, 13 (2013)1594.

[25] F. Caruso, Nanoengineering of Particle Surfaces., *Adv. Mater.*, 13 (2001) 11.

[26] (a) K. Naka, H. Itoh, Y. Tampo, Y. Chujo, Effect of Gold Nanoparticles as a Support for the Oligomerization of l-Cysteine in an Aqueous Solution Langmuir, 19, (2003) 5546, (b) Jingyue Fang, Shiqiao Qin, Xueao Zhang, Jiandong Sun, Jiadong Li, Hua Qin, Dongming Wu, Liang Fang, Shengli Chang, Fabrication and characterization of single electron tunneling device on Au/SiO2 nanocomposite films. *J. Nanosci. Nanotechnol*, 13 (2013) 878

[27] (a) S.S. Ghosh, P.M. Kao, A.W. McCue, H.L. Chappelle, Use of maleimide-thiol coupling chemistry for efficient syntheses of oligonucleotide-enzyme conjugate hybridization probes Bioconjugate Chem., 1(1990) 71.; (b) Gunasekaran Venugopal, Sang-Jae Kim, Observation of Nonvolatile Resistive Memory Switching Characteristics in Ag/Graphene-Oxide/Ag Devices. *J. Nanosci. Nanotechnol.* 12 (2012) 8522.

[28] J. Wang, Nanoparticle-based electrochemical DNA detection. *Anal. Chim. Acta*, 500, (2003) 247.

[29] (a) X. Michalet, F.F. Pinaud, L.A. Bentolila, J.M. Tsay, S. Doose, J.J. Li, S. Sundaresan, A.M. Wu, S.S. Gambhir, S. Weiss, Quantum dots for live cells, in vivo imaging, and diagnostics. *Science*, 307(2005) 538 (b) Vasilica Schiopu, Alina Matei, Adrian Dinescu, Mihai Danila, Ileana Cernica, Ce, Gd Codoped YAG Nanopowder for White Light Emitting Device. *J. Nanosci. Nanotechnol.* 12 (2012) 8836.

[30] Y. Weizmann, F. Patolsky, E. Katz, I. Willner, Amplified DNA Sensing and Immunosensing by the Rotation of Functional Magnetic Particles. *J. Am. Chem. Soc.*, 125(2003) 3452

[31] (a) H. T. Song, J.S. Choi, Y.M. Huh, S.J. Kim, Y.W. Jun, J.S. Suh, J.W.Cheon, Surface Modulation of Magnetic Nanocrystals in the Development of Highly Efficient Magnetic Resonance Probes for Intracellular Labeling. *J. Am. Chem. Soc.*, 127(2005) 9992 (b) E. Hourdakis, A. G. Nassiopoulou, Two-Terminal Charge-Trapping WORM Memory Device Using Anodic Aluminum Oxide. *J. Nanosci. Nanotechnol*, 12(2012) 7968.

[32] M. Mahmoudi, A. Simchi, M. Imani, U.O. Hafeli, Superparamagnetic Iron Oxide Nanoparticles with Rigid Cross-linked Polyethylene Glycol Fumarate Coating for Application in Imaging and Drug Delivery. *J. Phys. Chem. C* 1,(2009) 8142.

[33] (a) M.N. Rittner. Market Analysis of Nanostructured Materials. *Am. Ceram. Soc. Bull.*, 81(2002) 33. (b) Cheng-Hsien Lu, Chuan-An Cheng, Chia-Hua Ho, Kuan-Neng Chen, Effects of Bonding Technology and Thinning Process in Three-

Dimensional Integration on Device Characteristics. *J. Nanosci. Nanotechnol.* 12 (2012)8050.

[34] E. Oesterling, N. Chopra, V. Gavalas, X. Arzuag, E.J. Lim, R. Sultana, D. A. Butterfield, L. Bachas, B. Hennig, Alumina nanoparticles induce expression of endothelial cell adhesion molecules. *Toxicol. Lett*, 178(2008)160.

[35] (a) J.M. Laval, P.E. Mazeran, D. Thomas, Nanobiotechnology and its role in the development of new analytical devices. *Analyst* 125(1999)29; (b) Sujira Promnimit and Joydeep Dutta, Self-Organization of Colloidal Nanoparticles into Functional Pressure Sensing Device. *J. Nanosci. Nanotechnol.* 12 (2012) 8143.

[36] T. A. Desai, Micro- and nanoscale structures for tissue engineering constructs. *Med. Eng. Phys*, 22 (2000) 595.

[37] (a) L.A. Bauer, N.S. Birenbaum, G. J. Meyer, Biological applications of high aspect ratio nanoparticles. *J. Mater. Chem.* 14, 517-526, (2004); (b) Noorhana Yahya, Muhammad Hanis Zakariah, Synthesis and characterization of YBa2Cu3O7(Y123) via sol-gel method for development of superconducting quantum interference device magnetometer. *J. Nanosci. Nanotechnol*, 12 (2012)8147.

[38] L. Mazzola, Commercializing nanotechnology. *Nature Biotechnol*, 21 (2003) 1137.

[39] (a) X. Hu, S. Cook, P. Wang , H.-M. Hwang, In vitro evaluation of cytotoxicity of engineered metal oxide nanoparticles. *Sci. Total Environ.* 407 (2009) 3070; (b) A. Kathalingam, Jin-Koo Rhee. Fabrication and Characterization of Solution Processed *n*-ZnO Nanowire/*p*-Si Heterojunction Device. *J. Nanosci. Nanotechnol*, 12 (2012) 6948.

[40] S. Kim, M. Baek and S. J. Choi, Comparative cytotoxicity of Al2O3, CeO2, TiO2 and ZnO nanoparticles to human lung cells. *J. Nanosci. Nanotechnol*, 10 (2010) 3453.

[41] (a) A. Hanini, A. Schmitt, K. Kacem, F. Chau, S. Ammar ,J. Gavard, Evaluation of iron oxide nanoparticle biocompatibility. *Int. J. Nanomed.* 6(2011) 787.; (b) Marcos A. L. dos Reis, Augusto F. Saraiva, Manuel F. G. Vieira, Jordan Del Nero, Study of Ink Paper Sensor Based on Aluminum/Carbon Nanotubes Agglomerated Nanocomposites. *J. Nanosci. Nanotechnol.* 12 (2012)6955.

[42] A. Seidi and M. Ramalingam, Impact of Gradient Biomaterials on Interface Tissue Engineering. *J. Biomater, Tissue Eng.*, 2, (2012)89.

[43] Ahmad Umar, Yoon-Bong Hahn, *Metal Oxide Nanostructures and Their Applications*, American Scientific Publishers, USA, 1 (2009) 1.

[44] (a) S. G. Ansari, R. Wahab, Z. A. Ansari, Y. S. Kim, G. Khang, H. S. Shin, Effect of nanostructure on the urea sensing properties of sol–gel synthesized ZnO Sensors, *Actuators B*, 137 (2009)566.; (b) E. Esposito, C. Granata, A. Vettoliere, R. Russo, D. Peddis, M. Russo, Nano Superconducting Quantum Interference Device sensors for magnetic nanoparticle detection. *J. Nanosci. Nanotechnol.* 12 (2012)7468.

[45] (a) S. H. Kim, G. N. Dar, A. Umar, Growth of in-doped ZnO hollow spheres composed of nanosheets networks and nanocones: structural and optical properties.

J. Nanosci. Nanotechnol, 13 (2013) 4639. (b) V. F. Neto, J. Lencart, A. G. Dias, J. A. M. Santos, A. Ramos, C. Relvas, The Effect of Megavoltage Radiation on Polymeric Materials to be Used in Biomedical Devices, *J. Nanosci. Nanotechnol.* 12(2012)6779

[46] S. Al-Heniti, A. Umar, Structural, optical and field emission properties of urchin-shaped ZnO nanostructures. *J Nanosci Nanotechnol*, 13 (2013) 86.

[47] S. A. Khayyat, M. Abaker, A. Umar, Ahmad; M. O. Alkattan, N. J. Alharbi, S. Baskoutas, Synthesis and characterizations of Cd-doped ZnO multipods for environmental remediation application. *J Nanosci Nanotechnol*, 12 (2012) 8453.

[48] D. Liu, W. Wu, Y. Qiu, S. Yang, S. Xiao, Q. Wang, L. Ding, J. Wang, Surface functionalization of ZnO nanotetrapods with photoactive and electroactive organic monolayers. *Langmuir*, 24 (2008) 5052.

[49] A. K. Gupta, M. Gupta, Synthesis and surface engineering of iron oxide nanoparticles for biomedical applications. *Biomaterials*, 26 (2005) 3995.

[50] F. Witte, V. Kaese, H. Haferkamp, E. Switzer, A. Meyer-Lindenberg, C. J. Wirth and H. Windhagen, In vivo corrosion of four magnesium alloys and the associated bone response. *Biomaterials*, 26 (2005) 3557.

[51] Q. Yang, J. Sha, L. Wang, J. Wang , D. Yang, MgO nanostructures synthesized by thermal evaporation. *Mater. Sci. Eng.C.* 26 (2006) 1097.

[52] X. Gu, Y. Zheng, Y. Cheng, S. Zhong, T. Xi, In vitro corrosion and biocompatibility of binary magnesium alloys. *Biomaterials*, 30 (2009) 484.

[53] S. Ge, G. Wang, Y. Shen, Q. Zhang, D. Jia, H. Wang, Q. Dong , T. Yin, Cytotoxic effects of MgO nanoparticles on human umbilical vein endothelial cells in vitro IET. *Nanobiotechnol*, 5 (2011) 36.

[54] L. Bertinetti, C. Drouet, C. Combes, C. Rey, A. Tampieri, S. Coluccia, G. Martra, Surface characteristics of nanocrystalline apatites: effect of mg surface enrichment on morphology, surface hydration species, and cationic environments. *Langmuir*, 25 (2009) 5647.

[55] A. Fahmy, S. A. Cormier, Copper oxide nanoparticles induce oxidative stress and cytotoxicity in airway epithelial cells. *Toxicol. In Vitro* 23 (2009) 1365.

[56] W. Lin, Y. Xu, C. C. Huang, Y. Ma, K. B. Shannon, D. R. Chen ,Y. W. Huang, Toxicity of nano- and micro-sized ZnO particles in human lung epithelial cells. *J. Nanopart. Res*, 11 (2009) 25.

[57] W. Lin, I. Stayton, Y. W. Huang, X. D. Zhou , Y. Ma, Cytotoxicity and cell membrane depolarization induced by aluminum oxide nanoparticles in human lung epithelial cells A549. *Toxico. Environ. Chem.* 90 (2008) 983.

[58] W. Lin, Y. W. Huang, X. D. Zhou , Y. Ma, Toxicity of cerium oxide nanoparticles in human lung cancer cells. *Int. J. Toxicol*, 25 (2006) 451.

[59] C. K. Lai, M. B. Lai, S. Jandhyam, V. V. Dukhande, A. Bhushan, C. K. Daniels and S. W. Leung, Exposure to titanium dioxide and other metallic oxide nanoparticles

induces cytotoxicity on human neural cells and fibroblasts. *Int. J. Nanomedicine*, 3 (2008) 533.

[60] X. Q. Zhang, L. H. Yin, M. Tang ,Y. P. Pu, ZnO, TiO(2), SiO(2,) and Al(2)O(3) nanoparticles-induced toxic effects on human fetal lung fibroblasts. *Biomed. Environ. Sci.* 24 (2011) 661.

[61] M. Jugan, S. Barillet, A. Simon-Deckers, S. Sauvaigo, T. Douki, N. Herlin and M. Carrière, Cytotoxic and genotoxic impact of TiO2 nanoparticles on A549 cells. *J. Biomed. Nanotechnol.* 7 (2011) 22.

[62] JJ.Wang, BJ.Sanderson, Wang H. Cyto- and genotoxicity of ultrafine TiO2 particles in cultured human lymphoblastoid cells. *Mutat. Res*, 628 (2007) 99.

[63] S. J. Choi, J. M. Oh, J. H. Choy, Toxicological effects of inorganic nanoparticles on human lung cancer A549 cells. *J. Inorg. Biochem. 1*03 (2009) 463.

[64] J. Sun, S. Wang, D. Zhao, F. H. Hun, L. Weng ,H. Liu, Cytotoxicity, permeability, and inflammation of metal oxide nanoparticles in human cardiac microvascular endothelial cells: cytotoxicity, permeability, and inflammation of metal oxide nanoparticles. *Cell Biol. Toxicol*, 27 (2011) 333.

[65] J. Fischer, D. Pröfrock, N. Hort, R. Willumeit , F. Feyerabend, Improved cytotoxicity testing of magnesium materials. *Mater. Sci. Eng*, (2011) 176.

[66] H. S. Nalwa (ed.), *Encyclopedia of Nanoscience and Nanotechnology*, American Scientific Publishers, (2004/2011)1.

[67] American Cancer Society, *Cancer Facts and Figures 2010*, Atlanta, GA (2010).

[68] (a) Y. L. Zhao, H. S. Nalwa, Eds, *Nanotoxicology,* American Scientific Publishers, Los Angeles (2007); (b) S. Singh, H. S. Nalwa, Nanotechnology and health safety: Toxicity and risk assessments of nanostructured materials on human health. *J. Nanosci. Nanotechnol.*, 7(2007) 3048

[69] R. Singh, H. S. Nalwa, Medical Applications of Nanoparticles in Biological Imaging, Cell Labeling, Antimicrobial Agents, and Anticancer Nanodrugs. *J. Biomed. Nanotechnol.* 7(2011)489

[70] R. J. Aitken, M.Q. Chaudhry, A.B.A. Boxall, M. Hull, Manufacture and use of nanomaterials: current status in the UK and global trends. *Occup. Med.*, 56 (2006)300.

[71] A. Simon, B. Gouget, M. Mayne, N. Herlin, C. Reynaud, J. Degrouard, M. Carriere, In vitro investigation of TiO2, Al2O3, An nanoparticles and mutli-walled carbon nanotubes cyto- and genotoxicity on lung, kidney cells and hepatocytes, *Toxicol. Lett.* 36(2007) 172.

[72] S. Singh, Nanomedicine–Nanoscale Drugs and Delivery Systems. *J. Nanosci. Nanotechnol*, 10(2010) 7906

[73] Y. L. Zhao, H. S. Nalwa (eds.), *Nanotoxicology,* American Scientific Publishers, Los Angeles (2007).

[74] S. Singh, H. S. Nalwa, Nanotechnology and health safety: Toxicity and risk assessments of nanostructured materials on human health. *J. Nanosci. Nanotechnol*, 7 (2007) 3048.

[75] S. Prakash, M. Malhotra, W. Shao, C. Tomaro-Duchesneau , S. Abbasi, Polymeric nanohybrids and functionalized carbon nanotubes as drug delivery carriers for cancer therapy. *Adv. Drug Delivery Rev*, 63(2011) 1340.

[76] A. Saxena, C. Yashar, D. D. Taylor , C. Gercel-Taylor, Cellular response to chemotherapy and radiation in cervical cancer, *Am. J. Obstet. Gynecol.* 192 (2005) 1399.

[77] *WHO summary report on human papillomavirus and related cancers update 15* (2010) (http://apps.who.int/hpvcentre/statistics/dynamic/ico/country_pdf/IND.pdf)

[78] J. Huang, H. Li, W. Chen, G. H. Lv, X. Q. Wang, G. P. Zhang, K. Ostrikov, P. Y. Wang , S. Z. Yang, Dielectric barrier discharge plasma in Ar/O_2 promoting apoptosis behavior in A549 cancer cells. *Appl. Phys. Lett.* 99 (2011) 253

[79] K. Krishnamoorthy, J. Y. Moon, H. B. Hyun, S. K. Cho , S. J. Kim, Mechanistic investigation on the toxicity of MgO nanoparticles toward cancer cells. *J. Mater. Chem*, 22 (2012) 24610.

[80] K. M. Reddy, K. Feris, J. Bell, D. G. Wingett, C. Hanley , A. Punnoose, Selective toxicity of zinc oxide nanoparticles to prokaryotic and eukaryotic systems, *Appl Phys Lett.*, 90 (2007) 213902.

[81] N. Lewinski, V. Colvin, R. Drezek Cytotoxicity of nanoparticles, *Small*, 4 (2007) 26.

[82] A. Curtis, C. Wilkinson, Nantotechniques and approaches in biotechnology. *Trends in Biotechnology*, 19 (2001) 97.

[83] A. Vaseashta, D. Dimova-Malinovska, Nanostructured and nanoscale devices, sensors and detectors. *Science and Technology of Advanced Materials*, 6 (2005) 312.

[84] E. Klarreich, Biologists join the dots, *Nature*, 413 (2001) 450.

[85] B. Roszek, W. H. De Jong, R. E. Geertsma, *RIVM*, Bilthoven, The Netherlands 265001001, (2005).

[86] C. Leuschner, Challa SSR Kumar. *Nanofabrication Towards Biomedical Applications: Techniques, Tools, Applications, and Impact.* Editors: Challa SSR Kumar, J.Hormes, C. Leuschner, Wiley-VCH, 2005.

[87] Y. Kelly, M.A. Kim. Nanotechnology platforms and physiological challenges for cancer therapeutics. *Nanomedicine: Nanotechnology, Biology, and Medicine*, 3 (2007) 103.

[88] A. Thill, O. Zeyons, O. Spalla, F. Chauvat, J. Rose, M. Auffan, A.M. Flank, Cytotoxicity of CeO_2 Nanoparticles for *Escherichia coli.* Physico-Chemical Insight of the Cytotoxicity Mechanism. *Env. Sci. and Tech*, 40 (2006) 6151.

[89] H.H. Wang, C.S. Xie, W. Zhang, S.Z. Cai, Z.H. Yang, Y.H. Gui, Comparison of dye degradation efficiency using ZnO powders with various size scales. *J. Hazard. Mater.* 141 (2007) 652.

[90] J. Schulz, H. Hohenberg, F. Pflucker, E. Gartnera, T. Willa, S. Pfeiffera, R. Wepf, V. Wendel, H. Gers Barlag, K.P.Wittern, Distribution of sunscreens on skin, *Adv. Drug Deliv. Rev* 54 (2002) 163.

[91] R. Wahab, I. H. Hwang, Y.S. Kim, H. S. Shin, Photocatalytic activity of zinc oxide micro-flowers synthesized via solution method, *Chem. Engg. Jr.*, 168 (2011) 359

[92] R. Wahab, A. Mishra, S.I. Yun, Y.S. Kim, H.S. Shin. Antibacterial activity of ZnO nanoparticles prepared via non-hydrolytic solution route. *Appl. Microbiol. Biotech*, 87 (2010) 1917.

[93] R. Wahab, Y.S. Kim, A. Mishra, S.Il. Yun, H.S. Shin. Formation of ZnO Micro-Flowers Prepared via Solution Process and their Antibacterial Activity Nanoscale. *Res. Lett.* 5 (2010) 1675.

[94] S.G. Ansari, Z.A. Ansari, R. Wahab, Y.S. Kim, G.S. Khang, H.S. Shin. Glucose sensor based on nano-baskets of tin oxide templated in porous alumina by plasma enhanced CVD, *Biosens Bioelectron.* 23 (2008) 1838.

[95] S. P. Chang, C.W. Li, K. J. Chen, S. J. Chang, C. L. Hsu, T.J. Hsueh, H.T. Hsueh, ZnO-nanowire-based extended-gate field-effect-transistor pH sensors prepared on glass substrate. *Sci. Adv. Mater*, 4 (2012) 1174.

[96] T. Jansson, Zachary J., Clare-Salzler, T. D. Zaveri, S. Mehta, N. V. Dolgova, B. H. Chu, F. Ren, B. G.Keselowsky, Antibacterial effects of zinc oxide nanorod surfaces. *J. Nanosci. Nanotechnol.* 12 (2012) 7132.

[97] H. Algarni, M. M. El-Gomati, M. S. Al-Assiri, Low-Temperature Growth and Properties of Nanocrystalline Thin ZnO Nanosheet Interconnects on Zinc Foil. *Sci. Adv. Mater.* 4 (2012) 961.

[98] Y. Jang, J. Park, Y. Kim Pak, J. Jungho Pak, Immunosensor Based on the ZnO Nanorod Networks for the Detection of H1N1 Swine Influenza Virus. *J. Nanosci. Nanotechnol*, 12 (2012) 5173.

[99] P. Joshi, S. Chakraborti, P. Chakrabarti, S. P. Singh, Z. A. Ansari, M. Husain, and V. Shanker, *Sci. Adv. Mater.* 4 (2012) 173.

[100] Y. Wu, X. Li, J. H. Liu, Y. N. He, L. M. Yu, and W. H. Liu, ZnO nanomaterials based surface acoustic wave ethanol gas sensor. *J. Nanosci. Nanotechnol.* 12 (2012) 6505.

[101] Y. Nam, J. Park, Y. K. Pak, J. J. Park, Aptamer-based immunosensor on the ZnO nanorods networks. *J. Nanosci. Nanotechnol*, 12 (2012) 5547.

[102] K. Singh, Ahmad Umar, A. Kumar, G. R. Chaudhary, S. Singh, S. K. Mehta, Non-enzymatic glucose sensor based on well-crystallized ZnO nanoparticles. *Sci. Adv. Mater*, 4 (2012) 994.

[103] K. P. Jayadevan and T. Y. Tseng, One-dimensional ZnO nanostructures. *J. Nanosci. Nanotechnol*,12 (2012) 4409

[104] W. S. Chen, F.W. Chang, L. Selva Roselin, T. C. Ou, Production of Hydrogen by Partial Oxidation of Methanol Over ZnO-Promoted Copper Catalysts Supported on Rice Husk Ash. *Sci. Adv. Mater.* 4 (2012) 68
[105] Z. Luo, Q. Wu, J. Xue, Y. Ding, Selectively Enhanced Antibacterial Effects and Ultraviolet Activation of Antibiotics with ZnO Nanorods against *Escherichia Coli. J. Biomed. Nanotechnol*, 9 (2013) 69
[106] D. Bechet, P. Couleaud, C. Frochot, M.L. Viriot, F. Guillemin , M.B. Heyob, Nanoparticles as vehicles for delivery of photodynamic therapy agents. *Trends in Biotech*, 26(2008) 612.
[107] T. Jamieson, R. Bakhshi, D. Petrova, R. Pocock, M. Imani, A.H. Seifalian. Biological applications of quantum dots. *Biomaterials* 28(2007) 4717; (c) K.Y. Kim. *Nanomed: Nanotech, Biology and Medicine*, 3(2007) 103.
[108] R. Wahab, S. Dwivedi, Ahmad Umar, S. Singh, I. H. Hwang, H. S. Shin, J. Musarrat, A. A. Al-Khedhairy, Y. S. Kim, ZnO nanoparticles induce oxidative stress in Cloudman S91 melanoma cancer cells. *J. Biomed. Nanotechnol.*, 9 (2013) 441.
[109] S.J. Park, Y.C. Park, S.W. Lee, M.S. Jeong, K.N. Yu, H. Jung, J.K Lee, J.S. Kim, M.H. Cho, Comparing the toxic mechanism of synthesized zinc oxide nanomaterials by physicochemical characterization and reactive oxygen species properties. *Toxicology Letts*, 207 (2011) 197.
[110] S. Ostrovsky, G. Kazimirsky, A. Gedanken, C. Brodie. Selective cytotoxic effect of ZnO nanoparticles on glioma cells. *Nano Res.* 2 (2009) 882
[111] Z. Huang, X. Zheng, D. Yan, G. Yin, X. Liao, Y. Kang, Y. Yao, D. Huang, B. Hao, Toxicological effect of ZnO nanoparticles based on bacteria. *Langmuir*, 24 (2008) 4140.
[112] T. Puzyn, B. Rasulev, A. Gajewicz, X. Hu, T. P. Dasari, A. Michalkova, H. M. Hwang, A. Toropov, D. Leszczynska, J. Leszczynski, *Using nano-QSAR to predict the cytotoxicity of metal oxide nanoparticles*, 6 (2011) 175.
[113] Y.J. Jang, C. Simer, T. Ohm, Comparison of zinc oxide nanoparticles and its nano-crystalline particles on the photocatalytic degradation of methylene blue. *Mater. Res. Bull.* 41 (2006) 67.
[114] S. Makhluf, R. Dror, Y. Nitzan, Y. Abramovich, R. Jelinek, A. Geedanken, Microwave-Assisted Synthesis of Nanocrystalline MgO and Its Use as a Bacteriocide. *Adv. Funct. Mater,* 15 (2005) 1708.
[115] P.K. Stoimenov, R.L. Klinger, G.L. Marchin, K.J. Klabunde, Metal Oxide Nanoparticles as Bactericidal Agents. *Langmuir*, 18 (2002) 6679.
[116] S.M. Rozenzhak, M.P. Kadakia, T.M. Caserta TM, T.R. Westbrook, M.O. Stone, R.R. Naik, Cellular internalization and targeting of semiconductor quantum dots. *Chem. Commun. (Camb)*, 17 (2005) 2217.

[117] N. Jones, B. Ray, K.T. Ranjit, A.C. Manna, Antibacterial activity of ZnO nanoparticle suspensions on a broad spectrum of microorganisms. *Lett.*, 279 (2008) 71.

[118] Y. Liu, L. He, A. Mustapha, H. Li, Z.Q. Hu, M. Lin, Antibacterial activities of zinc oxide nanoparticles against Escherichia coli O157:H7. *J. Appl. Microbiol.*, 107 (2009) 1193.

[119] M. Banoee, S. Seif, Z.E. Nazari, P. Jafari-Fesharaki, H.R. Shahverdi, A. Moballegh, K.M. Moghaddam,A.R. Shahverdi, ZnO nanoparticles enhanced antibacterial activity of ciprofloxacin against Staphylococcus aureus and Escherichia coli. *J. Biomed. Mater. Res. Part B: Appl. Biomater.*, 93 (2010) 557.

[120] O. Yamamoto, M. Hotta, J. Sawai, T. Sasamoto, H. Kojima, Influence of Powder Characteristic of ZnO on Antibacterial Activity. *J. Ceram. Soc. Jpn*, 106 (1998) 1007.

[121] J. Sawai, O. Yamamoto, M. Hotta, H. Kojima, T. Sasamoto, antibacterial characteristics of heated zinc-oxide powders on escherichia-coli and staphylococcus-aureus. *Chem. Soc. Jpn – Chem. Indus. Chem. J.*, 9 (1998) 633.

[122] O. Yamamoto, Influence of particle size on the antibacterial activity of zinc oxide. *Int. J. Inorg. Mater.*, 3 (2001) 643.

[123] O. Yamamoto, M. Komatsu, J. Sawai, Z. Nakagawa, Effect of lattice constant of zinc oxide on antibacterial characteristics. *J. Mater. Sci. Mater. Med.*, 15 (2004) 847.

[124] J. Sawai, S. Shouji, H. Igarashi, A. Hashimoto, T. Kokugan, M. Shimizu, H. Kojima, Hydrogen peroxide as an antibacterial factor in zinc oxide powder slurry. *J. Ferment. Bioeng.*, 86 (1998) 521.

[125] O. Yamamoto, Y. Lida, Antifungal Characteristics of Spherical Carbon Materials with Zinc Oxide. *J. Ceram. Soc. Jpn*, 111 (2003) 614.

[126] J. Sawai, H. Igarashi, A. Hashimoto, T. Kokugan, M.Shimizu, Evaluation of growth inhibitory effect of ceramic powder slurry on bacteria by conductance method. *Jr. Chem. Engg. Japan*, 28 (1995) 288.

[127] J. Sawai, Quantitative evaluation of antibacterial activities of metallic oxide powders (ZnO, MgO and CaO) by conductimetric assay. *J. Microbiol. Methods*, 54 (2003) 177.

[128] X. Wang, F. Yang, W. Yang, X. Yang, A study on the antibacterial activity of one-dimensional ZnO nanowire arrays: effects of the orientation and plane surface. *Chem. Commun.*, 42 (2007) 4419.

[129] F. Sánchez-Jiménez, M V Ruiz-Pérez, J. L. Urdiales, M. A. Medina, Pharmacological potential of biogenic amine-polyamine interactions beyond neurotransmission., *Br. J. Pharmacol.* 170 (2013) 4.

[130] J. S. Johansen, A. K. Harris, D. J. Rychly, A. Ergul, Oxidative stress and the use of antioxidants in diabetes: Linking basic science to clinical practice. *Cardiovascul. Diabet*, 4 (2005) 5.

[131] Ömer Kesmez, Nadir Kiraz, Esin Burunkaya, H. Erdem Çamurlu, Meltem Asiltürk, Ertuğrul Arpaç, Effect of amine catalysts on preparation of nanometric SiO2 particles and antireflective films via sol–gel method. *J. Sol-Gel Sci. Technol.*, 56 (2010) 167.

[132] R. Rodrigo, M. Libuy, F. Feliú, D. Hasson, Molecular basis of cardioprotective effect of antioxidant vitamins in myocardial infarction. *BioMed Res. Int.*, 2013 (2013) 437613.

[133] H. J. Forman, M. Torres, Reactive oxygen species and cell signaling: respiratory burst in macrophage signaling. *Am. J. Respir. Crit. Care Med.*, 166 (2002)54.

[134] C. C. Wang, S. Wang, Q. Xia, W. He, J. J. Yin, P. P. Fu, J. H. Li, Phototoxicity of zinc oxide nanoparticles in HaCaT keratinocytes-generation of oxidative DNA damage during UVA and visible light irradiation. *J. Nanosci. Nanotechnol*, 13(2013) 3880.

[135] N. F. Wiernsperger, Oxidative stress as a therapeutic target in diabetes: revisiting the controversy. *Diabetes Metab*, 29 (2003) 579.

[136] J. P. Kehrer, L. G. Lund, Cellular reducing equivalents and oxidative stress. *Free Radi. Bio. Med.*, 17 (1994) 65.

[137] Y. W. Lin, Y. C. Chen, C. W. Wang, W. T. Chen, C.M. Liu, C. Y. Chen, H. T. Chang, Gold nanosponges: green synthesis, characterization, and cytotoxicity. *J Nanosci Nanotechnol.*, 13 (2013) 6566.

[138] H. Zhou, L Xiao, Y. Luo, J. H. Chen, J. H. Xu, Y. Zeng, M. Z.Zhong, Facile Synthesis of Monodispersed Fe_2O_3Nanoparticles and Its Cellular Uptake and Cytotoxicity Studies. *J Nanosci Nanotechnol.* 13 (2013) 6560.

[139] A. Mishra, R. Ahmad, V. Singh, M N Gupta, M Sardar, Preparation, characterization and biocatalytic activity of a nanoconjugate of alpha amylase and silver nanoparticles. *J Nanosci. Nanotechnol,* 13 (2013) 5028.

[140] H. M. Chiang, Q. Xia, X. Zou, C. Wang, S. Wang, B. J. Miller, P. C. Howard, J. J. Yin, F. A. Beland, H. Yu, P. P. Fu, Nanoscale ZnO Induces Cytotoxicity and DNA Damage in Human Cell Lines and Rat Primary Neuronal Cells. *J. Nanosci. Nanotechnol*, 12 (2012) 2126.

[141] N. K. Kaushik, H.S. Uhm, E.H. Choi, Oscillation of the magnetic moment in modulated martensites in Ni_2MnGa studied by *ab initio* calculations. *Appl. Phys. Lett.*, 100 (2012) 084106.

[142] R. Wahab, N. K. Kaushik, N. Kaushik, E. H. Choi, Ahmad Umar, S. Dwivedi, J. Musarrat, A. A. Al-Khedhairy, ZnO nanoparticles induces cell death in malignant human T98G gliomas, KB and non-malignant HEK cells. *J. Biomed. Nanotech*, 9 (2013) 1181.

[143] M.K. Patel, Md. Zafaryab, M. M. A. Rizvi, V. V. Agrawal, Z. A. Ansari, B. D. Malhotra, S. G. Ansari, Antibacterial and Cytotoxic effect of Magnesium Oxide nanoparticles on bacterial and human cell. *J. Nanoeng. Nanomanufac*, 3 (2013) 162.

[144] R. Wahab, Ahmad Umar, S. Dwivedi, K. J. Tomar, H. S. Shin,I. H. Hwang, ZnO Nanoparticles: Cytological Effect on Chick Fibroblast Cells and Antimicrobial Activities Towards *Escherichia Coli* and *Bacillus Subtilis. Sci. Adv. Mater.*, 5 (2013) 1571.

[145] Amit Kumar, Sumitra Arora, Navin Mogha, Salem S. Al-Deyab, Z. A. Ansari, S. G. Ansari, Glutathione Coated Zinc Oxide Nanoparticles: A Promising Material for Pesticide Detection. *Energy Environ. Focus*, 2 (2013) 101.

[146] Amit Kumar, Z. A. Ansari, H. Fouad, Ahmad Umar, S. G. Ansari, Oxidative Stress Control in *E. coli* and *S. aureus* Cells Using Amines Adsorbed ZnO. *Sci. Adv. Mater.* 6, (2014) 1236.

[147] T. Mosmann, Rapid colorimetric assay for cellular growth and survival: Application to proliferation and cytotoxicity assays. *J. Immunol. Meth.*, 65 (1983) 55

[148] R. Wahab, N.K. Kaushik, A.K. Verma, A. Mishra, I.H. Hwang, Y.B. Yang, H.S. Shin,Y.S. Kim. Fabrication and growth mechanism of ZnO nanostructures and their cytotoxic effect on human brain tumor U87, cervical cancer HeLa, and normal HEK cells. *J. Biol. Inorg. Chem.*, 16 (2011) 431.

[149] R. Wahab, Y. S. Kim, I. H. Hwang, H. S. Shin, A non-aqueous synthesis, characterization of zinc oxide nanoparticles and their interaction with DNA., *Synth. Metals*, 159 (2009) 2452.

[150] R. Wahab, S.G. Ansari, Y.S. Kim, H.K. Seo, G.S. Kim, G. Khang, H.S. Shin. Low temperature solution synthesis and characterization of ZnO nano-flowers. *Mater. Res. Bull*, 42 (2007) 1640.

[151] R. Wahab, S.G. Ansari, Y.S. Kim, G. Khang, H.S. Shin, Effect of hydroxylamine hydrochloride on the floral decoration of zinc oxide synthesized by solution method. *Appl. Surf. Sci.* 254 (2008) 2037

[152] P. H. Vajargaha, H. Abdizadeha, R. Ebrahimifarda, M.R. Golobostanfarda, Sol–gel derived ZnO thin films: effect of amino-additives. *Appl. Surf. Sci.*, 285 (2013) 732.

[153] J. R. Lakowicz, *Principles of Fluorescence Spectroscopy* 3rd ed., Springer, (2006) 278

[154] J. L. Bresloff, D.M. Crothers Equilibrium studies of ethidium--polynucleotide interactions. *Biochem.* 20 (1981) 3547

[155] P. Joshi, S. Chakraborti, P. Chakrabarti, D. Haranath, V. Shanker, Z. Ansari, S. P. Singh , V. Gupta, Role of surface adsorbed anionic species in antibacterial activity of ZnO quantum dots against Escherichia coli. *J. Nanosci. Nanotechnol.* 9, (2009)6427.

[156] R. K. Dutta, B. P. Nenavathu, M. K. Gangishetty, A.V. R. Reddy, Antibacterial effect of chronic exposure of low concentration ZnO nanoparticles on E. col. *Env. Sci. Health, Env. Engg.* 48 (2013) 871.

[157] R. Rohini, Yi Ting Chen, Roy T. W. Huang, G. H. Lin, I. J. B. Lin, *Sci. Adv. Mater*, 5 (2013) 216.

[158] R. Wahab, Y. S Kim, H. S. Shin, Synthesis, Characterization and Effect of pH Variation on Zinc Oxide Nanostructures. *Mater. Transactions*, 50 (2009) 2092.

[159] J. M. Yousef and E.N. Danial. In *Vitro* Antibacterial Activity and Minimum Inhibitory Concentration of Zinc Oxide and Nano-particle Zinc Oxide against Pathogenic Strains. *J. Health Sci*, 2 (2012) 38.

[160] R. Brayner, R. Ferrari-lliou, N. Briviois, S. Djediat, M. F. Benedetti, M.F. and F. Fievet, Toxicological Impact Studies Based on *Escherichia coli* Bacteria in Ultrafine ZnO Nanoparticles Colloidal Medium. *Nano Lett.* 6 (2006) 866.

[161] J. Doménech, A. Prieto, Stability of zinc oxide particles in aqueous suspensions under UV illumination. *J. Phys. Chem.*, 90 (1986) 1123.

[162] M. S. Hassan, T. Amna, O. B. Yang, M. H. El-Newehy, S. S. Al-Deyab, M. S. Khil, Smart copper oxide nanocrystals: synthesis, characterization, electrochemical and potent antibacterial activity. *Colloids Surf. B: Biointerfaces*, 97 (2012) 201.

[163] G. Fu, P. S. Vary, and C. T. Lin, Anatase TiO_2 Nanocomposites for Antimicrobial Coatings. *J. Phys. Chem. B* 109 (2005) 8889

[164] O. Yamamoto, J. Sawai, T. Sasamoto, Change in antibacterial characteristics with doping amount of ZnO in MgO–ZnO solid solution. *Int. J. Inorg. Mater*, 2 (2000) 451.

[165] O. Yamamoto, K. Nakakoshi, T. Sasamoto, H. Nakagawa, K. Miura, Adsorption and growth inhibition of bacteria on carbon materials containing zinc oxide. *Carbon*, 39 (2001)1643.

[166] K. H. Tam, A. B. Djurisic, Antibacterial activity of ZnO nanorods prepared by a hydrothermal method. *Thin Solid Films*, 516 (2008) 6167.

[167] Y. H. Leung, C. M. Chan, A. M. Ng, H. T. Chan, M. W. Chiang, F. C. Leung, W. K. Chan, Antibacterial activity of ZnO nanoparticles with a modified surface under ambient illumination. *Nanotechnology*, 23 (2012) 475703.

[168] A.M Rossi, P.R Burgmayer, *US Patent 5*, (1993)256.

[169] Lin W, Huang YW, Zhou XD, Ma Y, In vitro toxicity of silica nanoparticles in human lung cancer cells. *Toxicol. Appl. Pharmacol.* 217 (2006) 252.

[170] H.K. Yadav, K. Sreenivas, V. Gupta, S.P. Singh, R.S. Katiyar, Effect of surface defects on the visible emission from ZnO nanoparticles. *J. Mater. Res.*, 22 (2007) 2404.

[171] A. Nel, T. Xia, L. Madler, N. Li, Toxic Potential of Materials at the Nanolevel. *Science*, 311 (2006) 622.

[172] K. Donaldson, V. Stone, Current hypotheses on the mechanisms of toxicity of ultrafine particles. *Ann. Ist Super Sanità,* 39 (2003) 405.

[173] V. Sharma, R.K. Shukla, N. Saxena, D. Parmar, M. Das, A. Dhawan, DNA damaging potential of zinc oxide nanoparticles in human epidermal cells. *Toxicol. Lett.*, 185 (2009) 211.

[174] G. R. Martinez, A.P. Loureiro, S.A. Marques, S. Miyamoto, L.F. Yamaguchi, J. Onuki, E.A. Almeida, C.C. Garcia, L.F. Barbosa, M.H. Medeiros, P. Di Mascio, Oxidative and alkylating damage in DNA. *Mutat. Res.*, 544 (2003) 115.
[175] R.P. Schins, A.M. Knaapen, Genotoxicity of Poorly Soluble Particles. *Inhal. Toxicol.*, 19 (2007) 189.

In: Handbook of Remediation ...
Editors: Sadia Ameen, M. Shaheer Akhtar et al. ISBN: 978-1-53615-118-3

Chapter 9

NANOMATERIALS FOR LITHIUM–OXYGEN BATTERIES

Maria Christy[1,*], Anupriya Arul[2] and Kee Suk Nahm[1,2]
[1]Department of Mechanical Engineering, Hanyang University,
Seoul, Republic of Korea
[2]Department of Energy Storage & Conversion, Chonbuk National University,
Jeonju, Republic of Korea

ABSTRACT

The nature of alternative energy sources has changed considerably over time. At present, the major energy is produced by burning of fossil fuels or oils resulting in a high carbon dioxide emission. This is the major factor contributing to global warming, climate change etc. challenging our future other than weakening the world economy. The expeditious increase in the energy consumption coupled with the growing demand for clean energy has created the pressing need for efficient, clean, and renewable energy sources. In that context, batteries are best energy sources that are inherently simple in concept and have the potential to face the above-mentioned problems such as global warming and fossil fuel depletion in addition to strengthening the economy. Particularly, lithium-ion batteries have already taken over the portable electronic market with its high energy density. Lithium-air or -oxygen batteries are a step further ahead since these batteries could generate electricity by reacting the lithium metal and the oxygen from the air. The energy density of the lithium-air battery is theoretically equal to that of gasoline and ten times much more than that of a lithium-ion battery. Thus Lithium-oxygen (Li-O_2) battery with its largest theoretical energy densities among practical electrochemical couple is potentially considered to be the battery of the future mainly concerning electric vehicle applications. At the fundamental level, little was known about the reactions and processes that take place in the battery, representing a significant barrier to progress. So far, an advanced Li-air

* Corresponding Author's Email: mariaxty1@gmail.com.

cathode with a high capacity has been achieved with an average voltage of 2.33 V and a gravimetric capacity of 5813 mAh g^{-1} of carbon (without the use of catalysts) at a current density of 0.1 mA cm^{-2}. The progress has been slow due to the lack of suitable electrode materials and electrolytes, and the difficulties in mastering the interfaces between them. The battery requires a large chemical potential difference between the two electrodes by utilizing the reactants per exchanged electron as small as possible while ensuring that the electrolyte is not consumed in the chemistry of the battery. This chapter will discuss the overall progress of lithium-air batteries which include the basic structure, working principle, technical challenges, possible solution etc. The mechanisms of oxygen reduction and evolution that takes place on discharge and charge processes, respectively, are discussed in detail. The various reactions that take place in the cathode and electrolyte during discharge-charge mechanism are also considered. The parasitic reactions that occur on the cathode have been detailed describing the importance of suitable air-cathode catalysts. It also reviews the recent advances in understanding the chemistry and electrochemistry that govern the operation of the lithium–air battery. Insights on understanding the stability of the cathode and electrolyte and thereby the suitable design principles for better lithium–air battery are provided.

Keywords: lithium oxygen battery, capacity, nanomaterials, electrode, charge/discharge process

INTRODUCTION

"Energy is the single greatest challenge facing humanity", said the late Nobel laureate Richard Smalley. The drastic changes in the climatic conditions have increased the demand for clean energy from alternate energy resources. Emissions of heat–trapping gasses from human activity and atmospheric release of greenhouse gasses requires a carbon-free approach to energy generation. So, in the vision for clean, renewable and sustainable energy, batteries could be the next best possible energy resources. Batteries convert chemical energy directly into electrical energy with zero pollution. Specially, lithium ion battery technology tends to stay on top of the electrical power market dominating the portable electronic devices such as cell phones, laptops, cameras, etc. that are an essential part of everyday life. Recently lithium–ion batteries have replaced the gasoline and brought innovation to an era of clean energy since the invention of hybrid electric vehicles (HEV). In addition to its smaller size and lighter weight, lithium-ion batteries deliver performance that helps to protect the environment with features such as improved charge efficiency without memory effect. Yet, for HEV that runs on lithium–ion batteries, one of the biggest downsides is the limited range between recharging. Higher range electric vehicle (EV) means that the electric vehicles should run more kilometers before recharge. So it is necessary to develop a new generation battery to out-perform the limits of the current lithium-ion based battery. It is important to go beyond Li-ion technology to meet the growing energy storage needs of the future.

Lithium-oxygen (air) battery is said to possess one of the largest theoretical energy densities of any practical electrochemical couple. Figure 1 shows a comparison of common energy storage and conversion technologies based on their energy densities [1, 2]. From the Figure 1, it could be seen that a lithium–oxygen (Li–O_2) battery, in theory, could store energy as densely as a petrol engine which is more than ten times better than present days EV battery packs. In addition, Li-air technology uses O_2 derived from the air that is abundant and free, exhibiting highest theoretical specific energy (energy per unit mass) of any battery technology, 3,500 Wh kg^{-1}, making this battery extremely simple and low cost. Overall, by using lithium oxygen battery instead of lithium ion battery in EV just by changing the electrode in the lithium ion to air, it would be possible for a battery of the same size and weight to hold up to 10 times more energy, potentially harnessing the same energy density as gasoline [2-7]. This technology is a boon for energy storage and conversion since the transition from fossil fuel-based combustion engines to EV is mainly limited by the performance and the cost of the materials. Though the practical energy storage is still uncertain owing to various concerns, it has been estimated to deliver in the range of 800 to 1000 Wh kg^{-1} that would be sufficient to exceed 500 Km in EV without a recharge. So in the context of clean energy technology and the potential to replace gasoline at an even cheaper, simpler and better system, lithium oxygen batteries are expected to be future energy solution.

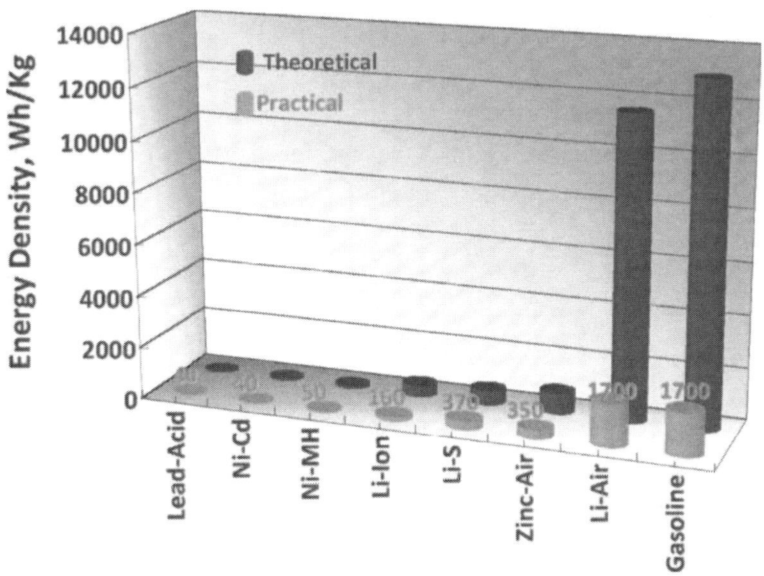

Figure 1. Energy density comparison chart. Reprinted with permission from Girishkumar, 2010, *J. Phys. Chem. Lett.* 1 (2010) 2193 © 2010 American Chemical Society.

The concept of Li-O_2 chemistry was initially proposed in 1976 which did not receive much attention until 1996 [8]. The first lithium air battery was reported by Abraham and Jiang in 1996 [2] who proposed a lithium air battery composed of non–aqueous polymer electrolyte, Li metal anode, and an oxygen cathode. Based on an inexhaustible and ubiquitous source of oxygen in the Li-air couple, they calculated the theoretical gravimetric energy density to be 11.14 kWh kg^{-1} and 5.2 kWh kg^{-1} including oxygen weight. However, experimentally utilizing a carbon air-cathode they achieved gravimetric capacities of ~ 1600 mAh g^{-1} in atmospheric air, and 1410 mAh g^{-1} in an oxygen atmosphere. The loss of capacity was attributed to the poor O_2 diffusion through the carbon-dense cathode. So clearly, although very large energy densities are theoretically achievable, there seem to be non-trivial challenges associated with the design of a practical Li-air battery. Read et al. [9–11] continued the development of Li-air technology by designing appropriate air cathodes and concluded that oxygen diffusion through the electrolyte and the electrolyte viscosity plays a major role in determining the cathode capacity. Dobley and Kuboki et al. [12, 13] also confirmed that oxygen diffusion, as well as, the porosity of the cathode plays a key role in determining the performance of the battery. Thus by the preliminary studies, it was evidently proved that the air cathode and its properties played a major role in the development of lithium-air batteries. In particular, an oxygen electrocatalyst involving in the fundamental reaction of Li-O_2 cell is one of the most challenging tasks in the improvement of the power density of Li-O_2 technology. Since then various researches have been focused on the improvement of the cathode performance by utilization of various electrocatalysts [2-7, 14-17]. On the other hand, the cycle life of the battery greatly depends on the electrolyte medium. Therefore, the development of new electrolytes with high conductivity, stability, and safety has been vigorous [18-21].

So, although Li–air batteries show excellence over many other rechargeable batteries, understanding of the underpinning chemistry and electrochemistry is still lacking. As discussed earlier, the crucial challenges that are faced by the Li-O_2 battery to become a reality involve the stability of the cathode towards reduced oxygen species, and the electrolyte solutions. Apart from that, the Li metal anodes do not deliver the necessary cycling efficiency because of the formation of dendrites. Lithium–air batteries unlike the name suggests, are currently limited to being operated under pure oxygen rather than ambient air due to the possible reaction of the discharge product with water and carbon dioxide in the ambient air that would render the battery irreversible. In the past decade, there has been tremendous progressing in understanding the fundamental science of Li–air technology, and many previous reports have led to the belief that Li–O_2 would inevitably be the energy of the future with a low rate capability, low capacity, and high polarization battery. In order to progress towards that and explore the practicality of the battery, all of these challenges need to be mastered first.

This chapter focuses on the progress and development of lithium-air batteries over the past decades. Specifically, the status and challenges and perspective of materials for non-aqueous Li−air batteries have been reviewed. The overall mechanism of the battery, the cathode degradation, importance of electrocatalysts, and the mechanisms of electrolyte etc. are discussed in detail. The significant challenges of lithium-oxygen batteries in the mechanisms of O_2 reduction on discharge and O_2 evolution on charge processes are reviewed and discussed with possible solutions.

WORKING PRINCIPLE OF LITHIUM-AIR BATTERY

A typical lithium air battery consists of a lithium metal anode, a carbonaceous air cathode, and an electrolyte medium as shown in Figure 2 [14]. Lithium-air batteries based on the electrolyte used can be classified into four types, namely, (i) non-aqueous [20], (ii) aqueous [22], (iii) hybrid [22] and, (iv) solid state system [23, 24]. Schematic representations of Li–O_2 cells based on the type of electrolytes are shown in Figure 3 [14].

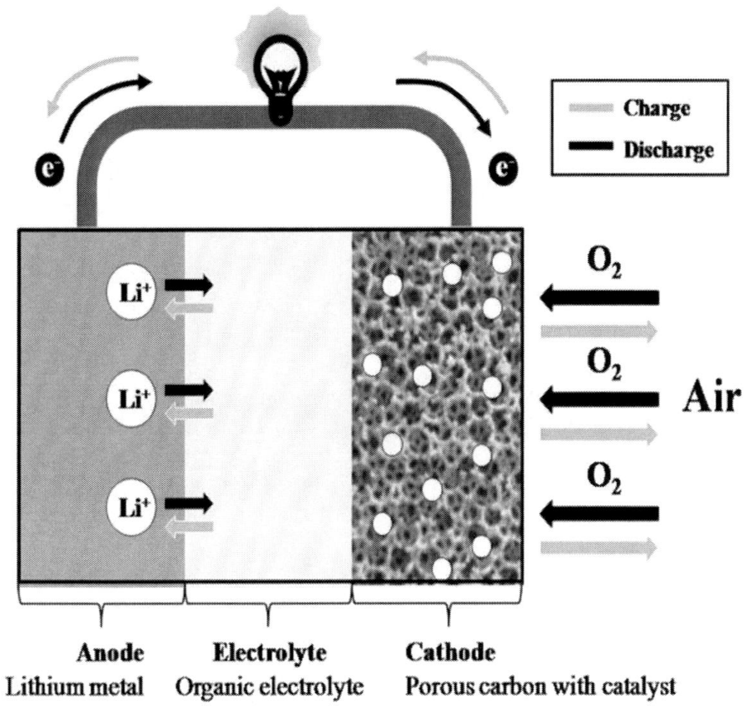

Figure 2. A typical lithium-air battery. Reprinted with permission from K. S. Nahm, 2012, *J. Electrochem. Sci. and Techn.* 3 (1) (2012) 14 © 2012 The Korean Electrochemical Society.

Aqueous and hybrid versions of electrolytes have been investigated by various research groups to understand their reaction mechanism [22]. Initially, Polyplus [25] developed the aqueous electrolyte version and Polyplus and Wang et al. [26] developed hybrid electrolyte version.

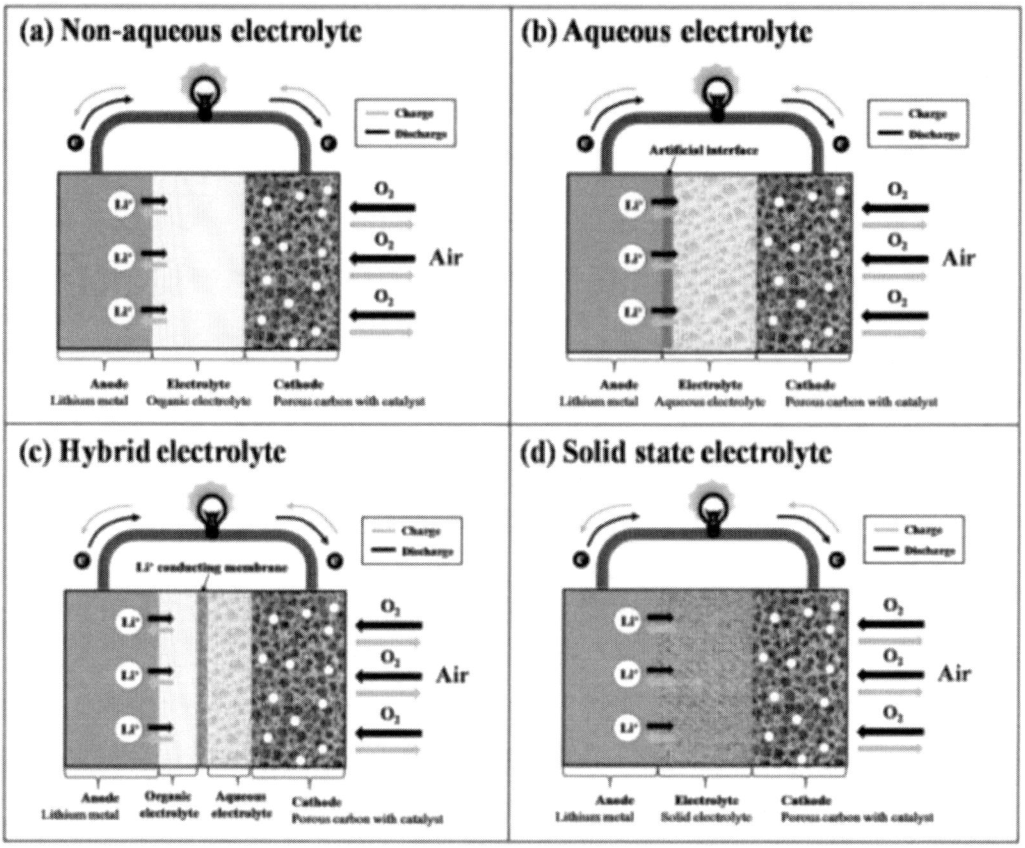

Figure 3. Lithium-air batteries based on the type of electrolyte used in the battery. Reprinted with permission from K. S. Nahm, 2012, *J. Electrochem. Sci. and Techn.* 3 (1) (2012) 14 © 2012 Korean Electrochemical Society.

The fundamental reaction mechanisms for the aqueous electrolytes could be represented as follows [14],

$$2Li + \frac{1}{2}O_2 + 2H^+ \rightarrow\ 2Li^+ +\ H_2O \qquad (1)$$

(in acidic electrolyte)

$$2Li + \frac{1}{2}O_2 + H_2O \rightarrow\ 2\ LiOH \qquad (2)$$

(in alkaline electrolyte)

The fundamental reaction mechanism for both aqueous and hybrid electrolytes are similar, as given in equations 1 and 2. The aqueous Li−air battery is not widely studied since its theoretical energy density is predicted to be lower than a non-aqueous system.

The hybrid electrolyte version consists of lithium metal anode, a non-aqueous electrolyte, superionic conducting glass ceramic membrane, aqueous electrolyte, and the air cathode. The solid electrolyte in hybrid cell has low Li+ ion conductivity and currently the hybrid and aqueous Li-air batteries are operated at a low current density of <1 mA cm^{-2}

and a limited power density of <50 mW cm^{-2} due to the high resistance associated with the solid electrolyte [14]. The main advantage of the aqueous electrolyte is that the discharge product is easily soluble in water which enables the fast diffusion of discharge product while the oxygen diffusion can also be enhanced resulting in better cycling efficiency. According to Manthiram et al. [22], aqueous and hybrid Li-air batteries are promising candidates for future EV and storage of electricity produced from renewable sources due to their large cell capacities, high cell voltages, and robust cell configurations in ambient atmospheres. However, the main challenge for the aqueous and mixed electrolytes is the lithium metal that reacts very aggressively with water. In detail, at the anode side, lithium dendrite growth and poor interfacial morphology are the major problems since accumulated dendrites will damage the solid electrolyte and the poor interfacial morphology will affect the cycling efficiency of the lithium-metal anode [20]. At the cathode side, the catalytic activity and durability of the bifunctional air electrode affect the overall efficiency and cycle life of the battery. The anode side problems could be solved by adjusting the electrolyte solvent (or polymer) and salts, introducing additives, and applying pressure on the whole cell to keep the cell components intact. Also, a lithium ion conducting membrane to protect the lithium reacting directly with water like an artificial solid electrolyte interface (SEI) would be necessary to protect the lithium metal anode in aqueous and hybrid systems [27]. The cathode side problems could be solved by selecting appropriate and compatible cathode/catalyst/electrolyte combinations and by filtering the incoming air to eliminate CO_2 from the air [20].

The all solid state battery chemistry has been investigated by several research groups [23, 24, 28-30] and the battery consists of lithium metal anode, a solid electrolyte, and an O_2 air cathode. The application of solid electrolytes to the Li-O_2 batteries can enhance the safety owing to their flame resistant properties. The high thermal stability of solid electrolytes prevents the thermal runaway of the batteries and concurrently allows them to operate effectively at a higher temperature. Higher temperature operation is beneficial to improve the cell performance since the Li^+ ion conductivity increases resulting in the activation of ORR and OER processes at the air electrode. A fully solid–state Li-air battery was recently reported by Kumar et al. [28]. First, they fabricated the solid-state Li-O_2 battery using $Li_yAl_xGe_{2-x}(PO_4)_3$ (LAGP) as a solid electrolyte and polymer electrolytes as a buffer layer to decrease the contact resistance between LAGP and the air electrode. Very recently, Li et al. [24], developed all solid state type battery without polymer materials for room temperature operation. Although some of them can reach a high Li^+ ion conductivity of 1×10^{-3} S cm^{-1} at room temperature, none of them show sufficient chemical stability in the complex chemical environment of hybrid and aqueous Li-air batteries. It is clear that the all-solid-state Li-O_2 cell has a significant scope, yet, the electrochemical performance, such as rate performance and a total capacity of this all-solid-state Li-O_2 cell, is still insufficient for practical use in the present stage. Possibly, highly conductive and thinner solid electrolytes could be used to reduce the internal resistance and improve the power performance.

The non-aqueous or fully aprotic system is the commonly used system wherein lithium–air battery comprises of lithium metal anode which plays a role as a lithium source, a carbon-based air-cathode, and an organic electrolyte medium. There have been various reaction mechanisms proposed for lithium-oxygen battery. Abraham et al. [2] suggested the following reaction mechanism for the aprotic cell:

$$Li + \frac{1}{2}O_2 \rightarrow \frac{1}{2}\ Li_2O_2 \qquad (3)$$

$$Li + \frac{1}{4}O_2 \rightarrow \frac{1}{2}\ Li_2O \qquad (4)$$

During the discharge process, the lithium ions from the anode move through electrolyte medium and meet with the atmospheric oxygen at air cathode and get oxidized resulting in the formation of lithium peroxide (Li_2O_2). The Li_2O_2, also known as the discharge product gets accumulated inside the pores of the porous electrode. The electrons from the anode are transported via an external circuit to the cathode while the lithium ions move through the electrolyte. As the electrons flow through the external circuit, the dissociated lithium ions flow through the electrolyte to the cathode via the mechanism of an electrochemical potential gradient. At the cathode, oxygen is reduced in either a two or four electron process to form lithium peroxide as shown in equations (3) and (4). The reversible cell voltage is Eo = 3.10 V for the reaction (3) and E_o = 2.91 V for reaction (4) calculated by using the standard Gibbs free energy of formation.

Later, Lu et al. [31] proposed the reversible cell voltages as the following equations:

$$2Li + O_2 \rightarrow\ Li_2O_2 \qquad (5)$$

$$4Li + O_2 \rightarrow 2Li_2O \qquad (6)$$

The reversible cell voltage, in this case, is Eo = 2.96 V for the reaction (5) and Eo = 2.91 V for reaction (6), and E is referred as Li/Li+. Eo = −ΔG/nF, where ΔG = Gibbs free energy of the reaction, n = number of electrons transferred and F = Faraday constant.

Laoire et al. [18] propose a mechanism in which oxygen is reduced at the cathode to form lithium superoxide as described by the following reaction:

$$O_2 + e^- + Li^+ \rightarrow\ LiO_2\ (3.0V) \qquad (7)$$

The mechanism states that, towards the end of the battery discharge when the reduction products have progressively filled the pores available for oxygen diffusion, the lithium superoxide further reacts via another one-electron reaction as indicated by the following step:

$$LiO_2 + e^- + Li^+ \rightarrow Li_2O_2\ (3.1V) \tag{8}$$

Furthermore, Mizuno et al. [32] have recently studied the electrochemical stability of carbonate-based liquid electrolytes on electrodes and have described the complicated cathode reactions involving O_2 radical. Lithium alkyl carbonate (RO–(C O)–OLi) and lithium carbonate (Li_2CO_3) were formed after discharge which suggests that the carbonate solvents highly influence the final discharge product. These suggested mechanisms for oxygen reductions at the cathode demonstrate the complexity and uncertainty of the electrochemical reactions; hence further investigations are required to clarify the reactions that take place during discharge. Based on the experimental observation, Bruce et al. [33] predicted ORR and OER pathways in Li^+ containing non-aqueous electrolyte as following; during ORR

$$O_2 + e^- \rightarrow O_2^- \tag{9}$$

$$O_2^- + Li^+ \rightarrow LiO_2 \tag{10}$$

$$LiO_2 \rightarrow Li_2O_2 + O_2 \tag{11}$$

During OER:

$$Li_2O_2 \rightarrow 2Li^+ + O_2 + e^- \tag{12}$$

They provided direct evidence that oxygen reduction in Li^+ -containing non-aqueous electrolyte first forms O_2^- that binds to Li^+ forming LiO_2 on the surface of the electrode which is unstable and disproportionate to more stable Li_2O_2 [15]. Upon charge or the OER process, the reactions do not involve LiO2; and the as-formed Li_2O_2 is directly decomposed to Li+ and O2 making it clear that the pathways for ORR and OER are different.

To summarize, during the discharge process, ORR takes place on the cathode resulting in the formation of lithium peroxide, Li_2O_2 (solid). On charge, the as-formed Li_2O_2 will be decomposed into 2Li and O_2 by OER. The recharge process of Li–O_2 battery requires higher potentials for the decomposition of the solid Li_2O_2. The ideal cathode reaction is generally expressed as follows:

$$2Li + 2e^- + O_2 \xleftrightarrow{\substack{Discharge/\\ Charge}} Li_2O_2 \tag{13}$$

Graphical representation of a typical charge/discharge curve of the lithium–oxygen battery is given in Figure 4. The difference between the discharge and charge voltage is called as overpotential, ΔV. For maximum energy density, the battery is expected to operate at low overpotential and relatively high current.

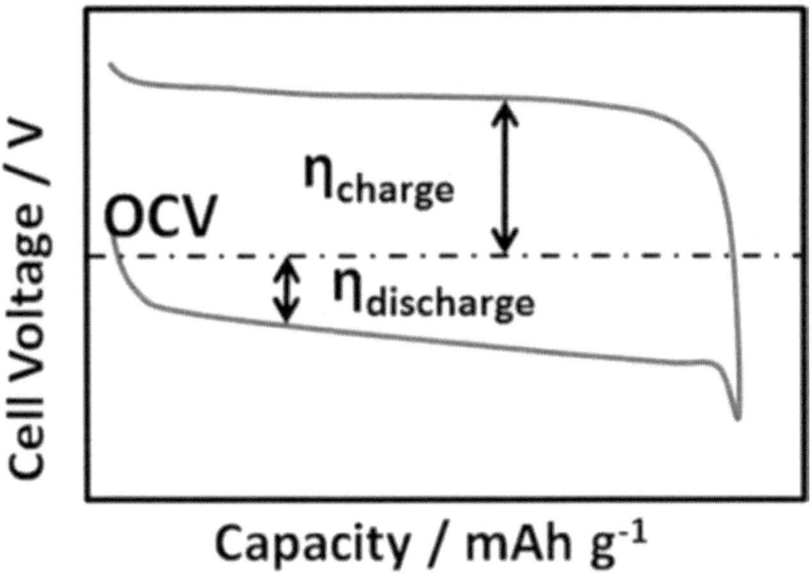

Figure 4. Graphical representation of a typical charge/discharge curve of the lithium–oxygen battery.

LIMITING FACTORS OF APROTIC LI-AIR BATTERIES

Energy Loss

One of the critical challenges that still prevent the implementation of the Li–air batteries is the huge energy loss for achieving a reversible Li–O_2 electrochemical process. Energy loss occurs due to the high charge–discharge polarization observed in the cell, that is, the large potential difference between the measured and the thermodynamically determined reversible potential (Erev=2.96 V, vs. Li/Li^+). Energy losses from charge/discharge process should be reduced for possible EV applications [34].

Over Potential

The discharge process depends on the current density and the overpotential. The standard potential for the discharge reaction, U_o, has been calculated as $U_o = 2.96$ V, using the Nernst equation [4]. The initial open circuit voltage (OCV) of a Li-O_2 cell is 3.1~3.3 V, which is significantly higher than the calculated U_o. The working voltage of the cell during discharge is approximately 2.6 – 2.7 V, which is significantly less than U_o. The difference between these is called the discharge overpotential η_{dis}. During recharge, the cell requires excessive energy to decompose the solid discharge product, Li_2O_2. So, during constant-current charging of the cell, the voltage increases to approximately 4.0 V. Hence the charge overpotential (η_{chg}) is significantly greater than the discharge overpotential (η_{dis}). It is important to reduce the overpotential and increase the electrical and thermodynamic efficiency of the battery. Overpotential, ΔV, can be clearly understood from Figure 4, where $\Delta V = \eta_{dis+} \eta_{chg}$.

Slow Reaction Kinetics

During the discharge/charge processes, the cathode catalyst offers active sites for electrochemical reactions such as formation and decomposition of Li_2O_2. So far, Pt and its alloys have been considered as best ORR catalyst and Ir, the best OER catalyst. However, even with Pt catalyst, ORR shows slow kinetics which is a major limitation of lithium air battery. Due to the sluggish kinetics of ORR and OER, poor capacity, rate capability, and round trip efficiency are obtained. So it is necessary to replace the Pt catalyst with an alternative low cost and high efficient catalyst in order to improve the reaction kinetics [34].

Unfavorable Reactions

Instead of the solid discharge products, sometimes, by–products other than Li_2O_2 are also formed in the air cathode due to the reaction between the electrolyte and the diffused oxygen that require further energy to be decomposed on recharge. As a result of electrochemical decomposition of carbonate species, the evolution of CO_2 instead of O_2 has been reported [34, 35]. In addition, the commonly used carbonate based electrolytes are highly susceptible to nucleophilic attack and easily decompose into lithium alkyl carbonates. So, unfortunately, discharge process mostly results in the decomposition of the electrolyte.

All the mentioned problems such as (1) energy loss, (2) overpotential, (3) slow kinetics and (4) unfavorable side reactions, can be suppressed by smoothing out the discharge/recharge processes by utilization of bifunctional (ORR and OER) electrocatalysts. The performance of the Li−O_2 battery mainly depends on the electrocatalysts regardless of its design (aprotic, aqueous, or solid state). Since the nature of the electrocatalyst plays a key role in the performance of the battery, it is necessary to identify the better suitable catalyst before focusing on other tasks, such as properties of the binder, and electrolyte. On the other hand, ethers, sulfones, sulfoxides and ionic liquids have been explored as alternative electrolytes for lithium-air batteries to overcome electrolyte decomposition.

IDEAL LITHIUM O_2 BATTERY

Anode Characteristics

Lithium is the lightest of all metals and has the greatest electrochemical potential and high theoretical reversible capacity 3,860 mAh/g. During discharge, lithium ions dissolve from the surface of the lithium metal anode and transfer to the air-cathode through the electrolyte medium. During charge, the lithium ions transfer back to the anode to be back onto the surface of the lithium metal anode evolving oxygen. In this mechanism, the inherent problem with the lithium metal anode is the dendrite formation that occurs during charging when the lithium metal reacts with the organic electrolyte forming a passive layer on their surface resulting in the non-uniform plating of lithium. This could lead to a short circuit between anode and cathode, resulting in cell death and also to serious safety problems due to localized hot spots. For the high energy density, efficiency and the safety of lithium air batteries, it is most important to control the lithium metal anode and electrolyte interface.

Till date researchers are trying to find out the best solution for the stability of lithium metal anode which includes gel type electrolytes, metallic lithium with protective layers that are composed of conductive ceramic, glass etc. The effect of protective layers on lithium metal anode has been investigated and the morphology and composition of SEI have been observed. Arie et al. [36] investigated the effect of diamond-like carbon coating on lithium metal anode by radio frequency plasma enhanced chemical vapor deposition. According to the study, the DLC coated lithium metal anode showed improved electrochemical properties with better specific capacity due to the relatively high sp3 carbon content that helped in the transport of lithium ions. This coating behaves as a thin layer on the lithium metals and prevents the lithium metal anode from direct contact with the electrolyte to suppress any dendrite formation. Earlier days, Ishikawa et al. [37] also investigated the effect of aluminum iodide (AlI3) and magnesium iodide (MgI_2) as additives in the electrolytes on lithium batteries, to form thin films of Li-Mg and Li-Al alloy layers at the electrode-electrolyte interface to suppress any dendrite formation. However, these compounds tend to exhibit more resistance to the lithium ions which leads to the low cycling efficiency of the battery. Likewise, Visco et al. [38] used a thin ceramic film of highly superionic lithium conductor, LISICON on lithium metal anode. The ceramic films created a barrier between lithium metal anode and the electrolyte and protected the anode from atmospheric contaminations. Coverage of the lithium metal electrode by protective lithium-ion conducting films has been proposed to address this issue. Yet, this approach cannot ensure total safety, since the protective films are usually fragile and may easily deteriorate or even break upon battery operation. Recently, a preliminary study of the behavior of lithium-ion-air battery where the common, unsafe lithium metal anode is replaced by a lithiated silicon−carbon composite is reported by scrosati et al. [39]. Lithium metal is, in fact, a very reacting electrode that needs to go a long way to ensuring safety and overcome serious safety hazards.

Electrolyte Characteristics

The practical development of lithium-oxygen (Li–O_2) batteries is mainly confined due to the limiting factors like poor cycle life and large overpotential of the battery. One of the factors causing this poor performance is the stability of the organic electrolytes against the superoxide anion radical, an intermediate phase formed during discharge. Depending on the electrolyte being used, the discharge product sometimes seems to vary resulting in the formation of the irreversible product, Li_2CO_3 instead of Li_2O_2 or evolution of CO_2 instead of O_2. Such problem results in the large overpotential and decomposition of the electrolyte, terminating the cell as a result. In that aspect, an ideal/suitable electrolyte is as essential as an electrocatalyst to improve the cycle life of the Li–O_2 battery. The challenge is to develop a stable electrolyte with better conductivity, solubility, stability and more importantly safety.

$LiPF_6$ had been used as the salt in Li-ion batteries for more than a decade for its unique properties like good ionic conductivity. However, they failed badly with the Li-O_2 system due to unwanted destructive reactions with Li-metal. Other lithium salts such as $LiClO_4$, $LiBF_4$, $LiAsF_6$, $LiPF_6$, or CF_3SO_3Li are commonly being used, where some of them are unstable with highly reactive Li metal anode. Amine et al. [40] studied the compatibility of lithium salts with non-aqueous solvents and found that the compatibility of lithium salts could significantly affect the stability of the whole electrolyte in a Li–O_2 battery. According to the report, three lithium salts, namely, LiPF6, Lithium bis (trifluoromethane sulfonyl) imide (LiTFSI) and $LiCF_3SO_3$, are compatible with tetraethylene glycol dimethyl ether (TEGDME) solvent. Later, four different salts, namely $LiCF_3SO_3$, $LiN(SO_2CF_3)_2$, LiTFSI), $LiClO_4$, and $LiPF_6$ were investigated in TEGDME solution under fixed capacity conditions by Hassoun et al. [41]. Among the four, the best behavior with the lowest interphase resistance and the highest lithium transference number was provided by $LiCF_3SO_3$ salt in TEGDME solution. LiTFSI salt has also been reported as a potentially viable alternative for $LiPF_6$ with its improved chemical and thermal stability, and more importantly safety compared to $LiPF_6$. Therefore, $LiCF_3SO_3$ and LiTFSI salts are commonly used in Li-O_2 battery system owing to their significant characteristics reported so far. Luntz and co-workers [4, 20, 42-44], also have extensively examined the stability of numerous solvent/salt combinations using differential electrochemical mass spectrometry (DEMS).

Table 1: Properties of electrolyte solvents

Solvents	conductivity, σ (mS/cm) Li+	DN (kcal/mol)	viscosity η (cP)	oxygen solubility (mM/cm^3)
dimethyl sulfoxide (DMSO)	2.11	28.9	1.948	2.10
acetonitrile (ACN)	14.39	14.1	0.361	8.1
tetrathylene glycol dimethyl ether (TEGDME)	1.16	16.6	4.05	4.43
1,2-dimethoxyethane (DME)	0.3	20.0	0.46	9.57

Like salts, solvents also play an important role in the electrolyte. For a non-aqueous electrolyte, solvents should hold low volatility, high conductivity, solubility, stability, high oxygen solubility and inert to superoxide radicals. In spite of not having enough low vapor pressure, alkyl carbonates were attractive electrolytes since most of the research on Li-metal anodes was conducted using such electrolytes. But, unlike Li-ion batteries, the oxygen-rich environment in Li–O_2 system dramatically influences the stability of carbonate-based electrolytes questioning their practicality. Also, organic carbonates are prone to nucleophilic attack by the superoxide radical. In that aspect, esters could be the next choice as electrolytes for Li–O_2 battery. According to Balaish et al. [21], the high dielectric constant and low viscosity of esters are the minimal criteria to ensure sufficient salt dissolution. However, the vapor pressures of common esters are found to be considerably high, and common esters are found to be too reactive toward Li metal resulting in low efficiency. The attachment of fluorine atoms to the ethereal carbon was considered as a possibility to suppress the nucleophilic substitution on the ethereal carbon, but it lowered the reaction barrier for the attack at the carbonyl atom. The instability of Esters in the presence of superoxide rules out the possibility of common organic esters to be suitable for the Li–air battery. Cormac et al. [18] initiated the studies of the redox reactions of oxygen in non-aqueous electrolytes to elucidate the roles of ion conducting salts and organic solvents on the mechanisms of the corresponding reactions.

Initially they started with acetonitrile and extended their investigation with solvents that are practically useful in Li-air batteries such as, anhydrous acetonitrile (MeCN), dimethyl sulfoxide (DMSO), 1, 2-dimethoxyethane (DME), and tetraethylene glycol dimethyl ether (TEGDME). According to their report, during cyclic voltammetry, the cathodic peak potential (*E*pc) shifts toward more negative potentials according to the order TEGDME < DME < MeCN < DMSO, indicating that the reduction of oxygen is hindered going from DMSO to TEGDME.

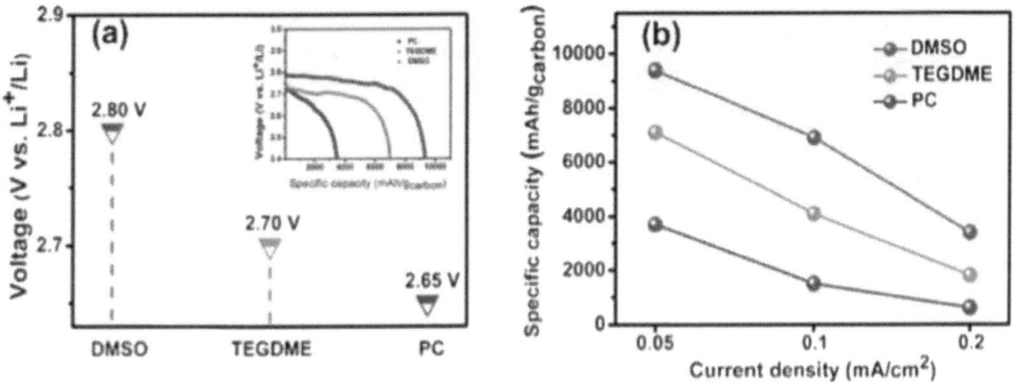

Figure 5. Li-O_2 battery performance using LiTFSI salt in TEGDME, DMSO and PC solvent electrolytes. Reprinted with permission from D. Xu, 2012, *Chem. Commun*. 48 (2012) 6948 © 2012 The Royal Society of Chemistry Ltd.

Table 1 gives the solvent properties of the four solvents as reproduced from Cormac et al. [18], which includes Li ion conductivity (σ), Donor number (DN), Viscosity (η) and oxygen solubility. A solvent's basicity is usually characterized by its donor number (DN), which has been reported for the solvents in the order MeCN (14.1) < TEGDME (16.6) < DME (20.0) < DMSO (29.8). Similarly, solvent acidity can be characterized by its acceptor number (AN), which in these solvents follows the order DME (10.2) < TEGDME (10.5) < MeCN (18.9) < DMSO (19.3) [18]. According to the detailed report by Loire et al, high DN solvents provide increased stability for the complex because of the modulated Lewis acidity of the hard acid resulting in an O_2/O_2^- reversible couple in the presence of Li^+. Recently, several groups have shown considerable interest on DMSO-based electrolytes which has demonstrated fair stability of the electrolytes [45-47]. Xu et al. [45], employed LiTFSI salt in DMSO solvent and investigated the Li-O_2 battery performance in comparison with LiTFSI in TEGDME and LiTFSI in PC electrolytes (refer Figure 5). The comparative analysis demonstrates DMSO to be desirable solvent for Li-O_2 batteries. However, DMSO decomposition is not negligible, especially with increased cycles where DMSO vapor pressure is not low enough to guarantee a long battery operation; it results in poor cell performances [21]. In addition, DMSO electrolytes interact with the Li-metal anode and do not demonstrate high Li-cycling efficiency. So, thus far, DMSO seems to be a less suitable solvent for a rechargeable Li–air cell which required further investigation. Ethers and glymes on the other hand, are considered good candidates as solvents for the non-aqueous Li–O_2 battery as they may have a fairly low vapor pressure. PG Bruce et al. compared ether-based electrolytes with organic carbonates, and found that ether-based electrolytes are more stable towards reduced O_2 species than the organic carbonates, since formation of Li_2O_2 was evident in ethers but organic carbonates. However, the study concluded that neither ether nor carbonates are suitable for Li-O_2 batteries. So far, TEGME has been a popular Li–air electrolyte solvent that possesses a room temperature vapor pressure of B105 bar. The main product at the first discharge in ether-based electrolytes is usually Li_2O_2, with negligible amounts of Li_2CO_3. Scrosati et al. [48] recently showed no discharge capacity degradation after 100 cycles at the high specific discharge and charge rate of 500 mAg c^{-1} (Refer Figure 6). This polar solvent also seems to have low volatility as opposed to lower molecular weight linear chain ethers, and displays a high lithium salt solubility and a fairly wide electrochemical window. The stability of TEGDME with various lithium salts requires further detailed investigation. Nevertheless, until now, the employed non-aqueous organic electrolytes cannot meet all of the requirements of rechargeable Li–air cell environment. Nevertheless, carbonate-based and ether-based electrolytes may offer adequate Li^+ transport properties, good Li-salt solubility and are compatible with the Li-anode compared to other possible electrolytes.

Air Cathode Characterization

The solid discharge products, Li_2O_2, formed on the cathode surface during discharge or ORR process, are non-conductive in nature. Hence a proper design of cathode for high-performance Li-O_2 battery is essential.

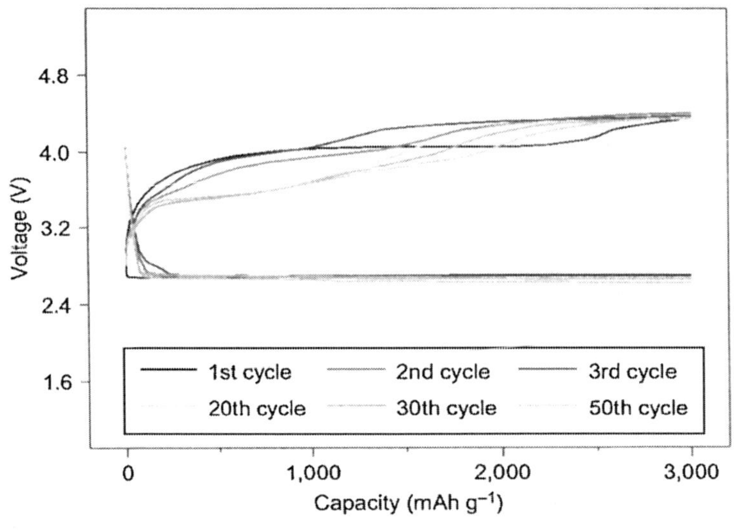

Figure 6. Discharge capacity at a high specific discharge and charge rate of 500 mA g c^{-1}. Reprinted with permission from H. G. Jung, 2012, *Nat. Chem.* 4(7) (2012) 579 © 2012 Macmillan publishers Ltd.

4.3.1 Carbon electrodes

Factors such as electrical conductivity and surface properties such as surface area and pore properties govern the electrochemical performance of Li-O_2 battery. Hence, nanostructured carbon materials have been attractive air cathode materials owing to their large surface area, high electrical conductivity, stability and low cost. Commercially available carbon materials such as carbon black (CB), Vulcan carbon (VC) with high surface area and electrical conductivity have been popular cathode materials. However, the relatively low capacity and stability makes them impractical. Potential carbon materials such as carbon nanotubes (CNT) and graphite nanofibers (GNF) have been greatly attractive electrode materials owing to their superior electrical and mechanical properties. According to Mitchell et al. [49], CNT and GNF exhibit improved capacity, reversibility and cyclability since their one-dimensional (1D) structure allows fast reaction kinetics. These 1D carbons are also famous for being a matrix or support for constructing nanocatalyst composites [50-53]. In the next level, two-dimensional (2D) carbons, graphene, have invited tremendous interests since 2010, when high-quality graphene became so easy to isolate [54-57]. Graphene or reduced graphene oxide (rGO) possess very high surface area, and their electrochemical properties could be tuned by stacking of the individual graphene sheets. The 1D and 2D carbon materials and their excellent properties led to the structural modification of multi-scale carbon materials into hierarchical 3D structured carbon materials with tunable large surface area and pore properties. Xiao et al. [58] reported excellent electrochemical activity of porous graphene which showed exceptionally high capacity without any catalysts.

Another effective technique to improve the carbon-based cathode materials is by doping heteroatoms into their structures. Carbon doping by boron, nitrogen, sulfur, etc. have been proved effective in improving the structural stability of the material and thereby increasing the round trip efficiency of the Li-O_2 battery when employed as cathode material [59-65]. The sp^2 carbon materials have abundant free-flowing π electrons that are too inert to be used directly in ORR. When the carbon is doped with nitrogen, the π electrons get activated by conjugating with the lone-pair of electrons from N-dopants. During ORR, O_2 molecules get reduced to the positively charged carbon atoms neighboring nitrogen. Nanostructured N-doped carbons such as nanotubes, nanobelts, nanorods, nanofibers, and nanosheets etc. have been widely utilized as air cathode materials for Li-O_2 batteries. Similarly, sulfur doped carbon (S-carbons) and boron doped carbon (B-carbons) have been reported to show better performance than nitrogen doped carbon (N-carbons) [66, 67]. Some researchers have even proposed to dope carbon with a combination of dopants like both B and N or S and N etc. to further improve the ORR catalytic activity, by the synergistic effects between the dopants [68-70].

Very recently, Lim et al. [71] reported a hierarchically woven porous carbon nanotube (CNT) fibril air electrode without any catalyst in Li-O_2 cell (Figure 7). According to the authors, the unique porous morphology of the air electrode significantly improved the cyclability of the Li−O_2 battery without catalyst even at high current density. The authors improved the porous structure of the cathode by embedding with catalysts producing catalyst embedded hierarchical air electrode which delivered even better performance [72]. On a step further, they [73] recently introduced hierarchical 3D carbon matrix electrode with redox mediators that delivered excellent stability and cyclability for Li-O_2 batteries. Later, Shui et al. [74] developed vertically aligned nitrogen doped coral like carbon fiber arrays that demonstrated high efficiency as carbon based oxygen electrode with very low overpotential (Figure 8).

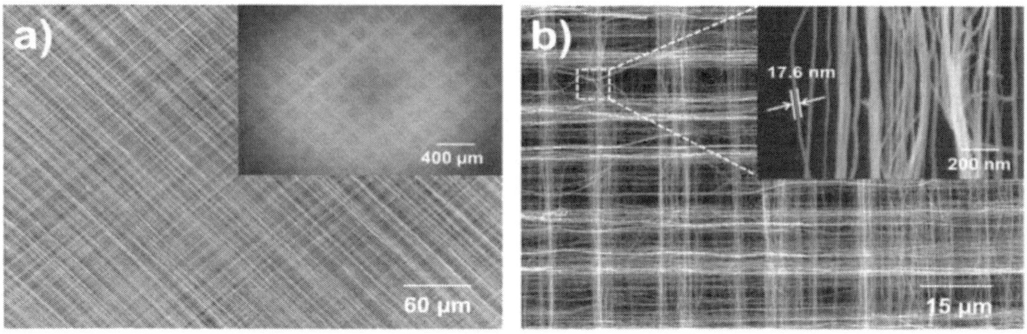

Figure 7. SEM images of the CNT fibril at a) low magnification (inset: large area image of the air electrode), and b) high magnification. Reprinted with permission from H. D. Lim, 2013, *Adv. Mater*. 25 (2013) 1348 © 2013 John Wiley & Sons, Inc.

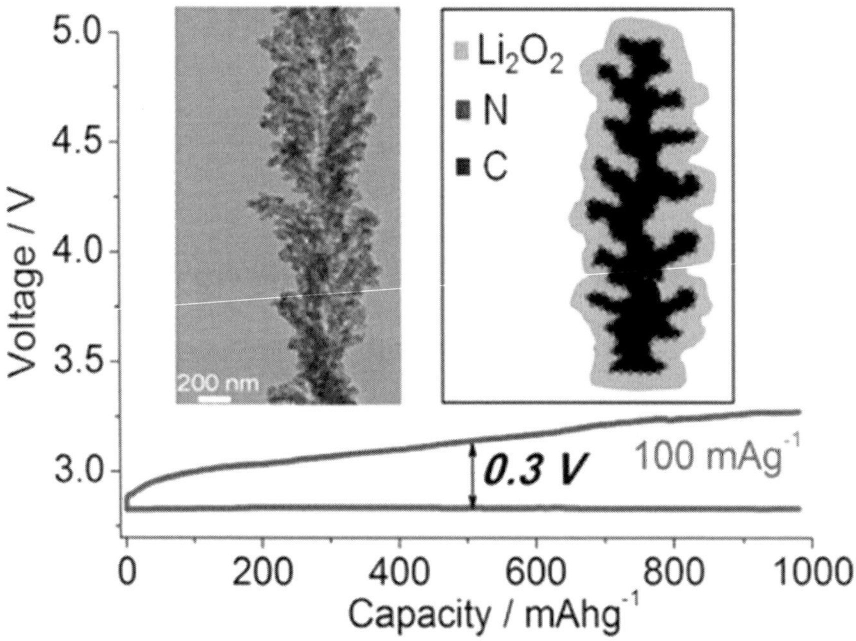

Figure 8. Hierarchical 3D carbon matrix electrode. Reprinted with permission from J. Shui, 2014, *ACS Nano 8* (3) (2014) 3015 © 2014 The American Chemical Society

Carbon-Free Electrodes

Developing carbon free cathode could mitigate the side reactions that arise from carbon-based cathodes and decrease the electrolyte decomposition during cycling. So far, precious and non-precious metals, transition metals, and metal oxides, carbides, nitrides, sulfides and their composite materials, have been investigated as a replacement for carbon cathode. Jung et al. [48] recently reviewed the oxygen electrodes for non-aqueous Li-O_2 batteries. According to the authors, carbon-free cathodes should possess certain physical apart from chemical stability. The physical attributes required of proper carbon-free cathodes are as follows: (1) high electrical conductivity; (2) high surface area; (3) mesoporosity; (4) high activity towards ORR and OER; and (5) low cost.

Peng et al, [75] introduced carbon-free nanoporous gold (NPG) cathode that exhibited high capacity retention even after 100 cycles. Though the precious materials possess the major physical properties mentioned above, they are scarce and very expensive. Chen et al, [76] proposed a core-shell structure with precious metal/non-precious metal oxide hybrid material like Au/Co_3O_4 and Au/MnO_2 etc. The unique nanostructure exhibited low overpotential owing to the highly conductive nanoporous Au core and a large number of reactive sites provided by the metal oxide layer surrounding Au. Alloying precious and non-precious metals or metal oxides could reduce the overall cost of the material. As a next step, bimetallic alloys such as 1-D AgPd-Pd nanotubes were reported by Luo et al. and others [77-79]. The Li-O2 battery showed better round trip efficiency with bimetallic nanotubes owing to the rapid connective network of nanotubes. Recently transition metal carbides have been investigated for their superior physicochemical properties. Thotiyl et al. [80] successfully demonstrated TiC cathode for $Li\text{-}O_2$ batteries which showed excellent capacity retention and greatly reduced overpotential. The oxidized TiOC phase developed on the surface of the cathode prevented it from nucleophilic attack, which in turn prevented any unwanted side reactions or electrolyte decomposition, resulting in the better performance of the cell even after 100 cycles. Molybdenum carbides have also been reported to be suitable carbon-free cathode materials. According to Kundu et al, [81] structural properties of oxide passivating layers on the carbide surface influences the stability of the carbides. Zhou's group introduces a conductive indium-tin-oxide supported on ruthenium nanoparticle (Ru/ITO) cathode. Compared to carbon cathodes, the Ru/ITO cathode exhibited reduced overpotential and high cyclability. It was evident that Ru nanomaterials possess high catalytic activity towards ORR and OER, in addition to good conductivity and surface properties. Followed by, highly conductive hollow RuO_2 has been investigated by Li et al. [82] and better rate capability of the cell was reported. Very recently, carbon-free Ruo2-TiO2 electrode for Li-air batteries has been proposed for Li-air batteries [83]

Carbon-free and binder-free also known as the "free-standing" electrode was introduced by Liao et al. [84] by depositing Ru on Ni foam by galvanic replacement method. The Li-O2 cell with free-standing cathode exhibited superior overall performance. Kim et al. [85] also proposed a free-standing electrode with Au nanoparticles on Ni nanowires by an electroplating method. Pt-Gd alloy coated in Ni foam by pulsed laser deposition have also been reported as a cathode for $Li\text{-}O_2$ battery. However, all the free-standing electrode materials require proper synthetic techniques. Recently, for the first time, free-standing electrodes were reported by simple chemical deposition reaction [86] that exhibited better capacity and efficiency. In addition to carbon-free cathode design, Riaz et al. [87] tailored the nanostructures for better performance. $NiCo_2O_4@MnO_2$

cathode, that is, ternary $NiCo_2O_4$ nanoneedles were grown on Ni foam current collector on which MnO_2 nanoflakes were deposited and used as carbon free cathode for Li-O_2 battery. The cell exhibited superior characteristics compared to conventional carbon cathodes.

The reversibility and cyclability of the Li-O_2 battery can be significantly improved using a stable cathode electrode. The side reaction arising from the cathode part, mainly cathode instability, results in the undesired reaction products and decomposition of electrolytes. So, carbon loading in carbon based cathodes should be reduced to avoid blocking of pores on the cathode surface. In addition carbon-free and free-standing cathodes could pave the way to mitigate the unwanted side reactions. However, lot more studies are required in order to achieve commercial status yet.

Electrocatalysts for Air Cathodes

The electrochemical reaction kinetics of oxygen is generally slow and complicated due to its strong irreversibility. There are two proposed mechanisms for the ORR: (i) a single step direct four-electron transfer mechanism and (ii) a two-by-two electron mechanism that involves two steps [14]. In the second mechanism, it goes through a hydroperoxide intermediate (HO_2^-) to complete the reaction where the intermediate could undergo disproportionation reaction and prevents the second step reaction. The base–stabilized intermediate could also degrade the carbon support, catalyst, and/or the membrane, resulting in poor ORR. Therefore, it is important to develop a bifunctional catalyst that promotes direct four–electron transfer mechanism. The electron transfer number per oxygen molecule (n) during ORR can be calculated from rotating ring-disk electrode (RRDE) experiment by using the following equation:

$$n = 4I_D/(I_D + (I_R/N)) \qquad (14)$$

where I_D, I and N are disk current, ring current and ring collection efficiency (here N = 0.37) respectively. The direct four-electron reduction involves an intermediate reaction of peroxide adsorption which is predominant in noble metal catalysts and transition metal oxides. In noble metal catalysts, like Pt, Ag, Au, Ir, Ru etc. oxygen is evolved from the corresponding metal oxides. Among them, Pt is the most suitable ORR catalyst, yet a poor electrocatalyst for OER. Similarly Ir, a suitable OER catalyst is a poor ORR catalyst. Hence, in order to achieve high efficiency, the reaction kinetics of both ORR and OER reactions, in other words, bifunctional catalytic activity should be improved.

Noble Metal Catalysts

Extensive research efforts were addressed on noble nanometals and their alloys including Pt, Au, Ru, Pd, Pt/Au, etc., to improve the electrokinetics of ORR and OER.

Horn et al. [88] reported the intrinsic oxygen reduction reaction (ORR) activity of polycrystalline palladium, platinum, ruthenium, gold, and glassy carbon surfaces in 0.1 M $LiClO_4$ 1,2-dimethoxyethane via rotating disk electrode (RRDE) measurements. According to the authors, the non-aqueous Li^+ ORR activity of these surfaces primarily correlates to oxygen adsorption energy, forming a "volcano- type" trend. The Li^+-ORR activity was found to be in the order of Pd > Pt > Ru ≈ Au > GC on bulk surfaces. According to the report, Au was the most active for ORR and Pt is the most active for OER among Pt, Au, and C in bulk and nanoparticle (NP) forms. So as a next step, a combination of both ORR and OER catalyst was analyzed by the authors. Bimetallic PtAu nanoparticles (NPs) were investigated and shown to strongly enhance the kinetics of ORR and OER in Li-O_2 batteries [89]. The effect of current density on the discharge and charge voltages of Li-O_2 cells with PtAu supported on carbon (C) was also investigated and PtAu/C exhibited bifunctional catalytic activity, where Au and Pt atoms were primarily responsible for ORR and OER kinetics in Li-O_2 cells, respectively.

In that trend we investigated bimetallic platinum-iridium nanocomposites supported on carbon and titanium oxides [90]. Platinum (Pt) is considered as a nearly ideal catalyst for ORR and alloying Pt with other metals (M), Pt/M is a viable approach to obtain bifunctional electrocatalytic activity. Iridium being the nearly ideal OER catalyst, Pt/Ir composites were chosen by the authors. Titanium oxides (TiO_2) had been one of the most broadly studied metal oxides in the field of energy conversion and storage due to its superior physicochemical properties, excellent stability, and low cost for synthesis and its being environmentally friendly and biocompatible. Another feasible approach to reduce the amount of noble metal catalysts is the usage of carbon supports that could provide a high surface area and stabilize the mass-based activity of the catalyst. Thus the Pt–Ir/C–TiO_2 bifunctional electrocatalysts were synthesized via a simple synthesis technique and high ORR and OER activities in both aqueous and non-aqueous electrolyte solutions were recorded. With the high cost and scarcity of precious metals, utilizing carbon or any other support and supportive non-precious metals could be a possible approach to minimal utilization of precious metals for maximum efficiency. Better ORR and

OER activities from the synthesized Pt–Ir/C–TiO_2 catalyst in both aqueous and non-aqueous electrolyte solutions were obtained, that was almost comparative to or much better than that of commercial Pt/C catalysts. The battery exhibited improved performance owing to the unique combination of the Pt–Ir/C–TiO_2 catalyst. Pt improved the reaction kinetics of ORR and Ir, OER, showing fast kinetics without any electrolyte decomposition in the Li–O_2 battery while C–TiO_2 support rendered a stable structure with better electronic conductivity.

Very recently, an air electrode was constructed by Lim et al. [72] based on hierarchically woven CNT fibrils with micro-mesh and nano-sized pores where the highly aligned CNT sheets are cross woven such that they formed a mesh, ensuring a homogenous and well-ventilated structure. The CNT fibrils were loaded with Pt nanoparticles, that demonstrated better OER characteristics and the authors explained that the three-dimensionally aligned pore structure of the Pt–CNT fibrils not only allowed facile accessibility of Li ions and oxygen but also effectively exposed catalysts to the discharge products, thereby providing an ideal condition for the catalytic reaction. Fig. 9 illustrates the scanning electron microscopic image (SEM) of Pt/CNT fibrils before and after discharge process.

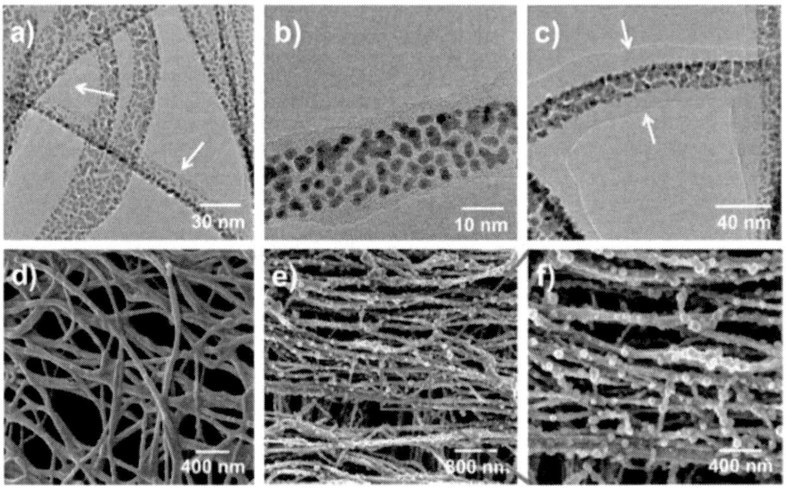

Figure 9. The scanning electron microscopic image (SEM) of Pt/CNT fibrils before and after discharge process. Reprinted with permission from H. D. Lim, 2013, *Energy Environ. Sci.* 6 (2013) 3570 © 2013 The Royal Society of Chemistry Ltd.

Non-Noble Metal Catalysts

The ORR mechanism in the non-aqueous system is different from conventional fuel cells and aqueous metal-air batteries since solid discharge product is formed in the non-aqueous Li-O_2 battery. So the non-precious metal catalysts are extensively investigated for application in the non–aqueous Li-O_2 batteries. Mainly, transition metal oxides such as MnO_x, Co_3O_4, Fe_3O_4 etc. have been investigated among which MnO_x is considered to be the most active ORR catalyst that could replace commercial Pt [91-99]. Few groups [100-103] compared all types of MnOx, such as α-MnO_2, β-MnO_2, bulk α-, β-, γ-, λ-MnO_2 and commercial Mn_2O_3 and Mn_3O_4 as well, and concluded that α-MnO_2 nanowires are the most effective catalysts for lithium–air battery with their crystal structure and high surface area. The catalytic activity of MnO_2 varies according to their crystalline structures, and Cao et al. [100] reported the sequence of the catalytic activity of MnOx as follows; β-MnO_2 < λ-MnO_2 < γ-MnO_2 < α-MnO_2 ≈ δ-MnO_2. The formation of different nanostructures of MnO_2 and the comparison between α-MnO_2 and δ-MnO_2 nanomaterials or their application in Li–O_2 battery was reported by Nahm et al. in 2014 [104]. Crystal structures of α- and δ-MnO_2 nanomaterials are shown in Figure 10 and the SEM images of nanourchin and birnessite shaped MnO_2 structures are shown in Figure 11. Both the synthesized MnO_2 nanomaterials are a suitable catalyst for the Li-air cathode and α-MnO_2 seem better than δ-MnO_2 in terms of stability and capacity. Among the two MnO_2 nanostructures each having different phase and structure, urchin shaped α-MnO_2 exhibited very good catalytic activity with very high capacity and stable cycling compared to layered birnessite δ-MnO_2 catalysts. The α-MnO_2 structure contains double chains of MnO_6 octahedra with the tunnel structure of (2×2) and (1×1) as shown in Figure 10. It is evident that the catalytic performance of α-MnO_2 is associated with the morphology, particle size, and their crystal structures with (2 x 2) tunnels. The difference in specific capacity has been explained in terms of their phase and crystal structures as follows. The large tunnel structures of α-MnO_2 favor the intercalation of ions in the lattice framework while its crystal defects and OH^- groups benefit to the adsorption of O_2 and dissociation of O–O bonds. Therefore, with a stronger O–O bond birnessite δ-MnO_2 shows poor activity resulting in a lower electrocatalytic activity in the cell performance.

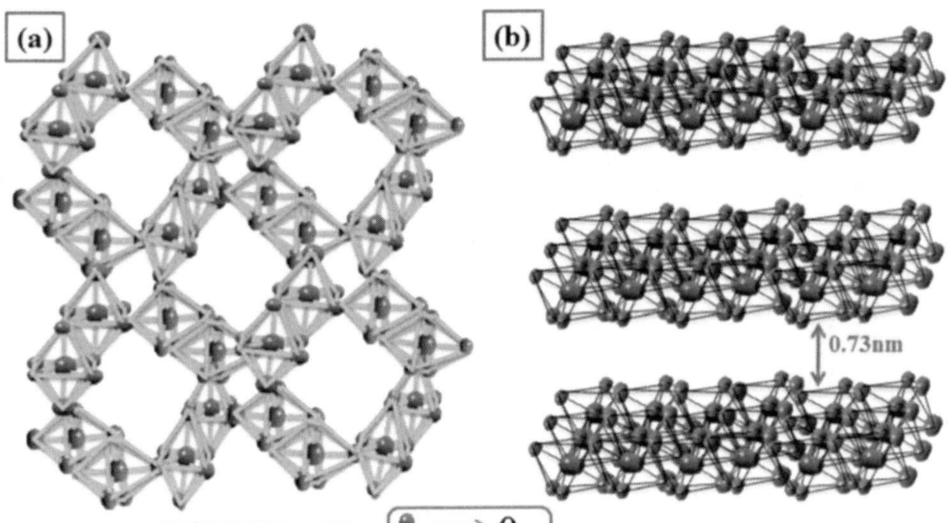

Crystallographic form	Size/Å
α (1X1) (2X2)	1.89, 4.6

Figure 10. Crystal structures of α– and δ–MnO_2 nanomaterials. Reprinted with permission from K. S. Nahm, 2014, *RSC Adv*. 4 (2014) 8973 © 2014 The Royal Society of Chemistry Ltd.

Similarly, other nonprecious metal and metal oxide ORR catalysts such as MnO_2, TiO_2, CoO, γ-MnOOH, Co_xMn, and $MnCo_2O_4$ etc. have been investigated for air-cathode applications [105-113]. Later, Jung et al. [114], reported the electrocatalytic property of metal-decorated MnO_2 catalysts and suggested that they could be used as an efficient bi-functional catalyst for rechargeable hybrid Li–O_2 batteries. Utilizing various metals to improve the conductivity and thereby the overall performance of the cathode is one of the desirable methods. So we synthesized Pd-deposited α-MnO_2 nanorods and observed better battery efficiency and reduced over potential with Pd-decorated α–MnO_2 electrocatalyst as shown in Figure 12 [115].

Figure 11. The SEM micrographs of urchin shaped α-MnO_2 and birnessite shaped δ-MnO_2 nanostructures Reprinted with permission from K. S. Nahm, 2014, *RSC Adv.* 4 (2014) 8973 © 2014 The Royal Society of Chemistry Ltd.

From Figure12, in the first cycle α-MnO_2 nanorod catalyst exhibit a maximum capacity of 3997 mAh g−1 with an overpotential (ΔV) of 1.36 V. Whereas, the first discharge curve of Pd-deposited α-MnO_2 nanorods reached a maximum capacity of 8526 mAh g^{-1} with a flat discharge plateau at 2.7 V. On charging, a maximum capacity of 8526 mAh g−1 at 4.3 V is obtained with maximum 100% reversibility. In addition, overpotential, ΔV, is 1.0 V which is much lower than the α-MnO_2 catalyst, favorable for the reversibility of the battery. It is clearly seen that the addition of Pd on α-MnO_2 has doubled the specific capacity and reduced the overpotential of the Li–air battery. A better performance of the Pd/α-MnO_2 catalyst was observed in aqueous phase as well [115]. The experiment results suggest that the high specific capacity, reversibility, and low overpotential can be attributed to the α-MnO_2 catalysts especially due to the addition of Pd on α-MnO_2. α-MnO_2 with its 2×2 crystal structure favored the movement of ions and OH^- groups helped in surface adsorption of O_2 and dissociation of O–O bonds, while Pd nanoparticles deposited on the surface of the MnO_2 nanorods improved the electrocatalytic activity towards ORR.

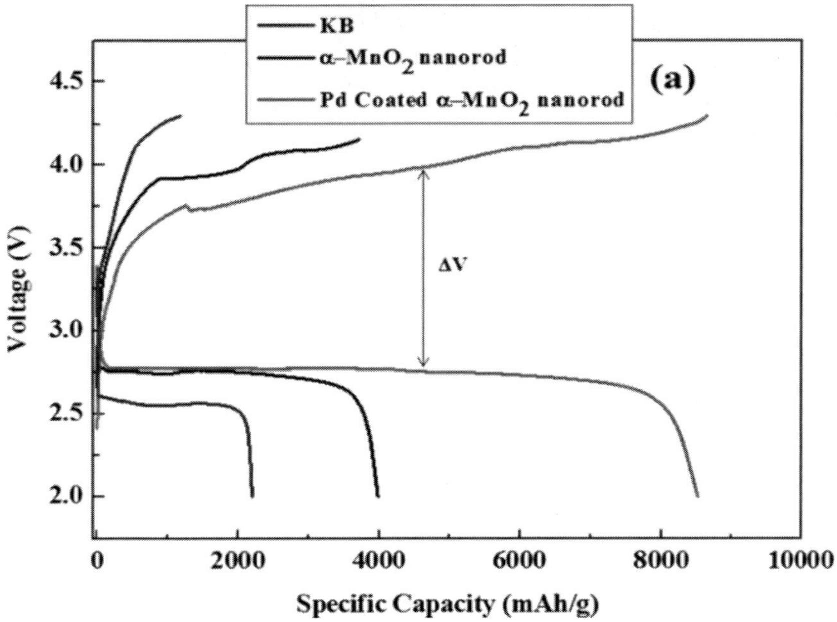

Figure 12. Charge–discharge profiles of Li–O_2 battery with Pd-deposited α-MnO_2 nanorod catalyst in comparison with α-MnO_2 nanorod catalyst and KB cathode without catalyst Reprinted with permission from K. S. Nahm, 2015, *J. Sol. St. Electrochem.* 19(5) (2015) 1501 © 2015 Springer International Publishing AG.

Another preferable way to improve the performance of MnO_2 materials is to combine or grow the α–MnO_2 catalysts on a highly conductive support. A few reports have been successfully shown the improved energy efficiency and cyclability of Li-O_2 battery using carbon supports like multi-walled CNT and reduced graphene oxide to grow MnO_2. We successfully developed α–MnO_2/NGNF nanomaterials with superior ORR activity as shown in Figure 13 [116]. The commercial graphite nanofibers exhibit rich fibrous structures with some organic and carbon residues. After N doping, however, the residues are completely removed and the N–GNF structure appeared smooth surfaced. Up on the development of α–MnO_2/NGNF, the α–MnO2 structures consisting of small nanorods are grown all over N–GNF nanostructures resulting in caterpillar-like structures (Fig. 13). The performance of caterpillar shaped a–MnO_2/N–GNF catalyst for ORR/OER in Li–O2 air cathode was evaluated and unlikely as the cycling number increased, the specific capacity dramatically increased, with a decrease in ΔV. The increase in discharge capacity upon cycling was because the ORR/OER activity of the catalysts that promoted the cell reaction near the active sites of the catalyst resulting in the increase in cycle performance. The α–MnO_2 in the air cathode significantly improved the reversible capacity of Li–O_2 batteries and exhibited low overpotential. The highly conductive nature of GNF connects the a–MnO_2 nanorods and N-doping in NGNF increased the surface properties of the electrode.

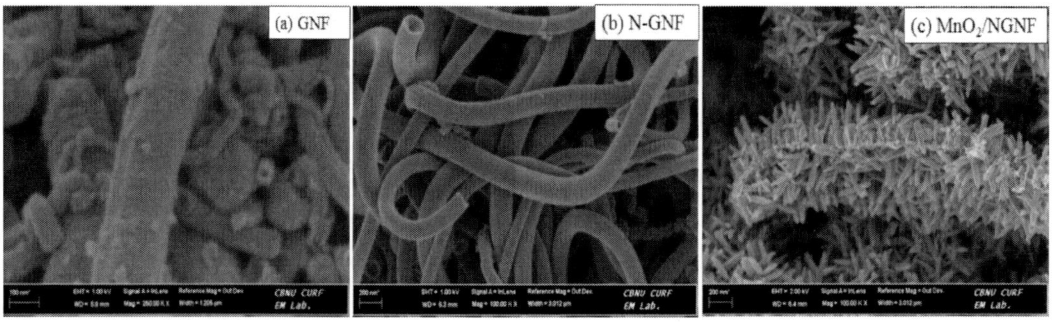

Figure 13. SEM micrographs of graphite nanofibers, nitrogen doped GNF and α–MnO_2/NGNF nanomaterials. Reprinted with permission from A. Zahoor, 2015, *Electrochimica Acta* 157 (2015) 299 © 2015 Elsevier B.V.

The collective effects of developed caterpillar shaped a–MnO_2/NGNF delivered better ORR and OER and proved to be a bifunctional electrocatalyst by exhibited very good rate capability with discharge capacity as high as 3000 mAh g^{-1} at even higher current densities 0.1, 0.2 and 0.3 mA cm^{-2}. Therefore, MnO_2 with highly conductive carbon support is proved to be a superior bifunctional electrocatalyst.

Later in a similar fashion, Byon et al. [117] reported the improved performance of lithium-air batteries by dispersing RuO_2 nanoparticles on carbon nanotube. Ru nanocrystals as OER catalyst for Li-O_2 batteries with lowest charge-discharge overpotential of about 0.37 V by Wang et al. [118] attracted many researchers. Similar results were also reported by Shao–Horn et al. [119] according to which RuO_2 exhibited excellent OER activities in both acidic and alkaline solutions. Hence, in an attempt to prepare an appropriate bifunctional catalyst, we combined MnO_2 and RuO_2 mixed oxides followed by Guo et al. [120] who prepared nanosized Mn–Ru binary oxides as effective bifunctional cathode electrocatalysts. Mn being the best to replace Pt in ORR and Ru, the superior OER catalyst, the combination of Mn and Ru would be ideal. Further, both optimization of composition and morphology control of the catalysts are necessary to enhance the electrocatalytic activity. So a unique 3D nanostructure of sea urchin shaped α–MnO_2 and RuO_2 (82:18) mixed oxides by simple low temperature hydrothermal technique was prepared and optimized for the application in air cathode of Li-O_2 battery [121]. The 3D nanostructure would provide large surface area and pores to facilitate the oxygen and electrolyte diffusions. The structure could also accommodate the reversible formation and decomposition of insoluble discharge products and enhance the ion transportation. The BET specific surface area of the synthesized α–MnO_2/RuO_2 mixed oxide was found to be 53.10 m^2 g^{-1} which is slightly higher than α–MnO_2 urchin (50 m^2 g^{-1}). The desorption–adsorption isotherm of the mixed oxide exhibited the characteristics of type IV, and the hysteresis loop indicated mesoporosity.

Mesopores could deliver a shorter ion transport pathway and provide more area for the triple–phase (solid–liquid–gas phases) region required for oxygen reduction. α–MnO_2/RuO_2 catalyst, facilitated the favorable formation and decomposition of Li_2O_2 when employed as air cathode catalyst in Li-O_2 battery. The 3D α–MnO_2/RuO_2 nanostructure is comparable to that of commercial Pt/C catalyst to catalyze ORR and better than Pt/C to catalyze OER.

Figure 14. FESEM images of α–MnO_2/RuO_2 hybrids at (b) (82:18), (c) (75:25), (d) (69:31), and (e) (50:50) ratios, and (a) α–MnO_2 and (e) RuO_2. Reprinted with permission from H. Jang, 2016, Electrochimica Acta 212 (2016) 701 © 2016 Elsevier B.V.

As a step further, the mixed oxide catalyst composition was optimized to achieve better electrocatalytic activity and stability, and to maximize the efficiency of Li–O_2 batteries. To achieve the best combination, three different combination viz. (75:25), (69:31) and (50:50) were synthesized and their physical and electrochemical properties were comparatively studied with respect to (82:18). The SEM image of sea urchin shaped α–MnO_2/RuO_2 mixed oxides and the mixed oxides prepared with different ratios are presented in Fig. 14 [121, 122]. The mixed oxide nanostructures have ability to catalyze ORR and OER with fast kinetics without electrolyte decomposition in both aqueous and non–aqueous electrolyte solutions. The bifunctional activity of the nanostructure catalyst is almost comparative to or much better than that of commercial Pt/C catalysts and α–MnO_2, owing to their intrinsic property of manganese oxide and the different O_2 adsorption kinetics and reaction pathway derived from the ruthenium oxide. Among the compositions investigated, α–MnO_2/RuO_2 (75:25) exhibited the best electrocatalytic activity which is better than that of commercial RuO_2 in OER. The cycling profiles of newly synthesized α–MnO_2/RuO_2 hybrids catalyzed Li–O_2 battery measured at 0.1 mA cm^{-2}, is given in Figure 15. The superior performance of the hybrid nanostructure is mainly attributed to its structural and compositional design which favors the electrochemical activity. Morphology–controllable synthesis proved necessary in the field of catalysis to improve the electrochemical properties by tailoring the morphology and surface structures in a nanoscale [121-125]. In particular, from the sequence observed from the above review of literatures, 3D nanostructures exhibit unique properties better suitable for catalysts. The high surface area–to–volume ratios in the 3D structure, large directions for charge transport, simultaneously constructing smooth pathways for extremely rapid charge/discharge processes and diffusion of electrolytes and oxygen, makes these structures viable for catalysts.

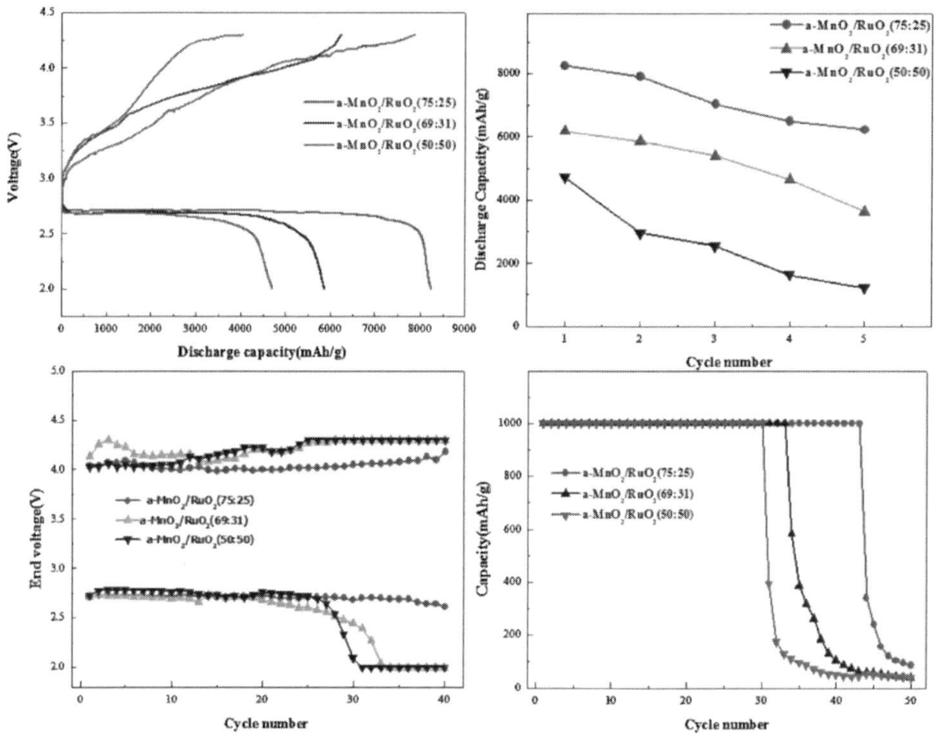

Figure 15. The cycling profiles of α–MnO_2/RuO_2 hybrids employed in Li–O_2 battery measured at 0.1 mA cm^{-2} [122]. Reprinted with permission from H. Jang, 2016, *Electrochimica Acta* 212 (2016) 701 © 2016 Elsevier B.V.

Other non-precious catalysts like phthalocyanine, have also been widely used as electrochemical catalysts for ORR in fuel cells and metal-air batteries [126-130]. Recently a series of non-precious metal catalysts like titanium nitride, and molybdenum based catalysts have been developed for Li-air battery system [131, 132]. A variety of conductive polymers, such as polyaniline (PANI), polypyrrole (PPy) and polythiophene (PTh), have also been found catalyze oxygen reduction reaction [133]. In order to achieve the rational design of efficient bifunctional catalyst to catalyze ORR and OER, it is necessary to gain critical insights into the detailed mechanisms.

CONCLUSION

Lithium-air batteries are perceived as the next generation battery owing to the demands of electrical energy of the world and to their superior theoretical capacity. However, there is still a long way towards practicality from where the present Li-O_2 technology stands.

While many non-precious and hybrid catalysts show promising progress, limitations still remain to develop a practical Li-air battery. Recently, vigorous researches and technological developments are carried out to reduce the limiting factors and improve the battery performances. Future research could emphasize on the following to achieve the goal;

1. To understand the electrochemical reaction mechanisms of Li-air batteries clearly
2. To develop appropriate interfaces since Li-O_2 battery involves three-phase reaction such as solid-liquid-gas.
3. To develop Li-ion conducting membrane to prevent the Li-metal anode reaction and dendrite formation
4. To improve the conductivity and connectivity of the catalysts and electrodes to improve the electron transfer.
5. To develop a rational design of catalyst to improve the ORR and OER performances since structure and morphology plays an important role.
6. To optimize the battery to operate at low current densities and extended periods of time, and also any passivation due to moisture contamination must be mitigated.

References

[1] D. Linden and T. Reddy, *Handbook of Batteries*, 3rd Edition, McGraw Hill Publishing, (2001).

[2] K. M. Abraham and Z. Jiang, A Polymer Electrolyte-Based Rechargeable Lithium/Oxygen Battery. *J. Electrochem. Soc.* 143 (1) (1996) 1.

[3] M. Armand and J. M. Tarascon, Building better batteries, *Nature* 451 (2008) 652.

[4] G. Girishkumar, B. D. McCloskey, A. C. Luntz, S. Swanson and W. Wilcke, Lithium−Air Battery: Promise and Challenges, *J. Phys. Chem. Lett.* 1 (2010) 2193.

[5] B. Scrosati and J. Garche, Lithium batteries: Status, prospects and future, *J. Power Sources* 195 (2010) 2419.

[6] P. G. Bruce, L. J. Hardwick and K. M. Abraham, Lithium-air and lithium-sulfur batteries, *MRS Bulletin* 36 (2011) 506.

[7] P. G. Bruce, S. A. Freunberger, L. J. Hardwick and J. M. Tarascon, Li–O_2 and Li–S batteries with high energy storage. *Nature Materials* 11 (2012) 19.

[8] E. L. Littauer and K. C. Tsai, Anodic Behavior of Lithium in Aqueous Electrolytes: I. Transient Passivation. *J. Electrochem. Soc.* 123 (1976) 771.

[9] J. Read, Characterization of the Lithium/Oxygen Organic Electrolyte Battery. *J. Electrochem. Soc.* 149 (9) (2002) A1190.

[10] J. Read, K. Mutolo, M. Ervin, W. Behl, J. Wolfenstine, A. Driedger and D. Fostera, Oxygen Transport Properties of Organic Electrolytes and Performance of Lithium/Oxygen Battery. *J. Electrochem. Soc.* 150 (10) (2003) A1351.

[11] J. Read, Ether-Based Electrolytes for the Lithium/Oxygen Organic Electrolyte Battery. *J. Electrochem. Soc.* 153 (1) (2006) A96.

[12] A. Dobley, C. Morein and K. Abraham, *208th ECS Meeting - Los Angeles, California*, (2005) Abs #823.

[13] T. Kuboki, T. Okuyama, T. Ohsaki and N. Takami, Lithium-air batteries using hydrophobic room temperature ionic liquid electrolyte. *J. Power Sources* 146 (2005) 766.

[14] A. Zahoor, M. Christy, Y. J. Hwang, and K. S. Nahm, Lithium air battery: alternate energy resource for the future. *J. Electrochem. Sci. and Techn.* 3 (1) (2012) 14.

[15] Y. Shao, S. Park, J. Xiao, J. G. Zhang, Y. Wang and J. Liu, Electrocatalysts for nonaqueous lithium–air batteries: status, challenges, and perspective. *ACS Catal.* 2 (5) (2012) 844.

[16] J. Wang, Y. Li, X. Sun, Challenges and opportunities of nanostructured materials for aprotic rechargeable lithium–air batteries, *Nano Energy* 2 (2013) 443.

[17] M. D. Bhatt, H. Geaney, M. Nolan and C. O. Dwyer, Key scientific challenges in current rechargeable non-aqueous Li–O_2batteries: experiment and theory, *Phys. Chem. Chem. Phys.* 16 (2014) 12093.

[18] C. O. Laoire, S. Mukerjee and K. M. Abraham, Influence of Nonaqueous Solvents on the Electrochemistry of Oxygen in the Rechargeable Lithium−Air Battery *J. Phys. Chem. C* 114 (19) (2010) 9178.

[19] Y. Chen, S. A. Freunberger, Z. Peng, F. Bardé and P. G. Bruce, Li–O_2 Battery with a Dimethylformamide Electrolyte *J. Am. Chem. Soc.* 134 (18) (2012) 7952.

[20] A. C. Luntz and B. D. McCloskey, Nonaqueous Li–Air Batteries: A Status Report, *Chem. Rev.* 114 (23) (2014) 11721.

[21] M. Balaish, A. Kraytsberg and Y. E. Eli, A critical review on lithium–air battery electrolytes, *Phys. Chem. Chem. Phys.* 16 (2014) 2801.

[22] A. Manthiram and L. Li, Hybrid and Aqueous Lithium-Air Batteries. *Advanced Energy Materials*, 5 (4) (2015) 1401302.

[23] B. Kumar, J. Kumar, R. Leese, J. P. Fellner, S. J. Rodrigues and K. M. Abraham, A Solid-State, Rechargeable, Long Cycle Life Lithium–Air Battery, *J. Electrochem. Soc.* 157 (2010) A50.

[24] F. Li, H. Kitaura and H. Zhou, The pursuit of rechargeable solid-state Li–air batteries. *Energy Environ. Sci.*, 6 (2013) 2302.

[25] S. J. Visco and Y. S. Nimon, *US Patent, 7, 645, 543* (2010).

[26] Y. Wang, and H. Zhou, A lithium-air battery with a potential to continuously reduce O_2 from air for delivering energy. *J. Power Sources* 195 (2010) 358.

[27] Q. C. Liu , J. J. Xu , S. Yuan , Z. W. Chang, D. Xu, Y. B. Yin, L. Li, H. X. Zhong, Y. S. Jiang , J. M. Yan and X. B. Zhang, Artificial Protection Film on Lithium Metal Anode toward Long-Cycle-Life Lithium–Oxygen Batteries. *Adv. Mater.* 27 (2015) 5241.

[28] B. Kumar, J. Kumar, R. Leese, J. P. Fellner, S. J. Rodrigues and K. M. Abraham, A Solid-State, Rechargeable, Long Cycle Life Lithium–Air Battery. *J. Electrochem. Soc.* 157 (1) (2010) A50.

[29] S. Rajendran, and T. Uma, Effect of ZrO_2 on conductivity of PVC–$LiBF_4$–DBP polymer electrolytes. *Mater. Lett.* 44 (2000) 208.

[30] S. Rajendran, P. Sivakumar, R.S. Babu, Investigation on polyvinylidene fluoride based gel polymer electrolytes. *Bull. Mater. Sci.* 29 (7) (2006) 673.

[31] Y. C. Lu, H. A. Gasteiger, M. C. Parent, V. Chiloyan, Y. S. Horn, The Influence of Catalysts on Discharge and Charge Voltages of Rechargeable Li–Oxygen Batteries. *Electrochem. Sol. St. Lett.* 13 (2010) A69.

[32] F. Mizuno, K. Takechi, S. Higashi, T. Shiga, T. Shiotsuki, N. Takazawa, Y. Sakurabayashi, S. Okazaki, I. Nitta, T. K., H. Nakamoto, H. Nishikoori, S. Nakanishi, Y. Kotani, H. Iba, Cathode reaction mechanism of non-aqueous Li–O2 batteries with highly oxygen radical stable electrolyte solvent. *J. Power Sources* 228 (2013) 47.

[33] Z. Q. Peng, S. A. Freunberger, L. J. Hardwick, Y. H. Chen, V. Giordani, F. Barde, P. Novak, D. Graham, J. M. Tarascon, P. G. Bruce, Oxygen reactions in a non-aqueous Li+ electrolyte. *Angew. Chem., Int. Ed.* 50 (28) (2011) 6351.

[34] Y. C. Lu, B. M. Gallant, D. G. Kwabi, J. R. Harding, R. R. Mitchell, M. S. Whittingham and Y. S. Horn, Lithium–oxygen batteries: bridging mechanistic understanding and battery performance. *Energy Environ. Sci.* 6 (2013) 750.

[35] K. N. Jung, J. Kim, Y. Yamauchi, M. S. Park, J. W. Lee and J. H. Kim, Rechargeable lithium–air batteries: a perspective on the development of oxygen electrodes. *J. Mater. Chem. A* 4 (2016) 14050.

[36] A. A. Arie and J. K. Lee, Electrochemical characteristics of lithium metal anodes with diamond like carbon film coating layer. *Diamond and Related Materials* 20 (3) (2011) 403.

[37] M. Ishikawa, S. Machino and M. Morita, Electrochemical control of a Li metal anode interface: improvement of Li cyclability by inorganic additives compatible with electrolytes. *J. Electroanal. Chem.* 473(1999) 279.

[38] S. J. Visco and Y. S. Nimon, *U.S. Patent, 7, 282, 295* (2007).

[39] J. Hassoun, H. G. Jung, D. J. Lee, J. B. Park, K. Amine, Y. K. Sun and B. Scrosati, A metal-free, lithium-ion oxygen battery: a step forward to safety in lithium-air batteries. *Nano Lett.* 12 (2012) 5775.

[40] P. Du, J. Lu, K. C. Lau, X. Luo, J. Bareño, X. Zhang, Y. Ren, Z. Zhang, L. A. Curtiss, Y. K. Sun and K. Amine, Compatibility of lithium salts with solvent of the non-aqueous electrolyte in Li–O_2 batteries. *Phys. Chem. Chem. Phys.* 15 (2013) 5572.

[41] G. A. Elia, J. B. Park, Y. K. Sun, B. Scrosati and J. Hassoun, Role of the Lithium Salt in the Performance of Lithium–Oxygen Batteries: A Comparative Study *Chem Electro Chem.* 1 (2014) 47.

[42] B. D. McCloskey, D. S. Bethune, R. M. Shelby, G. Girishkumar and A. C. Luntz, Solvents' Critical Role in Nonaqueous Lithium–Oxygen Battery. *Electrochemistry J. Phys. Chem. Lett.* 2 (10) (2011) 1161.

[43] B. D. McCloskey, R. Scheffler, A. Speidel, G. Girishkumar and A. C. Luntz, On the Mechanism of Nonaqueous Li–O_2 Electrochemistry on C and Its Kinetic Overpotentials: Some Implications for Li–Air Batteries. *J. Phys. Chem. C* 116 (45) (2012) 23897.

[44] B.D. McCloskey, D. S. Bethune, R. M. Shelby, T. Mori, R. Scheffler, A. Speidel, M. Sherwood, and A. C. Luntz, Hydrogen-Bond Dynamics in a Protic Ionic Liquid: Evidence of Large-Angle Jumps. *J. Phys. Chem. Lett.* 3 (2012) 3034.

[45] D. Xu, Z. L. Wang, J. J. Xu, L. L. Zhang and X. B. Zhang, Novel DMSO-based electrolyte for high performance rechargeable Li–O_2 batteries. *Chem. Commun.* 48 (2012) 6948.

[46] C. Liang, F. Wang, Y. Xu, J. Chen, D. Liu and Z. Luo, A stable electrolyte makes a nonaqueous Li–O_2 battery truly rechargeable. *New J. Chem.* 37 (2013) 2568.

[47] D. G. Kwabi, T. P. Batcho, C. V. Amanchukwu, N. O. Vitoriano, P. Hammond, C. V. Thompson and Y. S. Horn, Chemical Instability of Dimethyl Sulfoxide in Lithium–Air Batteries. *J. Phys. Chem. Lett.* 5 (2014) 2850

[48] H. G. Jung, J. Hassoun, J. B. Park, Y. K. Sun and B. Scrosati, An improved high-performance lithium–air battery. *Nat. Chem.* 4(7) (2012) 579.

[49] R. R. Mitchell, B. M. Gallant, C. V. Thompson and Y. S. Horn, All-carbon-nanofiber electrodes for high-energy rechargeable Li–O_2batteries. *Energy Environ. Sci.* 4 (2011) 2952.

[50] Y. Li, J. Wang, X. Li, J. Liu, D. Geng, J. Yang, R. Li and X. Sun, Nitrogen-doped carbon nanotubes as cathode for lithium–air batteries. *Electrochem. Commun.* 13 (2011) 668.

[51] J. Li, B. Peng, G. Zhou, Z. Zhang, Y. Lai and M. Jia, Partially cracked carbon nanotubes as cathode materials for lithium-air batteries. *ECS Electrochem. Lett.* 2 (2013) A25.

[52] J. Shui, F. Du, C. Xue, Q. Li and L. Dai, Vertically aligned N-doped coral-like carbon fiber arrays as efficient air electrodes for high-performance nonaqueous Li-O2 batteries. *ACS Nano,* 8 (2014) 3015.

[53] Y. Zhang, L. Wang, Z. Guo, Y. Xu, Y. Wang and H. Peng, High-Performance Lithium-Air Battery with a Coaxial-Fiber Architecture. *Angew. Chem. Int. Ed.* 55 (2016) 4487.

[54] Y. Wang and H. Zhou, To draw an air electrode of a Li–air battery by pencil. *Energy Environ. Sci.* 4 (2011) 1704.

[55] B. Sun, B. Wang, D. Su, L. Xiao, H. Ahn and G. Wang, Graphene nanosheets as cathode catalysts for lithium-air batteries with an enhanced electrochemical performance. *Carbon* 50 (2012) 727.

[56] Y. Li, J. Wang, X. Li, D. Geng, M. N. Banis, R. Li and X. Sun, Nitrogen-doped graphene nanosheets as cathode materials with excellent electrocatalytic activity for high capacity lithium-oxygen batteries. *Electrochem. Commun.* 18 (2012)12.

[57] H. Kim, H. D. Lim, J. Kim and K. Kang, Graphene for advanced Li/S and Li/air batteries. *J. Mater. Chem. A* 2 (2014) 33.

[58] J. Xiao, D. Mei, X. Li, W. Xu, D. Wang, G. L. Graff, W. D. Bennett, Z. Nie, L. V. Saraf, I. A. Aksay, J. Liu and J. G. Zhang, Hierarchically porous graphene as a lithium-air battery electrode. *Nano Lett.* 11 (2011) 5071.

[59] K. Gong, F. Du, Z. Xia, M. Durstock and L. Dai, Nitrogen-doped carbon nanotube arrays with high electrocatalytic activity for oxygen reduction. *Science* 323 (2009) 760.

[60] Y. F. Tang, B. L. Allen, D. R. Kauffman and A. Star, Electrocatalytic activity of nitrogen-doped carbon nanotube cups. *J. Am. Chem. Soc.* 131 (2009) 13200.

[61] L. Qu, Y. Liu, J.B. Baek and L. Dai, Nitrogen-Doped Graphene as Efficient Metal-Free Electrocatalyst for Oxygen Reduction in Fuel Cells. *ACS Nano* 4 (2010) 1321.

[62] S. Chen, J. Y. Bi, Y. Zhao, L. J. Yang, C. Zhang, Y. W. Ma, Q. Wu, X. Z. Wang, Z. Hu, Nitrogen-Doped Carbon Nanocages as Efficient Metal-Free Electrocatalysts for Oxygen Reduction Reaction. *Adv. Mater.* 24 (2012) 5593.

[63] E. Yoo, J. Nakamura, H. Zhou, N-Doped graphene nanosheets for Li–air fuel cells under acidic conditions. *Energy Environ. Sci.* 5 (2012) 6928.

[64] D. Yu, Q. Zhang, L. Dai, Highly Efficient Metal-Free Growth of Nitrogen-Doped Single-Walled Carbon Nanotubes on Plasma-Etched Substrates for Oxygen Reduction. *J. Am. Chem. Soc.* 132 (2010) 15127.

[65] Y. Shao, J. Sui, G. Yin, Y. Gao, Nitrogen-doped carbon nanostructures and their composites as catalytic materials for proton exchange membrane fuel cell. *Appl. Catal. B: Environ.* 79 (2008) 89.

[66] H. Wang, X. Bo, Y. Zhang, L. Guo, Sulfur-doped ordered mesoporous carbon with high electrocatalytic activity for oxygen reduction. *Electrochimica Acta* 108 (2013) 404.

[67] L. Yang, S. Jiang, Y. Zhao, L. Zhu, S. Chen, X. Wang, Q. Wu, J. Ma, Y. Ma and Z. Hu, Boron-doped carbon nanotubes as metal-free electrocatalysts for the oxygen reduction reaction. *Angew. Chem. Int. Ed.* 50 (2011) 7132.

[68] S. A. Wohlgemuth, R. J. White, M. G. Willinger, M. M. Titirici and M. Antonietti, A one-pot hydrothermal synthesis of sulfur and nitrogen doped carbon aerogels with enhanced electrocatalytic activity in the oxygen reduction reaction. *Green Chem.* 14 (2012) 1515.

[69] T. Wang, L. X. Wang, D. L. Wu, W. Xia, D. Z. Jia, Interaction between nitrogen and sulfur in co-doped graphene and synergetic effect in supercapacitor. *Scientific Reports*, 5 (2015) 9591.

[70] I. T. Kim, M. J. Song, Y. B. Kim, M. W. Shin, Microwave-hydrothermal synthesis of boron/nitrogen co-doped graphene as an efficient metal-free electrocatalyst for oxygen reduction reaction. *International Journal of Hydrogen Energy*, 41 (47) 2016 22026.

[71] H. D. Lim, K. Y. Park, H. Song, E. Y. Jang, H. Gwon, J. Kim, Y. H. Kim, M. D. Lima, R. O. Robles, X. Lep, R. H. Baughman, K. Kang, Enhanced power and rechargeability of a Li-O2 battery based on a hierarchical-fibril CNT electrode. *Adv. Mater.* 25 (2013) 1348.

[72] H. D. Lim, H. Song, H. Gwon, K. Y. Park,a J. Kim, Y. Bae, H. Kim, S. K. Jung, T. Kim, Y. H. Kim, X. Lepro, R. O. Robles, R. H. Baughman and K. Kang, A new catalyst-embedded hierarchical air electrode for high-performance Li–O_2 batteries. *Energy Environ. Sci.* 6 (2013) 3570.

[73] H. D. Lim, H. Song, J. Kim, H. Gwon, Y. Bae, K. Y. Park, J. Hong, H. Kim, Taewoo Kim, Y. H. Kim, X. Lepró, R. O. Robles, R. H. Baughman, K. Kang, Superior rechargeability and efficiency of lithium-oxygen batteries: hierarchical air electrode architecture combined with a soluble catalyst. *Angew. Chem. Int. Ed.* 53 (2014) 3926.

[74] J. Shui, F. Du, C. Xue, Q. Li and L. Dai, Vertically aligned N-doped coral-like carbon fiber arrays as efficient air electrodes for high-performance nonaqueous Li-O2 batteries. *ACS Nano 8* (3) (2014) 3015.

[75] Z. Peng, S. A. Freunberger, Y. Chen and P. G. Bruce, A reversible and higher-rate Li-O2 battery. *Science* 337 (2012) 563.

[76] L. Y. Chen, X. W. Guo, J. H. Han, P. Liu, X. D. Xu, A. Hirata and M. W. Chen, Nanoporous metal/oxide hybrid materials for rechargeable lithium–oxygen batteries. *J. Mater. Chem. A* 3 (2015) 3620.

[77] W. B. Luo, X. W. Gao, S. L. Chou, J. Z. Wang and H. K. Liu, Porous AgPd-Pd Composite Nanotubes as Highly Efficient Electrocatalysts for Lithium-Oxygen Batteries. *Adv. Mater.* 27 (2015) 6862.

[78] Y. C. Lu, Z. Xu, H. A. Gasteiger, S. Chen, K. HamadSchifferli and Y. S. Horn, Platinum-gold nanoparticles: a highly active bifunctional electrocatalyst for rechargeable lithium-air batteries. *J. Am. Chem. Soc.* 132 (2010) 12170.

[79] S. Zhang, X. Zhang, G. Jiang, H. Zhu, S. Guo, D. Su, G. Lu and S. Sun, Tuning nanoparticle structure and surface strain for catalysis optimization. *J. Am. Chem. Soc.* 136 (2014) 7734.

[80] M. M. Ottakam Thotiyl, S. A. Freunberger, Z. Peng, Y. Chen, Z. Liu and P. G. Bruce, A stable cathode for the aprotic Li-O2 battery. *Nat. Mater.* 12 (2013) 1050

[81] D. Kundu, R. Black, B. Adams, K. Harrison, K. Zavadil and L. F. Nazar, Nanostructured Metal Carbides for Aprotic Li-O2 Batteries: New Insights into Interfacial Reactions and Cathode Stability. *J. Phys. Chem. Lett.* 6 (2015) 2252.

[82] F. Li, D. M. Tang, T. Zhang, K. Liao, P. He, D. Golberg, A. Yamada and H. Zhou, Superior Performance of a Li–O_2 Battery with Metallic RuO_2Hollow Spheres as the Carbon-Free Cathode. *Adv. Energy Mater.* 5 (2015) 1500294.

[83] J. B. Park, I. Belharouak, Y. J. Lee and Y. K. Sun, *Abstract MA2016-03 408*, 2016.

[84] K. Liao, T. Zhang, Y. Wang, F. Li, Z. Jian, H. Yu and H. Zhou, Nanoporous Ru as a carbon- and binder-free cathode for Li-O2 batteries. *Chem Sus Chem* 8 (2015) 1429.

[85] S. T. Kim, N. S. Choi, S. Park and J. Cho, Optimization of Carbon- and Binder-Free Au Nanoparticle-Coated Ni Nanowire Electrodes for Lithium-Oxygen Batteries. *Adv. Energy Mater.* 5 (2015) 1401030.

[86] Y. Cui, Z. Wen and Y. Liu, A free-standing-type design for cathodes of rechargeable Li–O_2 batteries. *Energy Environ. Sci.* 4 (2011) 4727.

[87] A. Riaz, K. N. Jung, W. Chang, K. H. Shin and J. W. Lee, Carbon-, binder-, and precious metal-free cathodes for non-aqueous lithium-oxygen batteries: nanoflake-decorated nanoneedle oxide arrays. *ACS Appl. Mater. Interfaces*, 6 (2014) 17815.

[88] Y. C. Lu, H. A. Gasteiger and Y. S. Horn, Catalytic Activity Trends of Oxygen Reduction Reaction for Nonaqueous Li-Air Batteries. *J. Am. Chem. Soc.* 133 (2011) 19048.

[89] Y. C. Lu, Z. Xu, H. A. Gasteiger, S. Chen, K. H. Schifferli and Y. S. Horn, Platinum−Gold Nanoparticles: A Highly Active Bifunctional Electrocatalyst for Rechargeable Lithium−Air Batteries. *J. Am. Chem. Soc.* 132 (2010) 12170.

[90] A. Zahoor, M. Christy, Y. Kim, A. Arul, Y. S. Lee and K. S. Nahm, Carbon/ titanium oxide supported bimetallic platinum/iridium nanocomposites as bifunctional electrocatalysts for lithium-air batteries. *J. Sol. St. Electrochem.* 20 (5) (2016) 1397.

[91] M. K. Song, S. Park, F. M. Alamgir, J. Cho and M. Liu, Nanostructured electrodes for lithium-ion and lithium-air batteries: the latest developments, challenges, and perspectives. *Materials Science and Engineering: R: Reports* 72(11) (2011) 203.

[92] R. Cao, J. S. Lee, M. Liu and J. Cho, Recent Progress in Non-Precious Catalysts for Metal-Air Batteries. *Adv. Energy Mater.* 2 (2012) 816.

[93] H. Cheng and K. Scott, Carbon-supported manganese oxide nanocatalysts for rechargeable lithium–air batteries. *J. Power Sources* 195 (2010) 1370.

[94] D. Zhang, Z. Fu, Z. Wei, T. Huang and A. Yu, Polarization of Oxygen Electrode in Rechargeable Lithium Oxygen Batteries. *J. Electrochem. Soc.* 157 (2010) A362.

[95] R. S. Kalubarme, M. S. Cho, K. S. Yun, T. S. Kim and C. J. Park, Catalytic characteristics of MnO2 nanostructures for the O2 reduction process *Nanotechnology* 22 (2011) 395402.
[96] A. K. Thapa and T. Ishihara, Mesoporous α-MnO_2/Pd catalyst air electrode for rechargeable lithium–air battery. *J. Power Sources* 196 (2011) 7016.
[97] S. Ida, A. K. Thapa, Y. Hidaka, Y. Okamoto, M. Matsuka, H. Hagiwara and T. Ishihara, Manganese oxide with a card-house-like structure reassembled from nanosheets for rechargeable Li-air battery. *J. Power Sources* 203 (2012) 159.
[98] G. Q. Zhang, J. P. Zheng, R. Liang, C. Zhang, B. Wang, M. Au, M. Hendrickson and E. J. Plichta, α-MnO_2/Carbon Nanotube/Carbon Nanofiber Composite Catalytic Air Electrodes for Rechargeable Lithium-air Batteries. *J. Electrochem. Soc.* 158 (2011) A822.
[99] E. M. Benbow, S. P. Kelly, L. Zhao, J. W. Reutenauer, S. L. Suib, Oxygen Reduction Properties of Bifunctional α-Manganese Oxide Electrocatalysts in Aqueous and Organic Electrolytes. *J. Phys. Chem. C* 115 (2011) 22009.
[100] Y. L. Cao, H. X. Yang, X. P. Ai and L. F. Xiao, The mechanism of oxygen reduction on MnO_2-catalyzed air cathode in alkaline solution. *J. Electroanal. Chem.* 557 (2003) 127.
[101] A. Debart, A. J. Paterson, J. Bao and P. G. Bruce, Alpha-MnO2 nanowires: a catalyst for the O2 electrode in rechargeable lithium batteries. *Angew. Chem. Int. Ed. Engl.* 47(24) (2008) 4521.
[102] F. Cheng, J. Zhao, W. Song, C. Li, H. Ma, J. Chen and P. Shen, Facile controlled synthesis of MnO2 nanostructures of novel shapes and their application in batteries. *Inorg. Chem.* 45 (2006) 2038.
[103] F. Cheng, Y. Su, J. Liang, Z. Tao and J. Chen, MnO_2-Based Nanostructures as Catalysts for Electrochemical Oxygen Reduction in Alkaline Media. *Chem. Mater.* 22 (2010) 898.
[104] A. Zahoor, H. S. Jang, J. S. Jeong, M. Christy, Y. J. Hwang and K. S. Nahm, A comparative study of nanostructured α and δ MnO_2 for lithium oxygen battery application. *RSC Adv.* 4 (2014) 8973.
[105] A. Děbart, J. Bao, G. Armstrong and P. G. Bruce, An O_2 cathode for rechargeable lithium batteries: The effect of a catalyst. *J. Power Sources* 174 (2007) 1177.
[106] Y. Cui, Z. Wen and Y. Liu, A free-standing-type design for cathodes of rechargeable Li–O_2 batteries. *Energy Environ. Sci.* 4 (2011) 4727.
[107] M. Christy, H. Jang and K. S. Nahm, Cobaltite oxide nanosheets anchored graphene nanocomposite as an efficient oxygen reduction reaction (ORR) catalyst for the application of lithium-air batteries. *Journal of Power Sources* 288 (2015) 451.
[108] T. T. Truong, Y. Liu, Y. Ren, L. Trahey and Y. Sun, Morphological and crystalline evolution of nanostructured MnO2 and its application in lithium--air batteries. *ACS Nano* 6 (2012) 8067.

[109] G. Zhao, R. Mo, B. Wang, L. Zhang and K. Sun, Enhanced Cyclability of Li–O_2 Batteries Based on TiO_2 Supported Cathodes with No Carbon or Binder. *Chem. Mater.* 26 (2014) 2551.

[110] Y. Yang, Q. Sun, Y. S. Li, H. Li and Z. W. Fu, A CoO_x/carbon double-layer thin film air electrode for nonaqueous Li-air batteries. *J. Power Sources* 223 (2013) 312.

[111] L. Zhang, X. Zhang, Z. Wang, J. Xu, D. Xu and L. Wang, High aspect ratio γ-MnOOH nanowires for high performance rechargeable nonaqueous lithium–oxygen batteries. *Chem. Commun.* 48 (2012) 7598.

[112] J. Gomez, E. E. Kalu, R. Nelson, M. H. Weatherspoon and J. P. Zheng, Binder-free Co–Mn composite oxide for Li–air battery electrode. *J. Mater. Chem. A* 1 (2013) 3287.

[113] S. Ma, L. Sun, L. Cong, X. Gao, C. Yao, X. Guo, L. Tai, P. Mei, Y. Zeng, H. Xie, R. Wang, Multiporous $MnCo_2O_4$ Microspheres as an Efficient Bifunctional Catalyst for Nonaqueous Li–O_2 Batteries. *J. Phys. Chem. C* 117 (2013) 25890.

[114] K. N. Jung, A. Riaz, S. B. Lee, T. H. Lim, S. J. Park, R. H. Song, S. Yoon, K. H. Shin and J. W. Lee, Urchin-like α-MnO2 decorated with Au and Pd as a bi-functional catalyst for rechargeable lithium–oxygen batteries. *J. Power Sources* 244 (2013) 328.

[115] A. Zahoor, M. Christy, J. S. Jeon, Y. S. Lee and K. S. Nahm, Improved lithium oxygen battery performance by addition of palladium nanoparticles on manganese oxide nanorod catalysts. *J. Sol. St. Electrochem.* 19 (5) (2015) 1501.

[116] A. Zahoor, M. Christy, H. S. Jang, K. S. Nahm and Y. S. Lee, Increasing the reversibility of Li–O2 batteries with caterpillar structured α–MnO 2/N–GNF bifunctional electrocatalysts. *Electrochimica Acta* 157 (2015) 299.

[117] E. Yilmaz, C. Yogi, K. Yamanaka, T. Ohta and H. R. Byon, Promoting formation of noncrystalline Li2O2 in the Li-O2 battery with RuO2 nanoparticles. *Nano Lett.* 13 (2013) 4679.

[118] B. Sun, P. Munroe and G. Wang, Ruthenium nanocrystals as cathode catalysts for lithium-oxygen batteries with a superior performance. *Scientific Reports* 3 (2013) 2247.

[119] Y. Lee, J. Suntivich, K. J. May, E. E. Perry and Y. S. Horn, Synthesis and Activities of Rutile IrO2 and RuO2 Nanoparticles for Oxygen Evolution in Acid and Alkaline Solutions. *J. Phys. Chem. Lett.* 3 (2012) 399.

[120] K. Guo, Y. Li, J. Yang, Z. Zou, X. Xue, X. Lia and H. Yang, Nanosized Mn–Ru binary oxides as effective bifunctional cathode electrocatalysts for rechargeable Li–O_2 batteries. *J. Mater. Chem. A* 2 (2014) 1509.

[121] H. Jang, A. Zahoor, J. S. Jeon, P. Kim, Y. S. Lee and K. S. Nahm, Sea urchin shaped α-MnO2/RuO2 mixed oxides nanostructure as promising electrocatalyst for lithium–oxygen battery. *J. Electrochem. Soc.* 162 (3) (2015) A300.

[122] H. Jang, A. Zahoor, Y. Kim, M. Christy, M. Y. Oh, V. Aravindan, Y. S. Lee and K. S. Nahm, Tailoring three dimensional α–MnO 2/RuO 2 hybrid nanostructure as prospective bifunctional catalyst for Li–O 2 batteries. *Electrochimica Acta* 212 (2016) 701.

[123] F. Cheng, J. Zhao, W. Song, C. Li, H. Ma, J. Chen and P. Shen, Facile controlled synthesis of MnO2 nanostructures of novel shapes and their application in batteries. *Inorganic Chemistry* 45 (2006) 2038.

[124] Y. Xia, Y. Xiong, B. Lim and S. E. Skrabalak, Shape-Controlled Synthesis of Metal Nanocrystals: Simple Chemistry Meets Complex Physics? *Angew. Chem. Int. Ed.* 48 (2009) 60

[125] K. Song, D. A. Agyeman, J. Jung, M.R. Jo, J. Yang and Y. M. Kang, A Review of the Design Strategies for Tailored Cathode Catalyst Materials in Rechargeable Li-O_2 Batteries. *Israel Journal of Chemistry* 55 (2015) 458.

[126] J. Zagal and M. Pgez, Electrocatalytic activity of metal phthalocyanines for oxygen reduction. *J. Electroanal. Chem.* 339 (1992) 13.

[127] G. F. Dong, M. H. Huang and L. H. Guan, Iron phthalocyanine coated on single-walled carbon nanotubescomposite for the oxygen reduction reaction in alkaline media. *Phys. Chem. Chem. Phys.* 14 (2012) 2557.

[128] R. Cao, R. Thapa, H. Kim, X. Xu, M.G. Kim, Q. Li, N. Park, M. Liu and J. Cho, Promotion of oxygen reduction by a bio-inspired tethered iron phthalocyanine carbon nanotube-based catalyst. *Nat. Commun.* 4 (2013) 2076 (doi: 10.1038/ncomms3076).

[129] A. Abbaspour and E. Mirahmadi, Electrocatalytic activity of iron and nickel phthalocyanines supported on multi-walled carbon nanotubes towards oxygen evolution reaction. *Electrochimica Acta* 105 (2013) 92.

[130] A. Arul, M. Christy, M. Y. Oh, Y. S. Lee and K. S. Nahm, Nanofiber Carbon-Supported Phthalocyanine Metal Complexes as Solid Electrocatalysts for Lithium-Air Batteries. *Electrochimica Acta* 218 (2016) 335.

[131] P. He, Y. Wang and H. Zhou, Titanium nitride catalyst cathode in a Li–air fuel cell with an acidic aqueous solution. *Chem. Commun.* 47 (2011) 10701.

[132] S. Dong, X. Chen, K. Zhang, L. Gu, L. Zhang, X. Zhou, L. Li, Z. Liu, P. Han, H. Xu, J. Yao, C. Zhang, X. Zhang, C. Shang, G. Cui and L. Chen, Molybdenum nitride based hybrid cathode for rechargeable lithium–O_2 batteries. *Chem. Commun.* 47 (2011) 11291.

[133] V. G. Khomenko, V. Z. Barsukov and A. S. Katashinskii, The catalytic activity of conducting polymers toward oxygen reduction. *Electrochimica Acta* 50 (2005) 1675.

In: Handbook of Remediation ...
Editors: Sadia Ameen, M. Shaheer Akhtar et al.
ISBN: 978-1-53615-118-3

Chapter 10

NANOMATERIALS FOR THE THERMAL DECOMPOSITION OF AMMONIUM PERCHLORATE AND COMPOSITE SOLID PROPELLANTS

Jitendra Kumar Sharma, Pratibha Srivastava* and Gurdip Singh
Department of Chemistry, D.D.U. Gorakhpur University, Gorakhpur, U.P., India

ABSTRACT

Materials with diameter ranging from 1 to 100 nm have become widely utilized owing to their unique physicochemical properties. The synthesis protocols for effective nanomaterials in terms of sizes, shapes, morphologies and chemical compositions are the important aspect of nanotechnology. The remarkable size dependent physicochemical properties of nanoparticles (NPs) have fascinated and inspired research activity in this direction. This chapter describes several aspects on NPs synthesis, characterization, properties, and applications of metal oxides, ferrites, cobaltites, metals and metal alloys. The effect of these NPs on the thermal decomposition and combustion of ammonium perchlorate (AP) and composite solid propellants (CSPs) are also summarized. Influence of size effect of catalyst on thermal decomposition of AP and comparison of bulk and nanosized oxides is also discussed here. Due to small size and large surface area, NPs are highly efficient catalysts for the thermal decomposition of AP and CSPs in comparison with their bulk materials.

Keywords: nanomaterials, particles, catalytic decomposition, propellents, ammonium perchlorate

* Corresponding Author's Email: pratibhas1980@gmail.com.

1. INTRODUCTION

Nanotechnology is receiving a lot of attention of late across the globe. Nanotechnology is an emerging field that covers a wide range of disciplines, including the frontiers of chemistry, materials, medicine, electronics, optics, sensors, information storage, communication, energy conversion, environmental protection, aerospace and more [1, 3].

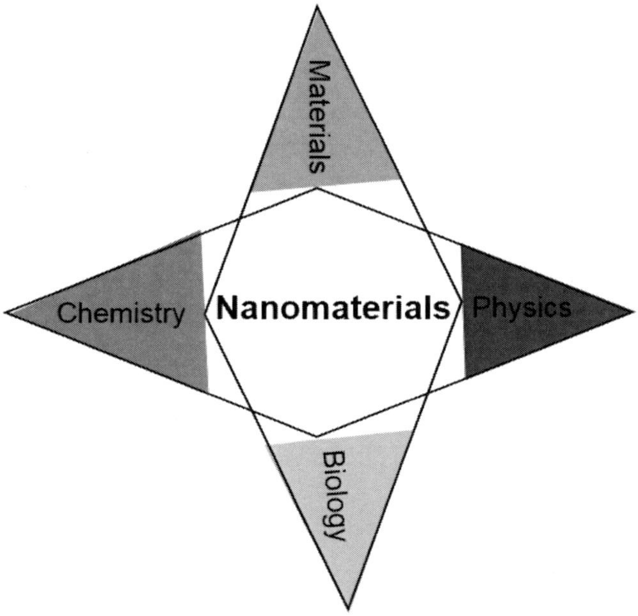

Figure 1. Relation of nanomaterials with other sciences.

Figure 1 shows that nanomaterials lie at the intersection of materials science, physics, chemistry, and for many of the most interesting applications also of biology and medicine. It focuses on the design, synthesis, characterization and application of materials and devices at the nanoscale [2, 4]. Nanomaterials are the foundation of nanotechnology and are anticipated to open new avenues to numerous emerging technological applications [1, 2].

The word 'nano' is derived from the Greek word 'nanos' or the Latin word 'nanus' which mean dwarf. Nanoparticles (NPs) are sized between 1 nm and 100 nm. The reduction in size of materials not only leads to its miniaturization but at a certain size scale, usually in nanometer-size, a drastic and dramatic change in the bulk properties of the material takes place. For example, an inert and noble bulk gold metal shows a size-dependent catalytic activity for CO oxidation at size < 5 nm. A drastic change in physical, chemical, optical, electrical, catalytic, magnetic properties of materials is observed at a size-scale of less than 100 nm [1, 4]. In general, the size at which nanomaterial displays properties different from that of the bulk is material dependent as well as property dependent i.e., different properties of a nanomaterial show transition from the bulk at different size scales. This size can be larger than 100 nm in case of some materials. For example, though one-dimensional (1D) nanomaterials like nanowires, nanorods and nanotubules have diameters in the range of a few nanometers but have lengths in the range of tens of microns.

These 1D nanomaterials have many new properties and functionalities that are different from the bulk materials, e.g., carbon nanotubes have properties, different from other allotropes of carbon, with a large number of promising applications in sensors, catalysis etc. Therefore, a NP is defined as a particle with at least one dimension in the size range of 1 nm to 100 nm. Nanostructures constitute a bridge between molecules and infinite bulk systems. Individual nanostructures include clusters, quantum dots, nanocrystals, nanowires, and nanotubes, while collections of nanostructures involve arrays, assemblies, and superlattices of the individual nanostructures [5, 6].

These unique properties depend on the atomic structure, size confinement, composition, microstructure, defects, and interfaces, all of which can be tailored by synthesis and other processes [1, 3, 7, 8].

The research on nanomaterials is highly interdisciplinary because it involves many different synthetic methodologies and characterization techniques. Many researchers are moving toward such molecular based synthesis and applying engineered nanomaterials to novel uses [9].

2. Nanomaterials

The use of materials that has the dimension of nanometers is not new; man has been utilizing nanosized materials for centuries. An ancient and beautiful example is the Roman Lycurgus Cup, which dates to the fourth century AD. When viewed in reflected light, it appears green but when light is transmitted through the glass, it appears red. This effect comes from minute particles (ca. 70 nm) of a silver-gold alloy dispersed in the glass.

The Chinese are known to have used Au NPs as an inorganic dye to introduce red colour into their ceramic porcelains for thousands of years. There is also a long history of using colloidal gold nanoparticles in medical applications. Colloidal gold NPs can be used to treat arthritis and to diagnose a number of diseases by observing the interaction of colloidal gold with spinal fluids obtained from patients. The first scientific report on the preparation and properties of colloidal gold was published by Michael Faraday in the middle 19th century. The colloidal gold NPs prepared by this method stayed stable for almost a century before being destroyed during World War II.

By the early 1940's, silica NPs, which are used as substitutes for carbon black, were synthesized in USA and Germany by precipitation method. Amorphous silica NPs have been applied in many every-day products, ranging from non-diary coffee creamer to sensors, drug carriers and optical fibers. In the 1960s and 1970s metallic NPs for magnetic recording tapes were developed. Feynman delivered a speech in 1959 entitled "There's Plenty of Room at the Bottom," which provided inspiration for the field of nanomaterials. Feynman commented that there were no fundamental physical reasons why materials could not be obtained by manipulating individual atoms, and he suggested that's calling issues would arise from the variation of the magnitude of various physical phenomena. However, Feynman's report was neglected in the first twenty years after it was published, due to the absence of instrument innovations. In 1981, the scanning tunneling microscope (STM), an instrument that allowed scientists to view the world from an atomic level, was developed by Gerd Binnig and Heinrich Rohrer at IBM Zurich Research Laboratory, for which they were awarded the Nobel Prize in Physics in 1986.

The evolvements of STM together with the invention of related instruments have enabled researchers to investigate a wider range of atomic-scale structures and properties with a greater depth. With the help of these novel instruments, even the manipulation of individual atoms and molecules is made possible. For example, IBM scientists had successfully manipulated individual xenon atoms on a nickel surface to spell out the company logo, which can be seen as a show of this great novel technology.

However, STM can only be used to observe the conductors and semiconductors. To extend the application, in 1986, Binnig, Calvin Quate and Christoph Gerber invented the first atomic force microscope (AFM), which provided a three-dimensional surface profile of the conductors, semiconductors and insulators.

The innovations of the experimental instruments have greatly accelerated the development of nanomaterials. In the mid-1980s, a new class of material, hollow carbon spheres was discovered, which were named buckyballs or fullerenes, in honour of the architect, Buckminster Fuller, who designed a geodesic dome with the similar geometry to that of the molecular structure of fullerenes [10]. Fullerenes, which is composed of 60 carbon atoms chemically bonded together in a ball-shaped molecule, has great potential applications in nanotechnology. Therefore, Kroto, Curl and Smalley, the people who first discovered fullerenes, were awarded the 1996 Nobel Prize in Chemistry.

The discovery of fullerenes has inspired research that led to the synthesis of carbon nanofibers, with diameters under 100 nm. In 1991, S. Iijima of NEC in Japan discovered single walled carbon nanotubes, by adding transition-metal catalysts into carbon in an arc discharge. Due to the excellent physical properties, carbon nanotubes and their nanocomposite have attracted more and more research attention in recent years.

3. BEHAVIOR OF NANOMATERIALS

Behavior serves as a model for nanomaterials and several inorganic, organic and biological materials, have existed in nature since the evolution of life started on the earth. Some examples are micro-organisms, fine-grained minerals in rocks and nanosized particles in bacteria and smoke (Figure 2). Also, in all natural processes, everything plays out in nanodimensions. Scientists are now attempting to imitate mother nature and create such materials synthetically.

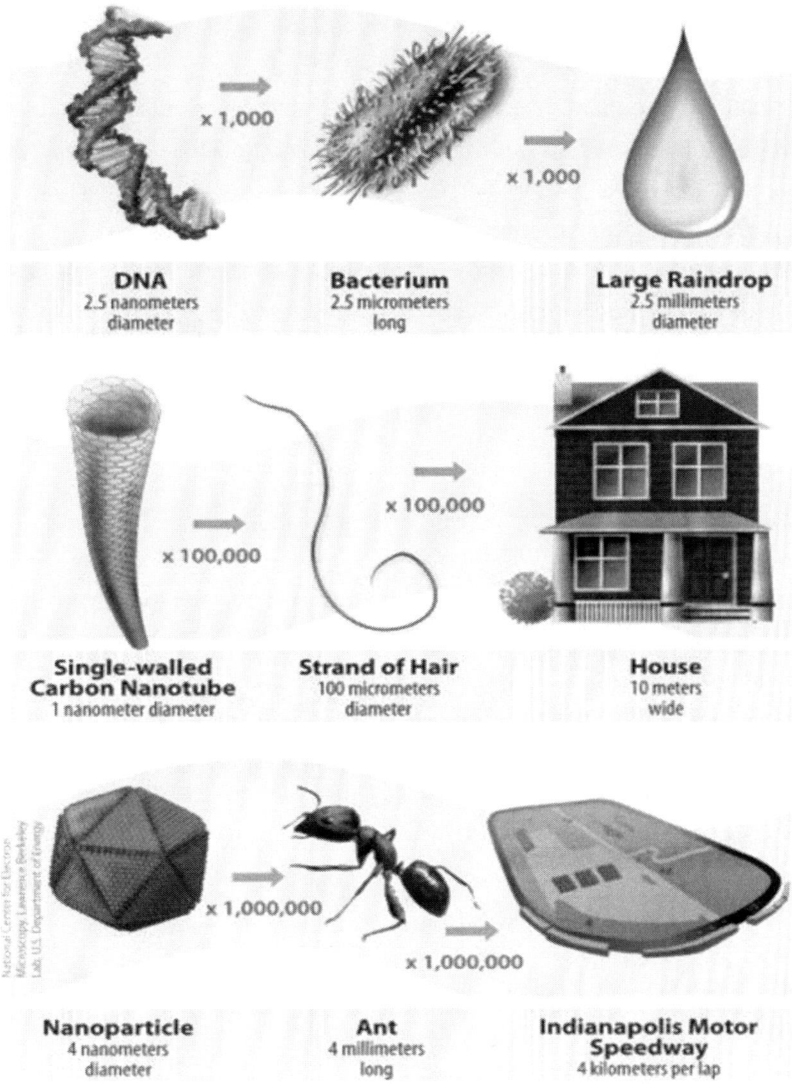

Figure 2. Nanomaterials: Various aspects [image courtesy of National Nanotechnology initiative].

Manmade nanostructured materials are of recent origin; their domain sizes have been engineered to precision at an atomic level simply by controlling the size of constituent grains or building blocks. Research on nanostructured materials began about two decades ago but did not gain much impetus until the late 1990s. Nanotechnology has now become a very active and vital area of research which is rapidly developing in industrial sectors and spreading to almost every field of science and engineering. It is expected that nanotechnology will grow to a multibillion dollar industry in the near future and will also become the most dominant technology of the twenty first century.

4. NANOPARTICLES

The physical and chemical properties of nanostructures are distinctly different from those of a single atom (molecule) and bulk matter with the same chemical composition. These differences between nanomaterials and the molecular and condensed-phase materials pertain to the spatial structures and shapes, phase changes, energetics, electronic structure, chemical reactivity and catalytic properties of large, finite systems and their assemblies. Some of the important issues in nanoscience relate to size effects, shape phenomena, quantum confinement and response to external electric and optical excitations of individual and coupled finite systems.

4.1. Surface Effects

For bulk materials, usually greater than one micron, the percentage of atoms at the surface is infinitesimally small compared to the total number of atoms of the material. As the size of the material is reduced, the ratio of surface atoms to inner atoms (or total atoms) starts increasing (Figure 3). Atoms on the surface of a bulk crystalline solid are principally located on basal planes, but they transform almost completely in edge and corner atoms with a decrease in size. Because of the low coordination number, edge and corner atoms are highly chemically reactive, catalytically active, highly polarizable and so on, in comparison with the atoms on basal planes.

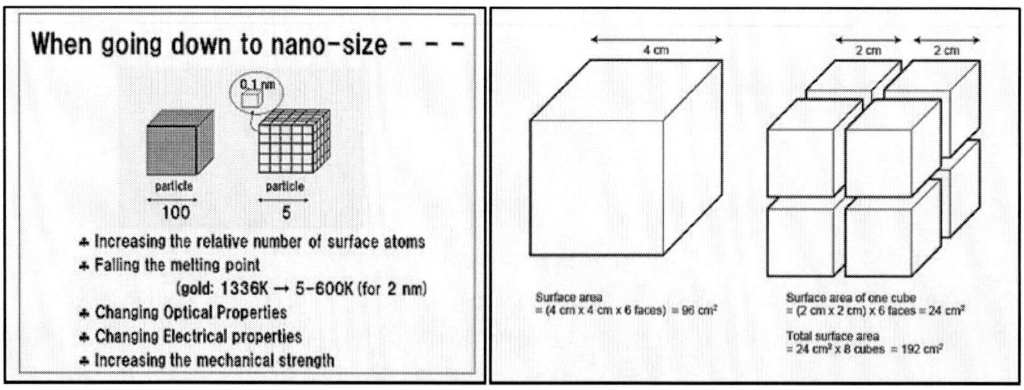

Figure 3. Schematic drawing showing nanoparticles have more surface area to volume than larger particles [image adapted from www.bbc.co.uk].

As a result, in the nanometer size range, the role of surface atoms starts becoming predominant in determining the properties of the material (surface effect) instead of the inner atoms. Therefore, higher proportion of the surface atoms results in some of the interesting and unexpected properties of NPs that are different from the bulk properties.

For example, bulk gold is an inert noble metal, but gold NPs are excellent catalyst with size dependent catalytic ability. The very high surface area to volume ratio provides a tremendous driving force for diffusion, especially at elevated temperatures. The large surface area to volume ratio also reduces the incipient melting temperature of NPs. Absorption of solar radiation in photovoltaic cells is much higher in materials composed of NPs compared to that in thin films of continuous sheets of material, i.e., the smaller the particles, the greater the solar absorption. Nanocrystals absorb then remit the light in different colours, the size of nanocrystal (in the Angstrom scale) determines the colour.

4.2. Quantum-Size Effect

The processes of diminishing the macroscopic size to nanometer range in one, two or all the three dimensions lead to the confinement of movement of conduction electrons in a very small space, which is comparable to their De Broglie wavelength. This leads to the formation of discrete energy levels in nanomaterials similar to atoms, unlike the continuous energy levels in the bulk materials. In case of semiconductor NPs, the band gap energy increases when exciton (a bound electron-hole pair produced by a photon) is confined to a length scale comparable or smaller to its Bohr radius. The optical absorption (and thus the NP color) and the emission spectra shifts toward the shorter wavelength (higher energy) and become sensitive to the size and shape of the particles.

4.3. Optical: Surface Plasmon Resonance (SPR)

Some metal NPs (Ag and Au) exhibit colours due to strong absorption bands in the visible region, and these colours are absent in the individual atoms as well as in the bulk. The basis of the light absorption by metal NPs is the collective coherent oscillation of the conduction band electrons induced by the incident photon. The electric field of an incoming light wave induces a polarization of the (free) conduction electrons with respect to the much heavier ionic core of a spherical NP. The positive charges in the particle are assumed to be immobile and the negative charges, that is, the conduction electrons, move under the influence of external fields. Therefore, a displacement of the negative charges from the positive ones occurs when the metallic NP is placed in an electric field. This results in a net charge difference at the NP boundaries which in turn gives rise to a linear restoring force to the system. As a consequence, a dipolar oscillation of the electrons is created (with a particular time period), and this is known as the surface plasmon oscillation.

For many metals, such as, Pb, In, Hg, Sn and Cd, the plasmon frequency lies in the UV part of the spectrum and NPs do not display strong colour effects. Hence, surface plasmon experiments are most commonly carried out with Cu, Ag and Au. The resonance frequency of this SPR is strongly dependent upon the size, shape, composition, interparticle interactions, dielectric properties and local environment of the NP. The unique surface plasmon absorbance features of metal NPs have been exploited for a wide variety of applications-colorimetric sensor, early detection and treatment of tumors, photo and electrochromic switching.

4.4. Catalytic Properties

Metal NPs show enhanced catalytic activity compared to that of the bulk metal, and it is dependent on the size, shape, composition, nature of stabilizing agent, crystal orientation etc. An inert and noble bulk gold metal shows a size-dependent catalytic activity for CO oxidation at size < 5 nm. Catalytic activity of the supported metal NP depends on the nature and electronic properties of the support. Metal NPs are fine tuners of interfacial properties at electrode-solution interface, and are widely used by electrochemists for decorating the surface of the conventional electrodes to produce electrocatalytic effect. Metal NPs have been used for the immobilization of many redox enzymes and bioelectrocatalysis of biomolecules (glucose, phenolic compounds, H_2O_2 etc.) has been observed. Metal NPs, especially gold and platinum, find widespread electrochemical applications in electrocatalysis and electrochemical sensing.

4.5. Novel Mechanical Properties

It is known that mechanical properties of a solid depend strongly on the density of dislocations, interface-to-volume ratio and grain size. An enhancement in damping capacity of a nanostructured solid may be associated with grain-boundary sliding or with energy dissipation mechanism localized at interfaces. A decrease in grain size significantly affects the yield strength and hardness. The grain boundary structure, boundary angle, boundary sliding and movement of dislocations are important factors that determine the mechanical properties of the nanostructured materials. One of the most important applications of nanostructured materials is in superplasticity, the capability of a polycrystalline material to undergo extensive tensible deformation without necking or fracture. Grain boundary diffusion and sliding are the two key requirements for superplasticity.

4.6. Unique Magnetic Properties

The magnetic properties of nano-size particles differ from those of bulk mainly in two points. The large surface-to-volume ratio results in a different local environment for the surface atoms in their magnetic coupling/interaction with neighbouring atoms, leading to the mixed volume and surface magnetic characteristics. Unlike bulk ferro-magnetic materials, which usually form multiple magnetic domains, several small ferromagnetic particles could consist of only a single magnetic domain.

In the case of a single particle being a single domain, the super paramagnetism occurs, in which the magnetizations of the particles are randomly distributed and they are aligned only under an applied magnetic field, and the alignment disappears once the external field is withdrawn. In ultra-compact information storage, for example, the size of the domain determines the limit of storage density.

Magnetic nanocrystals have other important applications such as in colour imaging [11], bioprocessing [12], magnetic refrigeration [13] and ferrofluids [14]. Metallic heterostructured multilayers comprised of alternating ferromagnetic and nonmagnetic layers such as Fe-Cr and Co-Cu have been found to exhibit giant magnetoresistance (GMR), a significant change in the electrical resistance experienced by current flowing parallel to the layers when an external magnetic field is applied [15]. GMR has important applications in data storage and sensors.

4.7. Crystal-Shape-Dependent Thermodynamic Properties

The large surface-to-volume ratio of nanocrystals greatly changes the role played by surface atoms in determining their thermodynamic properties. The reduced coordination number of the surface atoms greatly increases the surface energy so that atom diffusion occurs at relatively lower temperatures. The melting temperature of Au particles drops to as low as ~ 300°C for particles with diameters of smaller than 5 nm, much lower than the bulk melting temperature 1063°C for Au [16]. Nanocrystals usually have faceted shape and mainly enclosed by low index crystallographic planes. It is possible to control the particle shape, for example, cubic Pt nanocrystals bounded by {100} facets and tetrahedral Pt nanocrystals enclosed by {111} facets. The rod-like Au nanocrystals have also been synthesized, which are enclosed by {100} and {110} facets. The density of surface atoms changes significantly for different crystallographic planes, possibly leading to different thermodynamic properties.

5. Classification of Nanomaterials

Depending on the number of dimensions which are reduced to nanometer size range, nanomaterials can be defined as zero-dimensional (quantum dots), one-dimensional (nanowires, nanorods, and nanotubules), two-dimensional (quantum well, nanosheet, graphene) and three-dimensional (nanophase materials consisting of equated nanometer sized grains) as shown in Figure 4.

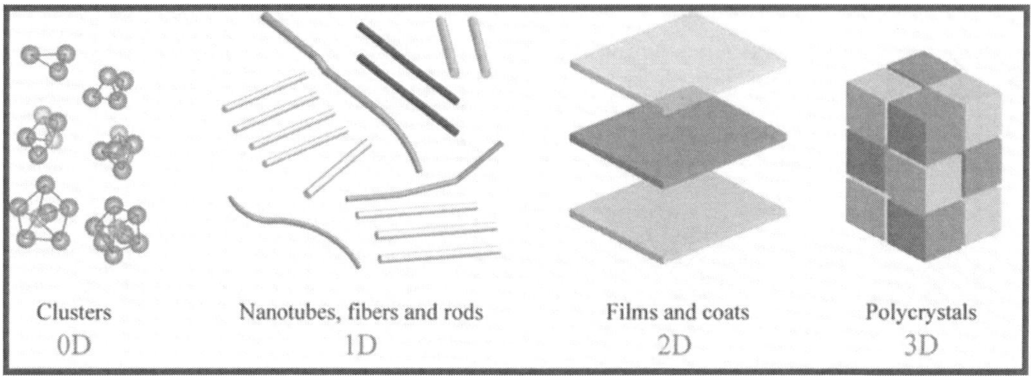

Figure 4. Classification of nanomaterials (a) 0D spheres and clusters, (b) 1D nanofibers,wires, and rods, (c) 2D films, plates, and networks, (d) 3D polycrystals [image adapted from eng. thesaurus.rusnano.com].

In zero dimensional nanomaterials, all the three dimensions are reduced to nanometer range. In one dimensional nanomaterials, two dimensions are in the nanometer range and one can remain large. In two-dimensional nanomaterials, one dimension is reduced to nanometer range and the other two dimensions remain large.

5.1. Nanoparticles

NPs are different from larger particles in many ways. First of all, NPs consists of less atoms and the fraction of the atoms located on the surface of a particle is higher. A high surface area to volume ratio will affect NPs reactivity but can also affect commonly physical stationary properties such as melting point and dielectric properties. It is known that particles with a diameter less than 50 nm, behave differently and below 10 nm it starts to be very difficult to predict material's physical properties. Physical properties that change with particle size is referred to as the "size effect" which involves quantum physics that allow new physical events to take place. Silicon is revolutionizing in its nanoparticle form.

5.1.1. Organic Nanoparticles

Organic NPs are carbon based and can take on many different molecular configurations. In addition to diamond, graphite and amorphous carbon, it has recently been discovered more allotropes of carbon. One important allotrope in nanoscience is the so called CNT (Carbon Nano Tube), which officially was discovered 1991 but papers on nanotubes have been published before that.

5.1.2. Carbon Nanotubes

CNTs are as the name inclines hollow tubes that usually have a diameter in nano range and a length in µm range, made out of carbon. The CNT is made from rolled graphene, a one atomic layer thick carbon sheet. The first method to produce CNTs was with an arc discharge process but today several methods to create CNTs exist with different physical properties. CNTs can vary a lot in size depending on processing method and multi-walled carbon nanotubes (MWCNT) can be created which are tubes with different diameters stacked inside each other. Graphene layers that constitute the CNTs can be rolled in different angels, so called chiral angels, which affect properties such as conductivity. It means that CNTs can be both conductive as well as semi conductive depending on what chiral angle it has, which generally is not easy to control during manufacturing. Approximately, one third of the tubes are conductive due to random distribution of chiral angles [17].

CNTs do get a lot of attention in scientific journals, partly because of their remarkable mechanical properties, with the highest tensile strength of all materials yet discovered. Today it is possible to buy CNTs, and they are used in applications such as for mechanical reinforcement in composites or as fillers in antistatic materials.

5.1.2.1. Amorphous Carbon

Carbon that does not have an ordered crystal structure but random distributed carbon atoms, is called amorphous carbon or in everyday language it is soot or coal. In material science amorphous carbon is mainly used to tint rubbers and make it more wear resistant. In these applications Carbon Black has been used for a long time, which is an amorphous carbon made from incomplete combustion of carbon rich products. This material has a very big surface area and due to its very small carbon particles it is suitable for tinting applications.

5.1.3. Fullerenes

Another newly discovered carbon allotrope is the fullerene that was discovered in 1985. It is closely related to the CNT but is shaped as a sphere instead of a tube. This particle can also be called Buckyball because of its football shape like structure; an extremely stable molecule with the roundest geometry of its size (Figure 5).

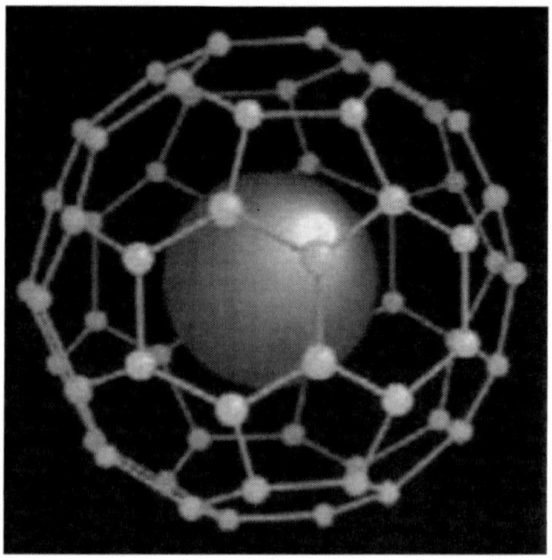

Figure 5. The structure of a C_{60} fullerene with a central metal atom [image adapted from www.georgehernandez.com].

5.1.4. Inorganic Nanoparticles

Even though inorganic NPs do not have the same extreme mechanical properties as the organic, it is still a very interesting field of science. Inorganic NPs are nanocrystals of metals or oxides.

5.2. Nanocrystals

Nanocrystals can have different shapes and sizes which largly affect their properties. However, particles larger than 50 nm are said to act similar to the bulk material, while particles smaller than that often have completely different properties. To create nano sized inorganic particles, there are plenty of processing methods or so called synthesis that can be used.

5.3. Nanocomposites

Many of the present nanocomposites are made out of dispersed NPs in a matrix of polymers, metals or ceramics. The particles can have different composition and shapes, which means that the variety of possible nanomaterials that can be made with help of NPs is vast. Creating a uniform nanocomposite is not an easy task. NPs always agglomerate when exposed to one another, which is not desirable in most applications. Therefore, the dispersion of the filler material, in a matrix, is a critical step in the manufacturing process of nanocomposites. Ways of dispersing NPs in a liquid can be done by using high energetic ultra sound or by treating the particles chemically so they repel each other. However, the intermolecular attraction force (van der Walls) is present at all surface interactions of small particles.

5.3.1. Polymer Matrix

A common high strength polymer composite is carbon fibre reinforced plastic (CFRP). The reason for its desirable material properties can be explained with the useful combination of flexibility and low weight of the polymer along with high strength fibres, which together constitutes a very light weight and strong material. The interaction between fibres and a polymer matrix is very important for the composites properties. Carbon nanotube reinforced plastics (CNTRPs), are proven to have higher strength than conventional polymer composites, which allows the possibility to create more lightweight composites, which is of great importance in many fields of engineering. Other possibilities with CNTRPs include the ability to use coating methods otherwise only suitable for conductive materials. One such method is powder coating, made possible due to the CNTs excellent electrical conductivity properties.

5.3.2. Elastomer Matrix

Elastomers in car tires has been used with nano fillers e.g., to increase strength, tint and improve friction properties. The old filler material used is often amorphous carbon or so called carbon black which is a nano material that exists naturally in for example, coal and soot. The largest application area for elastomeric nanocomposites is by far the tire industry where also, a lot of research is being made. Today, elastomers can get improved properties with other nano fillers such as CNTs or plate like nano particles (layered silicate and graphene) all with aim to decrease rolling resistance, increase braking friction and increase wear resistance in tires. Layered nano silicate, which is a clay mineral, does not only increase the strength of elastomers but it has also been found that they can achieve good gas barrier properties due to their shape. Therefore, these composites can be applicable in food packaging or in other applications such as oxygen, carbon dioxide, vapour or other light weight molecules should be secluded [18].

5.3.3. Ceramic Matrix

As for conventional ceramic composites, nano ceramic composites are often used in high temperature application due to its excellent heat resistance properties. For mechanical purposes, ceramic has a limited area of use because of brittleness. A lot of the research being made on the subject is to reduce the brittleness, which would turn ceramics into a material that could withstand both heat and tensile load. There are plenty of ways to increase fracture toughness in ceramics and some of these methods involve nano technology. It has been shown that nano sized carbides drastically can limit crack growth of Al_2O_3 which is a common ceramic. Furthermore plasma sprayed ceramic composites surfaces can be made with nano-sized grains. These surfaces can show lower friction, higher hardness and better wear resistance.

5.3.4. Metallic Matrix

A well-known metallic matrix composite is the hard metals consisting of cobalt reinforced with tungsten carbides. Yet, there is a wide range of particles that can be dispersed in metallic matrixes and when the particle size is decreased down to nano range, many properties can be improved. NPs in metal matrixes can also affect the grain growth behaviour, which then can be used to control grain shape and size in a metal. This makes it possible to influence many mechanical properties.

5.4. Nanocrystalline Materials

Nanocrystalline materials are materials with a grain size between 1-250 nm. These materials show different properties compared to those of bulk material with the same chemical composition. Since the size of the grains is so small, the grain boundary surface is vastly increased and more atoms are located on a grain boundary. It is recognised that these materials have better mechanical properties concerning hardness and tensile strength. Another effect of nano crystallinity is the decrease in elastic modulus which occurs for very small grains. There are multiple methods to create nano crystalline materials. One common way is to quench a liquid metal at a rate of at least around 106 K/s. At this rate, crystallization is suppressed and an amorphous structure is attained. When annealed, crystals starts to precipitate and if no further heating is applied a nano crystalline material is reached.

5.5. Nanoporous Materials

The difference between a normal porous material and a nano porous material is solely the size of the pores. A nano porous material has pores between 1-100 nm and is normally built up by ceramics, metals or polymers. Since the size of the pores is so small, these materials can be used as filters for molecules or other very small particles. Also the porousness vastly increases the materials surface area which can enhance the efficiency for catalysts. Today many new ways or producing these materials are explored and new areas of applications are also found rapidly.

5.5.1. Zeolites

Zeolites are a naturally occurring material that has a distinctive porous crystal structure (Figure 6). It is a crystalline hydrated alumina silicate with pores from 1-20 Å in diameter. Because of these very small and uniformed pores, zeolites can be used as selective separation membranes of chemicals, adsorbents or catalysts. Because of zeolite's great potential to separate chemical compounds on a molecular level, a lot of effort has been made to develop zeolites into membranes. This has pushed the development forward and zeolites are today used in separating processes in bio-ethanol production. One advantage of using zeolite membranes is that zeolites are more heat resistant, compared to polymers, which is another common selective membrane.

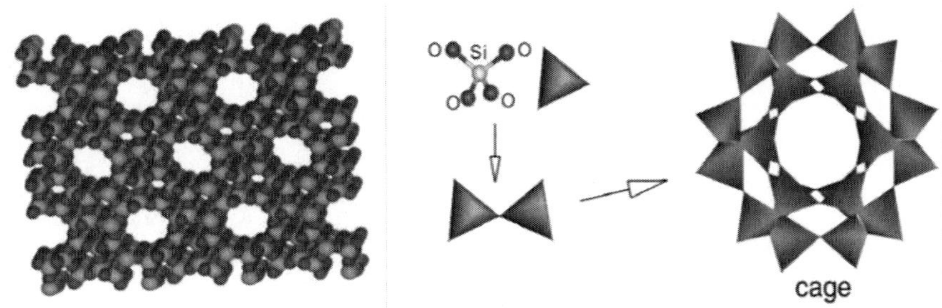

Figure 6. An open crystal structure of a zeolite.

5.5.2. Porous Polymers

Polymers are preferable made into nano porous material because of easier manufacturing methods compared to the zeolites. As the zeolites, polymers can act as separation membranes to, where molecule-size and chemical reactivity of the permeates determines if it can penetrate the membrane or not. This can be used to separate gases from each other or filter small particles. The efficiency of the membrane material is in many cases depending on the uniformity of pore sizes.

Porous nano polymers are also an important tool for the creation of nano fibers/wires where the polymer acts as a mold. The mold can be filled with a metallic material and create a network of metal in the pores of the polymer. If later on, the polymer is heated up to such temperature that it melts, what's left is a nano metal fibre network or so called a nanowire network.

6. NANOSTRUCTURED SURFACES

With nanotechnology, new surfaces can be made that can protect or change important properties on the surface of a material. In this section, some different kinds of nanostructured surfaces will be discussed in terms of composition and application areas.

6.1. Nanocrystalline Surfaces

The properties of nanocrystalline materials are very suitable for surfaces in mechanical applications. Nanocrystalline surfaces can be obtained by, for instance, different deposition methods. There are many variants of this technique and today these nano crystalline films can protect its underlying material from among other things, corrosion and erosion. Coatings for cutting tools where hardness, wear and heat resistance are critical properties, have also been developed with this technique. Electrodeposited nanocrystalline cobalt-based coatings have been tested as a replacement for normal chrome coatings that are strictly regulated due to its toxic composition. The result showed that the Cobalt based coatings had higher hardness and better wear and corrosion resistance [19].

6.2. Self-Cleaning Surfaces

Self-cleaning surface is today the accepted name for the lotus leaf imitation surface that has the property of staying clean by strongly repel water and dirt. A lotus leaf consists of a hydrophobic surface with nano/micro sized bumps that effectively decrease the contact area between a water droplet and the leaf. This property is called the Lotus Effect and has now been replicated by scientists. The Lotus leaf is super hydrophobic which means that a droplet has a contact angle higher than 150° degrees from a surface.

In order to produce hydrophobic surfaces, the underlying material must have low surface energy and to get super hydrophobic properties, the surface roughness must also be high. Both these requirements can be obtained with alkyl and fluorinated alkyl groups that possess a roughened surface. With this principal, many materials can become super hydrophobic, for instance, copper can get super hydrophobic when copper nanowires are grown on a copper substrate and oxidized. In the same manner, there are many more ways of creating super hydrophobic surfaces that can be used in application of self-cleaning applications but also for de-icing or make materials more corrosion resistant.

6.3. Nanofluids

It has recently been discovered that NPs that are dispersed in a cooling liquid can vastly increase the heat transfer performance. When NPs are mixed in a less conductive cooling liquid, they enhance the overall heat conductivity of the liquid. One effective way is mixing with ultrasound, but to keep the dispersion stable, surfactants must be used. Surfactants are molecules with one hydrophilic end and accumulate on surfaces in a suspension such as surfaces on particles. Nanofluids have also been successfully used as lubricant additives to efficiently reduce friction.

6.4. Dendrimers

These nanomaterials are nanosized polymers built from branched units. The surface of a dendrimer has numerous chain ends, which can be tailored to perform specific chemical functions.

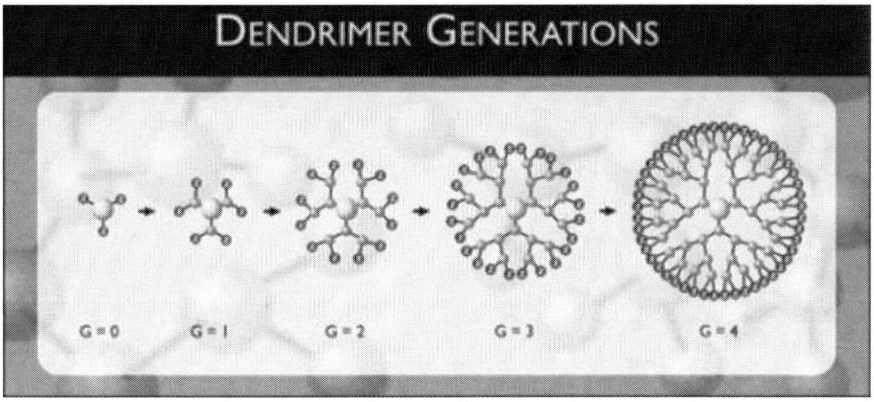

Figure 7. Dendrimer generations [ilustration courtesy of dendritic nanotechnologies, Inc.].

This property could also be useful for catalysis. Also, because three-dimensional dendrimers contain interior cavities into which other molecules could be placed, they may be useful for drug delivery. Dendrimers are unique macromolecules that have applications in developing strategies for nano-scale globular shapes with small cores. Dendrimers (Figure 7) generally possess multiple branches which can be used to carry a variety of agents fulfilling various functions at once.

The thickness of the polymers is in the range of several nanometers. These nanoobjects have gained interest because of their cylindrical structure, which provides new unique properties that could be applied within the emerging field of nanotechnology. For example, DNA/PAMAM (Polyamidoamine) dendrimer complex is used for DNA delivery to cell nucleus due to its high transfection efficiency and very low toxicity.

7. SYNTHESIS TECHNIQUES

Size, shape and composition-controlled monodisperse nanomaterials of metals, semiconductors, organic polymers, conducting polymers, inorganic materials, and their composites have been synthesized by different routes. Controlled synthesis of monodisperse nanomaterials with desired tunable properties is still a subject of great interest among the research communities and efforts are being continuously made to scale up the production for various applications. Generally, there are two methods for producing nanomaterials. One method is to generate particles of the correct size from scratch using crystallization, direct generation or other similar techniques, which are often referred to as "bottom up." In the other method, referred to as "top down," larger particles are reduced through milling to achieve the desired nanoscale particles.

The current fabrication technology for NPs is novel and versatile. Significant progress has been made over the last decade in understanding fundamental aspects of the synthesis of NPs, nanorods, etc. Various processing routes have been developed for the synthesis and commercial production of NPs including vapor, liquid, solid state processing routes, and combined methods.

7.1. Vapour Route

7.1.1. Physical Vapour Deposition

Physical vapour deposition (PVD) is a versatile synthesis method and capable of preparing thin film materials with structural control at the atomic or nanometer scale by careful monitoring the processing conditions. PVD involves the generation of vapour phase species either via evaporation, sputtering, laser ablation or ion beam. Inert gas condensation combined with thermal evaporation is commonly used to fabricate metallic and metal-oxide powders with a well-defined and narrow size distribution. This technique was originally introduced by Ganqvist and Buhrman [20] in 1976.

7.1.2. Chemical Vapor Deposition

Chemical vapour deposition (CVD) is a process where one or more gaseous adsorption species react or decompose on a hot surface to form stable solid products. More recently, Lai et al. [21] employed hot filament CVD to grow one dimensional SiC nanorods (Figure 8).

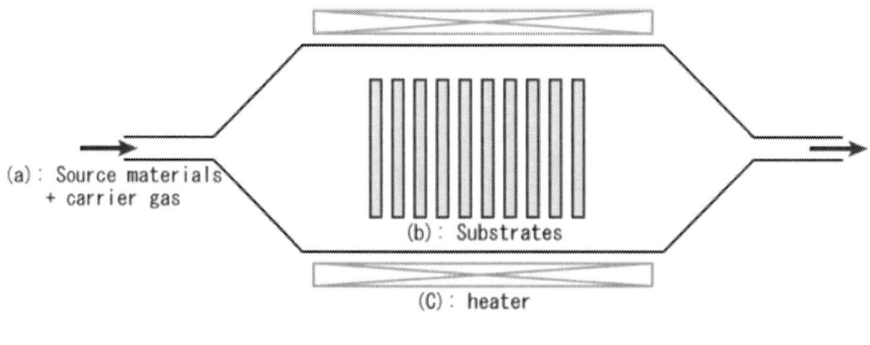

Figure 8. Chemical vapour deposition.

This route involves the atomization of chemical precursors into aerosol droplets that are dispersed throughout a gas medium. The aerosols are then transported into a heated reactor where the solution is evaporated or combusted to form ultrafine particles or thin films. This technique is versatile and inexpensive because of the availability of various low cost chemical solutions and has been used to produce a variety of high purity and non-agglomerated metal oxide NPs such as TiO_2, Al_2O_3, ZrO_2, V_2O_5 and Y_2O_3–ZrO_2 [22].

7.2. Liquid Route

7.2.1. Synthesis of Metallic Nanoparticles

Metallic core/shell NPs are synthesized by the reduction of metal salts. For the synthesis of metallic NPs, the synthesis media such as a microemulsion or aqueous phase, use of surfactants, and proper reducing agents, all play a key role. The solution phase reduction mechanism is used for the synthesis of different metal NPs, such as Au [23, 24] Ag [25] Co [26], Fe [27] and Ni [28, 29] from their respective metal salts. The choice of reducing agent is very important for controlling the reaction rate as well as the particle size. Based on the reducing agent, the synthesis routes are classified into the following groups.

7.2.2. Reduction by Sodium Borohydride

The most commonly used reducing agents are sodium borohydride and hydrazine for the reduction of metal ions from their respective salts. Sodium borohydride was discovered by Schlesinger in 1942 [30] and is an efficient water-soluble reducing agent used in both aqueous and organic media. Sodium borohydride has some distinct advantages for use as a reducing agent over other reducing agents:

it has low equivalent weight of 4.73 g/mol and 1 mol can supply eight electrons; (ii) it has high reducing power with a reducing potential -1.24 V vs. standard hydrogen electrode (SHE) at pH 14, decreasing to 0.48 V at pH 0; it can reduce metal salts in either aqueous or non-aqueous media at any pH.

Capping agents can also be used in this reducing media mainly for controlling the size of the particles as well as modifying the surface charge for incorporating a favourable shell coating. The fundamental chemistry of reduction by borohydride can be explained in terms of electron transfer from borohydride to metal ion. Depending on the pH of the medium, the reduction mechanism of borohydride is different. However, the overall reaction is a combination of reduction and hydrolysis of borohydride, which again depends on the reaction conditions, concentration of the reactants, mixing efficiency and kinetics of the reactions.

7.2.3. Reduction by Hydrazine

Hydrazine is another common reducing agent for metal NP synthesis from the metal salt, and it is comparatively weaker than borohydride. Therefore, hydrazine is used for the synthesis of metals that have higher reduction potentials than hydrazine. Because the standard reduction potential of metals normally varies from -1 to any higher positive value, hydrazine is mainly used in alkaline medium and the maximum reduction efficiency of hydrazine is at pH > 11.553, e.g., the standard reduction potential of Ni^{2+} and $Ni(OH)_2$ are -0.257 and -0.72 V, both of which are greater than that of hydrazine in alkaline media. Therefore, hydrazine is used for the synthesis of metals such as Ni [31, 32], Ag [33], Cu [34] and Pt [35] and bimetallic compounds such as Pd/Pt [36] and Pt/Ru [37]. The main advantage of using hydrazine is that the pH and temperature-dependent reducing ability of hydrazine makes the reduction rate more easily controllable.

7.2.4. Reduction-Transmetalation

Other than the more common reduction techniques, metallic particles can also be synthesized by using a redox-transmetalation method. It is an advanced process compared with the conventional reduction techniques. In this method, initially the core material is synthesized by a common reduction process using any suitable reducing reagent. Then the metal salt is added to the solution without any additional reducing agent. When the metal salts come into contact with the core surface, they are reduced by the surface atoms of the metal core and deposited on the surface of the core. As a result, some portions of the metal core are oxidized to metal salt and diffuse through the shell layer into the bulk solution. A redox reaction can then spontaneously precede under the favourable redox potentials between two metals. This technique has some advantages over normal reduction, such as a conventional reducing agent is not required, shell formation is spontaneous without any separate reaction being required, self-nucleation of the shell metal can be avoided, and in homogeneous growth of the shell can be prevented. However, the limitation of this method is that only those pairs of bimetallic core/shell NPs are possible where the shell metal has a high reduction potential compared with the core material. This method has been efficiently used for the selective formation of bimetallic core/shell structures, some common examples of which are Ni/Au [38], Co/Pt [39, 40], Ni/Ag [41], Co/Au, Co/Pd, Co/Cu [39] and Ag/Au [42].

7.3. Thermal Decomposition of Organometallic Compounds

High-temperature thermal decomposition of organometallic compounds is another suitable method for the synthesis of metal NPs especially for core particle synthesis. In this method, organometallic compounds are decomposed to metals at high temperature in the presence of a surfactant in non-aqueous media but in the absence of water and oxygen. The surfactant acts as a dispersing agent for controlling the particle size. A simple example of this method is the synthesis of cobalt from a octacarbonyldicobalt complex. In the presence of a surfactant molecule, $Co_2(CO)_8$ is decomposed to Co metal at 500°C. Organic surfactant molecules act as a dispersing and stabilizing agent. Adsorbed surfactant on the cobalt core surface can be removed by successive washings using anhydrous methanol and storage either in toluene or benzene. Freshly synthesized Co metal is highly reactive toward oxidation; therefore, contact with air or water should be avoided. Otherwise the surface layers of the cobalt will be oxidized to cobalt oxide. The metal carbonyls are highly toxic and relatively more expensive. Therefore, in order to improve the method, octacarbonyldicobalt is replaced by another organometallic precursor such as the Co(η^3-C_8H_{13}) (η^4-C_8H_{12}) [43] [bis(salicylidene) cobalt(II)] oleylamine complex [44]. By using the thermal decomposition method with an organometallic precursor of another metal such as Cu from copper oxalate [45], [bis(salicylidiminato)copper(II)] [$Cu(sal)_2$] [46], Au from gold acetate [$Au(OCOCH_3)_3$] [47] and Ni from nickel acetate [$Ni(OCOCH_3)_2$ $4H_2O$] [48] have all been synthesized. In the presence of a stabilizing agent, the mechanism of particle formation is the same as octacarbonyldicobalt complex decomposition. However, the main advantage of using another organic complex is that the process requires a lower temperature (~250°C) compared with the metal carbonyl complex (500°C). After separation of the metal, several washings with anhydrous alcohol are required to remove the stabilizing agent before coating with the required shell materials. This technique is more useful for the synthesis of metal/metal oxide core/shell NPs from the same metal compound. There is no additional step required for the synthesis of the shell.

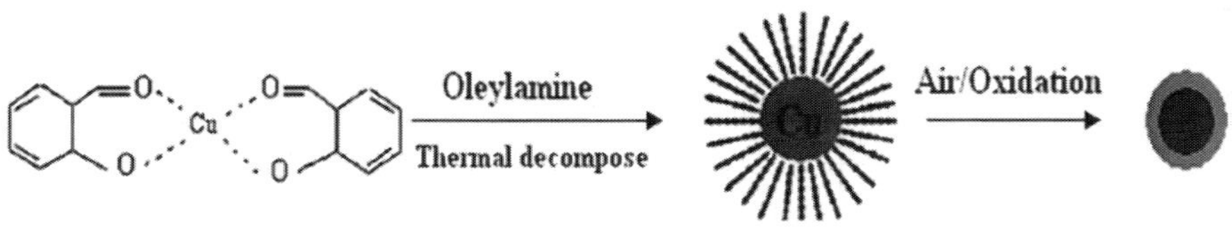

Figure 9. Schematic diagram for Cu/Cu_2O core/shell nanoparticle synthesis using the thermal decomposition of a metal organic complex.

After removal of the stabilizing agent from the core surface, surface atoms are oxidized to the corresponding metal oxide by atmospheric oxidation. A schematic diagram of Cu/Cu_2O core/shell NP formation by this technique is given in Figure 9.

7.4. Wire Electrical Explosion

Compared with all the other previously mentioned methods, metal NP synthesis by the wire electrical explosion (WEE) technique follows a totally different mechanism. In this method, a high voltage is applied through a thin wire in an inert gas atmosphere in a high pressure closed chamber. Under these conditions, a strong pulse current with a high current density is passed through the wire so that the wire becomes overheated and evaporates very quickly. The result is that different sized metal NPs are formed within that closed chamber because of the sudden explosion with a shock wave. In this process, the time required for thermal expansion is as short as several tens of microseconds. This process is not used conventionally for common industrial purposes, because it is not only very expensive but also impossible to use explicitly for different metals. It is mainly useful for those metals of high electrical conductivity that are easily available in the thin wire form. After particle formation, the particles should be collected carefully and kept in a closed vessel to prevent atmospheric oxidation. According to the available literature, this method is only used for the synthesis of metals such as Ni [49], Al [50] and Cu [51], core/shell alloys such as Cu/Zn [52] or Ti/Ni [53] and martensite stainless steel (MSS) (with alloy composition 72.1% Fe, 18.7% Cr, 6.7% Ni, 1.4% Mn, 0.8% Si, and 0.03% C) [54]. This technique is also used for the synthesis of bimetallic core/shells or metallic alloy formation. With respect to bimetallic core/shell NP synthesis, only those combinations that have a higher difference in saturation vapour pressure at a fixed temperature are possible. In this case, the particles having low saturation vapour pressure will form the core and the others form the shell after complete solidification of the core material. The core particles can also be coated in a separate step [49]. After separation of the core particles, a silica shell is coated in a non-aqueous media from TEOS.

8. Synthesis of Oxide Nanoparticles

Metal or metalloid oxides are another important class of compounds that have been extensively synthesized over the past decade as a core or shell in core/shell NPs. In general, liquid solid transformation is the more common transformation mode for metal oxide synthesis either as a core or as a shell than the gas-solid transformation. Among the different liquid-solid transformation methods, the sol-gel and co precipitation techniques are more acceptable methods for the synthesis of these metal or metalloid oxide NPs. Silicon and titanium oxides are preferentially synthesized by the sol-gel method. However, co precipitation techniques are mainly used for all the different types of oxides especially those metals from the hydroxide precipitate in alkaline pH. Other than these two, there are also other methods for metal oxide formation, and these will now be discussed.

8.1. Sol-Gel Method

The sol-gel process is a wet chemical technique widely used in material research related to science and technology applications, in particular for metal oxide NP synthesis. This process is a combination of two steps, hydrolysis and polycondensation of suitable molecular precursors (usually metal alkoxides or chlorides), which leads to the gradual formation of the solid-phase network (Figure 10).

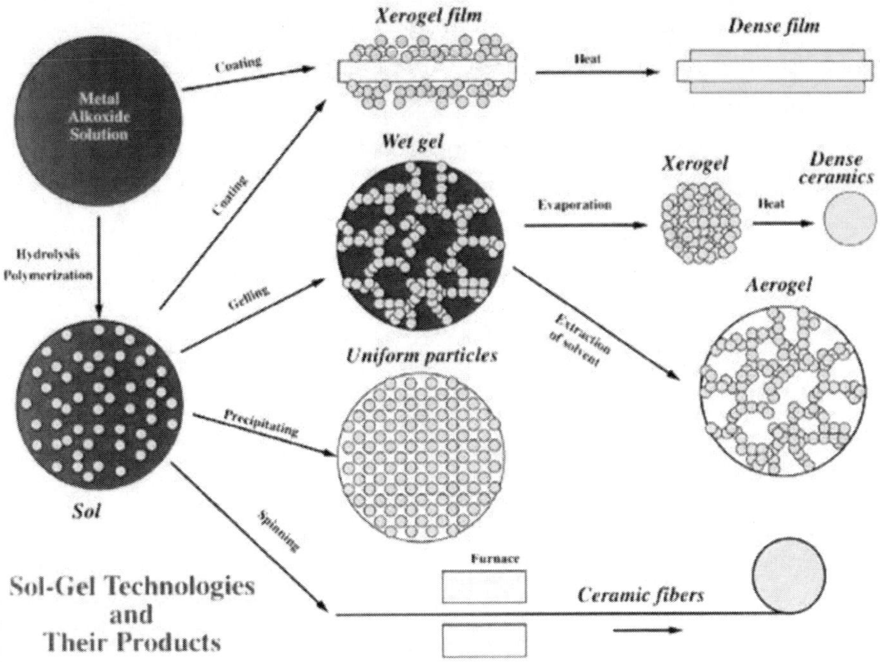

Figure 10. Various steps of sol-gel process for synthesis of nanomaterials [image adapted from sariyusriati.wordpress.com].

Different research groups used modified sol-gel methods for the synthesis of different metal oxides. Sol-gel processes are mainly classified into three different approaches:

- Gelation of solutions of a colloidal powder;
- hydrolysis and polycondensation of metal alkoxides or metal salt precursors followed by hypercritical drying of the gels;
- Hydrolysis and polycondensation of metal alkoxide precursors followed by aging and drying under ambient atmosphere. The last method is commonly used for the synthesis of metal oxides. These three approaches can be divided into seven steps, as describes below.
- Step 1. Mixing: For the first method, the suspension of colloidal powder or sol is formed by the mechanical mixing of colloidal particles in water at a suitable pH without any precipitation. For the other two methods, metal alkoxide or metal salt are used as a precursors, which are hydrolysed to form hydrated metal hydroxides in the presence of water, which subsequently undergo condensation to form an oxo bridge (-M-O-M-). When sufficient M-O-M bonds are formed in a region, then they cooperatively respond to form colloidal particles or sol. The size of the particles depends on the pH of the medium and the ratio of water to metal alkoxide or metal salt.
- Step 2. Casting: Since the sol is a low-viscosity liquid, it can be cast into a mold.

- Step 3. Gelation: With time, the colloidal particles and condensed species link together to form a three-dimensional network with increased viscosity, which finally becomes solid.
- Step 4. Aging: The aging of the gel involves maintaining the cast object for a period of time, completely immersed in liquid for polycondensation, syneresis, coarsening, and phase transformation.
- Step 5. Drying: The liquid is removed from the interconnected pore network during the drying process.
- Step 6. Dehydration or chemical stabilization: The surface metal hydroxide bonds are removed in this step and the material is converted into a stable ultra-porous solid with sufficient interconnected porosity, optical transparency and good mechanical strength.
- Step 7. Densification: After dehydration, solid materials are heated at high temperatures to increase density by reducing the internal porosity. During the last four steps, the shrinkage and densification occurs because of (1) capillary contraction, (2) condensation, (3) structural relaxation and (4) viscous sintering. In the first step, the rate of hydrolysis of metal or metalloid alkoxide or salt is affected by some physical parameters such as (i) pH, (ii) substituents, (iii) solvent, and (iv) water.

Aqueous phase sol-gel chemistry is a complex process, mainly because of the high reactivity of the metal oxide precursors and the double role of water as a ligand and solvent. In many cases, three types of reactions (hydrolysis, condensation, and aggregation) occur almost simultaneously, and these are also difficult to control individually, so slight changes in experimental conditions, result in altered particle morphologics. Furthermore, the as-synthesized metal oxides are often amorphous and it is difficult to retain full control over the crystallization process during any additional annealing step. Although, all these parameters can be controlled well enough for the preparation of bulk metal oxides, there is a big challenge in the case of NP synthesis by sol-gel methods. Therefore, to overcome these problems, researchers are trying to change the aqueous to a non-aqueous system. In non-aqueous systems, the exclusion of water as a continuous phase can minimize some of the limitations of the aqueous system. Water not only acts as the oxygen-supplying agent for the metal or metalloid oxide but also strongly influences particle size, shape, and surface and assembly properties. In specific cases, even the composition and crystal structure are affected. In 2007, Niederberger [55] explained the mechanism of the sol-gel process in non-aqueous systems. Different metal oxides, such as TiO_2, SiO_2-GeO_2, CeO_2, [56] and rare earth doped Al_2O_3, [57] are synthesized by this technique.

8.2. Co-Precipitation Technique

Precipitation is a most common type of reaction that can be used to prepare wide varieties of NPs. In a precipitation reaction, two or more water-soluble salts react with each other to form at least one water insoluble salt that precipitates out from the media. The solubility product of the precipitated compound is the most important parameter for such a precipitation reaction. In this technique, first the product will form in the liquid phase immediately after the completion of the reaction. When the concentration of the product crosses the solubility product value of that compound in the reaction media, particle formation will start. The mechanism of NP formation via precipitation is a combination of three separate steps one after another: (i) nucleation, (ii) growth and (iii) agglomeration.

However, there may be different types of reactions, such as acid-base precipitation reaction and redox precipitation reaction. In this process, reaction takes place in the liquid phase but the particles formed are in the solid state; for that reason it is also known as reactive crystallization and it also depends on several energy terms. The first step, the reaction in the liquid phase, is associated with the activation energy of the reaction. The rate of reaction is faster with a lower activation energy than for that of higher activation energy. After the completion of this reaction, embryos (consisting of few atoms) are generated in the bulk liquid phase. When the size of the embryos crosses, a critical radius (called a nucleus), then only the liquid state will be converted to solid. When the solid particles are formed in the bulk phase, a new solid liquid interfacial area is generated with a specific surface energy proportional to the square of its radius. In considering the bulk energy, because of the formation of a solid then the energy gain is proportional to the cube of the particle radius. These two energy terms are opposite in nature. The change in total energy is mainly the contribution of these two energies, but some other factor, such as motion of the particles in the media, defects in crystal energies and electrostatic contributions also contribute a little bit to the overall energy of the system. Normally nucleation is the faster step if the solubility product of the particle material is very low; in that case, growth rate will be controlled by the overall reaction rate. In the growth process, embryos form after the reaction diffuses to the nuclei surface from the bulk phase. Ultimately the growth rate depends on either the reaction rate or the diffusion of the molecules from the bulk phase to the nuclei surface, whichever is slow. The final step is agglomeration. Here small sized particles are converted to a larger size as a result of Oswald ripening with a decreasing specific solid liquid interface area to reduce the overall energy of the system. An alternate route for preparing Fe_3O_4 NPs has been reported by Gupta and Gupta [58]. Similarly, other core/shell NPs, such as $CaCO_3$/ SiO_2, can also be prepared by precipitation techniques using CaO and $Na_2SiO_39H_2O$ as the reactants. Both the core and shells were synthesized in the same media. By similar techniques, different metal or metalloid oxides such as ZnO, SiO_2, Fe_3O_4, and Y_2O_3 [59, 60] were synthesized either as a core or as a shell by different research groups.

8.3. Thermal Decomposition Technique

Thermal decomposition is another technique used for the decomposition of metal complex compounds to form metal oxides; it is also known as thermolysis. This method is used mainly for the synthesis of metal oxide as a core particle in the presence of air or oxygen. The associated reaction is normally endothermic in nature. This method is more common for the synthesis of transition metal oxides, because, this is only applicable to stable metal complexes. In general, because of the presence of a vacant d-orbital in the valence orbital, the transition metals can easily form metal complexes through the formation of a secondary bond between the metal and electron-donating ligands. In this process, initially a surfactant solution is used to stabilize the metal salts and then the surfactant-stabilized metal salt is mixed with a higher molecular weight organic compound to form an organometallic complex. Finally, the resulting organometallic complex is decomposed to form a metal oxide in a hydrophobic solvent in the presence of pure oxygen at high temperatures. After complete decomposition, reaction particles are separated by centrifugation and then washed with a suitable solvent. This method has been used for the synthesis of ZnO from [bis (2-hydroxyacetophenato) zinc(II)] [61] or [bis(acetylacetonato) zinc(II)]oleylamine complex, [62] zirconium oxide (ZrO_2) from zirconyl nitrate pentahydrate ($ZrO(NO_3)_2 5H_2O$) using two separate stabilizing agents, oleylamine and PEG, [63] or from bis-aqua,tris-2-hydroxyacetophenato zirconium(IV) nitrate [$Zr(HAP)_3(H_2O)_2](NO_3)$] as a precursor in oleylamine ($C_{18}H_{37}N$) and triphenylphosphine ($C_{18}H_{15}P$), [64] cobalt oxide (Co_3O_4) from Co(oxalate) [65] and [bis(2-hydroxyacetophenato) cobalt(II)], [66] and iron oxide (Fe_3O_4) from $Fe(acac)_3$ with 1,2-hexadecanediol, oleic acid, and oleylamine at elevated temperature [67]. The schematic diagram of zirconium oxide formation by the thermal decomposition method is shown in Figure 11.

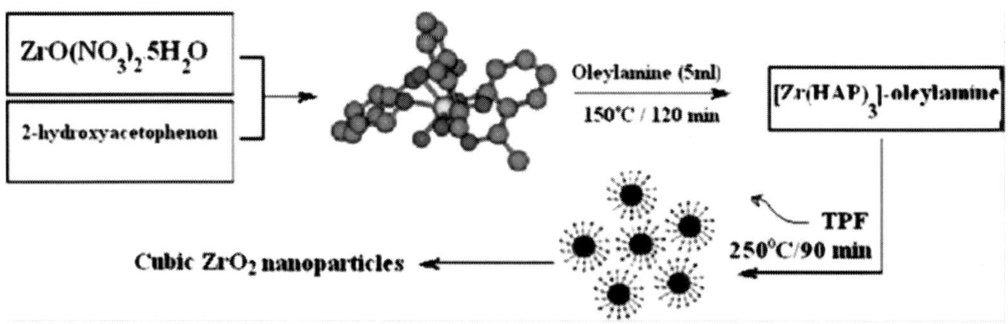

Figure 11. Schematic diagram of zirconium oxide formation by the thermal decomposition method [Reprinted with permission from (M. Salavati-Niasari et al. 2009), *Inorg. Chim. Acta* 2009; ©2009, Elsevier Publisher].

8.4. Solid Route

8.4.1. Mechanical Alloying/Milling

The mechanical alloying/milling process is a solid state processing technique and was originally developed by Benjamin of the International Nickel Company for the production of oxide dispersion strengthened (ODS) superalloys [68]. In this process, high energy ball milling is used for synthesizing a variety of materials like forming equilibrium and non-equilibrium phases.

The mechanical energy transfer to the powder particles in these mills takes place by shear action or impact of the high velocity balls with the powder. During the process, raw powder particles with a size of several microns experience severe plastic deformation, i.e., undergo a repetitive cold welding and fracturing mechanism. Vibratory mill, planetary mill, uniball mill and attritor mill are commonly used to perform mechanical alloying/milling. Due to mechanical limitations, these methods cannot produce very small particles. However, the advantage of this method is its relative simplicity, low cost, and the possibility to scale it up to tonnage quantities.

8.4.2. Mechanochemical Synthesis

This process involves mechanical activation of solid state displacement reactions in ball mill. Thus, mechanical energy is used to induce chemical reactions (Figure 12). The chemical precursors typically consist of mixtures of oxides, chlorides and/or metals that react either during milling or during subsequent heat treatment to form a composite powder consisting of the dispersion of ultrafine particles within a soluble salt matrix. The ultrafine particle is then recovered by selective removal of the matrix phase through washing with an appropriate solvent. This process has been used successfully to prepare NPs of a number of materials, including transition metals, oxide ceramics, and oxide semiconductors [69].

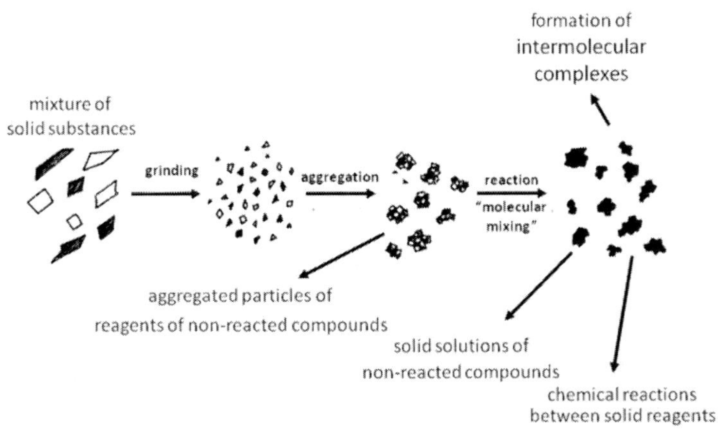

Figure 12. Various steps of mechanochemical synthesis [image adapted from book "The Complex World of Polysaccharides," 2012].

9. APPLICATIONS

9.1. Transparent Conductive Coatings

Coatings, pastes and inks made with nanomaterials can be more conductive due to the increased physical contact between the much smaller particles. Nanocoatings can also potentially be thinner than other kinds of coatings and may dry faster. The greatest potential impact for nanomaterial-based coatings is perhaps as functional inks and pastes, especially nanosilver inks, in high-end transparent conductive coatings. Nanomaterials may also be used to develop conductive coatings for electrodes in next-generation applications that require higher conductivity, lower processing costs and other performance features that only nanomaterials can provide. Not surprisingly, nanosilver is the most common nanomaterial being developed, but products with nanocopper and carbon nanotubes (CNTs) are also being produced, and coatings containing graphene or non-carbon nanostructures are at the research stage. Carbon nanotube coatings can be made transparent and are very strong and flexible.

9.1.1. Coatings for Electrodes

The most important emerging opportunities for conductive coatings are in contacts and electrodes for new types of electronics, optical devices, batteries and photovoltaic panels.

9.1.2. Antistatic Coatings

Chip bags, protective clothing and general antistatic coatings used mainly for optical applications. Tin oxide, for example, is often preferred for window coatings. Conductive polymers also find use in ESD protection systems for clean room environments.

9.2. Batteries

Conductive coatings in batteries provide two kinds of functionality: they are used as a core material in thin-film and printed batteries and in some batteries they can serve as a protective material that increases performance and/or enables the use of more aggressive chemistries. In general, these applications make use of lithium batteries and thus any new applications for conductive coatings in lithium batteries could eventually represent sizeable opportunities, because the volumes of lithium batteries shipped are already large and likely to continue to grow.

9.3. Fuel Cells

Conductive coatings are used in fuel cell applications for acid corrosion protection and electrical conductivity enhancement of bipolar metal plates in much the same way that they are in batteries, however, fuel cells remain in the early development stage. Most coatings are metallic (platinum, palladium, or ferrous nanopowder) and are applied to the electrodes to act as a catalyst. In the future, carbon nanotube coatings may have an important role to play in fuel cells.

9.4. Photovoltaics (PV)

Conductive coatings are used for the contacts and as reflecting layers on various types of solar panels. Thin-film PV (TFPV) producers mainly use ITO, as the transparent electrode. The cost of ITO, however, is driving interest in lower-priced alternatives, including zinc and tin oxide derivatives. Organic (OPV) and dye sensitive cell (DSC) photovoltaics claim niche applications (portable power, embedded power and certain building integrated photovoltaics (BIPV)) represents a significant source of value and opportunities.

9.5. Displays

ITO is the major conductive coating used in LCD display applications, which dominate the display market. Some other TCOs and polymers do find limited application and silver is used in PDP displays. The requirements for electrode materials are also shifting, however, as next-generation displays (e-paper, OLEDs, touchscreens, etc.) begin to take significant shares of the display business.

9.6. Solid-State Lighting

The phasing out of incandescent lighting across the developed world has intensified the interest in energy-efficient solid-state lighting (SSL), especially that using high-brightness (inorganic) or organic LEDS.

9.7. Magnetic Fluid Sealing

Ferrofluids or so-called magnetic liquids are suspensions of colloid magnetic particles stabilised by surfactants in liquid media [82]. Ferrofluids have been around since the 1960s when the uses were many, such as audio speaker coolants and high-end engineering seals. More recently, researchers are applying ferrofluids to cancer treatment and drug delivery systems. Cancer treatment has been proposed through methods utilizing the heating effect of alternating magnetic fields that the energy lost from such cycling. For drug delivery systems, magnetic drugs with a suitable surfactant are injected into the blood stream and manipulated with external magnetic fields for localizing treatment. These medical treatment advancements are only limited by the ability to create nontoxic, biologically compatible, magnetic particles.

9.8. Nanomaterials in Cosmetics

Nanomaterials in cosmetics are nanoemulsion, nanosomes and nanopigments. These nanomaterials are used for example in sun screens, skin creams and oral hygiene products. Nanoemulsions are macroscopic preparations which contain droplets of water and oil reduced to the nanometric size to increase the content of nourishing oil and to preserve the transparency and the lightness of the formula. Sometimes active ingredients that are fragile in air such as vitamins are protected and kept fresh inside nanometric bubbles (capsules) or nanosomes (liposomes). The active ingredients are released upon contact with the skin at the time of the application where the nanoemulsions and nanosomes.

Nanopigments are minerals already present in our natural environment, in the form of clay and sand for example. Titanium dioxide (TiO_2) is one of the best known. It is a totally insoluble, inert material. It is a reference of non-toxicity which is why we find it used widely in foodstuffs (colouring agent E171) as well as dental/oral hygiene products including toothpaste. Another well known is zinc oxide (ZnO). Both TiO_2 and ZnO are used in sunscreens because they are known to reflect and scatter UV light, and protect skin against adverse effects of UV light, including skin cancers. In sun creams titanium dioxide is present in the form of aggregates and agglomerates (size much greater than 100 nm) on the surface of the skin. Carbon black, an intense cosmetic colorant, can be used in the nano-form and is a good example of how reducing the pigment particle-size can alter the strength and opacity of colour.

9.9. Biomedical Applications

Magnetic NPs coated with biological molecules interact with or bind to a biological entity, thereby providing a controllable means of "Tagging" or addressing it. The magnetic NPs can be easily manipulated by an external magnetic field gradient. This 'action at a distance' combined with the intrinsic penetrability of magnetic fields into human tissue, opens up many applications involving the transport or immobilization of magnetic NPs or of magnetically tagged biological entities. In this way, they can be made to deliver a package, such as an anticancer drug to a targeted region of the body such as a tumor.

9.10. Spintronics

Similar to electrical charge, spin is another fundamental property of matter. While conventional electronic devices rely on the transport of electrical charge carriers, the emerging technology of spintronics employs the spin of electrons to encode and transfer information. Spintronics has the potential to deliver nanoscale memory and logic devices which process information faster, consume less power and store more data in less space. The extension of the hard disk capacities to the gigabyte and the terabyte ranges was the main achievement of spintronics by taking advantage of Giant Magneto-Resistance (GMR) and Tunnel Magneto-Resistance (TMR) effects which are effective only at the nano scale.

9.11. Electronics

In current research, the "hot topic" nanomaterial for potential electronic applications is graphene, the subject of the 2010 Nobel Prize in Physics. Graphene is a single atomic layer of graphite (an allotrope of carbon), in various shapes (e.g., "nanoribbons") and orientations. Graphene has amazing properties, in part, stemming from a band structure that exhibits linear, rather than the usual quadratic, dependence of energy on momentum (i.e., like a "relativistic" particle). Its potential uses include ultracapacitors [83], transparent conductive electrodes for PV (replacement for expensive indium tin oxide) [84], various forms of transistors, and many more. Although cost reduction would be the major benefit of using a nanomaterial in some cases, enhancement of material properties is generally the primary objective. Another electronics-industry example of the latter is improving the electrical and thermal conductivity of bonding materials, such as in packaging applications [85]. A long-pursued example is the use of NPs in nonvolatile memory [86].

9.12. Carbon-Based Nanosensors

In addition to the exceptional electrical properties of graphene and carbon nanotubes, their excellent thermal conductivity, high mechanical robustness and very large surface to volume ratio make them superior materials for fabrication of electromechanical and electrochemical sensors with higher sensitivities, lower limits of detection and faster response time. A good example is the carbon nanotube-based mass sensor that can detect changes in mass caused by a single gold atom adsorbing on its surface.

9.13. Carbon Nanotube Electronics

When a layer of graphene is rolled into a tube, a single-walled carbon nanotube (SWNT) is formed. Consequently, SWNTs inherit the attractive electronic properties of graphene but their cylindrical structure makes them a more readily available option for forming the channel in field-effect transistors. Such transistors possess an electron mobility superior to their silicon-based counterpart and allow for larger current densities while dissipating the heat generated from their operation more efficiently. During the last decade, carbon nanotube-based devices have advanced beyond single transistors to include more complex systems such as logic gates and radio-frequency components.

9.14. Cancer Treatment

Nanotechnology is being applied to cancer in two broad areas: the development of nanovectors, such as NPs, which can be loaded with drugs or imaging agents and then targeted to tumours, and high-throughput nanosensor devices for detecting the biological signatures of cancer (Figure 13). Combined, such technologies could lead to earlier diagnosis and better treatment for patients with cancer. Using iron-oxide NPs and a magnetic field to heat up cancer tumors has been shown to stimulate the immune system to fight cancer cells in other parts of the body. This methodology may be useful in preventing the spread of cancer cells, while other techniques are used to fight localized tumors.

The NPs are attracted to a protein present on many types of cancer tumors, resulting in a high rate of delivery of the chemotherapy drug to the tumors. Researchers are testing the use of chemotherapy drugs attached to nanodiamonds to treat brain tumors. The nanodiamond/chemotherapy drug combination stays in the tumor longer than the chemotherapy drug by itself, which should increase the effectiveness. Researchers are also testing the use of chemotherapy drugs attached to nanodiamonds to treat leukemia.

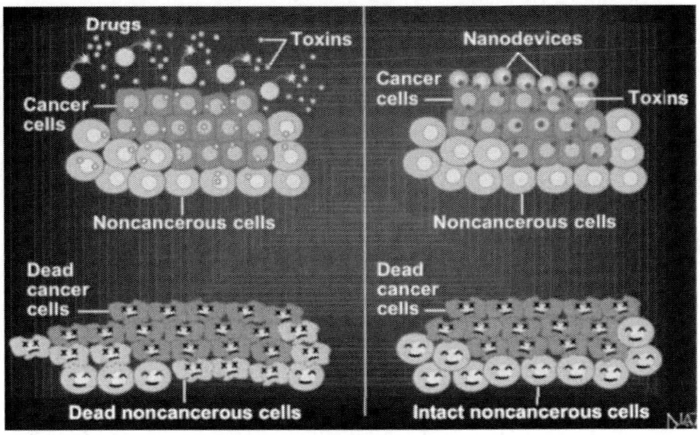

Figure 13. Nanomedicine can detect the cancer cells and deliver the toxin in a controlled, time-released manner [image adapted from www.cancer.gov].

9.15. Environmental Remediation

Nanomaterials offer the potential for the efficient removal of pollutants and biological contaminants. Nanomaterials in various shapes/morphologies, such as NPs, tubes, wires, fibres *etc.*, function as adsorbents and catalysts and their composites with polymers are used for the detection and removal of gases (SO_2, CO, NO_x, etc.), contaminated chemicals (arsenic, iron, manganese, nitrate, heavy metals, etc.), organic pollutants (aliphatic and aromatic hydrocarbons) and biological substances, such as viruses, bacteria, parasites and antibiotics. Nanomaterials show a better performance in environmental remediation than other conventional techniques because of their high surface area (surface-to-volume ratio) and their associated high reactivity. Recent advances in the fabrication of novel nanoscale materials and processes for the treatment of drinking water and industrial waste water contaminated by toxic metal ions, radionuclides, organic and inorganic solutes, bacteria, viruses and the treatment of air are highlighted (Figure 14).

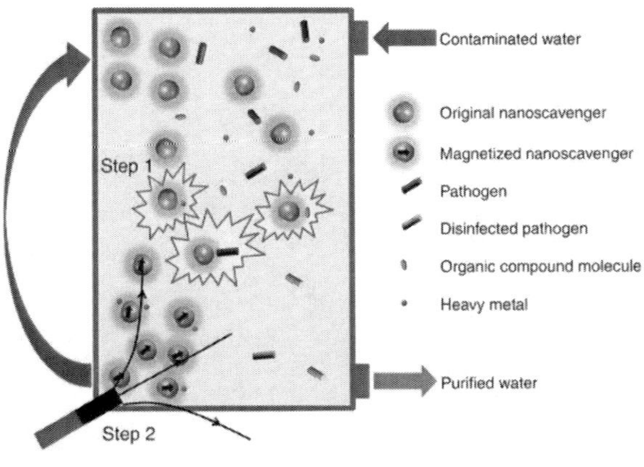

Figure 14. Schematic representation of new water purification process.

10. ROLE OF NANOMATERIALS IN THE THERMAL DECOMPOSITION OF AP AND CSPS

Generally, the bulk size transition and non-transition metal oxides, spinel oxides and metal fluorides have been widely used as catalyst to improve thermal decomposition of AP. But at their present size, these catalysts do not meet those objectives, as their addition beyond an optimum limit, leads to affect the process ability and the integrity of the propellant. Catalytic activity being a surface property generally increases with the increase of surface area of the catalysts. Hence, for achieving a large surface area, new classes of catalysts based on ultrafine particles (nanoparticles) were developed [88, 89]. With reduced particle size, these NPs have been found to be more efficient and active in thermal decomposition of AP than the commonly used bulk catalysts [90, 91]. Although, a lot of work on the decomposition of AP in the presence of NPs has been performed in the last decade, the amount of heat release, activation energy, and mechanism of decomposition of AP remain as key issues which depend mainly on particle size and particle agglomeration. Many workers have investigated the thermal decomposition of AP in the presence of NPs which has been discussed under following categories:

10.1. Nano Transition and Lanthanoid Metal Oxides

Metal oxides and their mixtures are most efficient and commonly used burn rate modifiers for CSPs. Several researchers have studied the effect of CuO NPs on the AP thermal decomposition and results show enhanced catalytic activity as compared to their bulk size. Xu et al. [92] demonstrated that CuO NPs exhibit a particular chemical reactivity due to their high concentration of dislocations and large surface areas. In addition, the different morphologies of NPs are expected to have different effects on the thermal decomposition of AP [93]. Author showed that the addition of nanosized CuO merges two exothermic peaks of AP, into one peak, greatly reducing the HTD peak and drastically increasing the apparent decomposition heat of AP.

The catalytic action of nickel oxide on thermal decomposition of AP is similar to the action of copper oxides. As compare to bulk size, the effect of NPs (~ 4 nm) of nickel oxide on thermal decomposition of AP was studied by Singh et al. [94]. According to their investigation catalytic activity enhanced ~4 times and decomposition rate of AP is increased. A comparison of catalytic effect of bulk size NiO to nanosize NiO in thermal decomposition of AP was studied by Wang et al. [95]. Author showed that complete decomposition of AP with nanosize NiO (~10nm) is done approximately at 350°C while in presence of bulk sized NiO, it occurs at 400°C. It was also shown that by addition of 2 wt % of NiO NPs to AP, the heat of the reaction increases from 580 to 1490 J/g.

Iron (III) oxide is a catalyst of thermal decomposition of AP mainly at high temperature (320–380°C). The efficiency of catalytic action of ferric oxide increases sharply when passing from micro-scale of the size of catalyst particles to the nano-size. Many properties of nanosized α-Fe_2O_3 such as high surface area, large number of surface atoms, oxygen vacancies on its surface enable it as an excellent catalyst. Above all, it is structurally simple, highly stable, easy to synthesize and inert to side reactions. Patil et al. [96, 97] have studied the catalytic effect of two different sizes of α-Fe_2O_3 (30 and 3.5 nm) nanoparticles on the thermal decomposition of AP. The 3.5 nm α-Fe_2O_3 particles showed pronounce effect in lowering the high temperature decomposition as 40°C (0.5 wt %), 35°C (1 wt %), 59°C (2 wt%) and 77°C (5 wt%) compared to pure AP. Such a marked reduction in high temperature exotherm is attributed to the presence of large number of active sites, higher surface area and smaller particle size of synthesized α-Fe_2O_3. Authors have also investigated that the thermal decomposition of AP is enhanced with increased percentage of Fe_2O_3.

Investigation of Xu. et al. [98] shows that different shape of additives also effect the thermal decomposition of AP. For nanorods, irregular particles and micro-octahedrons of α-Fe_2O_3, the temperature shift for high temperature exothermic peaks of AP was 48.9, 16.8, 10°C, respectively. Therefore, α-Fe_2O_3 nanorods exhibit better catalytic activity than other two samples, while the catalytic performance of irregular particles is slightly better than that of micro-octahedron. The particle with smaller size possess better performance and more active sites, which can further enhance the rate of heterogeneous decomposition of deprotonized $HClO_4$ gas on the surfaces of catalyst particles. Oxides of Cobalt are also studied as catalyst in thermal decomposition of AP [99]. When AP was mixed with CoO nanocrystals, the HTD process disappeared completely.

A comparative investigation on catalytic effect of synthesized NPs and commercial MgO on thermal decomposition of AP was done by Duan et al. [100]. The result shows MgO NPs has no effects on the crystallographic transition temperatures, but decreases the area of endothermic peak. The number of exothermic peak decreases from two peaks to one peak. The shapes of exothermic peaks also become narrow and sharp. The areas of exothermic peaks, which indicate the decomposition heat of AP, increase on a large scale. Comparing with commercial product, nanosized MgO showed clearly a good catalytic activity on the decomposition of AP following their decomposition heat value. The area of exothermic peak of nanosized MgO is about 20.1 units; but that of commercial is about 7.5 units. These results indicate that the catalytic decomposition effect on AP of nanosized MgO is better than that of commercial product.

The catalytic activity of mixed valent manganese oxide octahedral molecular sieve (OMS) on the thermal decomposition of AP also has been studied [101]. Results show that OMS plays a catalytic role on the thermal decomposition of AP. The decomposition temperature of AP has been decreased to 145°C and the heat of decomposition has been increased to 0.66 kJ/g with the addition of 2 wt% OMS. Such effects are mainly attributed to the mixed valence of Mn^{3+} and Mn^{4+} existing in OMS material and its high surface area.

Singh et al. [94] investigated the catalytic activity of MnO_2 NPs which enhanced the thermal decomposition of AP as compare to their bulk size. Catalytic effect of ZnO on thermal decomposition of AP also studied [102]. Investigation revealed that N-doped ZnO sample with peculiar morphology drives the thermal decomposition peak of AP decrease about 163°C with a strong heat of decomposition about 1.325 J/g, and the activation energy also decreases from 178.22 to 93.51 kJ/mol. The enhanced catalytic activity of N-doped ZnO sample can be attributed to oxygen vacancies and other defects induced by the doping of nitrogen. Singh et al. [103] have reported the synthesis of cadmium oxide (CdO) NPs and their characterization by XRD and TEM (Figure 3.28). The catalytic effect of CdO NPs on thermal decomposition of AP, HTPB and CSPs have been studied. Experimental results confirm that decomposition temperature of pure AP, HTPB and CSPs are lowered by incorporation of CdO NPs (Figure 15 and 16).

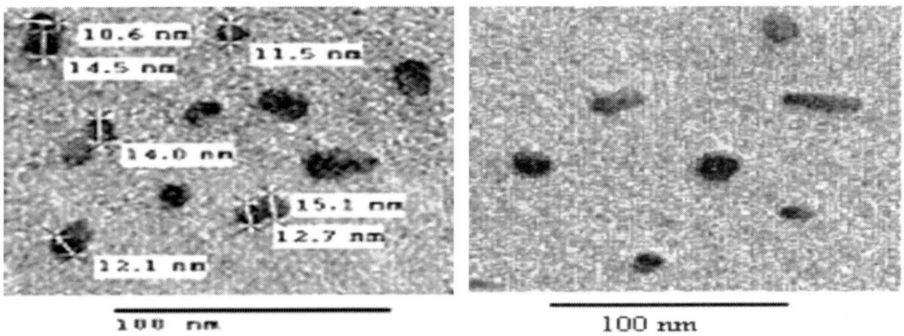

Figure 15. TEM images CdO NPs. [Reprinted with permission from (Singh et al. [103]), *Materials Science and Engineering* B 2011; ©2011, Elsevier Publisher].

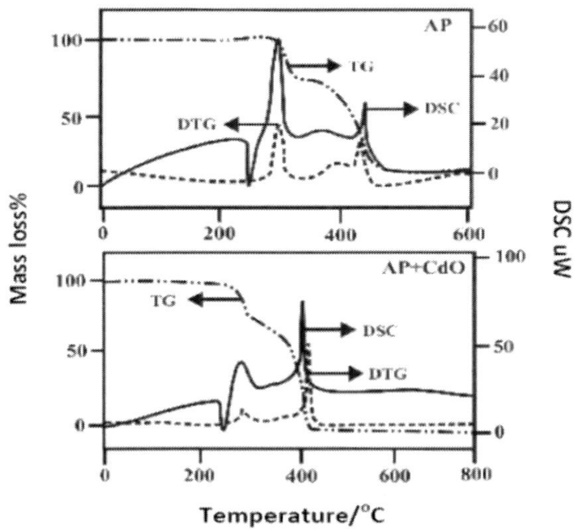

Figure 16. TG-DSC curves of AP with and without CdO NPs. [Reprinted with permission from (Singh et al. [103]), *Materials Science and Engineering* B 2011; ©2011, Elsevier Publisher].

The effect of 'p' type rare earth oxides with partially filled 'f' orbitals like La_2O_3, Pr_2O_3 and Nd_2O_3 on the thermal decomposition of AP has been studied [104]. DTA and TG results reveal that these oxides influence the thermal decomposition pattern of AP significantly and bring down the decomposition temperature substantially. Isothermal studies showed substantial acceleration of the decomposition of AP with the incorporation of these oxides as little as 0.05%. Isothermal data were analyzed using various kinetic models. The activation energy for the catalyzed decomposition of AP was significantly lower. The catalytic effect is explained on the basis wherein the 'p' type oxides act as conduits through metal cation in the electron-transfer mechanism for AP decomposition. Catalytic effect of NPs of perovskite-type oxides $LaMO_3$ (M = Fe, Co, Ni) were also studied [105].

Result showed that heat of decomposition of pure AP is 0.6 kJg^{-1} while in presence of $LaFeO_3$, $LaNiO_3$ and $LiCoO_3$ heat of decomposition of AP is 1.0, 1.1 and 1.4 kJ g^{-1} respectively. This shows that $LaCoO_3$ is obviously a more effective catalyst than $LaFeO_3$ and $LaNiO_3$.

The addition of $LaCoO_3$ lowers the decomposition temperature of AP and increases the speed of decomposition and heat of decomposition reaction. Catalytic activity of $LiCoO_3$ on the burning rate of AP-based propellants was also studied. Adding 1% $LaCoO_3$ nanocrystals to basic propellant; the burning rate of the modified propellant was increased around 8%.

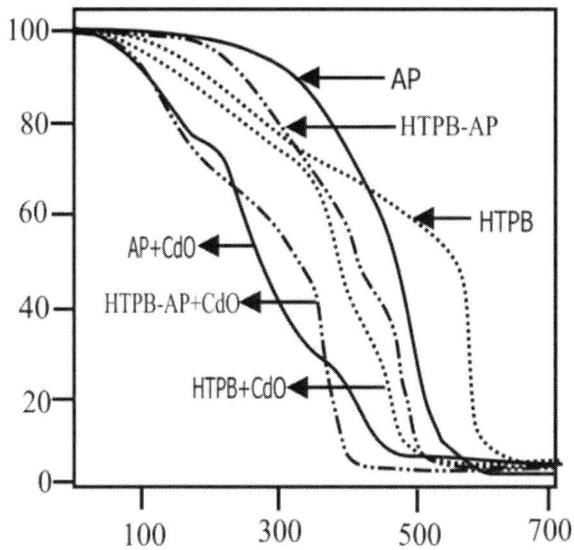

Figure 17. TG curves of AP, HTPB, HTPB-AP with and without CdO NPs. [Reprinted with permission from (Singh et al. [103]), *Materials Science and Engineering* B 2011; ©2011, Elsevier Publisher].

Singh et al. [106] have prepared rare earth metal oxide NPs (CeO_2, Pr_2O_3, and Nd_2O_3) by sol–gel route using $Ce(NO_3)_3.6H_2O$, $Pr(NO_3)_3.6H_2O$ and $Nd(NO_3)_3.6H_2O$ and citric acid as precursor materials. Powder X-ray diffraction, SEM and TEM are employed to characterize the size and morphology of rare earth metal oxide NPs (REMONPs). The particles are spherical in shape and the average particle size is of the order of 11-30 nm Figure 3.31). Their catalytic activity was measured on the thermal decomposition of AP and CSPs by TG-DSC. Addition of these REMONPs to AP led to shifting of the HTD peak toward lower temperature (Figure 17).

As the REMONPs were added to AP, the endothermic peak appears at 244°C, indicating that catalyst has no effect on the crystallographic transition temperature but noticeable changes were observed in first and second exothermic peaks. When Pr_2O_3 was added to the AP, the first exothermic peak shifted from 310°C to 294°C and second exothermic peak shifted significantly from 414°C to 374°C and when Nd_2O_3 was added to AP, the first exothermic peak shifted from 310°C to 294°C and second exothermic peak shifted from 414°C to 405°C but when CeO_2 was added to AP, the first exothermic peak shifted from 310°C to 289°C and the second exothermic peak shifted from 414°C to 407°C and the order of catalytic activity is $CeO_2 > Pr_2O_3 > Nd_2O_3$.

The data reported in Figure 18 clearly show that the rate of decomposition of AP and CSPs is also enhanced in presence of REMONPs having finer particle size and presence of positive hole in F-orbital. Thus, a large number of active sites would be available for the absorption of reactants as a consequence; the rate of reaction would be increased. Singh et al. [106] have studied the kinetics of thermal decomposition of AP with and without NPs using isothermal TG data.

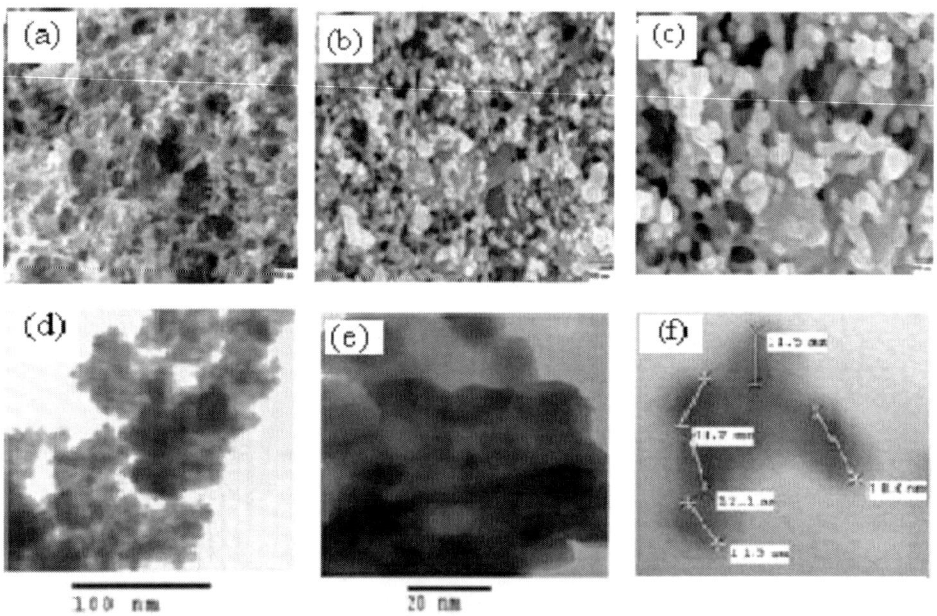

Figure 18. FE-SEM image of (a) CeO_2 (b) Pr_2O_3 (c) Nd_2O_3 and TEM image of (d) CeO_2 (e) Pr_2O_3 (f) Nd_2O_3. [Reprinted with permission from (Singh et al. [106]), *J. Therm. Anal. Calorim.*2013; ©2013, Springer International Publishing AG].

10.2. Kinetic analysis of isothermal TG data

The single-step kinetic equation (3.1) describes the kinetic analysis of thermal decomposition of a solid [107]

$$d\alpha/dt = k(T)f(\alpha) \tag{3.1}$$

where t, T, α, k(T) and f(α) denote the time, temperature, extent of conversion ($0 < \alpha < 1$), rate constant, and reaction model [107] respectively. This equation correlates the dependence of the reaction rate on the extent of reactions. The reaction model f(α) (Table 1) may take various forms.

Table 1. Set of reaction models applied to describe thermal decomposition of solids

S. No.	Reaction model	Differential form f(α)	Integral form g (α)	Rate-controlling process
1.	Power law	$4\,\alpha^{3/4}$	$\alpha^{1/4}$	-
2.	Power law	$3\,\alpha^{2/3}$	$\alpha^{1/3}$	-
3.	Power law	$2\,\alpha^{1/2}$	$\alpha^{1/2}$	-
4.	Power law	$2/3\,\alpha^{-1/2}$	$\alpha^{3/2}$	-
5.	One-dimensional diffusion	$1/2\,\alpha^{-1}$	α^2	One-dimensional diffusion
6.	Mampel (first order)	$1-\alpha$	$-\ln(1-\alpha)$	Random nucleation, one nucleus on each Particle
7.	Avrami-Erofeev	$4(1-\alpha)[-\ln(1-\alpha)]^{3/4}$	$[-\ln(1-\alpha)]^{1/4}$	Random nucleation
8.	Avrami-Erofeev	$3(1-\alpha)[-\ln(1-\alpha)]^{2/3}$	$[-\ln(1-\alpha)]^{1/3}$	Random nucleation
9.	Avrami-Erofeev	$2(1-\alpha)[-\ln(1-\alpha)]^{1/2}$	$[-\ln(1-\alpha)]^{1/2}$	Random nucleation
10.	Contracting sphere	$3(1-\alpha)^{2/3}$	$1-(1-\alpha)^{1/3}$	Three-dimensional
11.	Three-dimensional diffusion	$2(1-\alpha)^{2/3}[1-(1-\alpha)^{1/3}]^{-1}$	$[1-(1-\alpha)^{1/3}]^2$	Three-dimensional diffusion, Spherical symmetry
12.	Contracting cylinder	$2(1-\alpha)^{1/2}$	$1-(1-\alpha)^{1/2}$	Three-dimensional
13.	Prout-Tomkins	$\alpha\,(1-\alpha)$	$\ln(\alpha / 1-\alpha)$	Chain growth nuclei
14.	Ginstling-Brounshtein	$3/2[(1-\alpha)^{-1/3}-1]^{-1}$	$[1-(2\alpha/3)]-(1-\alpha)^{2/3}$	Three dimensional diffusion spherical symmetry

If we see the temperature dependence of k(T) then we have to substitute k(T) by Arrhenius equation in the above equation after which it takes the following form;

$$d\alpha/dt = A\exp(-E/RT).f(\alpha) \tag{3.2}$$

where A is pre exponential factor, E activation energy and R the gas constant.

Model Fitting Method

Rearrangement and integration of Eq. (3.1) for isothermal conditions gives

$$g_j(\alpha) = k_j(T)\,t \tag{3.3}$$

where $g(\alpha) = {}_0\int^{\alpha} [f(\alpha)]^{-1} d\alpha$ is the integrated form of the reaction model (Table 1).

The subscript j has been introduced to emphasize that substituting a particular reaction model in Eq. (3.3) results in evaluating the corresponding rate constant, which is determined from the slope of a plot of $g_j(\alpha)$ verses t. For each reaction model selected, the rate constants are evaluated at several temperatures T_i and Arrhenius parameters are determined using the Arrhenius Eq. (3.4) in its logarithmic form

$$\ln k_j (T_i) = \ln A_j - E_j/RT_i \tag{3.4}$$

The Arrhenius parameters were evaluated from isothermal TG experimental data by the model fitting method data is reported in the Table 2.

Table 2. Activation energy (E_a), Arrhenius parameters and correlation coefficient (r) for the isothermal decomposition of AP and AP with REMONPs

Model	AP		AP+CeO_2		AP+Pr_2O_3		AP+Nd_2O_3	
	E_a (kJmol^{-1})	r	E_a (kJmol^{-1})	r	E_a kJmol^{-1})	r	E_a (kJmol^{-1})	r
1	95.9	0.9609	64.4	0.9644	61.1	0.9711	60.6	0.9703
2	96.3	0.9612	65.6	0.9651	61.4	0.9715	60.9	0.9707
3	97.2	0.9598	66.3	0.9661	62.2	0.9692	61.7	0.9692
4	100.1	0.9634	68.7	0.9728	65.3	0.9762	64.8	0.9749
5	101.1	0.9636	69.8	0.9749	66.7	0.9775	66.2	0.9762
6	99.5	0.9630	68.3	0.9718	64.9	0.9756	64.3	0.9744
7	98.5	0.9633	66.9	0.9686	63.5	0.9732	63.0	0.9723
8	97.0	0.9616	66.2	0.9667	62.2	0.9724	61.6	0.9715
9	97.7	0.9620	66.8	0.9681	62.9	0.9733	62.4	0.9723
10	99.2	0.9629	68.1	0.9712	64.5	0.9752	64.0	0.9741
11	101.7	0.9634	79.7	0.9761	67.6	0.9783	67.1	0.9768
12	95.1	0.9629	67.9	0.9709	64.3	0.9750	63.8	0.9739
13	95.9	0.9607	65.5	0.9757	61.2	0.9712	60.7	0.9704
14	101.5	0.9635	70.4	0.9646	67.3	0.9780	66.8	0.9766

10.2. Isoconversional Method

This method allows the activation energy to be evaluated without making any assumptions about the reaction model. Additionally, the method evaluates the effective activation energy as a function of the extent of conversion which allows one to explore multistep kinetics. The basic assumption of the isoconversional method [108] is that the reaction model as defined in Eq. (3.1) is not dependent on temperature or heating rate. Under isothermal conditions, on combining Eqs. (3.3) and (3.4) we get,

$$-\ln t_{\alpha,i} = \ln [A_\alpha/g(\alpha)] - E_\alpha/RT_i \tag{3.5}$$

where E is evaluated from the slope of the plot of $-\ln t_{\alpha}i$ against Ti^{-1}.

The kinetics of the thermal decomposition of AP and AP+REMON (1%by wt) was evaluated using 14 mechanism based kinetic models discussed above. In the model fitting method, the kinetics is analysed by choosing a "best fit" model based on the value of the correlation coefficient 'r' close to 1. Among various values of r calculated for different models, the highest value of r for AP corresponds to model 5 while for the AP with CeO_2, Pr_2O_3 and Nd_2O_3 corresponds respectively to model 11, 11 and 14. Isothermal TG data best fitted for model 5, 11 and 14 proves the contention that the rate controlling step in the thermal decomposition may be respectively one dimensional diffusion, three dimensional diffusion; spherical symmetry and three dimensional diffusion; spherical symmetry. The corresponding values of E_a were reported for AP, AP+CeO_2, AP+Pr_2O_3 and AP+Nd_2O_3 are, respectively, 101.1, 67.2, 68.3 and 77.8 kJ mol^{-1} (Table 2). Kinetic analysis performed by the isoconversional method on thermogravimetric data as shown in Figure 19. The thermal decomposition of AP and AP+REMONs have initial overall activation energies are 123.94, 53.39, 55.45 and 59.63 kJ mol^{-1} respectively. These values changes with the extent of conversion to about 106.42, 65.11, 63.32 and 62.35 kJ mol^{-1} respectively at the end of reactions. Catalysed thermal decomposition of AP by REMONPs is also consistent with the fact that at each α, the value of E_a is lower as compared to the pure AP.

10.3. Ignition (D_i)/ Explosion Delay (D_E) Measurements

Freeman and Gordon [109] have suggested the following heat balance equation in order to evaluate the pre-ignition reactions.

$$dQ_H/dt = dH/dt - dq/dt \qquad (3.6)$$

where dQ_H/dt is the net rate of heat gain in the system. dH/dt is the rate of heat produced by pre-ignition reactions and dq/dt is the rate of heat dissipation.

Ignition will occur when,

$$Q_H = H' \qquad (3.7)$$

where H'= minimum amount of heat required to raise the temperature of the system to the point of ignition. From equation (3.7), it turns out that

$$Q_H = H' - q \qquad (3.8)$$

where H = total heat produced by pre-ignition reactions. The ignition would occur only if $H - q \geq H''$.

The total heat produced by the pre-ignition reactions must be greater than H' by the amount of heat dissipated. Thus

$$H = H' + q \tag{3.9}$$

The following equation was derived by Freeman and Gordon [174].

$$t_{id} = Ae^{\Delta H'/RT} \tag{3.10}$$

where ΔH* is the heat of activation and is approximately equal to activation energy (E_a).

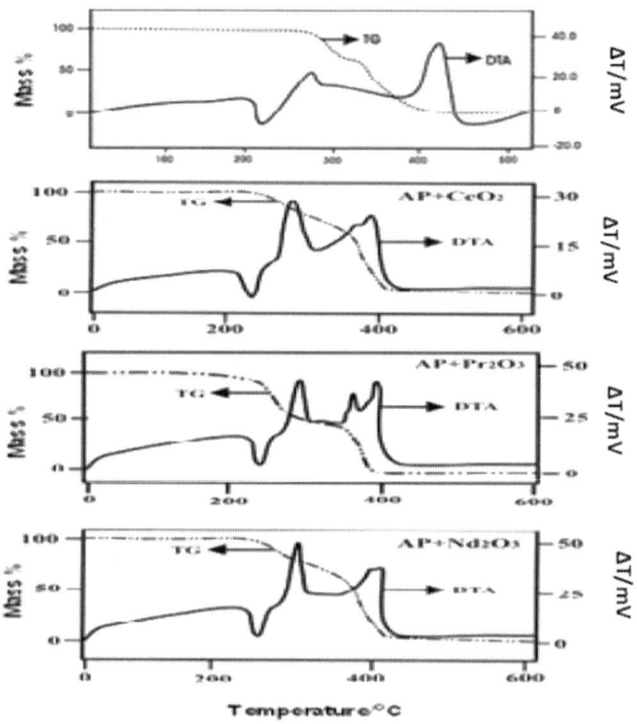

Figure 19. TG-DTA Thermograms of AP with and without REMONPs. [Reprinted with permission from (Singh et al. [106]), *J. Therm. Anal. Calorim.*2013; ©2013, Springer International Publishing AG].

D_i/D_E measurements have been carried out using tube furnace (TF) technique [110]. In these experiments, an ignition tube has been inserted manually into the TF and the time interval between the insertion of the sample tube into the TF and the moment of an audible explosion, noted with the help of a stop watch. Thus, this gives the value of D_i/D_E. The time taken for insertion of the ignition tube has been kept constant throughout the experiments. The D_i/D_E data were found to fit in the equation 3.11 [109].

$$D_i / D_E = A \exp (E_a / RT) \tag{3.11}$$

The activation energy for ignition in the temperature range of 360-420°C for AP and HTPB-AP with and without REMONs are reported in Table 3. The energies of activation for activation (E_a) are lower for AP+REMONs (28-30 kJ mol^{-1}) and AP+HTPB+REMONPs (21-27 kJ mol^{-1}) compared to AP (40.16kJ mol^{-1}) and HTPB+AP (37.73 kJ mol^{-1}). The ignition delay depends exponentially on temperature. The process of ignition [111] can never be treated as steady-state since it is a transient process prior to sustained combustion. Activation energies, calculated under different temperature ranges for isothermal kinetics and ignition delay measurements are different. From Table 3, it is clear that both ignition delay and activation energy for thermal ignition for AP is lowered by these NPs which infer high catalytic activity.

Table 3. Ignition delay (D_i), activation energy for ignition (E_a) and correlation coefficient of AP and AP+REMONPs

Sample	D_i (Sec.)					E_a	*r*(Correlation
	360±1°C	375±1°C	390±1°C	405 ±1°C	420±1°C	(kJ/mol)	coefficient)
AP	125	112	95	84	63	40.16	0.9785
AP+CeO_2 (1%)	72	67	55	49	47	28.43	0.9806
AP+Pr_2O_3	79	67	59	53	48	30.02	0.9967
AP+Nd_2O_3	85	73	68	55	52	30.82	0.9864
HTPB-AP	92	86	67	64	49	37.73	0.9724
HTPB-AP+CeO_2 (1%)	54	50	41	38	35	27.83	0.9860
HTPB-AP+Pr_2O_3 (1%)	57	53	46	42	37	26.65	0.9945
HTPB-AP+Nd_2O_3(1%)	59	55	49	45	42	21.43	0.9967

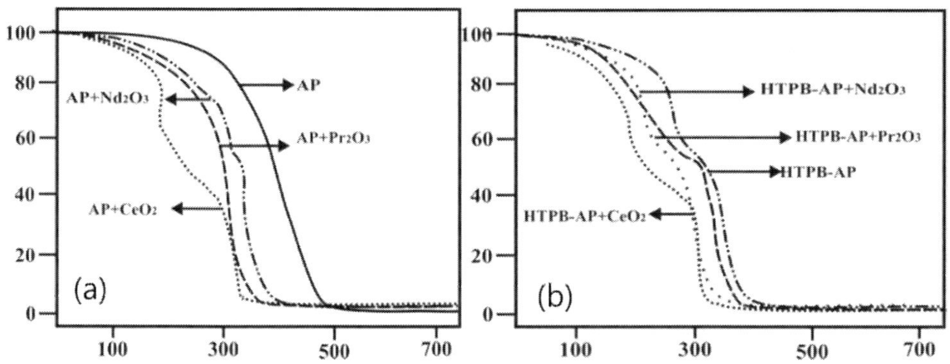

Figure 20. (a) Non-isothermal TG curves of AP and (b) CSPs with and without REMONPs. [Reprinted with permission from (Singh et al. [106]), *J. Therm. Anal. Calorim.*2013; ©2013, Springer International Publishing AG].

Burning rate of CSP, CSP+ CeO_2, CSP+ Pr_2O_3 and CSP+ Nd_2O_3 has been found to be 1.00, 1.68, 1.43 and 1.19. CSP+ CeO_2 imparts a two fold enhancement in burning rate and is the best catalyst followed by Pr_2O_3 and Nd_2O_3. Yu et al. studied [112] the catalytic effect of orthorhombic structural perovskite $NdCrO_3$ nanocrystals with size of 60 nm on thermal decomposition of AP. The results revealed that by adding 2% of $NdCrO_3$ NPs to AP decreased the temperature of thermal decomposition by 87° C and increased the heat of decomposition from 590 to 1073 J g^{-1}.

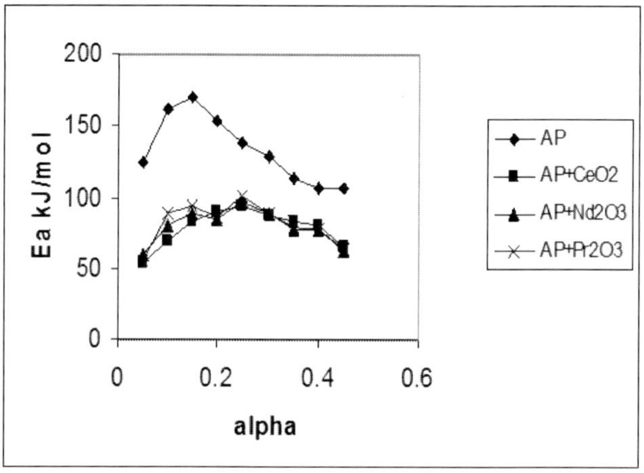

Figure 21. Plot of E_a versus α. [Reprinted with permission from (Singh et al. [106]), *J. Therm. Anal. Calorim.*2013; ©2013, Springer International Publishing AG].

10.4. Transition Metal Ferrites Nanomaterials

Ferrites (ferrospinels or spinel oxides) are of considerable interest due to their diverse applications in optical, electronic, catalytic, magnetic materials etc [113]. Ferrite spinels may also contain mixture of two divalent metal ions, in which ratio, of these divalent metal ions may vary, are called mixed ferrite. The cations distribution of mixed ferrite significantly affects the surface properties of ferrospinels making them catalytically active. Because of their small size and large number of cations, for co-ordination sites, nano sized ferriets are capable of enhancing the rate of chemical reactions. Singh et al. [114, 117] studied catalytic activity of binary, ternary and quaternary ferrites NPs in AP thermal decomposition. Ferrites affect both the LTD and HTD processes and further gasification of AP in presence of ferrite during HTD process not only begins early but also complete at lower temperature.

Singh et al. [118] synthesized nanoferrites of Mn, Co, and Ni with various morphologies such as nanorods, nanospheres and nanocubes by wet-chemical method under different synthesis conditions. These were characterized by XRD, FE-SEM, EDS, TEM and HR-TEM. The average size of MnF, CoF and NiF, estimated from the Scherrer equation, are 26.4, 27.4 and 31.4 nm respectively. The morphology of these three ferrites obtained under different synthetic conditions are presented in Figure 22 (a)–(c). From the

FE-SEM images, it is found that the morphologies of nanoferrites are extremely different shapes (with dimension 10–50 nm) of rod, sphere and cube respectively.

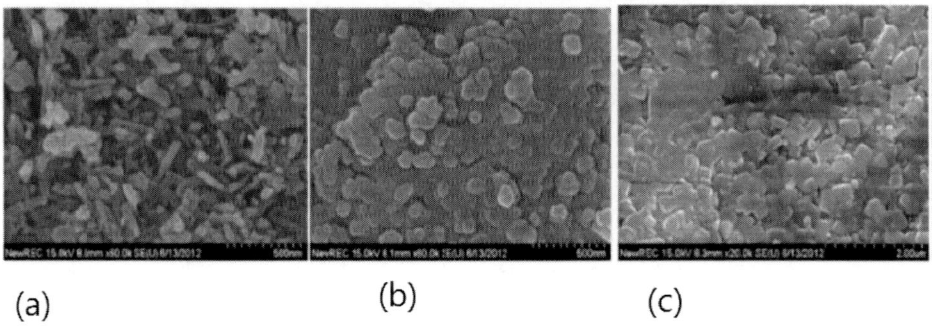

Figure 22. FE-SEM image of (a) MnF nanorods (b) CoF nanospheres (c) NiF nanocubes. [Reprinted with permission from (Singh et al. [118]), *Materials Research Bulletin* 2013; ©2013, Elsevier Publisher].

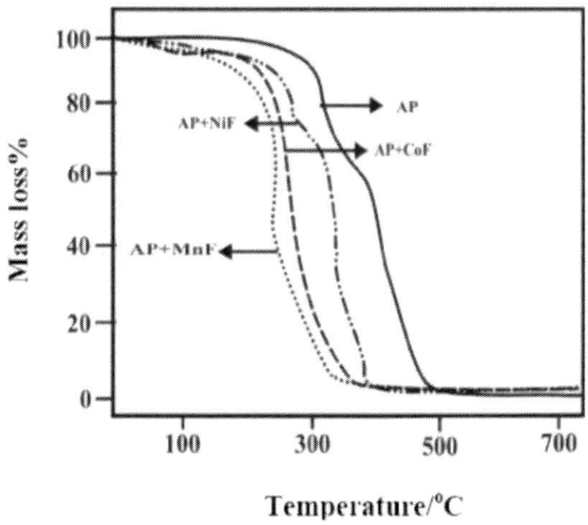

Figure 23. Non-isothermal TG curves of AP with and without MnF, CoF and NiF. [Reprinted with permission from (Singh et al. [118]), *Materials Research Bulletin* 2013; ©2013, Elsevier Publisher].

Burning rate of CSP with and without nanoferrites has been measured which show that burning rate is enhanced when nanoferrites are used as catalysts. Increase of burning rate might be due to the enhanced decomposition of HTPB or AP or CSP. These nanoferrites showed good catalytic activity in thermal decomposition of AP.

TG curve (Figure 23) for AP with nanoferrites confirm beyond doubt that nanoferrites affect both LTD and HTD of AP, to form gaseous products. When MnF, CoF and NiF were added to AP, the LTD shifted from 310^{0}C to 302^{0}C, 301^{0}C and 294^{0}C respectively and HTD significantly shifted from 430^{0}C to 335^{0}C, 344^{0}C and 357^{0}C respectively (Figure 24). MnF rods are expected to poses better catalytic properties on thermolysis of AP by predominantly exposing the reactive planes and greater number of welldefined active sites.

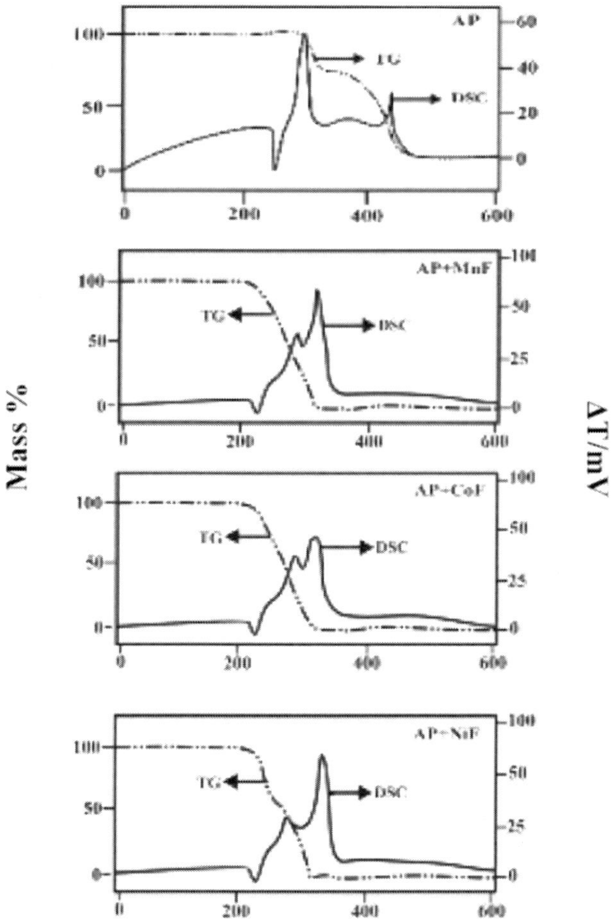

Figure 24. TG-DSC Thermograms of AP with and without nanoferrites. [Reprinted with permission from (Singh et al. [118]), *Materials Research Bulletin* 2013; ©2013, Elsevier Publisher].

Zhao and Ma [119] synthesize $CoFe_2O_4$ nanocrystallites by polyol-medium solvo thermal method and study their catalytic effect on the thermal decomposition of AP. The catalytic activity is found to be enhanced with the increase of content of $CoFe_2O_4$ nanocrystallites. The kinetic parameters such as the decrease in activation energy and increase in rate constant further confirmed the enhancement in the catalytic activity of AP. Based on the proton transfer process, a mechanism has been proposed for the thermal decomposition of AP in the presence of $CoFe_2O_4$ nanocrystallite.

10.5. Transition Metal Oxalates Nanomaterials

Metal oxalates are interesting candidates for nanostructured synthesis because these can be transformed to oxide or metals without losing the ordered structure [120]. A lot of works on the structure, properties and thermal behavior of metal oxalates are present in the literature [121]. Available evidence suggest that the oxalates of divalent transition metals are polymeric, with oxalate ion ($C_2O_4^-$) acting as a quadridentate bridging ligand and most such oxalates are isomorphous. The thermal decomposition of copper oxalate differs substantially from that of the other transition metal oxalates.

Ebrahim et al. [122] prepared nano-sized CuO, Co_3O_4 and $CuCo_2O_4$ powders via different methods such as co-precipitation, thermal decomposition of oxalate precursor and sol–gel, their catalytic activities towards thermal decomposition of AP were investigated using DSC technique. $CuCo_2O_4$ showed better catalytic activity than others and the thermal decomposition temperature of AP shift downward about 103°C.

Jia et al. [123] present a new preparation of zinc cobaltite ($ZnCo_2O_4$) nanorods via a hydrothermal-annealing method. Zinc cobalt oxalate ($ZnCo_2(C_2O_4)_3 \cdot 6H_2O$) nanorods as the precursor have been firstly synthesized by solvothermal method at 120°C using the mixed solvents as the reaction medium without the assistance of soft/hard template. Furthermore, the effect of zinc cobaltite ($ZnCo_2O_4$) nanostructures on the thermal decomposition of AP is investigated by DTA. The results show that the addition of zinc cobaltite ($ZnCo_2O_4$) nanorods to AP remarkably decreases the decomposition temperature. The as-prepared zinc cobaltite ($ZnCo_2O_4$) nanorods are promising as a high-performing ballistic modifier in AP-based CSPs.

Singh et al. [124] have prepared nano copper oxalate which shows 130^{0}C decrease in HTD exothermic peak where two exothermic peaks of AP were merge to give only one. Activation energy value calculated for AP and AP with copper oxalate NPs was found to be 140.5 and 122.2 kJ mol^{-1}, respectively.

10.6. Transition Metals Nanomaterials

Today, nano metals are applied in CSPs and become a research hot spot in worldwide scope. Owing to their small particle size, large specific surface area, electronic effects and lattice defects, nano-metals endow burning rate catalysts. Liu et al. [125] have studied the effect of nanometer Ni, Cu and Al metal powders on the thermal decomposition of AP and compared these results with their micron sized particles. The additions of nanometer Cu, Ni and Al powders lower the HTD of AP by 130.2°C, 112.9°C and 51.8°C respectively. This indicates that nanometer Cu, Ni and Al powders have an obvious catalytic effect on the main exothermic decomposition of AP. Among Ni, Cu and Al; Cu shows marked catalytic effect both on LTD and HTD of AP. In addition, the total DTA heat release of AP in presence of Ni, Cu and Al were 1.32, 1.20 and 0.90 kJ/g respectively. These are much larger than that of pure AP (0.44 kJ/g), indicating that the decomposition of AP is increased. Authors have also studied the effect of increasing content of metal nanopowders on the thermal decomposition of AP which showed enhanced catalytic activity by increasing the content.

As reported earlier [126] Cu nanopowder contains many defects over crystal lattice. Atoms on the defects are not saturated and tend to become steady by absorbing materials with surplus electrons onto its surface. The N atom of AP contains surplus electrons, so the N-X bond becomes weak and easy to break owing to the absorbing of the N atom on the surface of Cu atom. This is advantageous in the production of NH_3. It is also reported [126] that nitrogen oxides react easily with Cu:

$$\mathbf{4Cu + 2NO \rightarrow 2Cu_2O + N_2,}$$

$$\mathbf{2Cu_2O + 2NO \rightarrow 4CuO + N_2.}$$

Because the nitrogen oxides are produced both at the first and second decomposition step of AP, the nanometer Cu can accelerate the decomposition of AP by catalyzing the decomposition of nitrogen oxides. Because of both these effects, nanometer Cu powder shows high catalytic effect on the first and second decomposition step of AP.

The increase in the content of nanometal powders would enhance this effect. This may be due to more efficient heat transfer by metal particles in the mixture. Therefore, the results are that the first decomposition temperature of AP generally increases as the content of nanometal powders increase. In a comparison with nano metal powders, micro metal powders cover much less surface of AP particles and therefore do not produce a clear hindering effect. So with the same content of metal powders, micro metal powders slightly lower the first decomposition peak of AP while nano metal powders increase it.

Duan et al. reported [127] that Ni NPs have significant catalytic effect on the decomposition of AP. Adding Ni NPs (2–5 wt. %) in AP, its decomposition temperature decreases by 92–105^{0}C and its apparent decomposition heat increases by 787.1–796.1 J/g. Such effect is attributed to the large specific surface area and the active chemical nature of Ni NPs. The endothermic peak at 243.2^{0}C appeared in all samples containing Ni particles, indicating that the Ni species has little effect on the crystallographic transition temperature of AP. However, dramatic changes in the exothermic peak of AP decomposition have been observed after adding Ni particles. When the micrometric Ni powders was added to the AP, the first exothermic peak shifted from 330.6 to 328.2^{0}C, and the second exothermic peak shifted significantly from 440.6 to 410.0^{0}C. These results indicate the significant effect of Ni NPs on the decomposition temperature of AP. In addition, the relatively high content of Ni favors the further decreasing of the decomposition temperature of AP.

Ping et al. [128] have studied the thermal decomposition of AP in presence of Cu/CNT composite particles where HTD of AP lowered by 126.3^{0}C and the LTD peak disappeared. This work shows that the catalytic performance of Cu on the thermal decomposition of AP can be improved by compounding with CNTs. The effect of metal nanopowders (Al, Fe, W, Ni, Cu, and Cu-Ni alloys) on the decomposition of AP was studied by Gromov et al. [129] and it was found that the catalytic action appears in the case of Cu-Ni nanopowders.

The temperature of decomposition of energetic materials with the addition of metal nanopowders could be lowered by 161°C for AP.

The effect of cobalt NPs of different morphologies (ball-like, cauliflower-like and snowflake-like) on the thermal decomposition of AP is studied by Liu et al.[130]. They showed that the decomposition of AP at both low and high temperatures was significantly changed with the addition of cobalt NPs. The two exothermic decomposition peaks of the pure AP were immerged into one peak by addition of cobalt NPs. Based on the morphologies of the cobalt added, the maximum peak temperature for the decomposition of the AP, followed the decreasing order; ball-like (316.3°C), cauliflower-like (311.6°C), and snowflake like (284.3°C) cobalt. These observations clearly indicate that the cobalt nanocrystals were efficient to promote the decomposition of the AP and the snowflake-like cobalt showed the highest performance in the aspect of decreasing the decomposition temperature. The BET surface areas of the cauliflower-like and snowflake-like cobalt nanocrystals were measured to be of 7.23 and 1.65 $m^2 g^{-1}$, respectively. This indicates that the catalytic ability of the cobalt for decreasing the decomposition of AP is not proportional to its surface area. Thus, the high catalytic performance of cobalt nanocrystals may be explained as that a great deal of cobalt lattice defects were formed during the liquid-phase reduction. Based on the DSC curves, the apparent heat of decomposition of the pure AP was calculated to be of 0.437 kJ g^{-1}. When Co catalyst was added, the apparent decomposition heat increased to 0.884, 1.040, and 1.884 kJ g^{-1} for snowflake-like, cauliflower-like, and ball like Co, respectively.

Duan et al. [131] also reported similar results using Co NPs (42 nm, 2% in AP) and Co_3O_4 NPs (35 nm, ~2% in AP). They have decreased the thermal decomposition of AP by 145.01°C and 106.27°C, respectively. From the previous results, it is proved that, in the first exothermic decomposition step of AP, a solid decomposition reaction occurs to produce large amount of N_2O, O_2, Cl_2, H_2O, HCl and a small amount of NO [132, 133]. The second exothermic decomposition step of AP is a gas phase reaction to produce large amount of NO, O_2, Cl_2 and H_2O etc. [134, 135]. In many catalytic processes, it is proved that oxygen species is preferably absorbed by the metallic species, as compared with CO, H_2O etc., resulting in the following rapid reaction to form CoO. It is also normally observed that the metallic NPs are apt to burn or explode quickly as exposed to air. In one word, metallic Co_3O_4 is much unstable and active as compared with CoO. Notably, the interaction of metallic Co species with O_2 and Cl_2 to form CoO and $CoCl_2$ releases large amount of heat to be 237.9 kJ/ mol^{-1} and 313 kJ/mol^{-1}, respectively. These rapid exothermic reactions may accelerate the decomposition of AP during the first stage and when CoO is formed; the following catalytic reaction may be the same and still release large amount of heat [136].

The catalytic effect of Zn nano and micro powders on thermal decomposition of AP has been investigated by Song et al. [137]. The results show that both nano and micro Zn powders show similar excellent catalytic effect on the decomposition of AP, while the total heat releases of AP added by Zn nano powders are generally higher than those of AP added by Zn micro powders. Song et al. [138] have investigated the effect of Mg metal nanopowder on thermal decomposition of AP and result revealed that the addition of a small amount of Mg NPs remarkably catalyzes the decomposition process of AP by lowering the decomposition temperature and enhancing its heat output.

Singh et al. [139] reported the effect of Cu, Co, Ni, and Fe transition metal nanoparticles (TMNs). The addition of TMNs not only increases the percentage of mass loss of AP but lowers the HTD range of AP to form gaseous products. It is known that the surface area of TMNs is large due to their very small size and there are many reactive sites over the surface. Thus, during the exothermic decomposition of AP, TMNs can adsorb the gaseous reactive molecules on their surface and accelerate the reactions. However, TMNs may also be oxidized to metal oxides, which also serve as accelerants.

10.7. Nano Transition Bimetallic Alloys

In material science, the range of properties of metallic systems can be greatly extended by taking mixtures of elements to generate intermetallic compounds and alloys [140]. The rich diversity of compositions, structures and properties of metallic alloys has led to widespread applications in electronic and catalysis. To fabricate materials with well defined, controllable properties and structures on the nanometer scale, afforded by intermetallic materials have generated interest in bimetallic nanoalloys. Surface structures, composition and segregation properties of nanoalloys are of interest as they are important in determining chemical reactivity especially the catalytic activity. Moreover, nanoalloys are also of interest as they may display structures and properties which are distinct from those of the pure elemental cluster and bulk alloys. Nanoallloys have been prepared by low temperature synthetic pathway, co-decomposition, co-reduction methods etc. [140].

Singh et al. [140] investigated preparation, characterization and catalytic effect of Ni-Co, Ni-Cu, Ni-Zn nanoalloys while Srivastava et al. [141] studied the preparation, characterization and catalytic effect of Cu-Co, Cu-Fe, Cu-Zn nanoalloys on decomposition of AP and CSPs. Figure 25 shows SEM images of nanoalloys. TG-DSC thermograms (Figure 26) for AP with nanoalloys do confirm that nanoalloys affect both LTD and HTD of AP, addition of it, not only increases the mass loss of AP but also lower the HTD range of AP to form gaseous products. Computational calculation shows that activation energy for ignition for AP is lowered by the nanoalloys. Nanoalloys enhanced the burning rate 'r' of CSPs which is highest for Ni-Co near about two times of pure CSP and lowest for Ni-Zn nanoalloy. Most of the studies suggested that ballistic modifiers are active mainly in the condense phase at AP-binder interface [142]. Computational calculation shows that activation energy for ignition of CSPs lowered by the nanoalloys. Summarizing these results, it may be inferred that nanoalloy can be used as catalyst for AP and CSPs combustion.

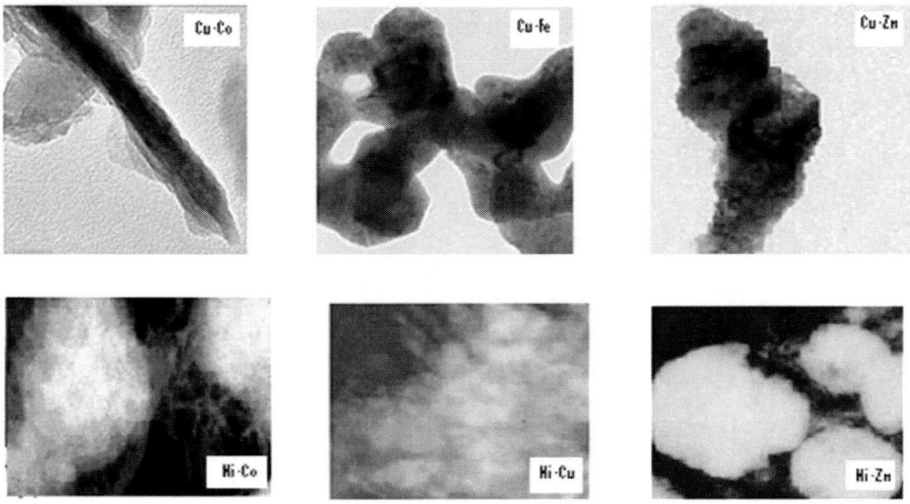

Figure 25. SEM images of nanoalloys. [Reprinted with permission from (Singh et al. [141]), *J. of Alloys and compounds* 2009; ©2009, Elsevier Publisher].

Dubey et al. [143] has synthesized MnCo, MnNi and MnZn nanocomposites via a polyol method. Particles of the nanocomposites varied in shape from spherical to cubic having an average size of the order of 9–18 nm (Figure 25). The nanocomposites showed very high activity as burning rate catalysts for CSPs. MnCo was the best catalyst among the three nanocomposites as it enhanced the burning rate of the CSP to more than three fold. These catalysts are active mainly in the condensed phase reactions as they have pronounced catalytic activity in the thermal decomposition of AP and CSPs. The catalytic activity of the nanocomposites is concentration dependant.

Better catalytic activity of MnCo composite was found not only during combustion of CSPs, but also during slow and rapid thermolysis of AP and the CSPs. Thermal sensitivity of AP was considerably increased with the mixing of nanocomposites to it. Catalytic effect of nanocomposites was clearly evident not only on the HTD process but also on the LTD process. MnCo was the best catalyst as the mass loss for AP + MnCo samples completed in a single step. HTD process not only began early but also completed early and thus resulted in a single step mass loss at a very low temperature (Figure 26).

MnNi was the poorer among the three nanocomposites in terms of catalytic activity. Interestingly, even MnNi was so active that thermal decomposition of AP was completed at 400°C when only 1 wt% MnNi was present as catalyst. Thermal decomposition of AP occurred at a lower temperature when the concentration of catalyst was increased in the mixture. However, mass % of the residue also increased with increasing concentration of the catalyst in the mixture. This is obvious as the metals or the corresponding metal oxides will remain as residue at the end of the thermal decomposition of AP. Thermal decomposition of AP started at a very low temperature of 260°C and was completed at 300°C, when 5 wt.% of MnCo was used as catalyst.

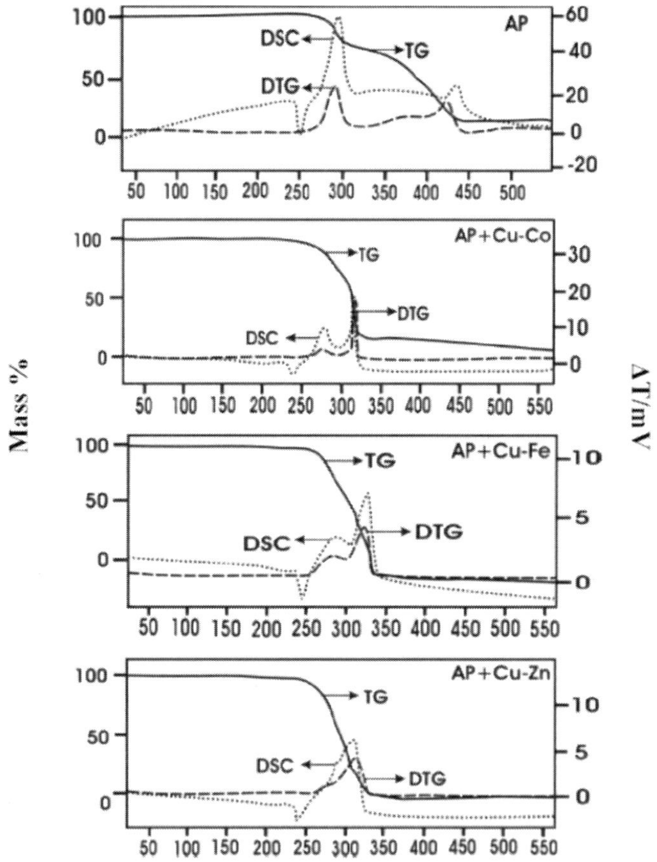

Figure 26. TG-DSC thermograms for AP with and without nanoalloys. [Reprinted with permission from (Singh et al. [141]), *J. of Alloys and compounds* 2009; ©2009, Elsevier Publisher].

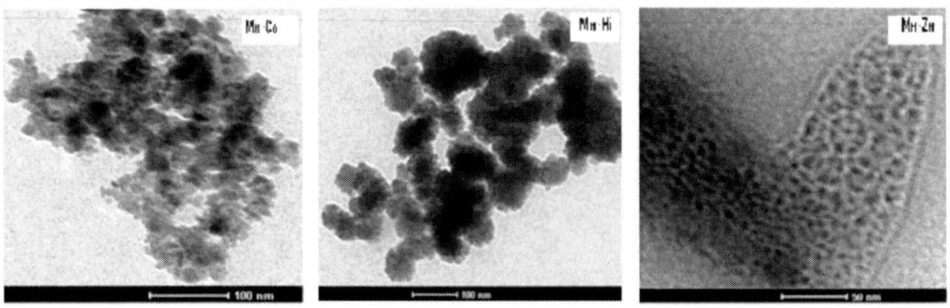

Figure 27. TEM images of nanocomposites [Reprinted with permission from (Singh et al. [141]), *J. of Alloys and compounds* 2009; ©2009, Elsevier Publisher].

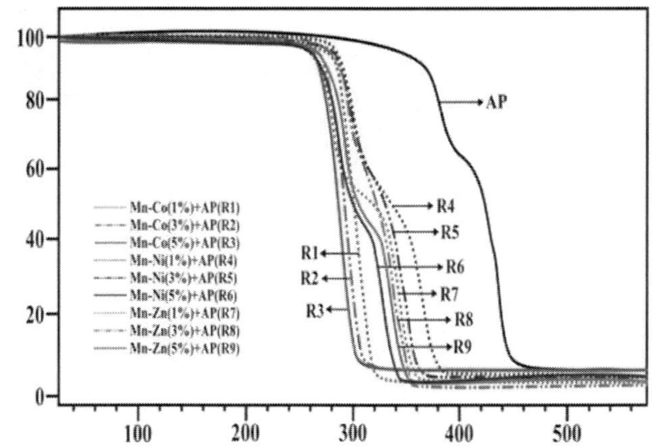

Figure 28a. TG Curve of AP with and without nanocomposites (1, 3 and 5% by wt.) [Reprinted with permission from (Singh et al. [141]), *J. of Alloys and compounds* 2009; ©2009, Elsevier Publisher].

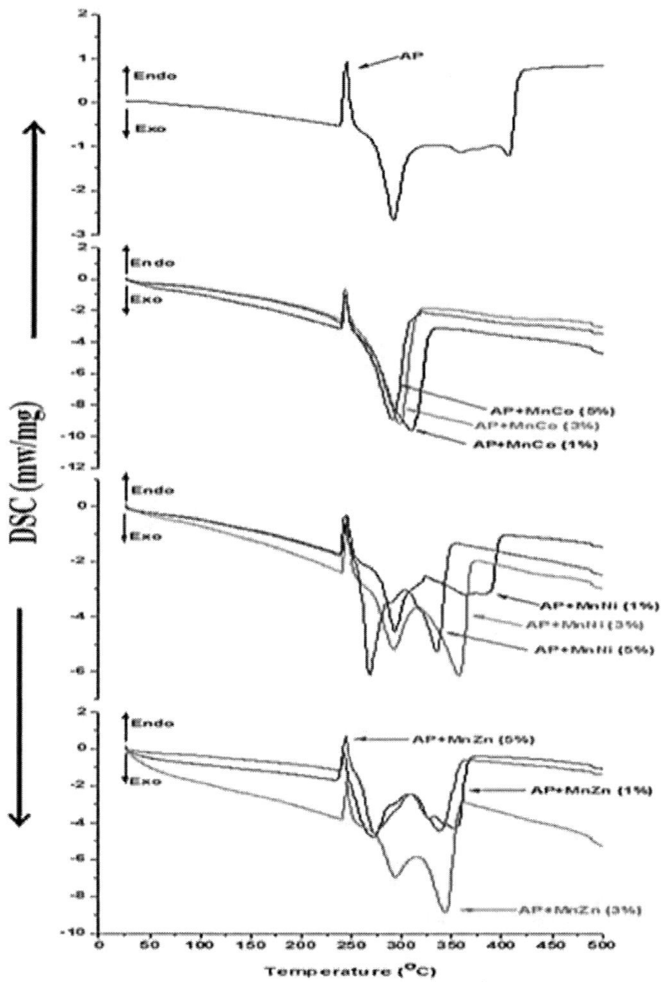

Figure 28b. DSC curve of AP with and without nanocomposites. [Reprinted with permission from (Singh et al. [141]), *J. of Alloys and compounds* 2009; ©2009, Elsevier Publisher].

An endothermic peak appeared at 244.4°C in the DSC thermal curve for pure AP that is shown in Figure 27. The first exothermic peak at 291°C is attributed to the LTD of AP. The second and main exothermic peak corresponding to HTD was observed at relatively higher temperature of 406°C. DSC thermal curves of AP in presence of nanocomposites showed noticeable changes in the decomposition pattern. From Figure 28 (a,b), it is clear that all the three nanocomposites have no significant effect on crystallographic transition temperature of AP. MnCo nanocomposite showed higher catalytic activity for the thermal decomposition of AP than MnNi and MnZn. Only a single exothermic peak appeared for thermolysis of AP when MnCo was present as catalyst.

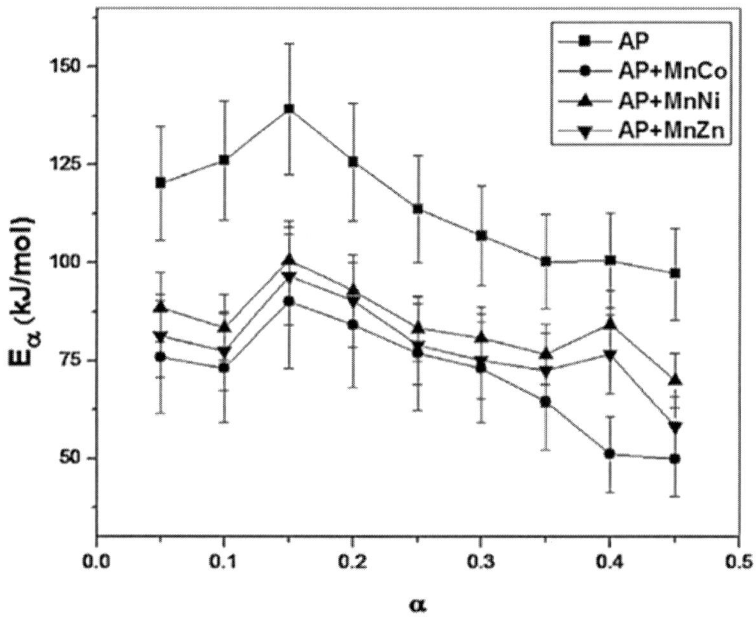

Figure 29. Variation of Ea with α for AP and AP + nanocomposites. [Reprinted with permission from (Singh et al. [141]), *J. of Alloys and compounds* 2009; ©2009, Elsevier Publisher].

Two exothermic peaks corresponding to LTD and HTD of AP were observed when MnNi and MnZn nanocomposites were present as catalysts, albeit the peaks were observed at lower temperature when compared to pure AP. Catalytic activity of the nano composites seems to be concentration dependent. Increasing the concentration of the nanocomposites in the mixture resulted in lowering of exothermic peaks.

Single exothermic peaks were observed at 310°C, 298.9°C and 291.3°C for AP + MnCo mixtures containing 1, 3 and 5% wt./wt. of MnCo respectively. When MnNi and MnZn nanocomposites were added to AP, the temperature shift for the peak corresponding to LTD of AP was 16.6–41°C and for HTD of AP was 49–93°C depending on the concentration of the catalyst. Enthalpy of exothermic decomposition for the catalyzed decomposition AP was higher than the enthalpy for pure AP. The enthalpy of decomposition was highest when MnCo was present as catalyst. Higher enthalpy of decomposition for catalyzed thermal decomposition of AP indicates the change in decomposition mechanism in presence of catalysts.

Kinetic parameters for isothermal decomposition were evaluated by a model-free isoconversional method using isothermal TG data. Values of activation energy (E_a) corresponding to different extent of conversion (α) were computed and the plots of E_a versus α for AP and AP + 1% wt./wt. nanocomposites are shown in Figure 28 (a,b). It is evident from (Figure 29) that E_a for AP is higher than E_a for AP + nanocomposites at all values of α. E_a for AP was initially ~120 kJ $mol^{-1}$and this value increased to ~140 kJ $mol^{-1}$when $\alpha = 0.15$ and then steadily decreased and reached a value of ~90 kJ mol^{-1}at $\alpha =$ 0.45. Variation of E_a with α indicates that thermal decomposition of AP is a complex interplay of multiple, competing processes and the rate limiting process varies with the extent of conversion. Vyazovkin and Wight used advanced isoconversional method to study the kinetics of thermal decomposition of AP [144]. They suggested that nucleation predominates during the initial values of α and ascending activation energy values at $\alpha =$ 0.1 indicates the dominance of growth of nuclei over nucleation. Dissociative sublimation that leads to porous AP and subsequent adsorption of NH_3 in the pores is the reason for slowing down of thermal decomposition of AP at $\alpha =$ ~0.3. Lowering of activation energy observed beyond $\alpha=$ ~0.15 is due to the transition from kinetically controlled decomposition process to the decomposition controlled by mass transfer [144].

The AP + nanocomposite samples showed similar trend of initial ascension in E_a with increasing α and dissention at higher values of α. However the values of E_a at all values of α was much lower than that for pure AP. Thus it appears that at initial values of α, nucleation and growth of nuclei predominates for the catalyzed samples. However, very low values of E_a observed at higher values of α is characteristic of mass transport limited reactions. High surface area of nanocomposites renders a lot of defective sites for the adsorption of the gaseous products of dissociative sublimation of AP (NH_3 and $HClO_4$). This renders the gas phase reactions to proceed to completion on the surface of the catalysts resulting in thermodynamically favored products such as Cl_2, HCl, H_2O, O_2, N_2O, and NO. MnCo appears to have more defective sites for the adsorption of gaseous products than the other two nanocomposites.

Chaturvedi et al. [145, 146] synthesized bimetallic nanoalloys (BMNs) of Co-Cu, Co-Fe, Co-Zn, Zn–Cu, Zn–Ni and Zn–Fe by hydrazine reduction of metal chloride in ethylene glycol. The effect of BMNs on the thermal decomposition of AP was investigated. Activation energy for HTD of AP with different BMNs (Co-Cu, Co-Fe, Co-Zn) was calculated using DSC by the Kissinger equation. The activation energy was calculated to be 69.52 kJ/mol for Co-Cu, which was lower than that of Co-Zn (124.45 kJ/mol) and Co-Fe (147.18 KJ/mol), indicating Co-Cu to be best catalyst. Moreover BMNs have found to be better and can be used as promoter for CSPs.

10.8. Transition Trimetallic Alloys Nanomaterials

The ability to design the size and composition of metal particles at the nanoscale could lead to improved catalytic properties. Powders of nickel, cobalt and copper are widely used because they possess good catalytic, electronic and magnetic properties. Investigations have indicated when metal was associated with another metal in trimetallic or alloy form, the properties of resulting material could be enhanced with respect to those of pure metals. There are variety of techniques for preparation of trimetallic alloys such as mechanical alloying [147], positive microemulsion [148], melt spinning [149] and electrodeposition [150].

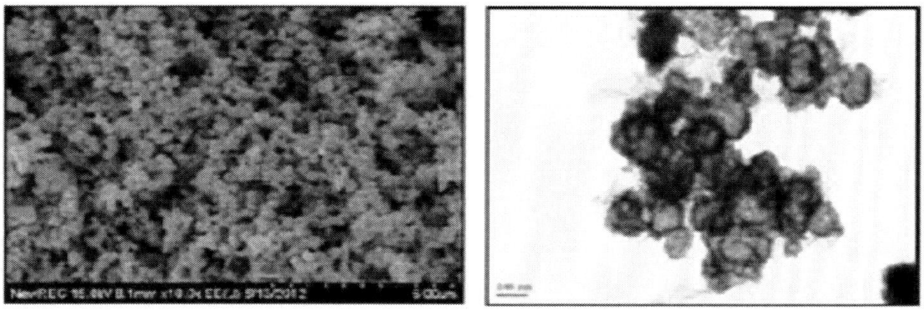

Figure 30. SEM and TEM images of TMN Co-Ni-Cu. [Reprinted with permission from (Singh et al. [152]), *J. of Alloys and compounds* 2013; ©2013, Elsevier Publisher].

Table 4. Activation energy (E_a) and correlation coefficient (r) for the isothermal decomposition of AP and AP with trimetallic alloy nanoparticles

Model	AP		AP+ Co-Ni-Cu	
	E_a (kJmol^{-1})	R	E_a (kJmol^{-1})	r
1	95.9	0.9609	74.1	0.9503
2	96.3	0.9612	74.3	0.9513
3	97.2	0.9598	74.8	0.9532
4	100.1	0.9634	76.2	0.9628
5	101.1	0.9636	76.8	0.9664
6	99.5	0.9630	75.9	0.9615
7	98.5	0.9633	74.6	0.9564
8	97.0	0.9616	74.9	0.9536
9	97.7	0.9620	75.8	0.9558
10	99.2	0.9629	77.3	0.9605
11	101.7	0.9634	75.7	0.9687
12	95.1	0.9629	75.7	0.9600
13	95.9	0.9607	74.2	0.9508
14	101.5	0.9635	77.1	0.9680

Reprinted with permission from (Singh et al. [152]), *J. of Alloys and compounds* 2013; ©2013, Elsevier Publisher.

Singh et al. [152] synthesized trimetallic nanoalloys (TMNs) Co-Ni-Cu (Figure 30) and study its catalytic activity in thermal decomposition of AP and CSP. Experimental results clearly show that burning rate is enhanced when TMNs are used as catalysts. Increase of burning rate might be due to the enhanced decomposition of AP or CSPs. The rate of decomposition and burning rate of CSPs is also enhanced in presence of TMNs. This is due to the fact that the freshly prepared TMNs have a finer particle size and a large number of defects and dislocations in the crystal lattice. Thus a large number of active sites would be available for the adsorption of reactants as a consequence; the rate of reaction would be increased.

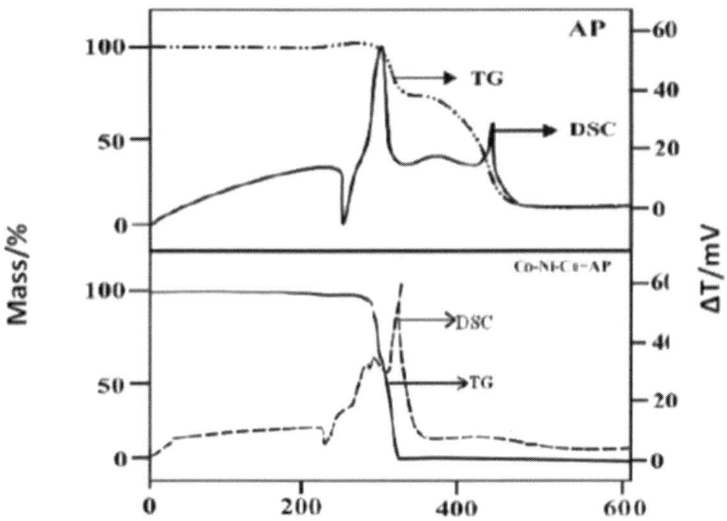

Figure 31. TG-DSC curve AP with and without TMN. [Reprinted with permission from (Singh et al. [152]), *J. of Alloys and compounds* 2013; ©2013, Elsevier Publisher].

The DSC curve for thermal decomposition of AP with and without TMN is shown in Figure 31, which clearly indicates that the LTD shifted from 310°C to 289°C and HTD from 430°C to 340°C after adding TMN in AP. The model fitting method has been used to evaluate kinetic parameters. Among the various values of r, calculated for different models (Table 3.4), the highest value of r for AP corresponds to model 5 and AP with TMNs corresponds to model 11. The corresponding values of E_a were reported for AP and AP with TMNs is 101.1 and 75.8 kJmol^{-1} respectively. Kinetic analysis performed by the isoconversional method on isothermal TG data also shows lower value of E_a at each α for AP+ (Co-Ni-Cu) than AP. The results of kinetic analysis clearly show that in the presence of TMNs catalyst, a noticeable decrease in the activation energy of thermal decomposition of AP occurred.

10.9. Transition Metal Cobaltite Nanomaterials

Recently, many studies have focused on the synthesis of nanomaterials, such as spinel cobaltite nanocrystals, which have attracted much attention because of their surface effect (large surface-to-volume ratio) and quantum confinement effects (size-dependent properties). These factors affect their physical and chemical properties. Binary spinel cobaltites find applications as materials in many fields, their main importance being as catalysts. In order to achieve high specific surface area, cobaltite synthesis by thermal treatment of coprecipitated precursors proves to be the most promising method [152].

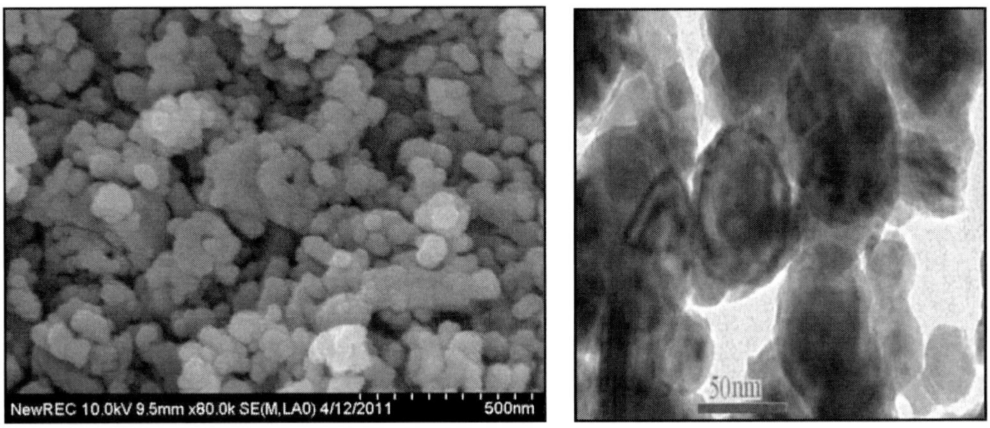

(a) (b)

Figure 32. SEM (a) and TEM (b) images of cadmium cobaltite NPs. [Reprinted with permission from (Singh et al. [154]), *J. Exper. Nanoscience* 2013; ©2013, Informa UK Limited, trading as Taylor & Francis Group].

Singh et al. [154] have synthesized and characterized cadmium cobaltite NPs (Figure 32). Its catalytic activity was investigated through the thermal decomposition of AP, NTO and HMX. The results of TG-DSC (Figure 33) on AP, NTO and HMX with and without cadmium cobaltite NPs (1% by wt) show that decomposition is occurred at lower temperatures with cadmium cobaltite NPs.

In case of AP first exothermic peak shifted from 310°C to 289°C and the second exothermic peak from 430°C to 305°C while DSC peak shifted from 272°C and 240°C for NTO and 280°C to 266°C for HMX. Singh et al. [154] also synthesized mixed transition metal oxide (MTMO) NPs of 3d-series ($NiCo_2O_4$, $CuCo_2O_4$ and $ZnCo_2O_4$) via co-precipitation method.

The thermolysis of AP, HTPB and CSPs was found to be catalyzed with MTMOs and the burning rate of CSPs was also enhanced. TG and ignition delay study demonstrated that the HTD of AP is catalyzed enormously by these catalysts and $CuCo_2O_4$ is found the best.

The catalytic effect of copper cobaltite, $Cu_xCo_{3-x}O_4$ spinel in the range ($0 \leq x \leq 3$) [155] on the thermal decomposition of AP has been studied. The results revealed that the addition of CuO ($x = 0.1$ or 0.3) led to an observable increase in the catalytic activity of Co_3O_4 towards the decomposition of AP, while the activity of CuO decreases at $x = 2.9$ and 2.7.

The creation of more holes within the p-type semiconducting catalyst is responsible for enhancing the decomposition of AP. In contrast, the catalysts containing copper cobaltite spinels are less active than the doped catalysts. The active redox sites established in these spinels, Co^{3+}/Co^{2+}; Cu^{2+}/Cu^{+} and Co^{3+}/Cu^{+}, are not as efficient as holes in promoting electron exchange with the perchlorate ions.

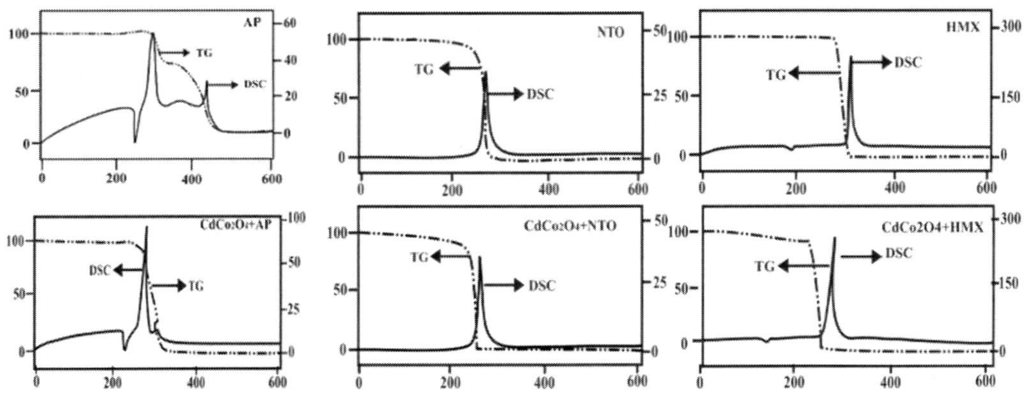

Figure 33. TG-DSC on AP, NTO and HMX with and without cadmium cobaltite NPs. [Reprinted with permission from (Singh et al. [154]), *J. Exper. Nanoscience* 2013; ©2013, Informa UK Limited, trading as Taylor & Francis Group].

10.10. Role of Biosynthesized Nanoparticles

Green chemistry is "the utilization of a set of principles that reduces or eliminates the use or generation of hazardous substances in the design, manufacture, and application of chemical products" [156]. The 12 principles of green chemistry are:

- Prevent Waste
- Atom Economy
- Less Hazardous Synthesis
- Design Benign Chemicals
- Benign Solvents & Auxiliaries
- Design for Energy Efficiency
- Use of Renewable Feed stocks
- Reduce Derivatives

- Catalysis (vs. Stoichiometric)
- Design for Degradation
- Real-Time Analysis for Pollution Prevention
- Inherently Benign Chemistry for Accident Prevention

Application of these principles has reduced the use of hazardous reagents and solvents, improved the material and energy efficiency of chemical processes, and enhanced the design of products for end of life. Employing these principles toward nanoscience will facilitate the production and processing of inherently safer nanomaterials and nanostructured devices.

Green nanoscience/nanotechnology involves the application of green chemistry principles to the design of nanoscale products, the development of nanomaterial production methods and the application of nanomaterials [157]. The approach aims to develop an understanding of the properties of nanomaterials, including those related to toxicity and ecotoxicity, and to design nanoscale materials that can be incorporated into high-performance products that pose little hazard to human health or the environment. It strives to discover synthesis/production methods that eliminate the need for harmful reagents and enhance the efficiency of these methods, while providing the necessary volume of pure material in an economically viable manner. It also provides proactive designs schemes for assuring the nanomaterials produced are inherently safer by assessing the biological and ecological hazards in tandem with design. Finally, it seeks applications of nanoscience that maximize societal benefit while minimizing impact on the ecosystem. In this way, green nanoscience guides materials development, processing, and application design throughout the life cycle, starting with raw materials election through end-of-life.

The techniques for obtaining NPs using naturally occurring reagents such as vitamins, sugars, plant extracts, biodegradable polymers, and microorganisms as reductants and capping agents could be considered attractive for nanotechnology. Among the reagents mentioned above, plant based materials seem to be the best candidates and they are suitable or large-scale 'biosynthesis' of NPs [158]. Plant parts such as leaf, root, latex, seed, and stem are being used for NPs synthesis. Greener synthesis of NPs provides advancement over other methods as it is simple, cost-effective, and relatively reproducible and often results in more stable materials (Scheme. 1).

Several research groups have successfully demonstrated the nanoscale synthesis of Ag, Au, and Pd NPs from various plant leaf extracts, for example, neem [159, 160], lemon grass [161], geranium leaves [162], Krishna Tulsi [163], Vinca rosea [164], hibiscus cannabinus [165], and so forth; plant fruit extract, for example, pear fruit extract [166], Prunus domestica fruit extract [167], Solanum lycopersicums extract [168], bark extract [169] and buds [170]. Medicinal plant extracts such as Aloe vera have been recently explored in the synthesis of metal oxides such as CuO [171], ZnO [172, 173], In_2O_3 [174],Fe_xO_y [175-177] and tin oxide [178] NPs.

Very recently Sharma et al. [180] have reported the biosynthesis of NiO NPs using C. gigantea leaf extract and its characterization by SEM, TEM and EDX (Figure 34). The biosynthesized NiO NPs possesses the aggregated spherical shape with a diameter of 20–50 nm. The catalytic behavior of the biosynthesized NiO NPs on the thermal decomposition of AP was demonstrated by TG-DSC experiment. TG curves for AP with and without green synthesized NiO NPs (1% by wt.) are shown in Figure 35.

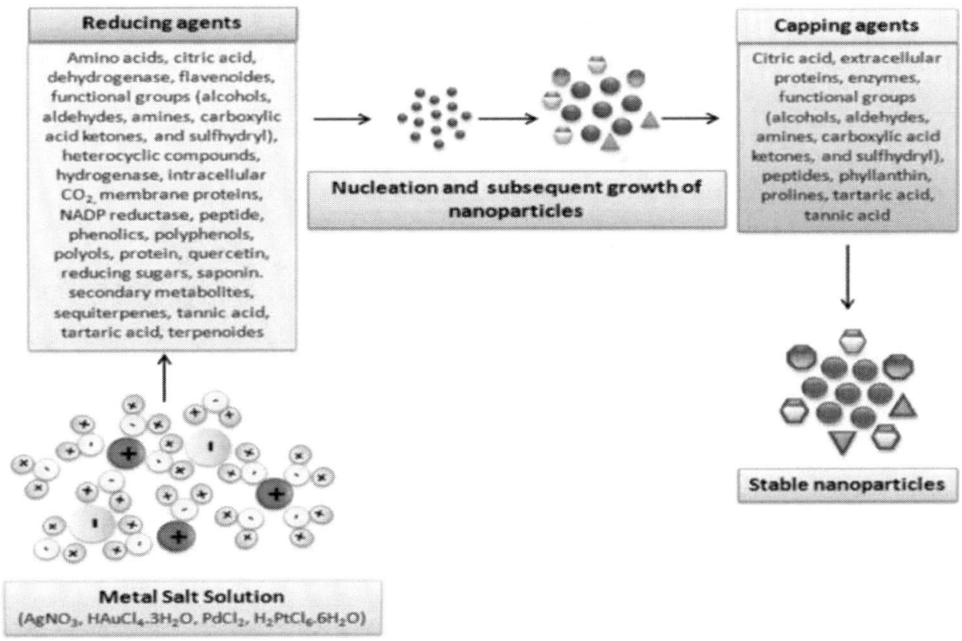

Scheme 1. Schematic diagram showing the mechanisms behind the green synthesis of NPs.

The thermal decomposition of AP was completed at 335°C when only 1wt% biosynthesized NiO NPs were present as catalyst. The DSC curves of AP with and without green synthesized NiO NPs (1% by wt.) shown in Figure 36 clearly indicates that the first and second exothermic peaks of pure AP were shift downward from 315°C to 280°C and 445°C to 352°C respectively by adding NiO NPs. This drastic decrease of 93°C in high temperature decomposition peak indicates the efficient catalytic effect of biosynthesized NiO NPs.

Burning rates of CSPs with and without biosynthesized NiO NPs (1% by wt.) were found to be 1.22 $mm.s^{-1}$ and 2.56 $mm.s^{-1}$ respectively. The burning rate of CSPs was increased to ~2.1 times when biosynthesized NiO NPs were present as catalyst. This increase in burning rate is mainly due to the enhanced condensed phase reactions during combustion of the propellants.

Kinetics of slow thermal decomposition was evaluated by isoconversional method using isothermal TG. Values of activation energy (E_a) corresponding to different extent of conversion (α) were computed and the plots of E_a versus α for AP with and without biosynthesized NiO NPs are shown in Figure 36. It is evident from Figure 37 that E_a for AP is higher than E_a for AP with biosynthesized NiO NPs at all values of α. Variation of E_a with α indicates that thermal decomposition of AP is a complex interplay of multiple, competing processes and the rate limiting process varies with the extent of conversion.

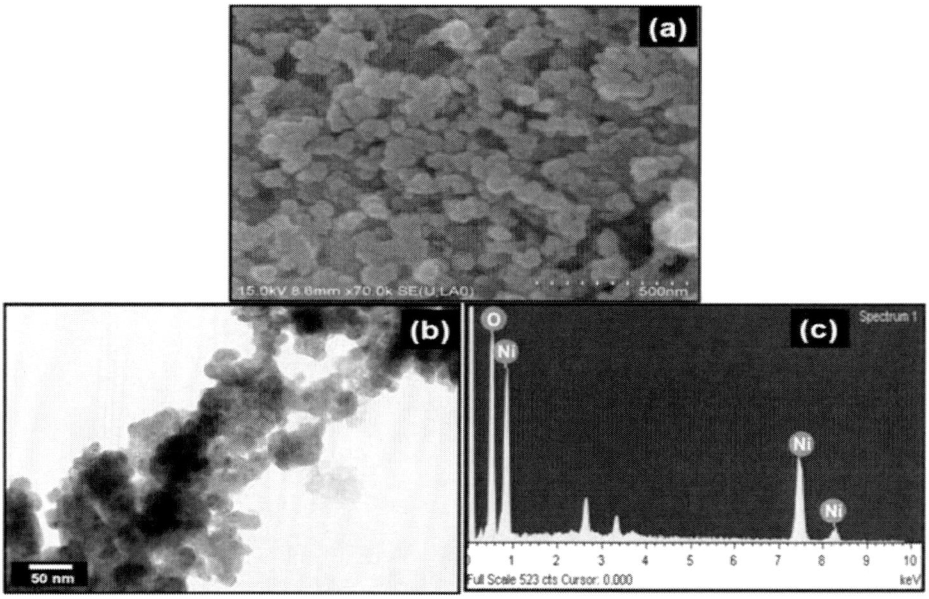

Figure 34. (a) FESEM image (b) TEM image and (c) EDX spectrum of biosynthesized NiO NPs. [Reprinted with permission from (Sharma et al. [180]), *Ceramic International* 2015; ©2015, Elsevier Publisher].

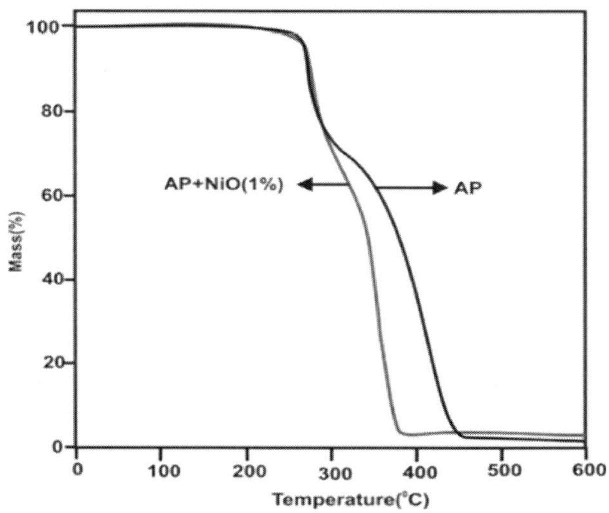

Figure 35. TG curves of AP with and without biosynthesized NiO NPs. [Reprinted with permission from (Sharma et al. [180]), *Ceramic International* 2015; ©2015, Elsevier Publisher].

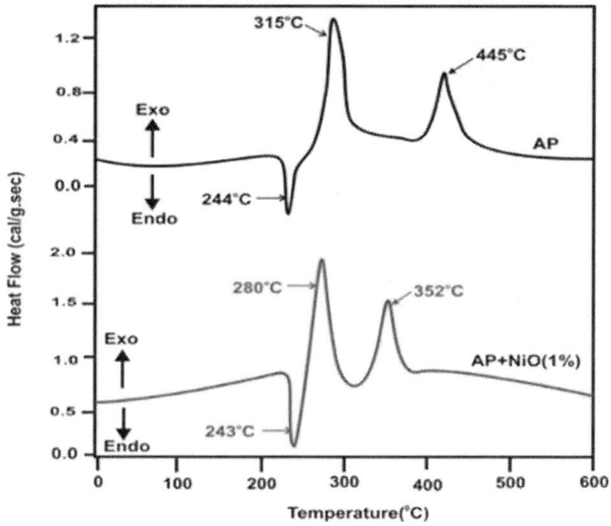

Figure 36. DSC curves of AP with and without biosynthesized NiO NPs. [Reprinted with permission from (Sharma et al. [180]), *Ceramic International* 2015; ©2015, Elsevier Publisher].

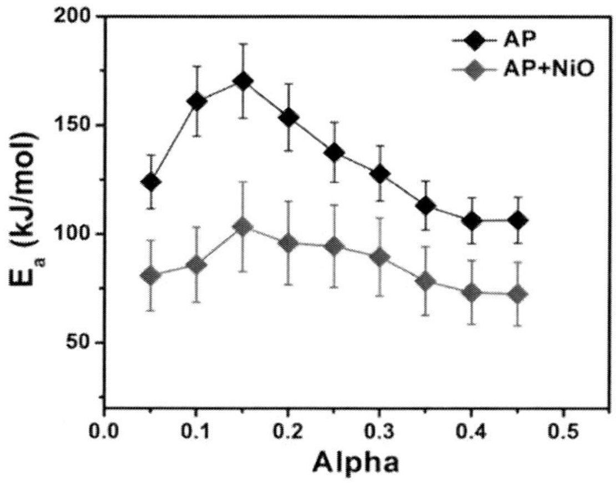

Figure 37. Variation of activation energy (Ea) with the extent of conversion (α) for AP with and without biosynthesized NiO NPs. [Reprinted with permission from (Sharma et al. [180]), *Ceramic International* 2015; ©2015, Elsevier Publisher].

Thus it appears that high value of activation energies at initial values of α, may be due to predominating nucleation and growth of nuclei. However, lowering of activation energy observed beyond $\alpha = 0.15$ is due to the transition from kinetically controlled decomposition process to the decomposition controlled by mass transfer.

Sharma et al. [181] have also reported the green synthesis of Co_3O_4 NPs using C. *gigantea* leaf extract (Figure 38) and its characterization by means of XRD, UV-Vis, SEM, EDX and TEM. In the XRD pattern of Co_3O_4 (Figure 39) all diffraction peaks can be readily indexed to a pure face centered cubic (fcc) structure (space group: Fd3m [227]) of Co_3O_4 cubic spinel and no diffraction peaks were seen arising from the possible impurity. The average crystallite size (C.S) of the green synthesized Co_3O_4 NPs was calculated to be

60 nm using the Debye–Scherer formula. UV-Vis spectrum of the green synthesized Co_3O_4 NPs presents two absorption bands in the wavelength ranges of i) 250 to 350 and ii) 400 to 600nm.

Figure 38. Photograph of C. *gigantea* plant (a), leaves extract (b), reaction mixture (c), as synthesized paste (d) and annealed Co_3O_4 NPs (e). [Reprinted with permission from (Sharma et al. [181]), *Materials Science and Engineering* B 2015; ©2015, Elsevier Publisher].

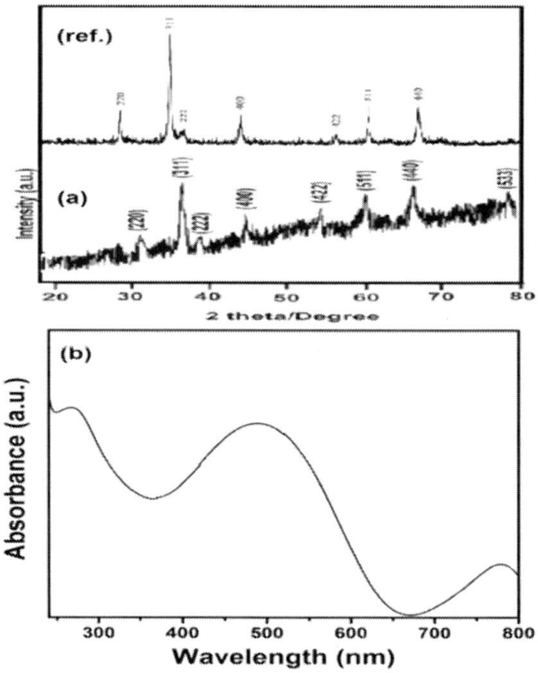

Figure 39. XRD pattern of authentic Co_3O_4 NPs (ref.) and green synthesized Co_3O_4 NPs (a), and UV-Vis spectroscopy of green synthesized Co_3O_4 NPs. [Reprinted with permission from (Sharma et al. [181]), *Materials Science and Engineering* B 2015; ©2015, Elsevier Publisher].

The appearance of first band near UV region is ascribed to the $O^{2-} \rightarrow Co^{2+}$ charge transfer process, and the second absorption band near green region is attributed to the $O^{2-} \rightarrow Co^{3+}$ charge transfer, confirming the p-type Co_3O_4 semiconductor. Another absorption band at 778 nm is also associated to the $O^{2-} \rightarrow Co^{3+}$ charge transfer in Co_3O_4

semiconductor. From UV-Vis spectrum, two direct band gap transitions in green synthesized Co_3O_4 NPs are estimated to ~2.53 eV and ~1.59 eV. FESEM, EDX spectrum, TEM and HRTEM images of green synthesized Co_3O_4 NPs were shown in Figure 40.

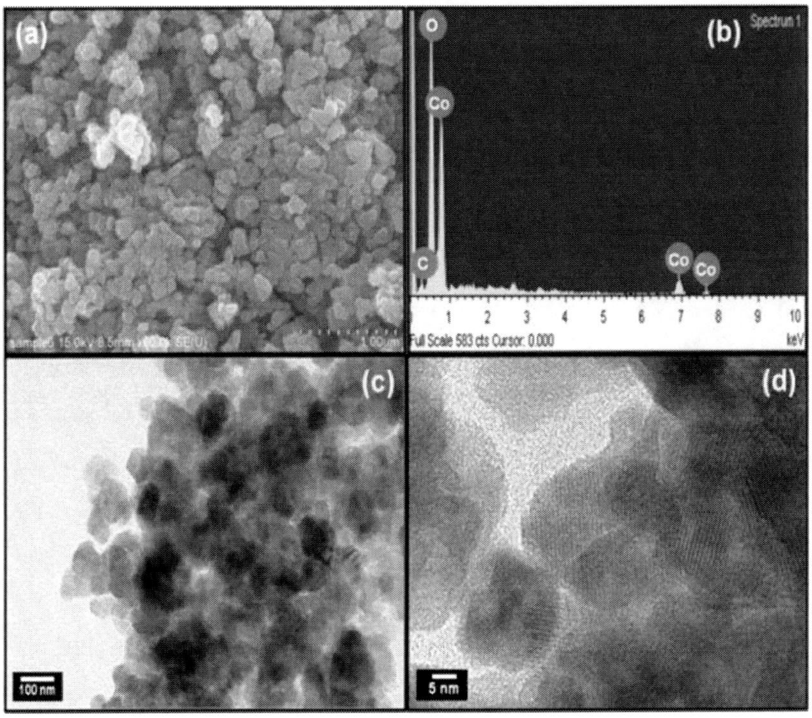

Figure 40. FESEM (a), EDX spectrum (b) TEM (b) and HRTEM (d) images of green synthesized Co_3O_4NPs. [Reprinted with permission from (Sharma et al. [181]), *Materials Science and Engineering* B 2015; ©2015, Elsevier Publisher].

FESEM image shows the nearly spherical shaped particles (~60-80 nm). From the TEM image, it is clear that the morphology of Co_3O_4 NPs is spherical shaped and the size ranges between 50-60 nm. EDX spectrum displays Co and O as the major elements. The element of C is the contaminant all around the places. No other elements can be detected, indicating the high purity of Co_3O_4 NPs.

The specific surface area of the green synthesized Co_3O_4 NPs is estimated to 46.7 m^2/g. which is higher than those of reported chemically synthesized Co_3O_4 NPs. In general, the large specific surface area of materials considerably produce a large number of active sites and higher number of adsorbed substrates for desired catalytic reaction. It is believed that the green synthesized Co_3O_4 NPs with surface area of 46.7 m^2/g could be potential catalytic materials for thermal decomposition of AP.

The catalytic effect of the green synthesized Co_3O_4 NPs on the thermal decomposition of AP was demonstrated by the TG-DSC experiment. TG curves for pure AP and AP with green synthesized Co_3O_4 NPs (1, 3 and 5 wt. %) and bulk Co_3O_4, shown in Figure 41 showed that two step decomposition process of AP was converted into a single step.

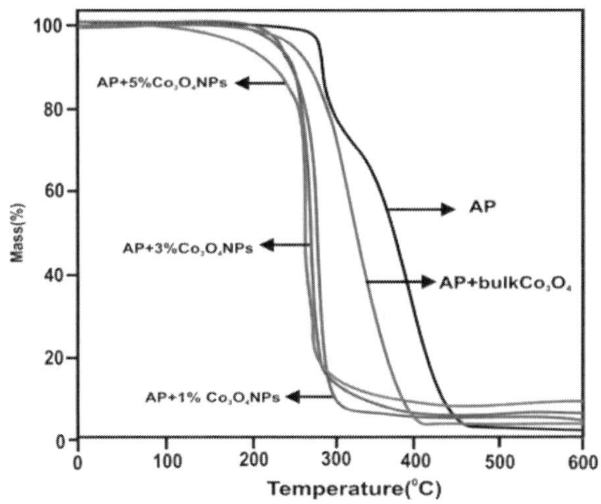

Figure 41. TG curves of pure AP, AP with green synthesized Co_3O_4 NPs (1%, 3% and 5%, by wt.) and AP with bulk Co_3O_4 (1% by wt). [Reprinted with permission from (Sharma et al. [181]), *Materials Science and Engineering* B 2015; ©2015, Elsevier Publisher].

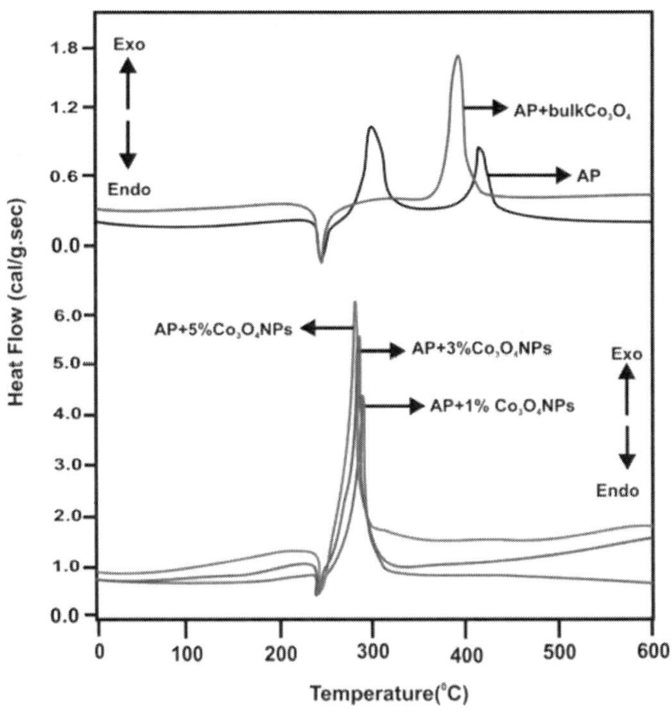

Figure 42. DSC curves of pure AP, AP with green synthesized Co_3O_4 NPs (1%, 3% and 5%, by wt.) and AP with bulk Co_3O_4 (1% by wt). [Reprinted with permission from (Sharma et al. [181]), *Materials Science and Engineering* B 2015; ©2015, Elsevier Publisher].

The presence of Co_3O_4 NPs (1 wt. %) resulted in considerable lowering (~250°C) of the decomposition temperature of AP. However, increasing the Co_3O_4 NPs loading, resulted in only minor decrease in the decomposition temperature. The DSC thermal curves for thermal decomposition of pure AP and AP with green synthesized Co_3O_4 NPs (1, 3 and

5 wt. %) and bulk Co_3O_4 were shown in Figure 42. The DSC thermal curve for thermal decomposition of pure AP showed three stages. However, the addition of bulk Co_3O_4 and Co_3O_4 NPs has apparently influenced the thermal decomposition of AP.

From (Figure 42), it is clear that bulk Co_3O_4 and Co_3O_4 NPs has no influence on crystallographic transition temperature of AP, while dramatic changes were observed in exothermic peaks in relatively high temperature region. The LTD and HTD peaks merged and a single exothermic peak was observed at 400°C for bulk Co_3O_4 and 292°C for green synthesized Co_3O_4 NPs (1 wt. %). Moreover, with increase in Co_3O_4 NPs loading, minor lowering of peak temperature was observed. However, the increase in Co_3O_4 NPs loading resulted in a considerable increase in heat release.

The increase in heat release can be attributed to the relatively increased condensed phase thermolysis of AP relative to dissociative sublimation and gas phase decomposition.

It is useful to compare the catalytic activity of the green synthesized Co_3O_4 NPs with classically synthesized Co_3O_4 NPs in the recent literature for thermal decomposition of AP are summarized in Table 5.

The catalytic activities are compared on the basis of peak temperature corresponding to the HTD (DSC) and final decomposition temperature (TG) of AP. Lower the value of the peak/decomposition temperature, higher is the catalytic activity. Green synthesized Co_3O_4 NPs seems to be the best catalyst.

Burning rates of CSPs and CSP with green synthesized Co_3O_4 NPs (1 wt. %) were found to be 1.22 $mm.s^{-1}$ and 2.95 $mm.s^{-1}$ respectively. The burning rate of CSPs increased ~2.5 times when green synthesized Co_3O_4 NPs were present as catalyst. This increase in burning rate is mainly due to the enhanced condensed phase reactions during combustion of the propellants.

Table 5. Catalytic activities of different Co_3O_4 NPs on the thermal decomposition of AP

Samples	Method of preparation of NPs	Final decomposition temperature of AP (TG)	HTD of AP (DSC)
AP+Co_3O_4 NPs (1wt%)	Green synthesis	300	292
AP+Co_3O_4 NPs (3wt%)	Sol-gel method	-	351
AP+Co_3O_4 NPs (2wt%)	Solvothermal method	297	300
AP+Co_3O_4 NPs (2wt%)	Solvothermal method	316	-
AP+Co_3O_4 NPs (2wt%)	Thermal decomposition method	342	-
AP+Co_3O_4 NPs (2wt%)	Thermal decomposition method	353	-
AP+Co_3O_4 NPs (2wt%)	Hydrothermal method	303	-

Reprinted with permission from Sharma et al. (181)], *Materials Science and Engineering* B 2015; ©2015, Elsevier Publisher.

Values of activation energy (E_a*) for AP, AP+ Co_3O_4, CSP and CSP+ Co_3O_4 were found to be 48.9, 25.9, 40.8 and 25.9 kJ/mol respectively. It is clear that E_a* for AP and

CSPs are lowered by adding green synthesized Co_3O_4 NPs. Values of E_a* for ignition generally tend to be very low and this is attributed to the predominance of mass transport processes during rapid heating.

The catalyzed samples showed much lower values of E_a* for ignition and this indicates that the green synthesized Co_3O_4 NPs are effectively adsorbing the gaseous products of dissociation of AP and thus promotes gaseous reactions.

CONCLUSION

There is continuous interest in the synthesis and application of metal or metal oxide nanomaterials, due to their large specific surface area and high activity in most catalytic processes. In contrast to metal and metal oxide nanomaterials, nano composites lead to the novel composition with unique properties and find applications in creating better technologies. For instance, nanomaterials have been proved effectively in improving the decomposition process of the AP, a key energetic material and the most common oxidizer in the rocket propellant. Propellants are one of the most important components to provide a driving force for rockets. A good solid propellant should have a stable burning rate and a low pressure exponent and for this purpose one of the best ways is to add a burning rate catalyst into the propellant. Nanometal oxides, pure metals, alloys, nano composites are considered as potential burning rate catalysts for CSPs. Studies on catalytic decomposition of AP and CSPs using nanomaterials emphasizes mass loss of AP and CSPs accelerates with time in the presence these nanomaterials. The burning rate of CSPs was also found to be enhanced.

REFERENCES

[1] T. Pradeep, *Nano: The Essentials: Understanding Nanoscience and Nanotechnology*, 2008.

[2] W. Meier, Nanostructure synthesis using surfactants and copolymers, *Current Opinion in Colloid & Interface Science* 4 (1999) 14.

[3] C. N. R. Rao, A. Muller, A. K. Cheetham, *The Chemistry of Nanomaterials*, Wiley-VCH, Weinheim, 2004.

[4] C. N. R. Rao, A. K. Cheetham, Science and technology of nanomaterials: current status and future prospects. *J. Mater. Chem*, 11 (2001) 2887.

[5] M. Lhadi, *Hybrid Nanocomposites for Nanotechnology. Electronic, Optical, Magnetic and Biomedical Applications*, 2009.

[6] P. D. Yang, *Chemistry of Nanostructured Materials*, 2004.

[7] J. Reithmaier, P. Petkov, W. Kulisch, C. Popov, *Nanostructured Materials for Advanced Technological Applications*, 2009.

[8] K. Bandyopadhyay, *Nano Materials*, New Age International (p) Limited, 2008.

[9] Pignataro, *Tomorrow's Chemistry Today Concepts in Nanoscience, Organic Materials and Environmental Chemistry*, 2008.

[10] P. Ravi, S. Dai, C. Wang, K. C. Tam, Fullerene containing polymers: a review on their synthesis and supramolecular behavior in solution. *J. Nanosci. & Nanotech.* 7 (2007) 1176.

[11] R. D. McMichael, R.D. Shull, L. J. Swartzendruber, L.H. Bennett, R.E. Watson, Magnetocaloric effect in superparamagnets. *J. Magn. Magnsm. Mater.*, 111 (1992) 29.

[12] I. Anton, I. de Sabata, L.Vékás, Application orientated researches on magnetic fluids, *J. Magn. Magnsm. Mater.*, 85 (1990) 219.

[13] M. N. Baibich, J. M. Broto, A. Fert, F. Nguyen Van Dau, F. Petroff, P. Etienne, G. Creuzet, A. Friederich, and J. Chazelas, Giant Magnetoresistance of (001)Fe/(001)Cr Magnetic Superlattices, *Phys. Rev. Letts.*, 61 (1988) 2472.

[14] P. Buffat, J.P. Borel, Size effect on the melting temperature of gold particles, *Phys. Rev. A*, 13 (1976) 2287.

[15] S. M. Prokes, Light emission in thermally oxidized porous silicon: Evidence for oxide-related luminescence, *Appl. Phys. Lett.*, 62 (1993) 3244.

[16] P. Alivisatos, Semiconductor Clusters, Nanocrystals, and Quantum Dots, *Science*, 271 (1996) 933.

[17] A. Grimes, Effect of purification of the electrical conductivity and complex permittivity of multiwall carbon nanotubes, *J. Applied Phys.*, 90 (2001) 4134.

[18] M. Heggli, *Elastomeric Nanocomposites, Research Plan*, Eidgenössische Technische Hochschule Zürich (2001).

[19] R.A. Prado, D. Facchini, N. Mahalanobis, F. Gonzalez, G. Palumbo, *Electrodeposition of nanocrystalline cobalt alloy coatings as a hard chroem alternative*. NAVAIR (2009).

[20] C. G. Granqvist, R. A. Buhrman, Ultrafine metal particles, *J. Appl. Phys.*, 47 (1976) 2200.

[21] H. L. Lai, N.B.Wong, Straight β-SiC nanorods synthesized by using C–Si–SiO_2, *Appl. Phys. Lett.*, 76 (2000) 294.

[22] Singhal, G. Skandan, N. Glumac, B. H. Kear, Minimizing aggregation effects in flame synthesized nanoparticles, *Scr. Mater.*, 4 (2001) 2203.

[23] Y. Lu, Y. Yin, Z. Li, Y. Xia, Synthesis and Self-Assembly of Au@SiO2 Core−Shell Colloids, *Nano Lett.*, 2 (2002) 785.

[24] W. Lu, B. Wang, J. Zeng, X. Wang, S. Zhang, J. Hou, Synthesis of Core/Shell Nanoparticles of Au/CdSe via Au−Cd Bialloy Precursor, *Langmuir*, 21 (2005) 3684.

[25] T. Ung, L. M. Liz-Marzan, P. Mulvaney, Controlled Method for Silica Coating of Silver Colloids. Influence of Coating on the Rate of Chemical Reactions, *Langmuir*, 14 (1998) 3740.

[26] V. Salgueiriño-Maceira , M. A. Correa-Duarte, M. Farle, M. A. López-Quintela, K. Sieradzki, R. Diaz, *Langmuir*, 22 (2006) 1455.

[27] X. Li, W. Zhang, Iron Nanoparticles: the Core−Shell Structure and Unique Properties for Ni(II) Sequestration, *Langmuir*, 22 (2006) 4638.

[28] B. B. Nayak, S. Vitta, A. K. Nigam, D. Bahadur, Ni and Ni–nickel oxide nanoparticles with different shapes and a core–shell structure, *Thin Solid Films*, 505 (2006) 109.

[29] C. C. Lee, D. H. Chen, Large-scale synthesis of Ni–Ag core–shell nanoparticles with magnetic, optical and anti-oxidation properties, *Nanotechnology*, 17 (2006) 3094.

[30] H. I. Schlesinger, H. I. Schlesinger, Herbert C. Brown, B. Abraham, A. C. Bond, N. Davidson, A. E. Finholt, J. R. Gilbreath, H. Hoekstra, L. Horvitz, E. K. Hyde, J. J. Katz, J. Knight, R. A. Lad, D. L. Mayfield, L. Rapp, D. M. Ritter, A. M. Schwartz, Irving Sheft, L. D. Tuck, and A. O. Walker, New Developments in the Chemistry of Diborane and the Borohydrides. I. General Summary, *J. Am. Chem. Soc.*, 75 (1953) 186.

[31] D. H. Chen, S. R. Wang, Protective agent-free synthesis of Ni–Ag core–shell nanoparticles, *Mater. Chem. Phys.*, 100 (2006) 468.

[32] J. W. Park, E. H.Chae, S. H. Kim, J. H. Lee, J. W. Kim, S. M. Yoon, J. Y. Choi, Preparation of fine Ni powders from nickel hydrazine complex, *Mater. Chem. Phys.*, 97 (2006) 371.

[33] H. H. Nersisyan, J. H. Lee, H. T. Son, C. W. Won, D. Y. Maeng, A new and effective chemical reduction method for preparation of nanosized silver powder and colloid dispersion, *Mater. Res. Bull.*, 38 (2003) 949.

[34] Y. Kobayashi, S. Ishida, K. Ihara, Y. T. M. Yasuda, S. Yamada, Synthesis of metallic copper nanoparticles coated with polypyrrole, *Colloid Polym. Sci.*, 287 (2009) 877.

[35] O. P. Yadav, A. Palmqvist, N. Cruise, K. Holmberg, Synthesis of platinum nanoparticles in microemulsions and their catalytic activity for the oxidation of carbon monoxide, *Colloids Surf. A*, 221 (2003) 131.

[36] M. L. Wu, D. H. Chen, T. C. Huang, Preparation of Pd/Pt Bimetallic Nanoparticles in Water/AOT/Isooctane Microemulsions, *J. Colloid Interface Sci.*, 243 (2001) 102.

[37] X. Zhang, K. Y. Chan, Water-in-Oil Microemulsion Synthesis of Platinum−Ruthenium Nanoparticles, Their Characterization and Electrocatalytic Properties, *Chem. Mater.*, 15 (2003) 451.

[38] D. Chen, S. Liu, J. Li, N. Zhao, C. Shi, X. Du, J. Sheng, Nanometre Ni and core/shell Ni/Au nanoparticles with controllable dimensions synthesized in reverse microemulsion, *J. Alloys Comp.*, 475 (2009) 494.

[39] W. Lee, M.G. Kim, J. Choi, J. Park, S. J. Ko, S. J. Oh, J. Cheon, Redox−Transmetalation Process as a Generalized Synthetic Strategy for Core−Shell Magnetic Nanoparticles, *J. Am. Chem. Soc.*, 127 (2005) 16090.

[40] J. I. Park et al., Characterization of Superparamagnetic "Core−Shell" Nanoparticles and Monitoring Their Anisotropic Phase Transition to Ferromagnetic "Solid Solution" Nanoalloys, *J. Am. Chem. Soc.*, 126 (2004) 9072.

[41] C. C. Lee, D. H Chen, Large-scale synthesis of Ni–Ag core–shell nanoparticles with magnetic, optical and anti-oxidation properties, *Nanotechnology*, 17 (2006) 3094.

[42] Y. Ji, S. Yang, S. Guo, X. Song, B. Ding, Z. Yang, Bimetallic Ag/Au nanoparticles: A low temperature ripening strategy in aqueous solution, *Colloids Surf.* A372 (2010) 204.

[43] J. Osuna, D.de Caro, C. Amiens, B. Chaudret, Synthesis, Characterization, and Magnetic Properties of Cobalt Nanoparticles from an Organometallic Precursor, *J. Phys. Chem.*, 100 (1996) 14571.

[44] M. Salavati-Niasari, F. Davar, M. Mazaheri, M. Shaterian, Preparation of cobalt nanoparticles from [bis(salicylidene)cobalt(II)]–oleylamine complex by thermal decomposition, *Magn. Mater.*, 320 (2008) 575.

[45] M. Salavati-Niasari, F. Davar, N. Mir, Synthesis and characterization of metallic copper nanoparticles via thermal decomposition, *Polyhedron*, 27 (2008) 3514.

[46] M. Salavati-Niasari, F. Davar, Synthesis of copper and copper(I) oxide nanoparticles by thermal decomposition of a new precursor, *Mater. Lett.*, 63 (2009) 441.

[47] J. Luo, Q. Fan, M.Suzuki , I.S. Suzuki, M. H. Engelhard, Y. Lin, N.Kim, J. Q. Wang, C. Zhong, Monodispersed Core−Shell Fe3O4@Au Nanoparticles, *J. Phys. Chem. B*, 109 (2005) 21593.

[48] E. A. Abdel-Aal, S. M. Malekzadeh, M. M. Rashad, A. A. El-Midany, H.El-Shall, Effect of synthesis conditions on preparation of nickel metal nanopowders via hydrothermal reduction technique, *Powder Technol.*, 171 (2007) 63.

[49] 49.W. Fu, H.Yanga, L. Chang, M. Li , H. Bala, Q.Yu, G. Zou, Preparation and characteristics of core–shell structure nickel/silica nanoparticles, *Colloids Surf. A*, 262 (2005) 71.

[50] A. Gromov, U. Forter-Barth, U. Teipel, Aluminum nanopowders produced by electrical explosion of wires and passivated by non-inert coatings: Characterisation and reactivity with air and water, *Powder Technol.*, 164 (2006) 111.

[51] Korshunov, M. Heyrovsky, Electrochemical behavior of copper metal core/oxide shell ultra-fine particles on mercury electrodes in aqueous dispersions, *J. Electroanal. Chem.*, 629 (2009) 23.

[52] Q. Wang, H. Yang, J. Shi, G. Zou, Preparation and characterization of nanocrystalline powders of Cu–Zn alloy by wire electrical explosion method, *Mater. Sci. Eng. A*, 307 (2001) 190.

[53] Y. Fu, C. Shearwood, Characterization of nanocrystalline TiNi powder, *Scr. Mater.*, 50 (2004) 319.

[54] W. Fu, H.Yang, Hari-Bala, S.Liu, M. Li, G. Zou, Preparation and magnetic characterization of core–shell structure stainless steel/silica nanoparticles, *Mater. Lett.*, 60 (2006) 1728.

[55] M. Niederberger, Nonaqueous Sol–Gel Routes to Metal Oxide Nanoparticles, *Acc. Chem. Res.*, 40 (2007) 793.

[56] H. Xiao, Z. Ai, L. Zhang, Nonaqueous Sol−Gel Synthesized Hierarchical CeO2 Nanocrystal Microspheres as Novel Adsorbents for Wastewater Treatment, *J. Phys. Chem. C*, 113 (2009) 16625.

[57] T. Yang, H. Wang, M. K. Lei, Phase transition of Er3+-doped Al2O3 powders prepared by the non-aqueous sol–gel method, *Mater. Chem. Phys.*, 95 (2006) 211.

[58] K. Gupta, M. Gupta, Synthesis and surface engineering of iron oxide nanoparticles for biomedical applications, *Biomaterials*, 26 (2005) 3995.

[59] D. Jiang, L .Cao, W. Liu , G. Su , H. Qu, Y. Sun , B. Dong, Synthesis and Luminescence Properties of Core/Shell ZnS:Mn/ZnO Nanoparticles, *Nanoscale Res. Lett.*, 4 (2009) 78.

[60] Y. L. Min, Y. Wan, S. H. Yu, Au@Y2O3:Eu3+ rare earth oxide hollow sub-microspheres with encapsulated gold nanoparticles and their optical properties, *Solid State Sci.*, 11 (2009) 96.

[61] M. Salavati-Niasari, F. Davar, Z. Fereshteh, Synthesis and characterization of ZnO nanocrystals from thermolysis of new precursor, *Chem. Eng. J.*, 146 (2009) 498.

[62] M. Salavati-Niasari, F. Davar, M. Mazaheri, Preparation of ZnO nanoparticles from [bis(acetylacetonato)zinc(II)]–oleylamine complex by thermal decomposition, *Mater. Lett.*, 62 (2008) 1890.

[63] M. Salavati-Niasari, M. Dadkhah, F. Davar, Synthesis and characterization of pure cubic zirconium oxide nanocrystals by decomposition of bis-aqua, tris-acetylacetonato zirconium(IV) nitrate as new precursor complex, *Inorg. Chim. Acta*, 362 (2009) 3969.

[64] M. Salavati-Niasari, M. Dadkhah, M. Davar, Pure cubic ZrO2 nanoparticles by thermolysis of a new precursor, *Polyhedron*, 28 (2009) 3005.

[65] M. Salavati-Niasari, N. Mir, F. Davar, Synthesis and characterization of Co3O4 nanorods by thermal decomposition of cobalt oxalate, *J. Phys. Chem. Solids*, 70 (2009) 847.

[66] M. Salavati-Niasari, Z. Fereshteh, F. Davar, Synthesis of cobalt nanoparticles from [bis(2-hydroxyacetophenato)cobalt(II)] by thermal decomposition, *Polyhedron*, 28 (2009) 1065.

[67] H. Zeng, J. Li, Z. L.Wang, J. P. Liu, S. Sun, Bimagnetic Core/Shell FePt/Fe3O4 Nanoparticles, *Nano Lett.*, 4 (2004) 187.

[68] J. S. Benjamin, Dispersion strengthened superalloys by mechanical alloying, *Metall. Trans.*, 1 (1970) 2943.
[69] L. M. Cukrov, T. Tsuzuki, P. G. McCormick, SnO2 nanoparticles prepared by mechanochemical processing, *Scr. Mater.*, 44 (2001) 1787.
[70] W. Friedrich, P. Knipping, M.V. Laue, *Acad.* Wissen Munich, 303 (1912) 6.
[71] M. A. Krivoglaz, *Theory of X-Ray and Thermal –Neutron Scattering by Real Crystals*, Plenum press, New York, (1969).
[72] L. Dan, *Fabrication and characterization of metal oxide nanostructures*, University of Hong Kong, July, (2007).
[73] N. D. Browning, M.F. Chisholm, S.J. Pennycook, Atomic-resolution chemical analysis using a scanning transmission electron microscope, *Nature*, 444 (2006) 235.
[74] H. H. Rose, Optics of high-performance electron microscopes, *Sci. Technol. Adv. Mater.*, 9 (2008) 014107.
[75] H. John Spence, *Experimental high-resolution electron microscopy*, New York: Oxford U. Press, (1988).
[76] D. L. Humphris, M. J. Miles, J. K. Hobbs, A mechanical microscope: High-speed atomic force microscopy, *App. Phys. Lett.*, 86 (2005) 034106.
[77] V. Crist, *Annotated Handbooks of Monochromatic XPS Spectra*, PDF of Volumes 1 and 2, published by XPS International LLC, Mountain View, CA, USA, (2005).
[78] J. Tersoff, D. R. Hamann, Theory of the scanning tunneling microscope, *Phys. Rev. B*, 31 (1985) 805.
[79] J. Bardeen, Tunnelling from a Many-Particle Point of View, *Phys. Rev. Lett.*, 6 (1961) 57.
[80] P. Zakharov, S. Bhat, P. Schurtenberger, F. Scheffold, Multiple-scattering suppression in dynamic light scattering based on a digital camera detection scheme, *Applied Optics*, 45 (2006) 1756.
[81] T. H. G Froerer, *Encyclopedia of Analytical Chemistry*, edited by R. A. Meyers, John Wiley & Sons Ltd., Chichester, UK, 17 (2000) 9209-9231.
[82] S Taketomi, S Chikazumi. *Magnetic Fluids: Principle and Application*, Tokio: Nokkan Kogyo Shinbun, (1988).
[83] M. D. Stoller, S. Park, Y. Zhu, J. An, R. S. Ruoff, Graphene-Based Ultracapacitors, *Nano Letters*, 9 (2008) 3498.
[84] S. Bae, H. Kim, Y. Lee, X. Xu, J. Park, Y. Zheng, J. Balakrishnan, T. Lei, H. R. Kim, Y. Il Song, Y. Kim, K. S. Kim, B. Özyilmaz, J. Ahn, B. H. Hong, S. Iijima, Roll-to-roll production of 30-inch graphene films for transparent electrodes, *Nature Nanotechnology*, 5 (2010) 574.
[85] D.Wakuda, K.S. Kim, K. Suganuma, Ag Nanoparticle Paste Synthesis for Room Temperature Bonding, *IEEE Trans. on Comp. and Packaging Tech.*, 33 (2010) 437.
[86] D. Tsoukalas, From silicon to organic nanoparticle memory devices, *Philosophical Trans. of the Royal Society*, 367 (2009) 4169.

[87] K. E. Drexler, *Nanosystems: Molecular Machinery, Manufacturing, and Computation.* Wiley (1992).

[88] E. Lucas, S. Decker, A. Khaleel, A. Seitz, S. Futlz, A. Ponce, W. Li, C. Carnes, K. J. Klabunde, Nanocrystalline Metal Oxides as Unique Chemical Reagents/Sorbents, *Chem. Eur. J.*, 7 (2001) 2505.

[89] S. Edelstein , R. C. Cammarata, *Nanomaterials Synthesis, Properties and Applications*, Second Edition Institute of Physics Publishing, London (1998), Vol. 1.

[90] P. R. Patil, S. K. Hait, V. N. Krishnamurthy, and S. S. Joshi, *5th International High Energy Materials Conference and Exhibition*, Hydrabad, India, November (2005), Vol. 111.

[91] P. S. Kapoor, P. Srivastava, G. Singh, Nanocrystalline transition metal oxides as catalysts in the thermal decomposition of ammonium perchlorate, *Propellants Explos. Pyrotechnics*, 34 (2009) 351.

[92] Y. Xu, D. Chen, X. Jiao, K. Xue; CuO microflowers composed of nanosheets: synthesis, characterization, and formation mechanism, *Mater. Res. Bull.*, 42 (2007) 1723.

[93] J. Wang, S. Hi, Z. Li, X. Jing, M. Zhang, Z. Jiang; Synthesis of chrysalis-like CuO nanocrystals and their catalytic activity in the thermal decomposition of ammonium perchlorate, *J. Chem. Sci.*, 121 (2009) 1077.

[94] Singh, I. P. S. Kapoor, S. Dubey, P. F. Siril; Preparation, characterization and catalytic activity of transition metal oxide nanocrystals, *J. Sci. Conf. Proce.*, 1 (2009) 11.

[95] Y. Wang, J. Zhu, X. Yang, L. Lu, X. Wang, Preparation of NiO nanoparticles and their catalytic activity in the thermal decomposition of ammonium perchlorate, *Thermochim. Acta*, 437 (2005).

[96] P. R. Patil, S. K. Hait, V. N. Krishnamurthy, S. S. Joshi, Characterisation of nano iron oxide and its effect on thermal decomposition of ammonium perchlorate, *International High Energy Material Conference and Exhibition, DRDL*, Hyderabad (2005).

[97] P. R. Patil, V. N. Krishnamurthy, S. S. Joshi, Differential scanning calorimetric study of HTPB based composite propellants in presence of nano ferric oxide, *Propellants Explos. Pyrotechnics*, 31 (2006) 442.

[98] Xu, X.wang L. Zhang; Selective preparation of nanorods and micro-octahedrons of Fe_2O_3 and their catalytic performances for thermal decomposition of ammonium perchlorate, *Powder Tech.*, 185 (2008) 176.

[99] L. Li, X. Sun, X. Qiu, J. Xu, G. Li, Nature of catalytic activities of CoO nanocrystals in thermal decomposition of ammonium percholorate, *Inorg. Chem.*, 74, 19 (2008) 8839,

[100] Duan, X. Yang, J. Chen, G. Huang, L. Lu, X. Wang, The catalytic effect of nanosized MgO on the decomposition of ammonium perchlorate, *Powder Tech.*, 172 (2007) 27.
[101] T. Fu, F. Liu, L.Liu, L.Guo, F. Li, Catalytic thermal decomposition of ammonium perchlorate using manganese oxide octahedral molecular sieve (OMS) *Cat. Communi.*, 10 (2008) 108.
[102] M. Zheng, Z.S. Wang, J. Q. Wu, Q. Wang, Synthesis of nitrogen-doped ZnO nanocrystalliteswith one-dimensional structure and their catalyticactivity for ammonium perchlorate decomposition, *J. of Nanoparticle Res* 12 (2010) 2211.
[103] G. Singh, I.P.S. Kapoor, Reena Dubey, Pratibha Srivastava, Synthesis, characterization and catalytic activity of CdO nanocrystals, *Materials Science and Engineering B*, 176 (2011) 121.
[104] D. V. Survase; D. B. Sarwade, E. M. Kurian, Effect of la_2o_3, pr_2o_3 and nd_2o_3 on the thermal decomposition of ammonium perchlorate, *J. of Energ. Mater.*, 19 (2001) 23.
[105] Y. Wang, X. Yang, L. Lu, X. Wang, Experimental study on preparation of $LaMO_3$ (M = Fe, Co, Ni) nanocrystals and their catalytic activity, *Them. Chem. Acta.*, 433 (2006) 225.
[106] S. Singh, P. Srivastava, I. P. S. Kapoor, G. Singh, Preparation, characterization, and catalytic activity of rare earth metal oxide nanoparticles, *J. Therm. Anal. Calorim.*, 111 (2013) 1073.
[107] M. E. Brown, D. Dollimore, A. K. Galway, *Comprehensive Chemical Kinetics*; Elsevier: Amsterdam, The Netherlands, (1960) Vol. 22.
[108] S.V. Vyazovkin, A unified approach to kinetic processing of nonisothermal data, *Int. J. Chem. Kinet.*, 28 (1996) 95.
[109] E.S. Freeman, S. Gordon, The Application of the Absolute Rate Theory to the Ignition of Propagatively Reacting Systems. The Thermal Ignition of the Systems Lithium Nitrate–Magnesium, Sodium Nitrate–Magnesium, *J. Phys. Chem.*, 60 (1956) 867.
[110] G. Singh, I.P.S. Kapoor, S.K. Vasudeva, *Indian J. Tech.* 29 (1991) 589.
[111] L.L. Bircumshaw, B.H. Newman, The thermal decomposition of ammonium perchlorate - I. Introduction, experimental, analysis of gaseous products, and thermal decomposition experiments, *Proc R Soc A.*, 227 (1954) 115.
[112] Z.Yu, Y. Sun, W.Wei, L.Lu, X.Wang, Preparation of NdCrO3 nanoparticles and their catalytic activity in the thermal decomposition of ammonium perchlorate by DSC/TG-MS, *J. Therm. Anal. Calorim.*, 97 (2009) 903.
[113] D.H. Chen, X. R. He, Synthesis of nickel ferrite nanoparticles by sol-gel method, *Res. Bulletin*, 36 (2001) 1369.
[114] G. Singh, I. P. S. Kapoor, S. Dubey, P. F. Siril, J. H. Yi, F. Q. Zhao, R.-Z. Hu, Effect of mixed ternary transition metal ferrite nanocrystallites on thermal decomposition of ammmonium perchlorate, *Thermochem. Acta*, 477 (2008) 42.

[115] P. Srivastava, I.P.S. Kapoor, G. Singh, Nanoferrites: Preparation, characterization and catalytic activity, *J. Alloys. Comp.*, 485 (2009) 88.

[116] G. Singh, I. P. S. Kapoor, S. Dubey, P. F. Siril, Kinetics of Thermal Decomposition of Ammonium Perchlorate with Nanocrystals of Binary Transition Metal Ferrites, *Propellants. Explos. Pyrotechnics*, 34 (2009) 72.

[117] G. Singh, I.P.S. Kapoor, R. Dubey, P. Srivastava, Preparation, characterization and catalytic behavior of $CdFe_2O_4$ and Cd nanocrystals on AP, HTPB and composite solid propellants, Part: 79, *Thermochim. Acta*, 511 (2010) 112.

[118] S. Singh, P. Srivastava, G. Singh, Nanorods, nanospheres, nanocubes: Synthesis, characterization and catalytic activity of nanoferrites of Mn, Co, Ni, Part-89, *Materials Research Bulletin*, 48 (2013) 739.

[119] S. Zhao, D. Ma, J. Nanomateri., 2010, Preparation of $CoFe_2O_4$ Nanocrystallites by Solvothermal Process and Its Catalytic Activity on the Thermal Decomposition of Ammonium Perchlorate, *J. of Nanomaterials*, Article ID 842816, 5 pages.

[120] L. Cristina Soare, P. Bowen, J. Lemaitre, H. Hofmann, Precipitation of Nanostructured Copper Oxalate: Substructure and Growth Mechanism, *J. Phys. Chem. B*, 110 (2006) 17763.

[121] D. Dollimore, The thermal decomposition of oxalates. A review, *Thermochim. Acta*, 117 (1987) 331.

[122] Ebrahim Alizadeh-Gheshlaghi Behrouz Shaabani Ali Khodayari Yashar Azizian-Kalandaragh Rahmatollah Rahimi, Investigation of the catalytic activity of nano-sized CuO, Co_3O_4 and $CuCo_2O_4$ powders on thermal decomposition of ammonium perchlorate, *Powder Technol.*, 217 (2012) 330.

[123] Z. Jia, D. Ren, Q. Wang, R. Zhu, A new precursor strategy to prepare $ZnCo_2O_4$ nanorods and their excellent catalytic activity for thermal decomposition of ammonium perchlorate, *Applied Surface Science*, (2013) 312.

[124] G. Singh, I.P.S. Kapoor, R. Dubey, P. Srivastava, Preparation, characterization and catalytic effects of copper oxalate nanocrystals, *J. Alloys and Compounds*, 513 (2012) 499.

[125] L. Liu, F. Li, Li. Tan, L. Ming, Y. Yi, Effects of Nanometer Ni, Cu, Al and NiCu Powders on the Thermal Decomposition of Ammonium Perchlorate, *Propellants Explos. Pyrotechnics*, 29 (2004) 34.

[126] Zhi, J, Feng-Qi, Z. Study on effects of nanometer metal powder on thermal decomposition of HMX, *J. Propu. Tech.*, 23 (2002) 258.

[127] H. Duan, X. Lina, G. Liua, L. Xua, F. Li, Synthesis of Ni nanoparticles and their catalytic effect on the decomposition of ammonium perchlorate, *J. Mater. Proce. Tech.*, 208 (2008) 494.

[128] C. Ping, F. Li, Z. Jian, J. Wei, Preparation of Cu/CNT Composite Particles and Catalytic Performance on Thermal Decomposition of Ammonium Perchlorate, *Propellants Explos. Pyrotechnics*, 31 (2006) 452.

[129] Gromov, Y. Strokova, A. Kabardin, A. Vorozhtsov, U. Teipel, Experimental Study of the Effect of Metal Nanopowders on the Decomposition of HMX, AP and AN, *Propellants Explos. Pyrotechnics*, 34 (2009) 506.

[130] Z. T. Liu, X. Li, Z. W. Liu, J. Lu, Synthesis and catalytic behaviors of cobalt nanocrystals with special morphologies, *J. Powder Tech.*, 189 (2009) 514.

[131] H. Duan, X. Lin, G. Liu, L. Xu, F. Li, Synthesis of Co Nanoparticles and Their Catalytic Effect on the Decomposition of Ammonium Perchlorate, *Chinese J. Chem. Engg.*, 16 (2008) 325.

[132] M. Foltz, J. Meinschein, Ammonium perchlorate phase transitions to 26 GPa and 700 K in a diamond anvil cell, *Mater. Lett.*, 24 (1995) 407.

[133] B.I. Kaidimov, V.S. Gavazova, A sharp change in the inhibiting effect of ammonia on the thermal decomposition of ammonium perchlorate as a result of its phase transition, *J. Inorg. Nucl. Chem.*, 36 (1974) 3848.

[134] P. M. W. Jacobs, G. S. Parasone, Mechanism of the decomposition of ammonium perchlorate, *Combus. Flame*, 13 (1969) 419.

[135] W. A. Rosser, S. H. Inami, Thermal decomposition of ammonium perchlorate, *Combustion and Flame*, 12 (1968) 427.

[136] L. Liu, F. Li, L. Tan, M. Li, Y. Yang, Effects of metal and composite metal nanopowders on the thermal decomposition of ammonium perchlorate (AP) and the ammonium perchlorate/hydroxyterminated polybutadiene (AP/HTPB) composite solid propellant, *Chin. J. Chem. Eng.*, 12 (2004) 595.

[137] M. Song, M. Chen, Z. Zhang, Effect of Zn Powders on the Thermal Decomposition of Ammonium Perchlorate, *Propellants Explos. Pyrotechnics*, 33 (2008) 261.

[138] M.-R Song, M. Chen, Z.-J. Zhan, Preparation and characterization of Mg nanoparticles, *Mater. Charac.*, 59 (2008) 514.

[139] G. Singh, S. K. Sengupta, I. P. S. Kapoor, S. Dubey, R. Dubey, S. Singh, Nanoparticles of Transition Metals as Accelerants in the Thermal Decomposition of Ammonium Perchlorate, Part 62, *J. Energ. Mater.*, 31 (2013) 165.

[140] G. Singh, I. P. S. Kapoor, S. Dubey, Bimetallic nanoalloys: Preparation, characterization and their catalytic activity, *J. of Alloys and Compounds*, 480 (2009) 270.

[141] P. Srivastva, G. Singh, I. P. S. Kapoor, R. Dubey, Synthesis, characterization and catalytic effect of bimetallic nanocrystals on the thermal decomposition of ammonium perchlorate, *Ind. J. Chem.* 49A, (2010) 1339.

[142] S.R. Chakravarthy, E.W. Price, R.K. Sigman, Mechanism of Burning Rate Enhancement of Composite Solid Propellants by Ferric Oxide, *J. Propul. Power*, 13 (1997) 471.

[143] R. Dubey, M. Chawla, P. F. Siril,, G. Singh, Bi-metallic nanocomposites of Mn with very high catalytic activity for burning rate enhancement of composite solid propellants, *Thermochim. Acta*, 572 (2013) 30.

[144] S. Vyazovkin, A.C.A. Wight, Kinetics of Thermal Decomposition of Cubic Ammonium Perchlorate, *Chem. Mater.*, 11 (1999) 3386.

[145] S. Chaturvedi, P. N. Dave, N. N. Patel, Thermal decomposition of AP/HTPB propellants in presence of Zn nanoalloys, *Applied Nanoscience*, 5 (2015) 93.

[146] S. Chaturvedi, P. N. Dave, N.N. Patel, Nano-alloys: Potential catalyst for thermal decomposition of Ammonium Perchlorate; Synthesis and Reactivity in Inorganic, Metal-Organic, and Nano-Metal. *Chemistry*; 44 (2) 258-262; 2014.

[147] B.N. Mondal, A. Basumallick, P.P. Chattopadhyay, Magnetic behavior of nanocrystalline Cu–Ni–Co alloys prepared by mechanical alloying and isothermal annealing, *J. Alloys Comp.*, 457 (2008) 10.

[148] M. Wen, Q. Liu, Y. Wang, Y. Zhu, Q. Wu, Positive microemulsion synthesis and magnetic property of amorphous multicomponent Co-, Ni- and Cu-based alloy nanoparticles, *J. Colloid Surf. A*, 318 (2008) 238.

[149] S. Zhanbo, Z. Yaomin, L. Xiaoyuan, S. Xiaoping, Compositions, magnetic properties and GMR performances of melt-spun Cu–Co–Ni ribbons, *J. Magn. Mater.*, 269 (2004) 341.

[150] S. Pane, E. Gomez, E. Valles, Magnetoresistive granular Cu–Co–Ni coatings prepared by electrodeposition, *J. Electroanal. Chem.*, 596 (2006) 87.

[151] S. Singh, P.Srivastava, G.Singh, Synthesis, characterization of Co–Ni–Cu trimetallic alloy nanocrystals and their catalytic properties, Part – 9, 1 *J. Alloys and Compounds*, 562 (2013) 150.

[152] D. Klissurski, E. Uzunova, Synthesis and features of binary cobaltite spinels *J. Mater Sci.*, 291 (1994) 285.

[153] S. Singh, P. Srivastava, I.P.S. Kapoor, G. Singh, Synthesis, characterisation and catalytic activity of cadmium cobaltite nanoparticles: part 87, *J. Exper. Nanoscience*, 10 (2013) 29.

[154] G. Singh, I. P. S. Kapoor, S. Dubey, Nanocobaltite: Preparation, Characterization, and their Catalytic Activity, *Propellants Explos. Pyrotechnics*, 36 (2011) 367.

[155] A.A. Said, R. Al-Qasmi, The role of copper cobaltite spinel, $Cu_xCo_{3-x}O_4$ during the thermal decomposition of ammonium perchlorate, *Thermochim. Acta*, 275 (1996) 83.

[156] P.T. Anastas, J.C. Warner, *Green Chemistry: Theory and Practice*; Oxford University Press: New York, (1998).

[157] L.C. McKenzie, J.E. Hutchison, Green Nanoscience, *Chim. Oggi*, 22 (2004) 30.

[158] S. Iravani, Green synthesis of metal nanoparticles using plants, *Green Chem.*, 13 (2011) 2638.

[159] S. P. Chandran, M. Chaudhary, R. Pasricha, A. Ahmad, M. Sastry, Synthesis of Gold Nanotriangles and Silver Nanoparticles Using Aloevera Plant Extract, *Biotechnol. Prog.* 22 (2006) 577.

[160] S. S. Shankar, A. Rai, A. Ahmad, M. Sastry, Controlling the Optical Properties of Lemongrass Extract Synthesized Gold Nanotriangles and Potential Application in Infrared-Absorbing Optical Coatings, *Chem. Mater.*, 17 (2005) 566.

[161] S. S. Shankar,, A. Rai, A. Ahmad, M. Sastry, B. Ankamwar, A. Singh, Biological synthesis of triangular gold nanoprisms, *Nat. Mater.*, 3 (2004) 482.

[162] S. S. Shankar, A. Ahmad, M. Sastry, R. Pasricha, Bioreduction of chloroaurate ions by geranium leaves and its endophytic fungus yields gold nanoparticles of different shapes, *J. Mater. Chem.*, 13 (2003) 1822.

[163] D. Philip, C. Unni, Extracellular biosynthesis of gold and silver nanoparticles using Krishna tulsi (Ocimum sanctum) leaf, *Physica E*, 43 (2011) 1318.

[164] V. S. Kotakadi, Y. S. Rao, S. A. Gaddam, T. N. V. K. V.; Prasad, A. V. Reddy, D. V. R. Gopal, Simple and rapid biosynthesis of stable silver nanoparticles using dried leaves of Catharanthus roseus. Linn. G. Donn and its anti microbial activity, *Colloids Surf. B: Biointerfaces*, 105 (2013) 194.

[165] M. R. Bindhu, M. Umadevi, Synthesis of monodispersed silver nanoparticles using Hibiscus cannabinus leaf extract and its antimicrobial activity, *Spectrochim. Acta, Part A: Molecular and Biomolecular spectroscopy*, 101 (2013) 184.

[166] G. S. Ghodake, N. G. Deshpande, Y. P. Lee, E. S. Jin, Pear fruit extract-assisted room-temperature biosynthesis of gold nanoplates, *Colloids. Surf. B: Biointerfaces*, 75 (2010) 584.

[167] P. Danthal, M. Mukhopodhyay, Prunus domestica Fruit Extract-Mediated Synthesis of Gold Nanoparticles and Its Catalytic Activity for 4-Nitrophenol Reduction, *Ind. Eng. Chem. Res.*, 51 (2012) 13014.

[168] M. Umadevi, M. R. Bindhu, V. Sathe, A Novel Synthesis of Malic Acid Capped Silver Nanoparticles using Solanum lycopersicums Fruit Extract, *J. Mater. Sci. Technol.*, 29 (2013) 317.

[169] M. Sathishkumar, K. Sneha, S. W. Won, C. W. Cho, S. Kim, Y. S. Yun, Cinnamon zeylanicum bark extract and powder mediated green synthesis of nano-crystalline silver particles and its bactericidal activity, *Colloids Surf., B: Biointerfaces*, 73 (2009) 332.

[170] D. Raghunandan, M. D. Bedre, S. Basavaraja, B. Sawle, S. Y. Manjunath, A. Venkataraman, Rapid biosynthesis of irregular shaped gold nanoparticles from macerated aqueous extracellular dried clove buds (Syzygium aromaticum) solution, *Colloids Surf. B: Biointerfaces*, 79 (2010) 235.

[171] S. Gunalan, R. Sivaraj, R. Venckatesh, Aloe barbadensis Miller mediated green synthesis of mono-disperse copper oxide nanoparticles: Optical properties, *Spectrochim. Acta, Part A*, 97 (2012) 1140.

[172] G. Sangeetha, S. Rajeshwari, R. Venckatesh, Green synthesis of zinc oxide nanoparticles by aloe barbadensis miller leaf extract: Structure and optical properties, *Mater. Res. Bull.*, 46 (2011) 2560.

[173] N. A. Samat, R. M. Nor, Sol–gel synthesis of zinc oxide nanoparticles using Citrus aurantifolia extracts, *Ceram. Int.*, 39 (2012) 545.

[174] S. Maensiri, P. Laokul, J. Klinkaewnarong, S. Phokha, V. Promarak, S. J. Seraphin, Indium oxide (In2O3) nanoparticles using Aloe vera plant extract: Synthesis and optical properties, *Optoelectron. Adv. Mater.*, 10 (2008) 161.

[175] R. Herrera-Becerra, C. Zorrilla, J. A. Ascencio, Production of Iron Oxide Nanoparticles by a Biosynthesis Method: An Environmentally Friendly Route, *J. Phys. Chem. C*, 111 (2007) 16147.

[176] R. Herrera-Becerra, J. L. Rius, C. Zorrilla, Tannin biosynthesis of iron oxide nanoparticles, *Appl. Phys. A, Mater. Sci. Process.*, 100 (2010) 453.

[177] S. Phumying, S. Labuayai, C. Thomas, V. Amornkitbamrung, E. Swatsitang, S. Maensiri, Aloe vera plant-extracted solution hydrothermal synthesis and magnetic properties of magnetite (Fe_3O_4) nanoparticles, *Appl. Phys. A: Mater. Sci. Process.*, 111 (2013) 1187.

[178] R. R. Gandhi, S. Gowri, J. Suresh, S. Selvam, M.J. Sundrarajan, Biosynthesis of Tin Oxide Nanoparticles Using Corolla Tube of Nyctanthes arbor-tristis Flower Extract, *J. Biobased Mater. Bioenergy*, 6 (2012) 204.

[179] J.K. Sharma, P. Srivastava, G. Singh, M. S. Akhtar, S. Ameen, Biosynthesized NiO nanoparticles: Potential catalyst for ammonium perchlorate and composite solid propellants, *Ceramic International*, 1 (2015) 1573.

[180] J.K. Sharma, P. Srivastava, G. Singh, M. S. Akhtar, S. Ameen, Green synthesis of Co_3O_4 nanoparticles and their applications in thermal decomposition of ammonium perchlorate and dye-sensitized solar cells, *Mat. Sci. Eng. B*, 193 (2015) 181.

ABOUT THE EDITORS

Professor Sadia Ameen obtained her Ph.D. in Chemistry, 2008, and then moved to Chonbuk National University. Presently she is working as Assistant Professor at Chonbuk National University. Her current research focuses on Dye-Sensitized Solar cells, perovskite Solar Cells, Organic Solar Cells, Inorganic Solar Cells, Chemical Sensors, Photocatalytic Degradation, Diodes and Optoelectronic Devices. She has achieved Gold Medal in academics, holder of Merit Scholarship and Merit Certificate for the best academic performances. She is recipient of Best Researcher Award, 2016. She has published more than 95 peer-reviewed papers, contributed book chapters, edited books and is inventor/co-inventor of patents.

Professor M. Shaheer Akhtar completed his Ph.D. in Chemical Engineering, 2008, from Chonbuk National University, South Korea. Presently, he is working as full time

Associate Professor at Chonbuk National University, South Korea. His research interest constitutes the photoelectrochemical characterizations of thin film semiconductor nanomaterials, composite materials, polymer based solid-state films, solid polymer electrolytes and electrode materials for dye-sensitized solar cells (DSSCs), hybrid organic-inorganic solar cells, small molecules based organic solar cells, and photocatalytic reactions.

Professor Hyung-Shik Shin received Ph.D. in the kinetics of initial oxidation A_l (111) surface from the Cornell University, USA, in 1984. Presently he is a Professor in the School of Chemical Engineering, Chonbuk National University, Korea. He has been promising researcher and visited several universities as visiting professor/invited speaker worldwide. He is an active executive member of various renowned programs/committee such as KiChE, copyright protection, KAERI, etc. He has extensive experience in electrochemistry, renewable energy sources, solar cells, organic solar cells, charge transport properties of organic semiconductors, inorganic-organic solar cells, biosensors, chemical sensors, nanopatterning of thin film materials, and photocatalytic degradation, etc.

INDEX

A

B

C

D

E

F

G

H

N

O

P

R

S

T

V

W

Z